Group Analysis
of
Differential Equations

Group Analysis of Differential Equations

L. V. Ovsiannikov

Institute for Hydrodynamics
USSR Academy of Sciences, Siberian Branch
Novosibirsk, USSR

Translation edited by

W. F. Ames

School of Mathematics
Georgia Institute of Technology
Atlanta, Georgia

1982

ACADEMIC PRESS

A Subsidiary of Harcourt Brace Jovanovich, Publishers

New York London
Paris San Diego San Francisco São Paulo Sydney Tokyo Toronto

ACADEMIC PRESS, INC.
111 Fifth Avenue, New York, New York 10003

United Kingdom Edition published by
ACADEMIC PRESS, INC. (LONDON) LTD.
24/28 Oval Road, London NW1 7DX

Library of Congress Cataloging in Publication Data

Ovsîannikov, L. V. (Lev Vasil′evich)
Group analysis of differential equations.

Translation of: Gruppovoĭ analiz differenťsial′nykh uravneniĭ.
Includes index.
1. Differential equations. 2. Transformation groups.
I. Ames, William F. II. Title. III. Series.
QA371.08713 515.3'5 82-4096
ISBN 0-12-531801-1 AACR2

PRINTED IN THE UNITED STATES OF AMERICA

82 83 84 85 9 8 7 6 5 4 3 2 1

Contents

Chapter IV **The Lie Theory**

Editor's Preface

Group analysis of differential equations combines the powers of analysis and algebra to develop an underlying structure of differential equations. The author L. V. Ovsiannikov has been a significant contributor to the development of this field during the past twenty-five years. It is especially useful in problems arising in physics, mechanics, geometry, numerical analysis, and other areas in which ordinary and partial differential equations play an important role.

The literal translation was ably carried out by Y. Chapovsky, a student of electrical engineering and mathematics at the Georgia Institute of Technology, with the assistance of Academic Press. The editing and checking of the mathematics took an enormous number of hours, but it was a worthy learning experience. The edited manuscript was prepared by Mrs. Joy Allen. The editor continues to enjoy the support of Academic Press. We are all indebted to them for publication of this important work.

WILLIAM F. AMES

Preface

Systematic investigation of continuous transformation groups was initiated in the second half of the previous century by the Norwegian mathematician Sophus Lie (1842–1899). These groups, now called Lie groups, represent a subject in which the algebraic group and topological structures are combined, both structures being interlinked by the continuity condition of the group multiplication operation. It may be said that the theory of Lie groups lies between two large mathematical sciences—algebra and analysis. In the works of Sophus Lie and his disciples this theory has also found profound and fruitful applications to various questions of geometry. Meanwhile, the essence of this fine and, to some extent, complete theory goes far beyond the limits of its application within mathematics proper. In fact, the continuous transformation groups are real objects of the world, and their impact may be judged from the way in which they form our conceptions of its physical structure. These include, for instance, the concepts of the homogeneity and isotropy of space and time, the dynamic similarity of phenomena, and Galilean and Lorentzian invariance.

Since the mathematical models of many phenomena of the real world are formulated in the form of differential equations, it becomes clear that their use in the general theory of differential equations is one of the most essential applications of Lie group theory. It is appropriate to note that the original aim of Sophus Lie was to create a theory of integrating ordinary differential equations similar to the Abelian theory of solving algebraic equations. In this connection he gave a definition and investigated the fundamental concept of the group admitted by the given system of differential equations. At present, the mathematical trend whose object is a common treatment of the Lie group of transformations and the differential equations admitted by these groups is called group analysis of differential equations.

One of the main problems of the group analysis of differential equations is to study the action of the group admitted by the given equation (system of equations) in a set of solutions to this equation. The action of the admitted group introduces into the set of solutions an algebraic structure which may be used to achieve several goals. These include, for example, a description of the general properties of all members of the family of solutions, the choice of certain classes of solutions that are simpler in some way than the general solution, and the generation of solutions on the basis of known ones.

For differential equations appearing in a mathematical model of a class of physical phenomena, the admitted group is usually not given together with the model. Therefore the purely technical task of finding the widest group admitted by this differential equation (system of equations) is essential. Another interesting and practically important problem consists in using the group analysis technique for the group classifications of differential equations. The solution of this problem is not only of purely mathematical interest but also has practical significance. The differential equations of mathematical physics, in many cases, contain parameters or functions that are determined experimentally and hence are not strictly fixed (these are called arbitrary elements). At the same time the equations of the mathematical model must be simple enough to be analyzed and solved successfully. The group approach enables the acceptance of (as the simplicity criterion) the requirement that the arbitrary element be such that, with it, the modeling differential equations admits a group having definite features or, in general, the widest group.

The problem of transformations for this differential equation is also interesting and useful. If the admitted group is known, it may be useful in the search for such transformations, as a result of which the differential structure of the transformed equation should be as simple and convenient as possible to facilitate finding concrete solutions.

The above-mentioned problems and other, more specific ones, form an extensive field for applying the theory and algorithms of the group analysis of differential equations. The methodological foundation for the development of operative techniques for this subject was already created in the works of Sophus Lie. The correspondence theorems between Lie groups and Lie algebras obtained by this scientist enable us to reduce complex nonlinear problems to simpler linear ones. This "linearization," which is typical of analysis but unusual for algebra, has been achieved by the creation of the infinitesimal investigation technique. Here the local action of the group of transformations is replaced by the action of linear differential operators—infinitesimal operators which form the Lie algebra. Of great importance is that the transition from Lie groups to Lie algebras completely retains the algebraic structure of the object being studied. So this transition makes it possible to obtain strong infinitesimal invariance criteria of various objects related to the continuous transformation group. But the finite transformations of the group (unlike the infinitesimal ones) are, generally speaking, restored implicitly as the solutions of some ordinary differential equations.

This book gives a systematic exposition of the theory of Lie groups and Lie algebras and its application to creating algorithms for solving the above-mentioned problems of the group analysis of differential equations. The book differs from existing volumes dealing with Lie group theory and its application in that this work has a number of features related to the initial assumptions and the manner of description.

The Lie groups, the manifolds, and other objects of analysis are considered only locally, in a manner similar to classical theory. The local approach makes the machinery flexible, easy to apply, and imparts the required generality, as far as the analysis of arbitrary differential equations is concerned. Of course, to make this theory informal, the local approach requires in many instances that careful consideration be given to the definitions of the mappings encountered (and arising). The relevant explanations are contained in Chapter I.

Further, the assumption of infinite differentiability of the involved mappings is a powerful simplifying factor. All the operations done on mappings and the definition of the mapping as a solution of an ordinary differential equation do not leave this class. Surely, due to the facts known for Lie groups, it might be proper to deal only with the assumption of their analyticity. But in this case one might assume that the differential equations being considered and the solutions to them are analytical too, but this is not needed. Besides, the analyticity assumption might make the theory more complicated since a proof of the convergence of the Taylor series that express any mappings would be required.

Another feature is that Lie theory is treated as the theory of Banach Lie groups and Lie algebras in that case when the parametric space is an arbitrary Banach space (generally, infinite dimensional), the differential equations mechanism being the basis. The representation space (whose transformations are the group elements) may also be considered as an arbitrary Banach space in the theory of transformation groups. However, a complete theory of invariants and invariant manifolds could not be obtained here, and hence finite dimensionality of the representation space or parametric space is assumed. Besides, in the case when the main object is the investigation of a differential equation, the representation space is usually finite dimensional.

The book presents a real group analysis of differential equations, i.e., Lie groups and Lie algebras where the Banach spaces and differential equations are considered over the field of real numbers. The field of complex numbers, required for solving some auxiliary algebraic problems, involves no difficulties.

The main feature of this manner of description is determined by the accepted vector (coordinate-free) form, in which the values and relationships are recorded, this form being indispensable in the case of infinite-dimensional Banach spaces. This form stresses, in the best possible way, that the vector values and relationships do not depend on any coordinate system. Accordingly, the geometric language of manifolds, their dimension, ranks of linear mappings, etc., is widely used. In particular cases (almost all examples) the coordinate form is also used, but always in a fixed coordinate system convenient for use in calculations. Systematic adherence to the vector concept involves the need to consider the change of values and relations during space transformations (instead of coordinate transformations) in which they have been determined. So the text contains these formulas which describe the relevant change of values and relations.

To unify the language of the theory of classical finite-dimensional and infinite-dimensional (Banach) Lie groups, the terminology generally accepted for the finite-dimensional case had to be slightly changed. Thus, the class of isomorphic (locally) Lie groups is considered instead of various coordinate systems in the given local Lie group, the concept of "canonical coordinates of second order" is replaced by that of "canonical group," etc. The terms "canonical group" or "canonical multiplication" are more descriptive since they emphasize the inner features inherent in the multiplication law.

Finally, the book gives a detailed description of the group analysis algorithms to which the general theory can be successfully brought. The description of algorithms contains illustrative examples in which the differential equations are taken, as a rule, from mechanics or, more exactly, from hydro–gas dynamics. Experts in hydromechanics will, naturally, draw more information from these examples than those who do not know this subject. However, it is not assumed that the reader knows hydrodynamics, and so all the special terms and designations are explained in the text.

Not all the constructions and problems of the group analysis are covered in this book. Only those results and algorithms that are well developed and therefore can be used in applications have been included. In this respect the book must be considered as an introduction to the subject rather than a thorough treatment. The following are some theoretical items that might be treated in a more extended work: the concept of imprimitivity and how it is applied to the classification of subgroups; Cartan's method in the theory of infinite Lie groups and Lie algebras; general features of invariance and methods of composing abstract determining equations; the problem of extension of Lie groups and Lie algebras; composition and classification of differential equations that are invariant with respect to the given group (or the given group representation); the problem of the canonical structure of infinite Lie groups; invariance and approximate modeling; group foundations of the theory of separation of variables for second-order linear equations.

Among the special problems not included in the book because of their narrow nature the following may be mentioned: Dickson's group classification of ordinary second-order differential equations, the supplement to S. A. Chaplygin's equation from gas dynamics, and partially invariant solutions of hydrodynamic problems with contact discontinuities. There are also many problems that have not yet been comprehensively solved and are in the course of development. These include: classification of Riemannian spaces with respect to groups of generalized motion, explanation of group foundations of the Huygens–Hadamard principle, group analysis of difference approximations of differential equations, applications in the theory of control, group analysis of differential equations by means of the Lie–Bäcklund groups, and a number of others.

The detailed contents depicts the subject matter of the book. Here only the description of the general structure is given. The book is based on the inductive method. The subject of Chapters I–III is the simplest concept, the one-parameter group with its tangential field of vectors (the operator). This is quite sufficient for developing the main technical means for the group analysis of differential equations—the criteria of invariance and continuation theory (Chapter I), for building the calculation algorithms of the full group and group classification (Chapter II), and for illustrating this technique by concrete nontrivial examples (Chapter III). The idea is that the choice of one-parameter transformation groups will be an aid in understanding the material for those who want to use it in their own problems. The algebraic structure is rather simple here and only at the end of Chapter II do the Lie algebra operators appear as objects resulting from calculations.

Chapter IV is indispensable for providing full understanding of the results obtained by the previously discussed methods and for promoting further development of the theory. As stated above, it contains a nonstandard treatment of the Banach Lie groups. This chapter deals mainly with algebraic problems, and only at its end is the complete theory of Lie group transformations presented.

Chapters V and VI comprise one of the main applications—the construction of partial solution classes for the given differential equation with a known admitted group. Considered sequentially are the classes of invariant solutions (Chapter V) and partially invariant ones (Chapter VI). Here the purely algebraic problem of the classification of subgroups for a given Lie group is applied, this being a problem for which there is not as yet any clear algorithm. The theory of differential invariants, treated in Chapter VII, is developed on an infinitesimal basis. Particularly interesting is that the main result—the theorem concerning the finiteness of the differential invariant basis—is also true for the infinite Lie groups.

Chapter VIII is aimed at creating an impression of the ways in which the methods of group analysis are used in special issues involving differential equations. It comprises problems which are interesting in themselves. These are: group classification of second-order linear equations, elements of the tangential transformation theory, invariance features of some boundary value problems, and foundations of conservation law theory for solving variational problems. The set of problems has been chosen with a view of providing definitions of more or less complete results, but these would be considerably wider in a more detailed description.

The Appendix comprises tables of full groups calculated by various authors for concrete differential equation systems of practical interest. This list demonstrates clearly the variety of applications for the group analysis method.

We hope that this book will draw the attention of many specialists and increase the circle of researchers in the field of group analysis of differential equations. It seems proper to introduce the main elements of group analysis in a university teaching program.

It is assumed that the reader has some knowledge and skill such as that within the university program of the first two courses of the Mechanics–Mathematics Faculty. The book is easily understood by those who know the concepts of group, ring, field, mapping, continuity, differentiability, as well as the concepts of vector and Banach spaces, and have some (minimal) knowledge in the theory of differential equations. References to *Analysis* encountered in the text (though rarely) are to J. Dieudonné's book *Foundations of Modern Analysis* (Academic Press, 1969–1978) and those to *Algebra* are to S. Lang's book *Algebra* (Addison-Wesley, 1965).

The text gives no references (with a few exceptions) to publications; priority issues are, as a rule, not discussed. References are listed at the end of individual chapters. The reference lists are preceded by brief comments, relating the sources with the main text.

Notation

SET-THEORETICAL SYMBOLS

$\times$	Product of Banach spaces
$\subset$	Set inclusion
$\in$	Element of a set
$\cap$	Intersection of sets
$\cup$	Union of sets
$\rightarrow$	Mapping symbol
$f: X \rightarrow Y$	f is a mapping of set X to set Y
$\otimes$	Tensor multiplication
$\oplus$	Direct sum of subspaces (subalgebras)
$\circ$	Composition of mappings
$\cdot$	Contraction of tensors ("scalar multiplication")
$\{\cdots\}$	Set, characterized by elements inside braces
$\{x \mid S\}$	Set of elements, possessing property S
$\langle\ \rangle$	Angular brackets for notation of images in linear and multilinear mappings
B^A	Set of all mappings $A \rightarrow B$

LETTER SYMBOLS

$X, Y, \cdots$	Banach spaces
$x, y, \cdots$	Vectors (points) of Banach spaces (in special cases—scalar coordinates of a point

$\|x\|$	Norm of a vector $x \in X$
X^n	Product of n copies of a space X
R	One-dimensional space (or field) of real numbers
$R^n(x)$	Space R^n; general vector is x
$R^2(t, x)$	Space R^2, where cartesian coordinates of a point are t and x
$\mathscr{L}(X)$	Space of continuous linear mappings $X \to X$
$\mathscr{L}(X; Y)$	Space of continuous linear mappings $X \to Y$
I_X	Identity mapping of space X
G	Group
$\mathrm{GL}(X)$	Group of all linear homeomorphisms of space X
L	Lie algebra
$C_k(V)$	Class of mappings of V that are continuously differentiable k times
∂	Symbol of an argument derivative, indicated either by index (∂_x) or before a symbol of mapping ($\partial f(x)$)
D	Total differentiation operator
$\xi \cdot \partial$	Infinitesimal operator, corresponding to a vector field (family of vector fields) ξ
$\delta_i^j, \delta^{ij}, \ldots$	Kronecker symbols
E	Differential equation (system)

Unconditionally, the rule of sum monomials, according to repeating index, is used without the sign of summation. Such summation is always assumed when the index is included once as a superscript and once as a subscript and the set of its values is clear from the context.

If the mapping f: $X \to Y$ is given and $x \in X$, $y \in Y$, then $x \to f(x)$ or $y = f(x)$ is called the *formula of the mapping f operation* and it is said that the mapping f "operates according to the formula" $x \to f(x)$ or $y = f(x)$. For example, the composition $v \circ u$ of mappings u: $X \to Y$ and v: $Y \to Z$ operates by the formula $(v \circ u)(x) = v(u(x))$.

Each subsection has its own enumeration. Theorems, lemmas, etc., are not enumerated individually because every subsection usually contains only one theorem, lemma, etc. A reference to Theorem 9 in Section 12 (§12), subsection 9, is designated as "Theorem 12.9". Formulas in a section are enumerated by a double index in parentheses, where the first index is a number of a section and the second is a number of a formula in the section; for example, (5.3) means formula 3 in Section 5. References to formulas in other sections are enumerated by a triple index; for instance, (7.3.1) means formula (3.1) in Section 7.

CHAPTER

I

One-Parameter Transformation Groups

Many elements of group analysis are based on the consideration of one-parameter transformation groups, to which this chapter is devoted. First ideas are presented by using the one-parameter continuous local Lie group concept of the local transformation of a Banach space (called group G^1).

Banach spaces are taken here as carriers of representations since differential equations are usually given on such spaces. Of course, it is inconvenient to be limited to transformations of any set or manifold onto itself. Under this limitation it is simple to operate with a local transformation only when it is unique. But, already in problems connected with the composition of some local transformations with different function domains, there are some difficulties. Here the continuous dependence of transformations on the group parameter removes the difficulty. Owing to the group parameter, a small translation of the transformation function domain occurs when the parameter undergoes a small change and a nonzero intersection of a mapping with an inverse image is guaranteed. Clearly, it is necessary to watch carefully for domain and function changes during the transformation. In theoretical studies this point is often missed, which makes the theory "formal," although actually there is no formality. The usual agreement concerning "forgetting terms" is made; in other words, the transformations so determined have the necessary properties on the whole Banach space. Also, it clearly suffices to consider only local transformation groups. Using this basis, a sufficiently general theory and efficient algorithms can be constructed.

The study of groups G^1 is closely connected with vector fields on Banach spaces. Here the concept of the tangent vector field to the group G^1 is essential. With the assistance of this (Lie) group, relatively simple ordinary differential equations can be studied. Introduced in this chapter is the concept of an infinitesimal operator (hereafter just "operator"), which is equivalent to the vector field concept that will be used throughout the book. The important theoretical fact is that there is a one-to-one correspondence of groups G^1 and their operators.

Furthermore, the fundamental concepts of invariant mappings and invariant manifolds, relative to transformations, are studied. Infinitesimal criteria of invariance are the most important, since they permit linearization of those problems. A new definition is presented by the introduction of a *universal invariance concept*, which, in spite of its simplicity and effectiveness, has not previously been introduced into the literature. The existence of a universal invariant for groups G^1 of an arbitrary Banach space of transformations is established in a very general form.

The *prolongation* theory is the principal element of group analysis of differential equations. It is connected with the necessity to consider the transformation of the product of two Banach spaces where the independent and dependent variables of the differential equations are cofactors. Here important properties about commutation and composition transformations, besides the working formulas of transformation continuity, are established. A calculated general form of the standard commutators, which are used in subsequent chapters, is also presented.

§1. DEFINITIONS AND EXAMPLES

1. Operations on and Representation of a Group

Let M be a set. A one-to-one mapping of the set M onto M is called a *transformation of the set M*. The ensemble $\tau(M)$ of all transformations of the set M forms a group, where the composition of mappings plays the part of a group operation and the identity transformation is designated I_M. The group $\tau(M)$ is called a *transformation group of the set M*.

Let G be a group. The group homomorphism $\pi: G \to \tau(M)$ is called a *representation of the group G to $\tau(M)$*. This means that the image $\pi(g)$ is a transformation of the set M for every element $g \in G$ and the relation

$$\pi(g_1) \circ \pi(g_2) = \pi(g_1 g_2) \tag{1.1}$$

is correct for any $g_1, g_2 \in G$.

The image $\pi(G)$ of the group G is a subgroup of the group $\tau(M)$, which is also called a *representation of the group G as a transformation group of the set M*.

The mapping $(x, g) \to \pi(g)(x)$ of the product $M \times G$ to the set M is called an *operation of the group G on the set M*. The definition of an operation determines a group structure on the set M. In future discussions, where it will not lead to a misunderstanding, the relation will be denoted by the same letter as a representation; thus $\pi(x, g) = \pi(g)(x)$.

A representation $\pi: G \to \tau(M)$ is called *exact* (or isomorphic) if the mapping π is an isomorphism. A representation π is exact if and only if $\pi(g) = I_M$ is correct only for a unit element e of the group G. A representation (or operation) π is called *trivial* if $\pi(g) = I_M$ for every element $g \in G$.

2. One-Parameter Group

In the sequel some Banach space X will play the role of the set M and the group will be that of all real numbers R under the group operation of addition, in the space X with the representation $f: R \to \tau(X)$. In this case the representation $f: R \to \tau(X)$ will be called *locally exact* if it is possible to find an interval $\Delta \subset R$, symmetric with respect to zero, such that for every $a \in \Delta$ the relation $f(a) = I_X$ is possible only if $a = 0$.

The *locally exact representation* $f: R \to \tau(X)$ is called a *one-parameter transformation group of the space X*.

In what follows the image of a point $a \in R$ in the representation $f: R \to \tau(X)$ will be denoted by the symbol f_a. This image is a transformation f_a of the space X for every $a \in R$.

This representation f of the group R on the space X is determined by the formula $f(x, a) = f_a(x)$. Thus, according to these definitions, the operation f defines a space X and a one-parameter group $f(R)$ of transformations f_a. From this it follows that the group properties of the representation $f(R)$ are equivalent to the following properties of the mapping f:

1° $f(x, 0) = x$ for any $x \in X$;
2° $f(f(x, a), b) = f(x, a + b)$ for any $a, b \in R, x \in X$;
3° symmetry relative to zero;

that is, for an interval $\Delta \subset R$, if $a \in \Delta$ and $f(x, a) = x$ for every $x \in X$, then $a = 0$.

Every mapping $f: X \times R \to X$ with properties 1°–3° defines a one-parameter transformation group $f(R)$ of the space X.

3. Continuity

Let Banach spaces Y and Z and an open set $U \subset Y$ be given. A mapping $u: Y \to Z$ is called a mapping of class $C_0(U)$, $C_k(U)$ $(k = 1, 2, \ldots)$, $C_\infty(U)$, or $C_A(U)$ if the mapping $u: U \to Z$ is continuous, continuously differentiable k times, infinitely differentiable, or analytic on U, respectively.

A group $f(R)$ is called a *continuous group of the class* C_k $(0 \leq k \leq \infty)$ or *class* C_A if the mapping $f: X \times R \to X$ belongs to the class $C_k(X \times R)$ $(0 \leq k \leq \infty)$ or to the class $C_A(X \times R)$, respectively.

Subsequently, a one-parameter continuous transformation group of a space X will be denoted by the symbol G^1. If it is necessary to emphasize that the group G^1 considered is generated by the mapping $f: X \times R \to X$, the symbol $G^1(f)$ will be used. The mapping f will be expressed by a formula of the type

$$x' = f(x, a).$$

4. Example: Translation Group

Let $x \in X$ be a fixed nontrivial vector. Specified by the formula

$$x' = x + ax_0 \qquad (x \in X, a \in R),$$

the mapping f generates a one-parameter continuous transformation group of the space X, which will be called the *translation group* (also called *translation* or *shift*) in the direction of the vector x_0. It is easy to verify that 2.1°–2.3° are satisfied.

5. Groups of Linear Homeomorphisms

Linear continuous invertible transformations of the space X generate the group $\mathrm{GL}(X)$ of linear homeomorphisms of the space X. The problem arises: "Find all continuous groups $G^1 \subset \mathrm{GL}(X)$ of given class C."

Every such group is determined by defining a representation $l: R \to \mathrm{GL}(X)$ and is generated by the mapping

$$x' = l(a)\langle x \rangle \qquad (x \in X, a \in R).$$

To satisfy properties 2.1°–2.3°, it is necessary that the relations $l(0) = I_X$ and $l(b) \circ l(a) = l(b + a)$ hold. In addition, the mapping f must be differentiable. In particular, the derivative $\partial l(0) = u$ exists and is a fixed linear mapping $X \to X$. Upon differentiating the relation $l(b + a) = l(b) \circ l(a)$ with respect to the parameter b, where $b = 0$, the relation

$$\partial l(a) = u \circ l(a) \tag{5.1}$$

is obtained.

Now the task is reduced to the construction of the solutions l of equation (5.1) with the initial condition $l(0) = I_X$. This is the Cauchy problem for the linear ordinary differential equation, the solution of which is well known: it is given by

$$l(a) = \exp(au) \qquad (a \in R).$$

This result solves the formulated task: One-parameter groups $G^1 \subset \mathrm{GL}(X)$ arise when a linear mapping u operates on the space $\mathscr{L}(X)$.

Further examples are discussed for the case of finite-dimensional X. If $\dim X = n$, then a square $n \times n$ matrix corresponds to a linear mapping $u \subset \mathscr{L}(X)$ in every cartesian coordinate system in which $x = (x^1, \ldots, x^n)$. Let $(u)^*$ be the transposed matrix. A mapping u is called *symmetric* if the matrix (u) is symmetric, that is $(u)^* = (u)$, and *antisymmetric* if the matrix (u) is antisymmetric, that is $(u)^* = -(u)$.

6. Example: Group of Dilatations

Let a linear mapping u be symmetric. Then a system of coordinates $x = (x^1, \ldots, x^n)$ can be found in which the matrix (u) is diagonal, that is, $(u) = \mathrm{diag}(\lambda^1, \ldots, \lambda^n)$. (A matrix all of whose elements are equal to zero except those on the main diagonal, which are equal to $\lambda^1, \ldots, \lambda^n$, is denoted this way.) Accordingly, the matrix $l(a) = \exp(au)$ is diagonal also, namely, $(l(a)) = \mathrm{diag}(\exp(\lambda^1 a), \ldots, \exp(\lambda^n a))$. That is why the mapping, generating a group G^1, will be like

$$x'^i = x^i \exp(\lambda^i a) \qquad (i = 1, \ldots, n).$$

For every $a \in R$, the resulting transformation of a space X is called a *dilatation*. So every symmetric mapping $u \in \mathscr{L}(X)$ defines some group of dilatations G^1 of the space X.

It is necessary to notice that the representation of group G^1, as a group of dilatations, depends on the way the coordinate system in the space X is chosen. Despite this, knowledge of a dilatation group is useful for applications to concrete differential equations, which are usually considered in a fixed coordinate system.

7. Example: Group of Rotations

Now let u be antisymmetric. Then the linear transformation $l(a)$ will possess the following property:

$$l(a)^* = \exp(au^*) = \exp(-au) = l(a)^{-1},$$

that is, $l(a)^* \circ l(a) = I$. Such transformations are called *orthogonal*.

In the case $n = 3$, orthogonal transformations are also *rotations*. Here for any antisymmetric transformation u there exists a vector $z \in X$ for which $u\langle z\rangle = 0$. Then $l(a)\langle z\rangle = \exp(au)\langle z\rangle = z$ for any $a \in R$ and a system of coordinates can be found, where $x = (x^1, x^2, x^3)$, such that $z = (0, 0, 1)$. In this system of coordinates the matrix $(l(a))$ has the form

$$(l(a)) = \begin{bmatrix} \cos a & -\sin a & 0 \\ \sin a & \cos a & 0 \\ 0 & 0 & 1 \end{bmatrix}.$$

The group G^1 of rotations around the axis (x^3), defined by the vector z and given by formulas

$$x'^1 = x^1 \cos a - x^2 \sin a, \qquad x'^2 = x^1 \sin a + x^2 \cos a, \qquad x'^3 = x^3,$$

corresponds to the matrix $(la))$. It can be seen from this example that it is essential to introduce the interval Δ into condition 2.3° in some nontrivial way. In this special case the interval $\Delta = (-\pi, \pi)$ can be taken.

8. Local Theory

For the goals of group analysis the class of one-parameter groups is too restricted. The point is that group analysis when connected with solutions of nonlinear problems gives rise to mappings f that are not generally defined on the whole space $X \times R$. An expansion of a class of one-parameter groups is achieved by restricting the set on which the generating mapping is considered. For that set the set $V \times \Delta$ is usually taken, where V is some open set in X and Δ is some symmetric (with respect to zero) interval from R. In accordance with this, for the mapping $f: X \times R \to X$ of the class $C_\infty(V \times \Delta)$ to satisfy properties 2.2° and 2.3° it is only required that $x \in V$ and $a, b, a + b \in \Delta$. When considering an expansion of a class of one-parameter groups, obtained as above, we are not interested in the properties of the transformation group outside the set V or for values of a outside the interval Δ. The theory, based on such an expansion of a class of one-parameter groups, is called a *local theory*.

Moreover, generally speaking, it is not required for the generating mapping f to be defined outside of the set $V \times \Delta$ in the local theory. Consequently, the problem arises to continue the mapping f, with properties 2.1°–2.3° only on set $V \times \Delta$, to some mapping $X \times R \to X$ with the same properties on the whole space. However, this interesting problem goes beyond the bounds of the local theory and will not be considered here.

It is worthwhile to note that it is always possible to continue the aforementioned mapping $f: V \times \Delta \to X$ to some mapping $\hat{f}: X \times R \to X$, which generates the one-parameter family of transformations of the space X, by means of the formula $\hat{f}_a(x) = f(x, a)$. Of course, this expansion is not unique. Nevertheless, this fact shows that in the local theory two essential properties of one-parameter groups hold. Its objects are the transformations of the space X; and their ensemble generates a one-parameter family.

So the local theory has to do with *local transformations of the space X*. The transformations of the open sets $V \subset X$, that is, mappings $v: V \to X$, for which $v(V)$ is open in X and the mapping $v: V \to v(V)$ is one-to-one, are called *local transformations*.

Let Δ be an interval in R which is symmetric with respect to zero. A mapping $f: V \times \Delta \to X$ is called a *one-parameter family of local transformations* of the space X if a particular mapping $f_a: V \to X$ is a transformation of an open set V, for every $a \in \Delta$.

9. Local Lie Group

The following definition introduces the object which forms the basis of the group analysis developed further on.

A local one-parameter Lie group of local transformations of the space X is a one-parameter family of the local transformations $f: V \times \Delta \to X$, which possess the following properties:

1° $f(x, 0) = x$ for any $x \in V$.
2° $f(f(x, a), b) = f(x, a + b)$ for any $a, b, a + b \in \Delta$, $x \in V$.
3° If $a \in \Delta$ and $f(x, a) = x$ for all $x \in V$, then $a = 0$.
4° $f \in C_\infty(V \times \Delta)$.

Comparison with the definition of a one-parameter group of transformations of the space X, given in Sections 1.2 and 1.3, shows that every such group is a local one-parameter Lie group of local transformations of the space X. That is why, in further exposition, the local one-parameter Lie group of local transformations of the space X will be denoted by the symbol G^1 also; and for short it will be called just the *group* G^1.

The absence of the sets V and Δ in the notation of the group G^1 is explained by the fact that they play a secondary role from the viewpoint of the local theory, and to display them is not essential. For example, if we consider f' as generating the mapping f on the subset $V' \times \Delta'$, where $V' \subset V$ and $\Delta' \subset \Delta$, then the mapping $f': V' \times \Delta'$ will possess all the properties 1°–4°. The groups generated by mappings f and f' are not different in the local theory. That is actually why the local theory operates with classes of indistinguishable groups. Although it would not be very difficult to give the definitions and to introduce groups G^1 as classes of indistinguishable groups to form the algorithm of the group analysis, there is no special necessity to do so.

The given definitions and the following reasons show that the mapping $f: V \times \Delta \to X$, generating the group G^1, can also be called a *local action* of an additive group R on the space X. In this case the group G^1 can be considered as a representation $f(\Delta) \subset \tau(X)$, where the local transformations $f_a: V \to X$ of the space X are elements. This terminology will be used further on.

10. Example: Projective Group

A one-parameter family of the local transformations $V \times \Delta \to R^2$ is considered on the plane $R^2(x, y)$, where V is a strip $\{(x, y) \mid |x| < 1, y \in R\}$ and $\Delta = (-1, 1)$, generated by the mapping

$$x' = \frac{x}{1 - ax}, \qquad y' = \frac{y}{1 - ax}.$$

It is easy to check that all properties 9.1°–9.4° are satisfied; therefore, the given mapping generates the group G^1. However, it is impossible to leave the limits of the set $V \times \Delta$ without violating at least one of the properties. This example shows that the expansion of the class of the one-parameter groups, achieved in the local theory, is essential. The given group G^1 is an example of the *projective transformation group* of the plane R^2.

11. Similitude of Groups

Let $\pi: M \times G \to M$ be an operation of a group G on a set M and $p: M \to M$ be some transformation of the set M. It is easy to check that the mapping $\pi_1: M \times G \to M$, defined by the formula

$$\pi_1(g) = p \circ \pi(g) \circ p^{-1}, \tag{11.1}$$

gives an operation of the group G on M. It is clear that the relation given by formula (11.1) is a theoretical-set criterion of an equivalence, denoted by $\pi_1 \sim \pi_2$.

If the operations π_1 and π_2 of the group G on the set M are equivalent, that is, they are in the relation $\pi_1 \sim \pi_2$, then the representations $\pi_1(G)$ and $\pi_2(G)$, in the group $\tau(M)$, are called *similar*.

This terminology is used on groups G^1 since each of them can be considered as a local representation of an additive group R into $\tau(X)$ (see Section 1.9). Because local transformations of the space X are elements of the group G^1 that precisely define the concept of the similarity of such groups, it is necessary to consider those open sets that define generating sets.

Let the mapping $f_i: V_i \times \Delta_i \to X_i$ generate a group G_i^1 $(i = 1, 2)$. The ensemble, $H(V_1, V_2)$, of every possible class C_∞ of transformation of the set V_1 onto the set V_2, that is, all one-to-one mappings $p: V_1 \to V_2$ for which $p \in C_\infty(V_1)$ and $p^{-1} \in C_\infty(V_2)$, is considered. Finally, let f_{ia} be elements of the group G_i^1 with representations $f_i(\Delta_i)$ $(i = 1, 2)$.

Groups G_1^1 and G_2^1 are called *similar* if the transformation $p \in H(V_1, V_2)$ and the interval $\Delta \subset \Delta_1 \cap \Delta_2$ exist so that the relation

$$f_{2a} = p \circ f_{1a} \circ p^{-1} \tag{11.2}$$

holds for every $a \in \Delta$.

12. Example

The group of dilatations G_1^1, as a representation $f_1(R)$ into $\tau(X)$, operates in the system of coordinates $x = (x^1, \ldots, x^n)$ according to the formula (see Section 1.6)

$$f^i_{1a}(x) = x^i \exp(a\lambda^i) \qquad (i = 1, \ldots, n)$$

and is considered on the set $V = \{x \mid x^i > 0\,(i = 1, \ldots, n)\}$. In the same coordinate system the transformation $p: V \to X$ is given by the formulas

$$p^i(x) = \ln x^i, \qquad p^{-1i}(x) = \exp x^i$$

$(i = 1, \ldots, n)$. By means of the transformation p, in formula (11.2), a similar group G_2^1 is obtained as the representation $f_2(R)$, whose elements are defined by the relation

$$x' = f_{2a}(x) = p \circ f_{1a} \circ p^{-1}(x).$$

A calculation gives

$$\begin{aligned} x'^i &= \ln((f_{1a} \circ p^{-1})^i(x)) = \ln(f^i_{1a}(p^{-1}(x))) \\ &= \ln(p^{-1i}(x) \exp(a\lambda^i)) = \ln((\exp x^i) \exp(a\lambda^i)) \\ &= x^i + a\lambda^i, \end{aligned}$$

that is, in vector form $x' = x + a\lambda$, where $\lambda = (\lambda^1, \ldots, \lambda^n)$ is a fixed vector of the space X. For this reason the group G_2^1 is the translation group of the space X. So it has been shown that the *dilatation group of the space X is similar to the translation group of the space X.*

13. Tangent Vector Field

Let the group G_2^1 generated by the mapping $f: V \times \Delta \to X$ be given. For every *fixed* $x \in V$ the partial mapping $f_x: \Delta \to X$ defines some curve in the space X, passing through the point x, that is a one-dimensional manifold $f_x(\Delta)$, given parametrically by the equation $x' = f_x(a) = f(x, a)$. The curve $f_x(\Delta)$ is called the *orbit of the point* x.

The vector $\xi = \partial f_x(0)$ is a tangent to the orbit $f_x(\Delta)$ at the point x. Thereby, the vector field $\xi: V \to X$, operating according to the formula

$$\xi(x) = \partial_a f(x, 0), \tag{13.1}$$

is defined on V. This vector field will be called a *tangent vector field of the group* G^1.

The connection between the group G^1 and its tangent vector field turns out to be closer than that given by (13.1). In fact, a group is defined by its tangent vector field. This fact will be established later using the special differential equation of the group G^1.

14. Lie Equation

Differentiating the relation 9.2° with respect to the parameter b at the point $b = 0$, and using (13.1), results in

$$\partial_b f(f(x, a), 0) = \xi(f(x, a)) = \xi \circ f(x, a),$$

so that the new relation

$$\partial_a f(x, a) = \xi \circ f(x, a) \tag{14.1}$$

is obtained.

This relation together with $f(x, 0) = x$ (property 9.1°) shows that the partial mapping $f_x: \Delta \to X$ is a solution of the Cauchy problem for the ordinary differential equation

$$\partial_a x' = \xi(x'), \qquad x'(0) = x, \tag{14.2}$$

with the desired mapping $x': R \to X$. Equation (14.2) is called the *Lie equation*. A simple example of a concrete Lie equation has already been seen in relation (5.1).

In case of a finite-dimensional space X (n dimensions) equation (14.2) becomes a system of n scalar ordinary differential equations

$$\frac{dx'^i}{da} = \xi^i(x'^1, \ldots, x'^n), \qquad x'^i(0) = x^i \qquad (i = 1, \ldots, n), \tag{14.3}$$

in the coordinate system $x = (x^1, \ldots, x^n)$, where $\xi^i(x^1, \ldots, x^n)$ are the components of the vector field $\xi = (\xi^1, \ldots, \xi^n)$ evaluated at the point x.

The Lie equation serves as one of the basic instruments of the group analysis. As will be shown later, a one-to-one correspondence between the groups G^1 and their tangent vector fields is established with the aid of the Lie equation. Just this fact is decisive in all applications of Lie group theory (including differential equations) because it permits the formation of precise algorithms for determination and application of the group properties of the considered objects. Because of the transition from the group G^1 to its tangent vector field many problems are linearized. This makes it possible to study problems effectively with group analysis.

15. Examples

Tangent vector fields of the groups G^1, considered in Sections 1.4–1.7 and 1.10, can be found directly using formula (13.1).

For translation groups (see Section 1.4), $\xi = x_0$ is, obviously, a constant vector field. The family of constant vector fields is obtained when x_0 covers the whole space X.

For the group G^1 of linear homeomorphisms of the space X, generated by the mapping $x' = \exp(au)\langle x\rangle$ (see Section 1.5),

$$\xi(x) = u\langle x\rangle \tag{15.1}$$

is obtained, where $u \in \mathscr{L}(X)$ is the constant linear mapping $X \to X$. When u covers the whole space $\mathscr{L}(X)$, formula (15.1) gives the family of linear vector fields on the space X.

The dilatation group of the (finite) n-dimensional space $X = R^n(x^1, \ldots, x^n)$ is given by formulas $x'^i = x^i \exp(\lambda^i a)$ (see Section 1.6). For its tangent vector field, written as $\xi = (\xi^1, \ldots, \xi^n)$ in the same coordinate system, $\xi^i = \lambda^i x^i$ is obtained for every $i = 1, \ldots, n$ (no summation on i); or in vector form,

$$\xi(x) = (\lambda^1 x^1, \ldots, \lambda^n x^n). \tag{15.2}$$

When the numbers λ^i cover all the set independently of each other, formula (15.2) gives an n-parameter family of dilatation tangent fields of the n-dimensional space X, in the coordinate system $x = (x^1, \ldots, x^n)$.

For the group of rotations with respect to the (x^3) axis case (see Section 1.7), formula (13.1) gives $\xi^1 = -x^2$, $\xi^2 = x^1$, $\xi^3 = 0$, or in vector form

$$\xi(x) = (-x^2, x^1, 0), \tag{15.3}$$

for the vector $\xi = (\xi^1, \xi^2, \xi^3)$ in the coordinate system $x = (x^1, x^2, x^3)$.

Calculation of the vector $\xi = (\alpha, \beta)$ gives $\alpha = x^2$, $\beta = xy$, or in vector form

$$\xi(x, y) = (x^2, xy), \tag{15.4}$$

for the projective group in the plane $R^2(x, y)$ (see Section 1.10). Here, as opposed to the previous examples, the tangent vector field ξ is a quadratic one (second power) relative to the coordinates of the point (x, y).

16. Similitude of Tangent Fields

When the transformation from the group $G_1^1(f_1)$ to its similar group $G_2^1(f_2)$ takes place by means of the transformation p (see Section 1.11), the tangent vector field ξ_1 of the group G_1^1 transforms into the tangent vector field ξ_2 of the group G_2^1. To ascertain the corresponding formula of the transformation $\xi_1 \to \xi_2$ it is necessary to differentiate the relation (11.2) with respect to the parameter a at the point $a = 0$. From definition (13.1) and using the relation $f(x, 0) = x$, there results

$$\xi_2(x) = \partial p(p^{-1}(x))\langle \xi_1(p^{-1}(x))\rangle,$$

or in simpler form,

$$\xi_2(\bar{x}) = \partial p(x)\langle \xi_1(x)\rangle, \qquad \bar{x} = p(x). \tag{16.1}$$

The formula (16.1) will be called a *formula of similitude of vector fields.*

For example, when the transformation from the group of dilatations to its similar one by the transformations p from Section 1.12 occurs, the derivative $\partial p(x)$ in matrix form will be

$$(\partial p(x)) = \operatorname{diag}\left(\frac{1}{x^1}, \ldots, \frac{1}{x^n}\right).$$

This is why formula (16.1) gives

$$\xi_2(x) = \partial p(x)\langle(\lambda^1 x^1, \ldots, \lambda^n x^n)\rangle = (\lambda^1, \ldots, \lambda^n) = \lambda,$$

which is the constant tangent vector field of the translation group.

§2. INFINITESIMAL OPERATOR

1. Problem: To Construct a Group

In connection with the definition in Section 1.13, the question arises: How is the group G^1 defined by its tangent vector field ξ? A more general form of the question leads to the following problem: The vector field $\xi: V \to X$ of class $C_\infty(V)$ is given; it is required to find out whether there is a group G^1, for which the given ξ is its tangent vector field.

The solution of this problem is based on the use of the Lie equations (1.14.2).† From the theory of ordinary differential equations it is known that if the given vector field $\xi: V \to X$ is continuously differentiable on V, then the solution for the Cauchy problem (1.14.2) exists and is unique for any $x \in V$. This solution, defined and continuously differentiable in some neighborhood of the point $(x, 0) \in V \times R$, gives a mapping $f: V \times R \to X$, operating according to the formula $x' = f(x, a)$.

2. Lemma

Lemma. *Let $x' = f(x, a)$ be a solution of problem* (1.14.2) *with vector field $\xi: V \to X$ which is continuously differentiable on V. Then a sphere $B(x_0, r) \subset V$, a symmetric interval $\Delta \subset R$, and a number $v > 0$ can be found for any point $x_0 \in V$ such that the inequality*

$$|f(x, a) - x| \geq |a|\,|\xi(x)| - v|a|^2 \tag{2.1}$$

holds for every $(x, a) \in B(x_0, r) \times \Delta$; and the mapping $f_a: V \to X$ is a one-to-one mapping of the sphere $B(x_0, r)$ onto $f_a(B(x_0, r))$, for any $a \in \Delta$.

Proof. By virtue of the continuity of the mapping f and the relation $f(x, 0) = x$, for any $r_1 > 0$ it is possible to choose a number $r < r_1$ and an interval $\Delta \subset R$ so that the inequality $|f(x, a) - x_0| < r_1$ holds for all $(x, a) \in B(x_0, r) \times \Delta$. Then by virtue of the assumption about the vector field ξ it is possible to choose a number r_1 such that both mappings ξ and $\partial\xi$ are bounded on the sphere $B(x_0, r_1)$. Under these conditions the second derivative, calculated from the relation (1.14.2),

$$\partial_a^2 f(x, a) = \partial\xi(f(x, a))\langle\xi(f(x, a))\rangle,$$

will be bounded on the set $B(x_0, r) \times \Delta$. Therefore the mapping ψ, from the Taylor formula

$$f(x, a) = x + a\xi(x) + a^2\psi(x, a), \tag{2.2}$$

will be bounded on the same set, for example, $|\psi(x, a)| < v$. The inequality (2.1) now follows directly from formula (2.2).

Later, the derivative $g(a) = \partial f_a(x_0)$, obtained by the differentiation of relations (1.14.2),

$$\partial g(a) = \partial\xi(f_a(x_0)) \circ g(a), \qquad g(0) = I_X,$$

will be shown to exist as a solution of the Cauchy problem. Therefore, the mapping $g: \Delta \to \mathscr{L}(X)$ is continuous for the above defined interval Δ. As is

† Citation of equations in other sections will include the section number as a prefix.

known, the set of linear homeomorphisms of the Banach space X is open in $\mathscr{L}(X)$. So by decreasing the interval Δ, if necessary, it is possible to assume that the linear mapping $g(a)\colon X \to X$ is a linear homeomorphism of the space X. This ensures that such an $r > 0$ and an interval Δ can be found for which the mapping $f_a\colon B(x_0, r) \to f_a(B(x_0, r))$ will be one-to-one for all $x \in \Delta$, because of the sufficient condition for the local homeomorphism (inverse function theorem), known from analysis. ■

This lemma shows that the mapping $f\colon B(x_0, r) \times \Delta \to X$, constructed as a solution of the Cauchy problem (1.14.2), generates a one-parameter family of local transformations of the space X (see Section 1.8).

3. Lie Theorem

Theorem. *Let a vector field $\xi\colon V \to X$ of the class $C_\infty(V)$, $\xi(x_0) \neq 0$ for $x_0 \in V$, be given. Then the mapping f, constructed as a solution of the problem* (1.14.2) *with this ξ, generates the local one-parameter Lie group for which the vector field ξ is its tangent vector field.*

Proof. Since it is established in Section 2.2 that the mapping f, obtained as the solution of problem (1.14.2), is a one-parameter family of local transformations of the space X, it remains to make sure that conditions 1°–4° of the definition in Section 1.9 are satisfied.

To establish that condition 1° is true is trivial. To check condition 2°, the sphere $B(x_0, r)$ and interval Δ defined in the lemma in Section 2.2 are chosen. Let $a, b, a + b \in \Delta$ and $x \in B(x_0, r)$. The notation $y = f(x, a)$, $u = f(y, b)$, and $v = f(x, a + b)$ will be used. Mappings u and v are considered as functions of the parameter b when x and a are fixed. Since $u = f(f(x, a), b)$, it is necessary to establish the relation $u = v$. The mapping $x' = u$ satisfies equation (1.14.2) with the initial condition $u(0) = y$. The mapping $x' = v$ satisfies the same equation with the initial condition $v(0) = f(x, a) = y$. Consequently, both of the mappings u and v are solutions of the same Cauchy problem, and by virtue of its uniqueness they are coincidental.

To check condition 3°, the inequality (2.1) from the lemma in Section 2.2 is used. That inequality ensures that the relation $f(x, a) = x$, with $a \neq 0$, is possible only for such $(x, a) \in B(x_0, r) \times \Delta$ for which $|a|v \geq |\xi(x)|$. Since the condition $|\xi(x_0)| \neq 0$ holds, it is possible to choose a sufficiently small sphere $B(x_0, r)$ so that the inequality, $|\xi(x)| > \frac{1}{2}|\xi(x_0)|$, holds for all $x \in B(x_0; r)$. Then it follows that for $a \neq 0$ the relation $f(x, a) = x$ is possible only for those values of a which satisfy the inequality $|a| \geq |\xi(x_0)|/(2v)$. Hence, if the interval Δ is so small that $|\xi(x_0)|/(2v) \notin \Delta$, then the relation $f(x, a) = x$, for $a \in \Delta$ and $x \in B(x_0, r)$, is possible only for $a = 0$.

Finally, the condition 4° is satisfied because it is known from the theory of ordinary differential equations that if the right-hand side of the equation belongs to the class C_∞, then all of its solutions also belong to this class. The last statement of the Lie theorem, that the group G^1 obtained as a result of the solution of problem (1.14.2) has the given vector field ξ as its tangent vector field, follows directly from the differential equations (1.14.3), when $a = 0$, and from the definition in Section 1.3. ■

4. Example of Group Construction

One of the applications of the Lie theorem is to use it for constructing a group G^1 from the given tangent vector field. In the finite-dimensional case, this construction is accomplished by means of the integration of the common equation system (1.14.3). This may not be a simple problem. But such integration can be done in many cases which are practically important.

A simple example is given by a vector field in the three-dimensional space $R^3(t, x, u)$, given by the formula $\xi = (0, t, 1)$. In this case the corresponding system of differential equations (1.14.3) has the form

$$\left.\begin{aligned} dt'/da &= 0, &\quad t'(0) &= t;\\ dx'/da &= t', &\quad x'(0) &= x;\\ du'/da &= 1, &\quad u'(0) &= u; \end{aligned}\right\},$$

and these can be integrated easily. The solution of the Cauchy problem is

$$t' = t, \qquad x' = x + at, \qquad u' = u + a.$$

In mechanics, where t is time, x a space coordinate, and u a velocity, the resulting group G^1 is called the *group of Galilean translations along the* (x) *axis*.

5. Correspondence of Groups and Vector Fields

The Lie theorem, together with the definition in Section 1.13, establishes the desired correspondence between groups G^1 and vector fields ξ. The question arises whether this is a one-to-one correspondence. It is clear that one defined group G^1 corresponds to the given vector field ξ. This follows from Lie's theorem. But conversely, this is not unique since the same group G^1 corresponds to the vector fields ξ and $\lambda\xi$ for any $\lambda \in R$, $\lambda \neq 0$.

Actually, if the group G^1 is generated by the mapping f, which has transformations $x' = f(x, a)$ belonging to G^1, then the mapping f^λ, defined by the formula $f^\lambda(x, a) = f(x, \lambda a)$, has all the properties required by the definition given in Section 1.9. The function f^λ also generates some group

G^{1^λ}. But the groups G^1 and G^{1^λ} are not distinguishable because they are sets of the same transformations of the space X.

It can be shown that the ambiguity of the correspondence between groups G^1 and vector fields ξ is exhausted by the above argument. Actually, because of the definition given in Section 1.9, this ambiguity occurs as a result of the possibility that different additive structures can be introduced on R. But if the transformation $\mu: R \to R$ introduces a new additive structure on R, then the relation

$$\mu(a + b) = \mu(a) + \mu(b)$$

must hold for any $a, b \in R$. As we know, every nontrivial continuous solution of this functional equation has the form $\mu(a) = \lambda a$, where $\lambda \in R$. This is that collection of additive structures, which was previously obtained.

These results demonstrate that a *one-to-one correspondence exists between groups G^1 and vector fields ξ, up to a random nonzero numerical multiplier*. In the sequel this correspondence will be denoted $G^1(\xi)$.

6. Group Operator

Let $f: V \times \Delta \to X$ be a mapping generating a group $G^1(\xi)$ and let Y be a Banach space. Consider the values of differentiable mappings $F: X \to Y$, which orbit a point $x \in V$, which are equal to $F \circ f(x, a)$. By virtue of the definition in Section 1.13, the rate of change of those values at the point $a = 0$ is given by the formula

$$\partial_a(F \circ f(x, a))|_{a=0} = \partial F(x)\langle \xi(x) \rangle.$$

In the case where $Y = R$ [that is, $F(x)$ is a scalar], the right side can be rewritten as $\xi \cdot \partial F(x)$ (the scalar product of vectors ξ and ∂F is denoted by a point $\cdot$), that is, as a result of an operation by the linear differential operator $\xi \cdot \partial$ on the mapping F. It is convenient to write this definition in general, by means of the following definition:

A linear differential operator $\xi \cdot \partial$, operating on the mapping $F: X \to Y$ according to the formula

$$(\xi \cdot \partial)F(x) = \partial F(x)\langle \xi(x) \rangle, \tag{6.1}$$

is called an *infinitesimal operator* of the group $G^1(\xi)$. The word "infinitesimal" means "arbitrarily small." Its usage is a tribute to the classic theory, which often operated with arbitrarily small and arbitrarily large values. With the aid of infinitesimal operators the change in the main linear part of the mapping F, along the orbit of the point x, is written

$$F \circ f(x, a) = F(x) + a(\xi \cdot \partial)F(x) + O(a^2). \tag{6.2}$$

In the case of an n-dimensional space X when the coordinate system is $x = (x^1, \ldots, x^n)$, $\xi = (\xi^1, \ldots, \xi^n)$, the operator ∂ is $\partial = (\partial_1, \ldots, \partial_n)$, where $\partial_i = \partial/\partial x^i$. Then the infinitesimal operator will be written in the form

$$\xi \cdot \partial = \xi^i \partial_i = \xi^i \frac{\partial}{\partial x^i}. \tag{6.3}$$

In this case the infinitesimal operator acts on the mapping $F: X \to Y$ to give the formula

$$(\xi \cdot \partial)F = \xi^i \partial_i F = \xi^i \frac{\partial F}{\partial x_i}. \tag{6.4}$$

In the case of a finite-dimensional space, in which a vector representation is employed, the components of the vector field ξ will be called *components of the infinitesimal operator* $\xi \cdot \partial$. In the sequel, the adjective infinitesimal will be dropped and the form *group operator* will be used, unconditionally.

The equivalence between groups G^1 and vector fields established in Section 2.4 naturally suggests an equivalence between groups G^1 and their operators. This equivalence between groups G^1 and their operators is one-to-one if it is assumed that the infinitesimal operator is exact up to a nonzero numerical multiplier.

7. Examples

Usually the name that characterizes the group is given to the group operator. The following infinitesimal operators correspond to the tangent vector fields considered in Section 1.15:

Translation operator: $\xi \cdot \partial = x_0 \cdot \partial = x_0^i \partial_i$.
Linear transformation operator: $\xi \cdot \partial = u\langle x \rangle \cdot \partial = u_j^i x^j \partial_i$.
Dilatation operator: $\xi \cdot \partial = \sum_{i=1}^n \lambda^i x^i \partial_i$.
Rotation about axis (x^3) *operator*: $\xi \cdot \partial = -x^2 \partial_1 + x^1 \partial_2$.
Projective transformation operator: $\xi \cdot \partial = x^2 \partial_x + xy \partial_y$.

Here $\partial_i = \partial/\partial x^i$, and in the last example $\partial_x = \partial/\partial x$, $\partial_y = \partial/\partial y$.

Operators of rotation about the (x^1) and (x^2) axes are obtained from the rotation operator about the (x^3) axis by the circular permutation of indices $1 \to 2 \to 3 \to 1$:

$$(x^1): \quad \xi \cdot \partial = -x^3 \partial_2 + x^2 \partial_3;$$

$$(x^2): \quad \xi \cdot \partial = -x^1 \partial_3 + x^3 \partial_1.$$

8. Operator Invariance

An important property of the infinitesimal operator is that it is invariant with respect to a transformation from a group G^1 to any of its similar groups. If the group $\bar{G}^1(\bar{\xi})$ is obtained from the group $G^1(\xi)$ by a transformation of similitude $p: X \to X$ defined by means of the relation $\bar{x} = p(x)$, then the tangent vector fields of these groups are connected by equation (1.16.1), which is rewritten here as

$$\bar{\xi}(\bar{x}) = \partial p(x)\langle \xi(x) \rangle, \qquad \bar{x} = p(x). \tag{8.1}$$

If the differentiation operator $\bar{\partial}$, which operates on mappings $\bar{x} \to u(\bar{x})$ (that is, $\bar{\partial}$ is differentiation with respect to the variable $\bar{x}$), is introduced, then the transformed operator $\xi \cdot \partial$ will be written as $\bar{\xi} \cdot \bar{\partial}$. Calculating the result of the operator $\xi \cdot \partial$ acting on the mapping $F: X \to Y$, for which $F(x) = \bar{F}(\bar{x})$, we find

$$\begin{aligned}(\xi \cdot \partial)F(x) &= \partial_a(\bar{F} \circ p \circ f)(x, 0) = \bar{\partial}\bar{F}(\bar{x})\langle \partial p(x)\langle \xi(x) \rangle\rangle = \bar{\partial}\bar{F}(\bar{x})\langle \bar{\xi}(\bar{x}) \rangle \\ &= (\bar{\xi} \cdot \bar{\partial})\bar{F}(\bar{x}) = (\bar{\xi} \cdot \bar{\partial})F(x).\end{aligned}$$

From this it follows that both operators $\xi \cdot \partial$ and $\bar{\xi} \cdot \bar{\partial}$ operate on any mapping $F: X \to Y$ in the same way. Because of this we write the relation

$$\xi \cdot \partial = \bar{\xi} \cdot \bar{\partial}, \tag{8.2}$$

which expresses the invariant property of the operator with respect to the similitude transformation $p: X \to X$.

In fact, we need not connect this property with the groups G^1 if we accept equation (8.1) as a definition of vector similitude. It is useful to have in mind that the differentiation operators ∂ and $\bar{\partial}$ considered as vectors of the conjugate space X^* are connected by the formula

$$\partial = \partial p^* \langle \bar{\partial} \rangle \tag{8.3}$$

when the transformation $\bar{x} = p(x)$ holds. The star designates the conjugate linear mapping. With the aid of equation (8.3) the invariant property (8.2) of the infinitesimal operator is easily found to be

$$\bar{\xi} \cdot \bar{\partial} = \partial p\langle \xi \rangle \cdot (\partial p^*)^{-1}\langle \partial \rangle = ((\partial p)^{-1} \circ \partial p)\langle \xi \rangle \cdot \partial = \xi \cdot \partial.$$

For example, for the dilatation operator $\xi \cdot \partial = \sum_{i=1}^n \lambda^i x^i \partial_i$ and transformation p, given by $\bar{x}^i = \ln x^i$,

$$\partial p(x) = \operatorname{diag}\left(\frac{1}{x^1}, \ldots, \frac{1}{x^n}\right).$$

Because of these, the relations

$$\bar{\xi} = \operatorname{diag}\left(\frac{1}{x^1}, \ldots, \frac{1}{x^n}\right)\langle(\lambda^1 x^1, \ldots, \lambda^n x^n)\rangle = (\lambda^1, \ldots, \lambda^n),$$

$$\bar{\partial} = \operatorname{diag}(x^1, \ldots, x^n)\langle(\partial_1, \ldots, \partial_n)\rangle = (x^1\partial_1, \ldots, x^n\partial_n)$$

hold and relation (8.2) is nicely exemplified.

§3. INVARIANTS AND INVARIANT MANIFOLDS

1. Continuation of Group Representation

Let G be a group and M, N be sets. It is further assumed that representations $\pi_M: G \to \tau(M)$ and $\pi_N: G \to \tau(N)$ are given (see Section 1.1). The set N^M of all types of mappings $u: M \to N$ is considered. It is possible to construct a new representation π_{NM} of the group G in the set $\tau(N^M)$ using the given representations π_M and π_N. This new representation, derived from the representations π_M and π_N, called a *continuation of the pair of representations* (π_M, π_N) *to the representation* π_{NM}, is defined by the equation

$$\pi_{NM}(g)(u) = \pi_N(g)^{-1} \circ u \circ \pi_M(g) \qquad (g \in G). \tag{1.1}$$

According to equation (1.1), to every mapping $u: M \to N$ and every element $g \in G$ there is a new congruent (one-to-one) mapping $u': M \to N$ operating by means of

$$u'(x) = \pi_{NM}(g)(u)(x) = \pi_N(g)^{-1}(u(\pi_M(g)(x))). \tag{1.2}$$

Particular cases of equation (1.2) are obtained when one of the representations π_N or π_M is the identity. In further considerations that case of π_N being the identity, where equation (1.2) becomes

$$u'(x) = u(\pi_M(g)(x)), \tag{1.3}$$

is especially important.

2. Concomitants and Invariants

An element $x_0 \in M$ is called a *fixed element* with respect to the representation $\pi: G \to \tau(M)$ if the relation $\pi(g)(x_0) = x_0$ holds for any $g \in G$.

A fixed element of the continuation representation $\pi_{NM}: G \to \tau(N^M)$ is called a *concomitant* of the group G with respect to the representation pair

(π_M, π_N). Because of equation (1.2) the mapping $u: M \to N$ is a concomitant of the group with respect to the pair (π_M, π_N) if the relation

$$\pi_N(g)(u(x)) = u(\pi_M(g)(x)) \tag{2.1}$$

holds for any $g \in G$, $x \in M$.

In the case when the representation π_N is trivial, the concomitant $u: M \to N$ is called an *invariant mapping or invariant of the group G relative to its representation π in the set $\tau(M)$*. Because of equation (1.3) the mapping $u: M \to N$ is an invariant of the group G relative to its representation π if

$$u(x) = u(\pi(g)(x)) \tag{2.2}$$

for any $g \in G$, $x \in M$. The term "concomitant" means "consistent joint invariant," that is, an invariant which is consistent [in the sense of relation (1.1)] under the joint operation of the pair (π_M, π_N).

3. Invariants of Lie Groups

Here the study of invariants of the group G^1 as the local representation $f: R \to \tau(X)$, described in Section 1, begins. In what follows, when it does not harm the discussion, the formulations will not specifically assert that the whole theory is local. For example, the mapping f, generating the group G^1, will be written as a mapping $f: X \times R \to X$. And the domain of the vector field ξ will not always be specified, and so on.

The mappings $F: X \to Y$ are considered, where Y is a Banach space and it is assumed that the group R operates on Y trivially. According to definition (2.2), *the mapping F is an invariant of the group G^1, generated by the mapping $f: X \times R \to X$, if the relation*

$$F(f(x, a)) = F(x) \tag{3.1}$$

holds for any $(x, a) \in X \times R$.

The study of invariants is based on a lemma in which it is assumed that $\xi \cdot \partial$ is an infinitesimal operator of the group $G^1(\xi)$ and the designation $x' = f(x, a)$ is used.

Lemma. *For any mapping $F: X \to Y$ of the class $C_1(X)$ and for any $(x, a) \in X \times R$ the relation*

$$\partial_a F(x') = (\xi \cdot \partial) F(x') \tag{3.2}$$

holds.

Proof. Relation (3.2) arises from the Lie equation (1.14.2) and the definition of an infinitesimal operator (2.6.1), whence

$$\partial_a F(x') = \partial F(x')\langle \partial_a x' \rangle = \partial F(x')\langle \xi(x') \rangle = (\xi \cdot \partial) F(x'). \quad \blacksquare$$

4. Criterion of Invariance

In the following theorem the necessary and sufficient condition that some mapping $F: X \to Y$ is an invariant of the group $G^1(\xi)$ is presented.

Theorem. *The mapping $F: X \to Y$ of class $C_1(X)$ is an invariant of the group $G^1(\xi)$ if and only if the relation*

$$(\xi \cdot \partial)F(x) = 0 \tag{4.1}$$

holds for any $x \in X$.

Proof. If F is an invariant, then equation (3.1) is satisfied. Differentiation of the right-hand side of (3.1) with respect to the parameter a gives exactly the left side of equation (4.1) because of (3.2).

Conversely, if equation (4.1) is satisfied, then because of (3.2) the relation $\partial_a F(x') = 0$ holds. This means that the vector $F(x') = F(f(x, a)) \in Y$ is independent of the parameter a and hence equal to its value when $a = 0$, that is, $F(x') = F(x)$. Consequently F is an invariant of the group $G^1(\xi)$ because of the definition (3.1). ■

5. Universal Invariant

If $F: X \to Y$ is an invariant of the group G^1 and Z is a Banach space (where R also operates trivially), then the composition $\Phi \circ F$ is an invariant of the same group G^1 for any mapping $\Phi: Y \to Z$.

The invariant $J: X \to Y$ of the group G^1 is called its *universal invariant* if for any Banach space Z and any invariant $F: X \to Z$ there exists a mapping $\Phi: Y \to Z$ such that $F = \Phi \circ J$.

Because of its definition it is possible to obtain all possible invariants of the group G^1 with the aid of the universal invariant. Therefore, finding the universal invariant is the main problem of invariant theory.

It is natural to assert that the universal invariant of the group G^1 is smooth since it must be a *local mapping of the class C_∞*.

Theorem. *For any group $G^1 = f(R) \subset \tau(X)$ and Banach space Y a universal invariant $J: X \to Y$, of the class C_∞, exists.*

Proof. Let ξ be the tangent vector field of the group G^1. A point $x_0 \in X$ is selected at which the vector $e = \xi(x_0) \neq 0$ and the space X is represented as a direct sum, $X = T \oplus Y$, where T is a one-dimensional space of vectors te ($t \in R$), and Y is a direct complement of T in X. Then for vectors $x, x_0 \in X$ the relations

$$x = te + y, \quad x_0 = t_0 e + y_0$$

hold, where the number $t \in R$ and the vector $y \in Y$ are uniquely defined by the vector x.

Owing to this representation any mapping $u: X \to Z$ can be considered as a mapping $R \times Y \to Z$, operating according to the formula $z = u(t, y)$. With this the tangent vector field ξ of the group G^1 and the mapping f generating it are expressible as

$$\xi(x) = c(t, y)e + \eta(t, y), \qquad f(x, a) = p(t, y, a)e + \phi(t, y, a).$$

In addition, because of the definition of the vector $e = \xi(x_0)$, the relations

$$c(t_0, y_0) = 1, \qquad \eta(t_0, y_0) = 0$$

hold. Further, let $x' = f(x, a) = t'e + y'$. Then

$$t' = p(t, y, a), \qquad y' = \phi(t, y, a). \tag{5.1}$$

In this designation the group property $f(x', b) = f(x, a + b)$ takes the form

$$p(t', y', b) = p(t, y, a + b), \qquad \phi(t', y', b) = \phi(t, y, a + b). \tag{5.2}$$

The Lie equation (1.14.2) is then rewritten as the system of equations

$$\partial_a t' = c(t', y'), \qquad t'(0) = t,$$

$$\partial_a y' = \eta(t', y'), \qquad y'(0) = y.$$

Let $\varepsilon = |a| + |t - t_0| + |y - y_0|$. For ε sufficiently small the inequality $c(t', y') > \frac{1}{2}$ holds because of the relation $c(t_0, y_0) = 1$. Therefore, the first of the relations (5.1) can be solved for the parameter a as $a = \alpha(t, y, t')$. Consequently, it is possible to exclude the parameter a as an "argument" of the mapping ϕ and to form a new mapping $\psi: R \times Y \times R \to Y$, operating according to the formula

$$\psi(t, y, t') = \phi(t, y, a) = \phi(t, y, \alpha(t, y, t')). \tag{5.3}$$

Finally, using the value $t' = t'_0$, the mapping $J: R \times Y \to Y$ is obtained defined by the formula

$$J(t, y) = \psi(t, y, t_0). \tag{5.4}$$

It must be verified that the mapping J is the universal invariant of the group G^1. First, it will be shown that J is an invariant. Since the relation $\phi(t, y, b) = \psi(t, y, p(t, y, b))$ is an identity relative to the variables t, y, b, it holds for the values (5.1), that is,

$$\phi(t', y', b) = \psi(t', y', p(t', y', b)),$$

or because of the group property (5.2),

$$\phi(t, y, a + b) = \psi(t', y', p(t, y, a + b)).$$

On the other hand, because of definition (5.3), the identity

$$\phi(t, y, a + b) = \psi(t, y, p(t, y, a + b))$$

is satisfied. Comparison of these two identities leads to the relation

$$\psi(t', y', p(t, y, a + b)) = \psi(t, y, p(t, y, a + b)),$$

where b is a free parameter since t' and y' depend only on t, y, and a. At the expense of a selection of the parameter b, it is possible to satisfy the relation $p(t, y, a + b) = t_0$ when ε is sufficiently small. This occurs because of the equivalence to the relation $a + b = \alpha(t, y, t_0)$. Moreover, $\alpha(t_0, y_0, t_0) = 0$ because of the initial condition $p(t, y, 0) = t$. But, if $p(t, y, a + b) = t_0$, then, according to definition (5.4), the last relation implies $J(t', y') = J(t, y)$ for all sufficiently small values of the parameter a, or $J(x') = J(x)$, which is the same thing. So J is an invariant.

On careful examination of the process of constructing the invariant J, it is easy to determine that J is a class $C_\infty(V)$ mapping of some neighborhood V of the point x_0.

It is left to prove that the constructed invariant J is universal. First let us show that the mapping $J_1: X \to X$, which is represented as the mapping $J_1: R \times Y \to R \times Y$, operating according to the formula

$$J_1(t, y) = (t, J(t, y)),$$

is a one-to-one mapping of some neighborhood V of the point x_0. For that it is enough to show that the derivative $\partial_y J(t_0, y_0)$ is a homeomorphism of $Y \to Y$, and then the derivative $\partial J_1(x_0)$ will be a homeomorphism of $X \to X$. A calculation gives

$$\partial_y J(t, y) = \partial_y \phi(t, y, a) + \eta(t', y') \otimes \partial_y \alpha(t, y, t_0).$$

It is easy to verify that $\partial_y \alpha(t_0, y_0, t_0) = 0$ using the definition of the function α and the Lie equation. Therefore, for a sufficiently small ε the second term will be small compared to the norm of the space $\mathscr{L}(Y)$. As far as the first term is concerned it converges to I_Y as $a \to 0$. So for a sufficiently small a, the derivative $\partial_y J(t, y)$ differs from I_Y by only a small amount and therefore is a homeomorphism of the space Y.

Consequently, the correspondence between J_1 and V is one-to-one and because of that the inverse mapping $J_1^{-1}: X \to X$ exists. If $F: X \to Z$ is any invariant of the group G^1, then the mapping $\Phi_1 = F \circ J_1^{-1}: X \to Z$ is defined, under which the relation $F = \Phi_1 \circ J_1$ holds, that is,

$$F(t, y) = \Phi_1(t, J(t, y)).$$

Next it is shown that the value of $\Phi_1(t, y)$ is independent of t. Since F and J are invariants, the relation $F(t', y') = \Phi_1(t', J(t', y'))$ can be rewritten in the form $F(t, y) = \Phi_1(t', J(t, y))$; and upon comparison with the initial expression it follows that $\Phi_1(t', J) = \Phi_1(t, J)$ for any $a \in R$. For every t there is an a for which $t' = t_0$, so $\Phi_1(t, J) = \Phi_1(t_0, J)$. From this the mapping $\Phi: Y \to Z$, operating by the formula $\Phi(y) = \Phi_1(t_0, y)$, is defined so that the relation $F = \Phi \circ J$ holds. The universality of the invariant J is established. ■

6. Finite-Dimensional Case

If $X = R^n$, then the construction of the universal invariant in the proof of the theorem in Section 3.5 shows that the space Y, to which the universal invariant J maps the space X, is a finite-dimensional space, namely, $Y = R^{n-1}$. So in this case J has a form

$$J = (J^1, \ldots, J^{n-1}), \tag{6.1}$$

where every component $J^k: X \to R$ is a *scalar invariant*. Obtained in this way, scalar invariants have the important property of functional independence.

Scalar invariants $J^k: X \to R$ $(k = 1, \ldots, n)$ are called *functionally dependent* if there exists a not identically constant mapping $\Phi: R^m \to R$ such that

$$\Phi(J^1(x), \ldots, J^m(x)) = 0.$$

Otherwise, scalar invariants are called *functionally independent*.

If for the invariants (6.1) the mapping $\Phi: Y \to R$, for which $\Phi \circ J(x) = 0$, can be found, then, as is shown in the proof of the theorem in Section 3.5, the relation $\Phi_1 \circ J_1(x) = 0$ holds also, where $\Phi_1(t, y) = \Phi(y)$ and $J_1(x) = (t, J(t, y))$. But the mapping $J_1: X \to X$ is invertible, from which it follows that $\Phi_1(t, y) = \Phi(y) = 0$ for all $y \in Y$. Hence the scalar invariants (6.1) are functionally independent.

Another important property of the set of scalar invariants (6.1) is that any invariant of the group G^1 is expressible in terms of them. Of course, the universal invariant J is expressible in terms of the invariants (6.1); and with its aid, it is possible to get all possible invariants of the group G^1 according to the definition in Section 3.5.

The set of functionally independent scalar invariants of the group, with which it is possible to express any of its invariants, is called a *functional basis of the invariants of the group* G^1.

It makes sense to talk about the functional basis of the invariants only in the case of finite-dimensional spaces X. From the above arguments it is clear that the idea of the universal invariant is a natural generalization of the concept of functional basis for the invariants of the group G^1 of the transformations of any Banach space X.

Because of criterion (4.1), construction of the scalar invariants of the group $G^1(\xi)$, in the case $X = R^n$ with $\xi = (\xi^1, \ldots, \xi^n)$, is reduced to finding the solutions of the linear partial differential equation of the first order

$$\xi^i \partial_i F = 0 \qquad (\partial_i = \partial/\partial x^i). \tag{6.2}$$

As is well known, this problem can be solved by constructing solutions of the system of ordinary differential equations

$$\frac{dx^1}{\xi^1(x)} = \cdots = \frac{dx^n}{\xi^n(x)}. \tag{6.3}$$

7. Examples of Invariants

Universal invariants of the groups G^1 with reduced infinitesimal operators (2.7) are represented in terms of functional bases of scalar invariants by means of formula (6.1).

Group of translations in R^n. The operator is $x_0^i \partial_i$ and the system (6.3) becomes

$$\frac{dx^1}{x_0^1} = \cdots = \frac{dx^n}{x_0^n}.$$

If it is assumed that $x_0^n \neq 0$, then the first integrals have the form $c^j = x_0^n x^j - x_0^j x^n$ $(j = 1, \ldots, n-1)$ and the basis of the invariants is found to be

$$J = (x_0^n x^1 - x_0^1 x^n, \ldots, x_0^n x^{n-1} - x_0^{n-1} x^n).$$

Group of linear transformations in R^n. The operator is $u_j^i x^j \partial_i$. If an auxiliary parameter t is inserted in the differential system (6.3), then the system can be rewritten as a homogeneous system of linear ordinary differential equations with constant coefficients

$$dx^i/dt = u_j^i x^j \qquad (i = 1, \ldots, n).$$

The procedure to solve such systems is well known. Let

$$x_k(t) = (x_k^1(t), \ldots, x_k^n(t)) \qquad (k = 1, \ldots, n)$$

be the fundamental system of its solutions, and let $\phi_j^i(t)$ be elements of the matrix, inverse to the matrix $(x_k^i(t))$. If it is assumed that the equation

$$x^j \phi_j^n(t) = 1$$

is solved for t and its solution is $t = \alpha(x)$, then the basis of invariants will have the form

$$J = (x^i(\phi_i^1 \circ \alpha)(x), \ldots, x^i(\phi_i^{n-1} \circ \alpha)(x)).$$

Group of dilatations in R^n. The operator is $\sum_{i=1}^n \lambda^i x^i \partial_i$, and system (6.3) becomes

$$\frac{dx^1}{\lambda^1 x^1} = \cdots = \frac{dx^n}{\lambda^n x^n}.$$

Without loss of generality, it can be assumed that $\lambda^n = 1$. Then first integrals are $c^j = x^j(x^n)^{-\lambda^j}$ $(j = 1, \ldots, n-1)$. Consequently, the basis of invariants is

$$J = (x^1(x^n)^{-\lambda^1}, \ldots, x^{n-1}(x^n)^{-\lambda^{n-1}}).$$

Group of rotations in R^3. The operator of rotation in R^3 around the (x^3) axis is $-x^2\partial_1 + x^1\partial_2$ and the system (6.3) becomes

$$\frac{dx^1}{-x^2} = \frac{dx^2}{x^1} = \frac{dx^3}{0}.$$

The first integral is clear: $c^1 = x^3$. The second integral, found by integration of the equation $x^1\,dx^1 + x^2\,dx^2 = 0$, is $c^2 = (x^1)^2 + (x^2)^2$. Therefore, the basis of invariants is

$$J = (x^3, (x^1)^2 + (x^2)^2).$$

Projective group in $R^2(x, y)$. The operator is $x^2\partial_x + xy\partial_y$ and the system (6.3) consists of one equation,

$$\frac{dx}{x} = \frac{dy}{y},$$

with the obvious first integral $c = y/x$. The basis of invariants consists of one invariant, $J = y/x$.

8. Theorem about Similarity

The similarity formula of the tangent vector fields (1.16.1), when a transition from the group G_1^1 to its similar form G_2^1 occurs, can be considered, with given vector fields ξ_1, ξ_2, as an equation for the mapping $p\colon X \to X$, transforming the field ξ_1 to the field ξ_2. In particular, the following problem is examined: For a given vector field $\xi\colon X \to X$ and some nonzero constant vector field $\eta \in X$, ascertain whether there exists a diffeomorphism $p\colon X \to X$ such that

$$\partial p(x)\langle \xi(x) \rangle = \eta \tag{8.1}$$

for all $x \in X$. This problem is posed in "general terms" but must be considered in the local theory format. It is found that it can be solved with the aid of the universal invariant.

Theorem. *Let a nonzero vector $\eta \in X$ and a vector field $\xi: V \to X$, of the class $C_\infty(V)$, be given such that $\xi(x_0) \neq 0$ for $x_0 \in V$. Then there exists a diffeomorphism $p: V_0 \to X$ of some neighborhood V_0 of the point x_0 which is such that the identity* (8.1) *holds.*

Proof. The notation and quantities introduced in the proof of the theorem in Section 3.5 are used here. The universal invariant J of the group $G^1(\xi)$ is the mapping $X = T \oplus Y \to Y$ and the equation $J(t, y) = z$ can be solved uniquely for y, with any z taken from a sufficiently small neighborhood of the point $z_0 = J(x_0)$. Let this solution be given by the mapping $y = \Phi(t, z)$. The expansion of the vector field $\xi(x) = c(t, y)e + \eta(t, y)$ is taken, where $e = \xi(x_0)$ and it is further assumed that $\tilde{c}(t, z) = c(t, \Phi(t, z))$. Then $\tilde{c} = 1$ at the point (t_0, z_0) and is different from zero in the neighborhood V of this point. Consequently, the mapping $v: X \to R$, of the class $C_\infty(V)$, is defined and is calculated by means of

$$v(t, z) = \int_{t_0}^{t} \frac{d\tau}{\tilde{c}(\tau, z)}.$$

Now consider the mapping $k: X \to X$, of the class $C_\infty(V)$, which is defined by means of

$$k(t, y) = ev(t, J(t, y)) + J(t, y).$$

Clearly, k is a homeomorphism of the neighborhood V_0 of some point x_0 contained in an open set of the space X. Indeed, for any vector $r = ea + b \in X$ the relation

$$\partial k\langle r\rangle = e((\partial_t v)a + \partial_z v \cdot ((\partial_t J)a + \partial_y J\langle b\rangle)) + ((\partial_t J)a + \partial_y J\langle b\rangle)$$

holds, where $a \in R$ and $b \in Y$. Because of the identity $\partial_z v(t_0, z) = 0$ it follows that the equation $\partial k\langle r\rangle = s$, for all $s \in X$, has a unique solution. The remainder of this portion of the proof follows directly from the implicit function theorem.

When $r = \xi$ and it is recalled that J is an invariant of the group G^1, the previous relation becomes

$$\partial k\langle \xi\rangle = e(\partial_t v)c = e$$

because of the definitions for v and $\tilde{c}$. Finally, with the linear transformation $l \in \mathrm{GL}(X)$, which transforms the vector e to the vector η, there is generated the transformation $p = l \circ k: X \to X$, of the class $C_\infty(V_0)$, for which the relation (8.1) is satisfied. ■

9. Consequence

(a) *All vector fields (which are local) are similar on the given space X.* Indeed, because of the theorem just proved, every vector field is similar to a constant vector field; and a relation of similarity is a set-theoretical criterion of equivalence, as is easy to verify.

(b) All groups G^1 of (local) transformations of the space X are similar. Indeed, without difficulty it is easy to verify that the similar groups G^1 correspond to similar vector fields.

In particular, any group G^1 is similar to the group of translations. This property of the groups G^1 is remarkable and its use permits simplification of the number of statements because of the similar transformation.

In the finite-dimensional case, when $X = R^n(x^1, \ldots, x^n)$, we can always regard the considered group G^1 to be similar to the group of translations in the coordinate x^n, which has the generating mapping

$$x'^j = x^j (j = 1, \ldots, n-1), \qquad x'^n = x^n + a.$$

The tangent vector field of this group is

$$\xi = (0, \ldots, 0, 1),$$

and its infinitesimal operator is

$$\xi \cdot \partial = \partial_n = \partial/\partial x^n.$$

The universal invariant $J: R^n \to R^{n-1}$ of this group is given by the formula

$$J(x) = (x^1, \ldots, x^{n-1}).$$

10. Invariant Manifolds

Let X be a Banach space and $\omega: R^k \to X$ be a mapping of the class $C_\infty(R^k)$, which is a one-to-one mapping of R^k onto $\omega(R^k)$. The image $\Omega = \omega(R^k) \subset X$ is called a *manifold of dimensionality k (or k-dimensional manifold)* in the space X.

In the local theory, some open set $V \subset R^k$ is taken instead of R^k. For example, a sphere $B_\rho \subset R^k$ is selected with its center at zero and of radius ρ. With the aid of the mapping $\omega: V \to X$, of the class $C_\infty(V)$, mapping V onto $\omega(V)$ in one-to-one correspondence, the local k-dimensional manifold $\Omega = \omega(V)$ in the space X is found (the adjective "local" is dropped). When the manifold Ω is defined in this way it is called a *parametric definition* and R^n is the *space of parameters*.

Let $G^1(f)$ be a group of transformations of the space X. A manifold Ω is called an *invariant manifold of the group $G^1(f)$* if an orbit $f_x(\Delta)$ of any point $x \in \Omega$ belongs to Ω. In other words, the manifold $\Omega \subset X$ is invariant relative to the group $G^1(f)$ if it, as a whole, is fixed in X and transforms into itself under every transformation $f_a \in G^1(f)$.

If $\Omega = \omega(V)$ is an invariant manifold of the group $G^1(f)$, then the group $G^1(f)$ can be considered as a group of transformations of the set Ω. This means that the additive group R operates on Ω. Because of the one-to-one nature of the mapping $\omega\colon V \to X$, this operation f of the group R on Ω *induces* some operation ϕ of the group R on the space of parameters R^k. That is, the group $G^1(\phi)$ of the transformations of the space R^k appears. Evidently, the groups $G^1(f)$ and $G^1(\phi)$, generated by mappings f and ϕ, are connected by the relation

$$f(\omega(t), a) = \omega(\phi(t, a)) \qquad (t \in R^k). \tag{10.1}$$

A comparison of (10.1) with (2.1) shows that the mapping ω is a concomitant relative to the pair of operations (f, ϕ).

With the aid of the tangent vector field $\eta\colon V \to R^k$ of the group $G^1(\phi)$, the differentiation of the relation (10.1) with respect to a, at the point $a = 0$, gives rise to an infinitesimal analogue of that equation

$$\partial\omega\langle\eta\rangle = \xi \circ \omega, \tag{10.2}$$

which expresses a necessary condition for invariance of the manifold Ω [where ξ is the tangent vector field of the group $G^1(f)$]. As the next theorem shows, equation (10.2) gives an infinitesimal criterion for invariance of the manifold Ω.

Theorem. *The manifold* $\Omega = \omega(V)$ *is an invariant manifold of the group* $G^1(\xi)$ *if and only if there exists a vector field* $\eta\colon V \to R^k$ *such that the mapping* $\omega\colon V \to X$ *satisfies relation* (10.2).

Proof. The necessity of relation (10.2) has previously been established. The sufficiency is proven with the aid of the result that if the relation (10.2) is fulfilled, then the generating mappings f and ϕ, of the groups $G^1(\xi)$ and $G^1(\eta)$, respectively, are connected with relation (10.1). With $t' = \phi(t, a)$ this result follows from the Lie equation (1.14.2) and (10.2), which give the relations

$$\partial_a \omega(t') = \xi(\omega(t')), \omega(t')|_{a=0} = \omega(t).$$

On the other hand, the vector $x' = f(\omega(t), a)$ satisfies the Lie equation

$$\partial_a x' = \xi(x'), x'|_{a=0} = \omega(t).$$

Hence, both vectors x' and $\omega(t')$ are solutions of the same Cauchy problem. Consequently, because of the uniqueness of the solution, $x' = \omega(t')$; that is, relation (10.1) is satisfied. But this relation means that $x = \omega(t) \in \Omega$ goes into $x' = \omega(\phi(t, a)) \in \Omega$ as the result of any transformations f_a; that is, the manifold Ω is invariant relative to the group $G^1(\xi)$. ■

11. Regularly Assigned Manifolds

In applications, a so-called *implicit assignment of a manifold* is often used instead of a parameter assignment. In further considerations it is more essential to consider manifolds in a finite-dimensional space X. Implicit assignment is studied only for this case in what follows.

Let the dimensionality of the space X be equal to n and $s \leq n$. For any continuously differentiable mapping $\psi: X \to R^s$, the concept of rank of this mapping on an open set $V \subset X$ is introduced. To do this the derivative $\partial\psi(x)$ is considered, which is a linear mapping of $X \to R^s$. The rank of the matrix $(\partial\psi(x))$ of this linear mapping (which does not depend on the choice of the coordinate system, as is well known) is called the *rank of the mapping* ψ *at the point* x. We say that *the mapping* ψ *has a given rank on the set* V *if its rank is the same at all points* $x \in V$.

If the mapping ψ has rank q on V, then the number q is called the *rank of the mapping* ψ *on the set* V.

As is known from differential geometry, any mapping $\psi: V \to R^s$, of the class $C_1(V)$, having rank on V with $\psi(x_0) = 0$ for some $x_0 \in V$, generates (locally, in the neighborhood of the point x_0) some manifold $\Psi \subset X$ which is the set of those points $x \in V$ for which the equality

$$\psi(x) = 0 \tag{11.1}$$

holds. This is called the *equation for the manifold* Ψ. An assignment of a manifold by an equation is called its implicit assignment. Equation (11.1) is called *regular*, and the manifold Ψ assigned by it a *regularly assigned manifold*, if the rank of the mapping ψ on V is equal to s. In this case the dimensionality of the manifold Ψ is equal to $n - s$.

In the coordinate system $x = (x^1, \ldots, x^n)$, equation (11.1) is a system of s equations

$$\psi^j(x^1, \ldots, x^n) = 0 \qquad (j = 1, \ldots, s), \tag{11.2}$$

and the condition of regularity becomes that the Jacobi matrix $(\partial_i \psi^j)$ has the rank s everywhere in the neighborhood V of the point x_0.

Lemma. *If the manifold* Ψ *is regularly assigned by the equations* (11.2), *then there exists a mapping* $\bar{x} = q(x)$, *which is a diffeomorphism of some neighborhood of the point* x_0 *and such that the manifold* $\overline{\Psi} = q(\Psi)$ *is assigned by the equations*

$$\bar{x}^j = 0 \qquad (j = 1, \ldots, s). \tag{11.3}$$

Proof. Without loss of generality, we can assume that the rank of the minor of the matrix $(\partial_i \psi^j)$ is contained within the first s columns. (It is always possible to obtain this by a renumbering of the coordinates x^i.) Then a mapping q which satisfies all of the conditions is given by

$$\bar{x}^j = \psi^j(x), \qquad \bar{x}^l = x^l \qquad (j = 1, \ldots, s; l = s + 1, \ldots, n). \quad \blacksquare$$

12. Criterion of Invariance

According to the definition of invariance of a manifold, given in Section 3.10, the manifold Ψ, assigned by equation (11.1), is invariant with respect to the group $G^1(f)$ if for any $x \in \Psi$, $a \in R$, $\psi(f(x, a)) = 0$. Differentiation of this equality with respect to the parameter a, evaluated at the point $a = 0$, and using the notation (2.6.1), results in a necessary condition for invariance of the manifold (11.1)

$$(\xi \cdot \partial)\psi\,|_\Psi = 0, \tag{12.1}$$

where $\xi \cdot \partial$ is the operator of the group G^1 and the symbol $|_\Psi$ is used instead of the words "on the manifold Ψ" and means that the equality is satisfied for all points $x \in \Psi$. It turns out that equation (12.1) is also a sufficient condition of invariance for manifolds which are regularly assigned by equation (11.1).

Theorem. *The manifold $\Psi \subset X$, regularly assigned by equation* (11.1), *is invariant relative to the group $G^1(\xi)$ if and only if equation* (12.1) *is satisfied.*

Proof. Because of the previous result, it is only necessary to prove that if (12.1) holds, then the manifold Ψ is invariant. Because of this goal, the property of invariance of the condition (12.1) relative to a similarity is used. Indeed, if $\bar{x} = p(x)$ is a diffeomorphism of some neighborhood of the point $x_0 \in \Psi$, and also $\bar{\psi}(\bar{x}) = \psi(x)$, then the equation $\bar{\psi}(\bar{x}) = 0$ assigns the manifold $\bar{\Psi} = p(\Psi)$ and the group $G^1(\xi)$ transforms into its similar group $G^1(\bar{\xi})$. With this, because of (2.8.2), equation (12.1) transforms into

$$(\bar{\xi} \cdot \bar{\partial})\bar{\psi}(\bar{x}) = 0$$

for all $\bar{x} \in \bar{\Psi}$. If invariance of $\bar{\Psi}$ relative to $G^1(\bar{\xi})$ follows from this, then the manifold Ψ is invariant relative to $G^1(\xi)$.

Therefore, because of the regularity of the assignment of Ψ by equation (11.1) and the lemma in Section 3.11, it is sufficient only to prove the required statement for the manifold Ψ, assigned in the coordinate system $x = (x^1, \ldots, x^n)$ by equation (11.3). Let $\xi = (\xi^1, \ldots, \xi^n)$ and $f = (f^1, \ldots, f^n)$

in the same coordinate system. Then condition (12.1), for the manifold (11.3), will have the form

$$\xi^j(0, \ldots, 0, x^{s+1}, \ldots, x^n) = 0 \qquad (j = 1, \ldots, s). \tag{12.2}$$

Let $x'^i = f^i|_\Psi = f^i(0, \ldots, 0, x^{s+1}, \ldots, x^n, a)$. Then the Lie equation can be written as two subsystems of equations

$$\partial_a x'^j = \xi^j(x'^1, \ldots, x'^s, x'^{s+1}, \ldots, x'^n), \qquad x'^j(0) = 0 \qquad (j = 1, \ldots, s),$$

$$\partial_a x'^l = \xi^l(x'), \qquad x'^l(0) = x^l \qquad (l = s + 1, \ldots, n).$$

Now it can be seen that, because of condition (12.2), the function $x'^j \equiv 0$ $(j = 1, \ldots, s)$ is a solution of the first subsystem. Since the solution of the Cauchy problem for the Lie equations is unique, for all $x \in \Psi, x'^j = f^j(x, a) = 0$ $(j = 1, \ldots, s)$ will hold for any $a \in R$, that is, $x' \in \Psi$. According to the definition in Section 3.10 this means that the manifold (11.3) is invariant relative to the group $G^1(\xi)$. ■

13. Examples

The following examples illustrate different peculiarities of an invariant manifold and a criterion of the invariance concept. In these the groups G^1 of transformations of the plane $R^2(x, y)$, defined by their infinitesimal operators of the form $\xi \cdot \partial = \alpha(x, y)\partial_x + \beta(x, y)\partial_y$, and manifolds (curves), defined by $\psi(x, y) = 0$, are considered.

(a) Let $\xi \cdot \partial = \partial_x + \partial_y$ be an operator of translation and $\psi = x - y$. Here $(\xi \cdot \partial)\psi \equiv 0$ and the manifold $\psi = 0$ is invariant because this ψ is an invariant of the group $G^1(\xi)$. This situation is general in two ways.

First, any manifold Ψ, defined by equation (11.1), in the left-hand side of which there is an invariant ψ, is an invariant manifold. This fact is obvious.

Second, any invariant manifold Ψ, which is not singular, for the group of transformations, can always be defined by equation (11.1), on the left side of which there is an invariant ψ. This fact will be studied completely in Chapter V.

(b) Let $\xi \cdot \partial = x\partial_x + y\partial_y$ be an operator of dilatation and $\psi = x - y$. Here $(\xi \cdot \partial)\psi = x - y$ and condition (12.1) is satisfied, although ψ is not an invariant of the group $G^1(\xi)$. The universal invariant of this group is $J = y/x$ and with its aid the manifold $\psi = 0$ is defined by the equation $J = 1$ everywhere, except at the special point $x = y = 0$.

(c) Let $\xi \cdot \partial = \partial_y$ be an operator of translation and $\psi = y^2$. Then $(\xi \cdot \partial)\psi|_{\Psi=0} = 2y|_{y=0} = 0$ and condition (12.1) holds. But the manifold $y^2 = 0$ is not invariant relative to the transformations of the group $G^1(\xi)$, which are $x' = x$, $y' = y + a$. This is not a contradiction with the theorem in Section 3.12 since the manifold $y = 0$ is not regularly defined by the equation $y^2 = 0$.

§4. THE CONTINUATION THEORY

1. Spaces of Polylinear Mappings

Let X and Y be Banach spaces. The mapping $\phi: X^k \to Y$ is called *polylinear* (*k-linear*) if it is continuous at zero and if, for every l, $1 \leq l \leq k$, the equality

$$\begin{aligned}\phi\langle x_1, \ldots, x_{l-1}, ax_l + a'x_l', x_{l+1}, \ldots, x_k\rangle \\ = a\phi\langle x_1, \ldots, x_{l-1}, x_l, x_{l+1}, \ldots, x_k\rangle \\ + a'\phi\langle x_1, \ldots, x_{l-1}, x_l', x_{l+1}, \ldots, x_k\rangle\end{aligned}$$

is satisfied for any vectors $x_1, \ldots, x_k, x_l' \in X$ and any numbers $a, a' \in R$.

The union of all polylinear mappings $\phi: X^k \to Y$ generates a Banach space, denoted by the symbol $\mathscr{L}^k(X; Y)$. The linear mappings $X \to Y$, generating the Banach space $\mathscr{L}(X; Y)$, are a special case of polylinear mappings.

The polylinear mapping $\phi: X^k \to Y$ is called *symmetric* if the equality

$$\phi\langle x_{r_1}, x_{r_2}, \ldots, x_{r_k}\rangle = \phi(x_1, x_2, \ldots, x_k\rangle$$

holds for any vectors $x_1, \ldots, x_k \in X$ and for any permutations $(r_1, \ldots, r_k)$ of the indices $(1, \ldots, k)$. Symmetric k-linear mappings $\phi: X^k \to Y$ also generate a Banach space, which will be denoted by the symbol $\mathscr{L}_s^k(X; Y)$. Obviously, $\mathscr{L}_s^1(X; Y) = \mathscr{L}(X; Y)$. An example of a symmetric polylinear mapping consists of all derivatives of the mapping $u: X \to Y$ of the class $C_\infty(X)$. As is known from analysis, the derivative $\partial^k u(x)$ is a symmetric k-linear mapping $X^k \to Y$, whose value on the set of the vectors $x_1, \ldots, x_k \in X$ is $\partial^k u(x)\langle x_1, \ldots, x_k\rangle$. With this the formula of differentiation

$$\partial(\partial^k u(x)\langle x_1, \ldots, x_k\rangle)\langle x_{k+1}\rangle = \partial^{k+1}u(x)\langle x_1, \ldots, x_k, x_{k+1}\rangle$$

is satisfied. In what follows, if it will not lead to a misunderstanding, the derivative $\partial^k u(x)$ will be denoted by the symbol $\partial^k u$ (the x is understood).

2. Continuations of Space

The space $Y_k = \mathscr{L}_s^k(X; Y)$ will be called the kth *continuation of the space Y with respect to X*. Any vector of the space Y_k will be denoted by $\underset{k}{y}$.

Let $Z = X \times Y$. The space $Z_1 = Z \times Y_1$ is called the *first continuation of the space* Z. Continuations of higher orders will be defined inductively: $Z_k = Z_{k-1} \times Y_k$ $(k = 2, 3, \ldots)$. The space Z_k is called the kth *continuation of the space* Z. Any vector of the space Z_k will be denoted by $\underset{k}{z}$. If $x \in X$, $y \in Y$, then this vector is a set

$$\underset{k}{z} = (x, y, \underset{1}{y}, \ldots, \underset{k}{y}). \tag{2.1}$$

In a finite-dimensional case, when $X = R^n$, $Y = R^m$, $Z = R^{n+m}$, the component expression of the considered vectors is

$$x = (x^1, \ldots, x^n), \qquad y = (y^1, \ldots, y^m),$$

$$\underset{1}{y} = (y_1^1, \ldots, y_i^l, \ldots, y_n^m), \qquad \ldots, \qquad \underset{k}{y} = (\ldots, y_{i_1 \ldots i_k}^l, \ldots),$$

where indices $i, i_1, \ldots, i_k$ take values from 1 to n, independently, and the index l has values $1, \ldots, m$. With this, the quantities $y_{i_1 \ldots i_k}^l$ are to be symmetric relative to the indices $i_1, \ldots, i_k$ (that is, they are independent of their permutations). The dimensionality of the space Y_k is $\dim Y_k = m\binom{n+k-1}{k}$.

3. Continuation of Differentiation Operators

In accordance with (2.1), a differentiation operator ∂ of the mappings $Z_k \to P$ is represented by a set of differentiation operators of the mappings of spaces $X, Y, Y_1, \ldots, Y_k$; in particular,

$$\partial = (\partial_x, \underset{0}{\partial}, \underset{1}{\partial}, \ldots, \underset{k}{\partial}),$$

where $\underset{0}{\partial} = \underset{y}{\partial}$. A *differential operator*

$$D = \partial_x + \underset{1}{y} \cdot \underset{0}{\partial} + \underset{2}{y} \cdot \underset{1}{\partial} + \cdots + \underset{k}{y} \underset{k-1}{\partial} + \cdots \tag{3.1}$$

is called an *operator of exact differentiation*. Formally, the sum can be considered as being continued to infinity because an operator of exact differentiation will be used only for mappings $F: Z_k \to P$, with finite k, on which it operates according to the formula

$$DF = \partial_x F + \underset{1}{y} \cdot \partial_y F + \underset{2}{y} \cdot \underset{1}{\partial} F + \cdots + \underset{k+1}{y} \cdot \underset{k}{\partial} F. \tag{3.2}$$

In some cases, together with the operator of exact differentiation (3.1), it is convenient to use truncated operators of exact differentiation

$$\underset{0}{D} = \partial_x, \qquad \underset{1}{D} = \partial_x + \underset{1}{y} \cdot \partial_y, \qquad D_2 = \partial_x + y_1 \cdot \partial_y + y_2 \cdot \underset{1}{\partial}, \ldots, \tag{3.3}$$

which are defined by the recurrence relation

$$\underset{k+1}{D} = \underset{k}{D} + \underset{k+1}{y} \cdot \underset{k}{\partial} \qquad (k = 0, 1, 2, \ldots). \tag{3.4}$$

An operator of differentiation $\underset{k}{\partial}$ should be considered as a vector of the conjugate space $\overset{k}{Y}{}_k^*$. In the finite-dimensional case, using notation given in Section 4.2, its components are partial derivatives

$$\underset{k}{\partial}{}_l^{i_1 \ldots i_k} = \partial^k / \partial y^l_{i_1 \ldots i_k}.$$

The operator $\underset{k+1}{y} \cdot \partial_k$ is a vector of the conjugate space X^*. In the finite-dimensional case, the jth component is a sum (k is fixed)

$$y^l_{i_1 \ldots i_k j}(\partial^k / \partial y^l_{i_1 \ldots i_k})$$

over all values $l = 1, \ldots, m$ and over all possible combinations $(i_1, \ldots, i_k)$ of the numbers $(1, 2, \ldots, n)$. All together there are $\dim Y_k$ terms in this sum.

Accordingly, the operator of exact differentiation is also a vector of a conjugate space X^*. In the finite-dimensional case, the jth component is denoted by the symbol D_j (not to be confused with the symbol $\underset{j}{D}$!).

4. Continuation of Mappings

Every mapping $u: X \to Y$ of the class $C_\infty(X)$, operating according to $y = u(x)$, can be continued to the mapping $X \to Y_k$. This continuation is obtained with the aid of differentiation operators ∂_x^k given by derivatives $\partial_x u: X \to Y_1, \partial_x^2 u: X \to Y_2, \ldots$, which operate by the formula

$$\underset{k}{y} = \partial_x^k u(x) \qquad (k = 1, 2, \ldots).$$

Thereby, every mapping $u: X \to Y$ is continued to the mapping $\underset{k}{u}: X \to Z_k$. The *continued mapping* $\underset{k}{u}$ operates by means of the formula

$$\underset{k}{u}(x) = (x, u(x), \partial_x u(x), \ldots, \partial_x^k u(x)) \qquad (k = 0, 1, 2, \ldots). \tag{4.1}$$

In particular, $\underset{0}{u}(x) = (x, u(x))$ is a "zero continuation" of the mapping $u: X \to Y$ to the mapping $\underset{0}{u}: X \to Z$.

It is useful to remark that because of definitions (2.1) and (4.1) for mappings $v: Z_k \to P$ and $u: X \to Y$, the composition $v \circ \underset{k}{u}: X \to P$, operating by the formula $(v \circ \underset{k}{u})(x) = v(\underset{k}{u}(x))$, can be generated. Differentiation of such compositions is realized with the aid of the operator of exact differentiation (3.1).

Lemma. *For mappings $\underset{k}{v}: Z_k \to P$ and $u: X \to Y$, the differentiation of a composition is given by*

$$\partial_x(v \circ \underset{k}{u})(x) = (\underset{k}{D}v) \circ \underset{k+1}{u}(x) \qquad (k = 0, 1, 2, \ldots). \tag{4.2}$$

Proof. Using the rule for differentiation of a composition there results

$$\begin{aligned}\partial_x(v \circ \underset{k}{u})(x) &= \partial_x v(\underset{k}{u}(x)) + (\underset{0}{\partial} v(\underset{k}{u})) \circ (\partial_x u(x)) \\ &\quad + (\underset{1}{\partial} v(\underset{k}{u})) \circ (\partial_x^2 u(x)) + \cdots + (\underset{k}{\partial} v(\underset{k}{u})) \circ (\partial_x^{k+1}(u(x)) \\ &= ((\partial_x + \underset{1}{y} \cdot \underset{0}{\partial} + \underset{2}{y} \cdot \underset{1}{\partial} + \cdots + \underset{k+1}{y} \cdot \underset{k}{\partial})v) \circ \underset{k+1}{u}(x) = (\underset{k}{D}v) \circ \underset{k+1}{u}(x).\end{aligned}$$

■

5. Continuation of the Transformation

In the notation of (4.2) the transformation $h: Z \to Z$ of class $C_\infty(Z)$, given by $z' = h(z)$, is considered. Since $Z = X \times Y$, this formula can be written in component form as $z' = (x', y')$, where

$$x' = f(x, y), \qquad y' = g(x, y). \tag{5.1}$$

It is assumed that the following condition holds: The linear mapping $\partial_x f(z): X \to X$ is a homeomorphism of the space X. In the local theory the satisfaction of this condition is necessary only for the points $z \in W$, where W is an open set in Z, and in that set this condition is written as

$$\partial_x f(z) \in GL(X) \qquad (z \in W). \tag{5.2}$$

If condition (5.2) is satisfied and if $B_\rho \subset Y_1$ is a sphere with center zero and radius ρ, then a ρ can always be found, perhaps by restricting W, such that for every $\underset{1}{z} \in W_1 = W \times B_\rho$ the linear mapping $\underset{1}{D}f(\underset{1}{z}) = \partial_x f(z) + \underset{1}{y} \cdot \partial_y f(z)$ will also be a homeomorphism of the space X. Further considerations are with respect to the points $z \in W_1$.

The transformation h is continued to the transformations $\underset{s}{h}: Z_s \to Z_s$ $(s = 1, 2, \ldots)$, where the first two components are (5.1) and the remainder is given by

$$\underset{k}{y}' = \underset{k}{v}(\underset{k}{z}) = \underset{k}{v}(x, y, \underset{1}{y}, \ldots, \underset{k}{y}) \qquad (k = 1, 2, \ldots), \tag{5.3}$$

where mappings $\underset{k}{v}: Z_k \to Y_k$ are defined by the recurrence formulas

$$\underset{1}{v} = (\underset{1}{D}g) \circ (\underset{1}{D}f)^{-1}, \qquad \underset{k}{v} = (\underset{k}{D}\,\underset{k-1}{v}) \circ (\underset{1}{D}f)^{-1} \qquad (k = 2, 3, \ldots). \tag{5.4}$$

The resulting transformation $\underset{s}{h}: Z_s \to Z_s$ operates by the formula $\underset{s}{z}' = \underset{s}{h}(\underset{s}{z}) = (f, g, \underset{1}{v}, \ldots, \underset{s}{v})$, which is called the sth *continuation of the transformation h.*

At first glance, the formulas (5.4) seem artificial, but they have a simple analytic structure. Let the mapping $u: X \to Y$, defined by the equation $y = u(x)$, be given. Using the transformation (5.1), the mapping u transforms into a new mapping u', given by $y' = u'(x')$. This new mapping is obtained by the elimination of x from the equations

$$x' = f(x, u(x)), \qquad u' = g(x, u(x)). \tag{5.5}$$

Relations (5.4) operate in such a way that they generate values of the derivatives $\partial_{x'}^k u'(x')$ in terms of values of the derivatives $\partial_x^l u(x)$. In the following lemma this property is stated in the notation of (4.1).

Lemma. *The identities*

$$\partial_{x'}^k u'(x') = \underset{k}{v} \circ \underset{k}{u}(x), \qquad x' = f \circ \underset{0}{u}(x) \qquad (k = 1, 2, \ldots) \tag{5.6}$$

hold for the mappings (5.4).

Proof. From the definition of the mapping u' and using equations (5.5), the identity

$$\begin{aligned}\partial_{x'} u'(x') \circ [\partial_x f(\underset{0}{u}(x)) + \partial_y f(\underset{0}{u}(x)) \circ \partial_x u(x)]\\ = \partial_x g(\underset{0}{u}(x)) + \partial_y g(\underset{0}{u}(x)) \circ \partial_x u(x)\end{aligned}$$

is obtained, where $x' = f(\underset{0}{u}(x))$. Because of the obvious relation

$$\partial_x f(\underset{0}{u}(x)) + \partial_y f(\underset{0}{u}(x)) \circ \partial_x u(x) = (\underset{1}{D} f) \circ \underset{1}{u}(x)$$

and one similar to it for the mapping g, the identity can be rewritten as

$$(\partial_{x'} u'(x')) \circ (\underset{1}{D} f \circ \underset{1}{u}(x)) = \underset{1}{D} g \circ \underset{1}{u}(x).$$

On the other hand, if equation (5.4) is rewritten as $\underset{1}{v} \circ \underset{1}{D} f = \underset{1}{D} g$ and its composition is taken with $\underset{1}{u}(x)$, there results $(\underset{1}{v} \circ \underset{1}{u}(x)) \circ (\underset{1}{D} f \circ \underset{1}{u}(x)) = \underset{1}{D} g \circ \underset{1}{u}(x)$. Upon comparison with the preceding result, it follows that $\partial_{x'} u'(x') = \underset{1}{v} \circ \underset{1}{u}(x)$, which establishes (5.6) for $k = 1$.

Now (5.6) is established, for general k, by induction. Using the results of the lemma of Section 4.4, equation (5.6) can be written as

$$(\partial_{x'}^{k+1} u'(x')) \circ (\underset{1}{D} f \circ \underset{1}{u}(x)) = (\underset{k+1}{D} \underset{k}{v}) \circ \underset{k+1}{u}(x).$$

On the other hand, if equation (5.4) is rewritten, with k replaced by $k + 1$, it becomes $\underset{k+1}{v} \circ \underset{1}{D} f = \underset{k+1}{D} \underset{k}{v}$, and the composition of this with $\underset{k+1}{u}(x)$ gives

$$(\underset{k+1}{v} \circ \underset{k+1}{u}(x)) \circ (\underset{1}{D} f \circ \underset{1}{u}(x)) = (\underset{k+1}{D} \underset{k}{v}) \circ \underset{k+1}{u}(x).$$

Comparison with the preceding result establishes (5.6) with $k + 1$ instead of k. ■

6. Main Feature of Continuation

Let h and $\bar{h}$ be two transformations of the space $Z = X \times Y$. For group analysis it is of fundamental importance to establish the equality of two continuations of the continuous space Z_k, namely, the kth continuation of the composition $\underset{k}{\bar{h} \circ h}$ and the composition of the kth continuations $\underset{k}{\bar{h}} \circ \underset{k}{h}$. Although this equality is almost obvious, because of its extraordinary importance for all further work, it is worth further study and proof in detail.

Lemma. *The continuation of the compositions of two transformations is equal to the composition of the two continuations of the transformations, that is,* $\underset{k}{\bar{h} \circ h} = \underset{k}{\bar{h}} \circ \underset{k}{h}$.

Proof. Let $h = (f, g), \bar{h} = (\bar{f}, \bar{g})$, and $\underset{k}{z'} = \underset{k}{h}(\underset{k}{z})$. First it is established that the identity

$$\underset{k}{D}F'(\underset{k}{z}) = \underset{k}{D}F(\underset{k}{z'}) \circ \underset{1}{D}f(\underset{1}{z}), \tag{6.1}$$

where $F'(\underset{k}{z}) = F \circ \underset{k}{h}(\underset{k}{z}) = F(\underset{k}{z'})$, holds for any mapping $F: Z_k \to P$. To establish this, it is necessary to note that because of definition (5.3), the operators $\underset{s}{\partial}$ operate on the mapping $\underset{k}{z} \to F'(\underset{k}{z})$ by means of

$$\underset{s}{\partial}F'(\underset{k}{z}) = \sum_{\sigma=-1}^{k} \underset{\sigma}{\partial}F(\underset{k}{z'}) \circ \underset{s\sigma}{\partial}\underset{\sigma}{v}(z),$$

where the notation $\underset{-1}{\partial} = \partial_x$ is used. If it is assumed that $\underset{0}{y} = I_X$, then because of this result and (5.4), we obtain

$$\begin{aligned}
\underset{k}{D}F'(\underset{k}{z}) &= \sum_{s=-1}^{k} \underset{s+1}{y} \cdot \underset{s}{\partial}F'(\underset{k}{z}) \\
&= \sum_{\sigma=-1}^{k} \underset{\sigma}{\partial}F(\underset{k}{z'}) \circ \sum_{s=-1}^{k} \underset{s+1}{y} \cdot \underset{s\sigma}{\partial}\underset{\sigma}{v}(z) \\
&= \sum_{\sigma=-1}^{k} \underset{\sigma}{\partial}F(\underset{k}{z'}) \circ \underset{k+1}{D}\underset{\sigma}{v}(z) \\
&= \sum_{\sigma=-1}^{k} \underset{\sigma+1}{y'} \cdot \underset{\sigma}{\partial}F(\underset{k}{z'}) \circ \underset{1}{D}f(\underset{1}{z}) \\
&= \underset{k}{D}F(\underset{k}{z'}) \circ \underset{1}{D}f(\underset{1}{z}),
\end{aligned}$$

that is, (6.1) is correct.

Introduction of the components of the product $Z_k = Z \times Y_1 \times \cdots \times Y_k$ leads to the representation

$$\bar h_k \circ h_k(z_k) = \bar h_k(z'_k) = (\bar h(z'), \bar v_1(z'_1), \ldots, \bar v_k(z'_k)),$$

where the mappings $\bar v_s$ are defined by the formulas (5.4) using the mapping $\bar h = (\bar f, \bar g)$. Let the analogous representation of the continuation of the composition be

$$\bar h_k \circ h_k(z) = (\bar h \circ h(z), \bar{\bar v}_1(z_1), \ldots, \bar{\bar v}_k(z_k)).$$

Since $\bar h(z') = \bar h \circ h(z)$, to prove the lemma it is necessary to establish the correctness of the equalities

$$\bar{\bar v}_s(z_s) = \bar v_s(z'_s) \qquad (s = 1, 2, \ldots, k), \tag{6.2}$$

where $z'_s = h_s(z_s)$.

Let $\bar{\bar f}(z) = \bar f(z')$, $\bar{\bar g}(z) = \bar g(z')$, whence $\bar h \circ h = (\bar{\bar f}, \bar{\bar g})$. Because of the identity (6.1), the equalities

$$D_1 \bar{\bar f}(z_1) = D_1 \bar f(z'_1) \circ D_1 f(z_1),$$

$$D_1 \bar{\bar g}(z_1) = D_1 \bar g(z'_1) \circ D_1 f(z_1)$$

hold. From these and definition (5.4) it follows that

$$\begin{aligned} \bar{\bar v}_1(z_1) &= D_1 \bar{\bar g}(z_1) \circ [D_1 \bar{\bar f}(z_1)]^{-1} \\ &= D_1 \bar g(z'_1) \circ [D_1 \bar f(z'_1)]^{-1} \\ &= \bar v_1(z'_1), \end{aligned}$$

which is (6.2) for $s = 1$. If it is assumed that (6.2) is correct for $s = k - 1$, then because of (5.4) and (6.1) it follows that

$$\begin{aligned} \bar{\bar v}_k(z_k) &= D_k \bar{\bar v}_{k-1}(z_k) \circ [D_1 \bar{\bar f}(z_1)]^{-1} \\ &= D_k(\bar v_{k-1}(z'_{k-1})) \circ [D_1 \bar{\bar f}(z_1)]^{-1} \\ &= [D_k \bar v_{k-1}(z'_k) \circ D_1 f(z_1)] \circ [D_1 \bar f(z'_1) \circ D_1 f(z_1)]^{-1} \\ &= D_k \bar v_{k-1}(z'_k) \circ [D_1 \bar f(z'_1)]^{-1} \\ &= \bar v_k(z'_k), \end{aligned}$$

that is, (6.2) holds for $s = k$. ■

7. Continuation Group

Let G^1 be a Lie group of transformations of the space $Z = X \times Y$, generated by the local operator $h: Z \times R \to Z$. As a result of a continuation of every transformation $h_a \in G^1$ to the transformation $\underset{k}{h_a}$ of the space Z_k there is obtained a one-parameter family of transformations generated by the mapping $\underset{k}{h}: Z_k \times R \to Z_k$.

The mapping $\underset{k}{h}$ is an operation of the additive group R on the space Z_k. Indeed, because of the lemma in Section 4.6, the relation $\underset{k}{h_a} \circ \underset{k}{h_b} = \underset{k}{h_{a+b}}$ corresponds to $h_a \circ h_b = h_{a+b}$, that is, the definition of the operation in Section 1.1 is satisfied.

As a result of this fact it is easy to check that the mapping $\underset{k}{h}$ possesses all of the properties of a mapping generating a local continuous one-parameter group of local transformations of the space Z_k, as described in the definitions of Section 1.9. In particular, the set V can be formed, for which all transformations $\underset{k}{h_a}$ are defined and belong to the class $C_\infty(V)$ for all $a \in \Delta$, while keeping in mind the remark of Section 4.5 made because of condition (5.2).

So the continuation mapping $\underset{k}{h}: Z_k \times R \to Z_k$ generates a group $G^1(\underset{k}{h})$ of local transformations of the space Z_k. This group is denoted by $\underset{k}{G^1}$ or $\underset{k}{G^1}(h)$ and called the *continuation group*, obtained by the kth continuation of the group $G^1(h)$.

Later, the mapping $h = (f, g)$, generating the group G^1 of transformations of the space $Z = X \times Y$, will be given by formulas of the type

$$x' = f(x, y, a), \qquad y' = g(x, y, a), \tag{7.1}$$

and the kth continuation $\underset{k}{h}$ of h, with space Y_k, will be given by

$$\underset{k}{y}' = \underset{k}{v}(x, y, \underset{1}{y}, \dots, \underset{k}{y}, a), \tag{7.2}$$

where the mappings v are defined by f and g in formulas (5.4).

Example. The group of translations in $Z = X \times Y$ is generated by the mapping $h(x, y, a) = (x, y) + (\lambda, \mu)a$, where $x, \lambda \in X$, $y, \mu \in Y$, $a \in R$, and λ and μ are constant vectors. Here $f(x, y, a) = x + \lambda a$, $g(x, y, a) = y + \mu a$, and equation (3.2) gives $\underset{1}{D} f = I_X$, $\underset{1}{D} g = \underset{1}{y}$. Consequently, $\underset{1}{v} = \underset{1}{y}$ and since

$D\underset{k}{y} = \underset{k+1}{y}$ for any k, according to (3.2), it follows that $\underset{k}{v} = \underset{k}{y}$ from (5.4). As a result, for all $a \in R$, the continuous transformations have the form

$$(x', y', \underset{1}{y}', \ldots, \underset{k}{y}') = (x + \lambda a, y + \mu a, \underset{1}{y}, \ldots, \underset{k}{y}),$$

that is, all transformations $\underset{k}{h}_a$ of the group $\underset{k}{G}^1$ leave invariant all components $\underset{1}{y}, \ldots, \underset{k}{y}$ and actually lead to transformations of the space Z. In other words, the continuation mappings $\underset{k}{h}$ trivially operate on the space $Y_1 \times \cdots \times Y_k$. Expressing this fact, sometimes it is said that the group of translations "is not continuous."

Remark. If the groups $G^1(h)$ and $\bar{G}^1(\bar{h})$ are similar (see Section 1.11), that is, transformations h_a and $\bar{h}_a$ are related by the equation $\bar{h}_a = p \circ h_a \circ p^{-1}$ by means of some transformation p of the space Z, then because of the lemma in Section 4.6 the continuation transformations will be connected by the formula $\underset{k}{\bar{h}}_a = \underset{k}{p} \circ \underset{k}{h}_a \circ \underset{k}{p}^{-1}$. Consequently, the continuation groups $\underset{k}{G}^1$ and $\underset{k}{\bar{G}}^1$ are also similar, and for them the continuation transformation $\underset{k}{p}$ is the similarity transformation.

8. Continuation of Infinitesimal Operator

Let ζ be a tangent vector field for the group $G^1(h)$ of transformations of the space $Z = X \times Y$. As ζ is a mapping $Z \to Z$, it is possible to assume $\zeta = (\xi, \eta)$, where ξ is the component of the vector ζ in X and η is its component in Y. If the mapping h, generating the group, is given by formulas (7.1), then these components are calculated from

$$\xi(x, y) = \partial_a f(x, y, 0), \qquad \eta(x, y) = \partial_a g(x, y, 0) \tag{8.1}$$

because of equation (1.13.1). According to this notation the infinitesimal operator of the group $G^1(\zeta)$ has the form

$$\zeta \cdot \partial = \xi \cdot \partial_x + \eta \cdot \partial_y. \tag{8.2}$$

The tangent vector $\underset{k}{\zeta}$ of the continuation group $\underset{k}{G}^1(h)$ is written in the form

$$\underset{k}{\zeta} = (\xi, \eta, \underset{1}{\eta}, \ldots, \underset{k}{\eta}), \tag{8.3}$$

where η_s $(s = 1, \ldots, k)$ are the components of the vector $\underset{k}{\zeta} \in Z_k$ in the continuation space Y_k. All components $\underset{k}{\eta}$ $(k = 1, 2, \ldots)$ are expressed by means of ξ, η with the aid of the so-called *formulas of continuation of the vector field* $\zeta = (\xi, \eta)$.

From (7.2) and the definition in Section 1.13 the relation $\underset{k}{\eta} = \partial_a \underset{k}{v}(\underset{k}{z}, 0)$ is easily obtained and employed to deduce the formulas of continuation. Since the operator of total differentiation D commutes with differentiation with respect to the parameter a, the recurrence formula (5.4), rewritten as $\underset{k+1}{v} \cdot \underset{1}{D} f = \underset{k+1}{D}\underset{k}{v}$, becomes the relation

$$\underset{k+1}{\eta} + \underset{k+1}{y} \cdot \underset{1}{D}\xi = \underset{k+1}{D}\underset{k}{\eta}$$

after differentiation with respect to the parameter a, setting $a = 0$, and using $\underset{k}{v}(\underset{k}{z}, 0) = \underset{k}{y}$. From this, assuming that $\underset{0}{\eta} = \eta$, the recurrence formula for the continuation of a vector field is found to be

$$\underset{k+1}{\eta} = \underset{k+1}{D}\underset{k}{\eta} - \underset{k+1}{y} \cdot \underset{1}{D}\xi \qquad (k = 0, 1, 2, \ldots). \tag{8.4}$$

In some applications it is convenient to use a formula equivalent to (8.4),

$$\underset{k}{\eta} = D^k(\eta - \underset{1}{y} \cdot \xi) + \underset{k+1}{y} \cdot \xi, \tag{8.5}$$

where D^k is the kth power of an operator of total differentiation, (3.1).

Corresponding to the representation (8.3) of a tangent vector for a continuation group $\underset{k}{G^1}(\zeta)$ is the infinitesimal operator

$$\underset{k}{\zeta} \cdot \partial = \xi \cdot \partial_x + \eta \cdot \partial_y + \eta_1 \cdot \underset{1}{\partial} + \cdots + \underset{k}{\eta} \cdot \underset{k}{\partial}. \tag{8.6}$$

The operator (8.6) is called a *continuation operator* or kth *continuation of the operator* (8.2).

9. Standard Commutators

Let a sequence of mappings $\underset{k}{\sigma}: Z_k \to Z_\infty$ of class $C_\infty(Z_k)$ $(k = -1, 0, 1, 2, \ldots)$ be given, where $Z_{-1} = Z_0 = Z$ and Z_∞ denotes the infinite continuation of the space Z (union of all the Z_k). The infinite operator

$$\sigma \cdot \partial = \sum_{k=-1}^{\infty} \underset{k}{\sigma} \cdot \underset{k}{\partial} \qquad (\underset{-1}{\partial} = \partial_x, \underset{0}{\partial} = \partial_y) \tag{9.1}$$

generated with the aid of this sequence is considered (the sum is formal).

Let $\tau \cdot \partial$ be another infinite operator, analogous to (9.1). The *commutator* of these operators is an infinite operator defined by

$$[\tau \cdot \partial, \sigma \cdot \partial] = ((\tau \cdot \partial)\sigma - (\sigma \cdot \partial)\tau) \cdot \partial$$

$$= \sum_{k=-1}^{\infty} ((\tau \cdot \partial)\underset{k}{\sigma} - (\sigma \cdot \partial)\underset{k}{\tau}) \cdot \underset{k}{\partial}. \tag{9.2}$$

Some commutators, called *standard* in the sense that they possess an explicit expression, are considered below. They are generated for operators of total differentiation $\underset{k}{D}$ or for the infinite operator of total differentiation (3.1)

$$D = \sum_{k=-1}^{\infty} \underset{k+1}{y} \cdot \underset{k}{\partial},$$

where $\underset{0}{y} \cdot \underset{-1}{\partial} = \partial_x$, and the infinite operator (8.2)

$$\underset{\infty}{\zeta} \cdot \partial = \sum_{k=-1}^{\infty} \underset{k}{\eta} \cdot \underset{k}{\partial},$$

where $\underset{-1}{\eta} \cdot \underset{-1}{\partial} = \xi \cdot \partial_x, \underset{0}{\eta} \cdot \underset{0}{\partial} = \eta \cdot \partial_y$. First the expression for the fundamental standard commutator is obtained.

Lemma. *The equality*

$$[D, \underset{\infty}{\zeta} \cdot \partial] = D\xi \cdot D \tag{9.3}$$

holds.

Proof. Because of (8.4),

$$\underset{k}{D\eta} - (\underset{\infty}{\zeta} \cdot \partial)\underset{k+1}{y} = \underset{k+1}{\eta} + \underset{k+1}{y} \cdot \underset{1}{D\xi} - \underset{k+1}{\eta} = \underset{k+1}{y} \cdot \underset{1}{D\xi}.$$

Therefore, according to the definition of the commutator (9.2),

$$[D, \underset{\infty}{\zeta} \cdot \partial] = \sum_{k=-1}^{\infty} (\underset{k}{D\eta} - (\underset{\infty}{\zeta} \cdot \partial)\underset{k+1}{y})\underset{k}{\partial} = \sum_{k=-1}^{\infty} (\underset{k+1}{y} \cdot \underset{1}{D\xi}) \cdot \underset{k}{\partial}$$

$$= \underset{1}{D\xi} \cdot \sum_{k=-1}^{\infty} \underset{k+1}{y} \cdot \underset{k}{\partial} = D\xi \cdot D,$$

where the next to the last equality holds because of $(k+1)$ symmetry of the linear mapping $\underset{k+1}{y}$ and the last equality is true because $\xi: Z \to X$. ■

The other standard commutators are deduced from (9.3). For example, if the operator $\underset{\infty}{\zeta} \cdot \partial$ is changed to the operator $\underset{k}{\zeta} \cdot \partial$, then all sums are finite and (9.3) gives a standard commutator in the form

$$[\underset{k+1}{D}, \underset{k}{\zeta} \cdot \partial] = \underset{1}{D\xi} \cdot \underset{k+1}{D}. \tag{9.4}$$

Then with the aid of (9.3), rewritten as

$$D(\underset{\infty}{\zeta}, \partial) = (\underset{\infty}{\zeta} \cdot \partial)D + D\xi \cdot D,$$

another commutator is calculated to be

$$\begin{aligned}[D^2, \underset{\infty}{\zeta} \cdot \partial] &= D^2(\underset{\infty}{\zeta} \cdot \partial) - (\underset{\infty}{\zeta} \cdot \partial)D^2 = D((\underset{\infty}{\zeta} \cdot \partial)D + D\xi \cdot D) - (\underset{\infty}{\zeta} \cdot \partial)D^2 \\ &= (D(\underset{\infty}{\zeta} \cdot \partial))D + D^2\xi \cdot D + D\xi \cdot D^2 - (\underset{\infty}{\zeta} \cdot \partial)D^2 \\ &= D^2\xi \cdot D + 2D\xi \cdot D^2.\end{aligned}$$

In an analogous way there is obtained without difficulty the following expression for a standard commutator (verified by induction with respect to k),

$$[D^k, \underset{\infty}{\zeta} \cdot \partial] = \sum_{l=0}^{k-1} \binom{k}{l} D^{k-l}\xi \cdot D^{l+1}, \tag{9.5}$$

of which (9.3) is a particular case (when $k = 1$). These calculated standard commutators will be used in Sections 19 and 24.

10. Finite-Dimensional Case

If $X = R^n$, $Y = R^m$, and $Z = R^{n+m}$, then the notations of Section 4.2 are employed, and it is further assumed that $\xi = (\xi^1, \ldots, \xi^n)$ and $\eta = (\eta^1, \ldots, \eta^m)$. In this case the operator (8.2) is rewritten as

$$\zeta \cdot \partial = \xi^i \partial_i + \eta^k \underset{0}{\partial_k}, \tag{10.1}$$

where $\partial_i = \partial_{x^i}$ and $\underset{0}{\partial_k} = \partial_{y^k}$. The vector $\underset{k}{\eta}$ has scalar components $\eta^l_{i_1 \cdots i_k}$, symmetric relative to permutations of the lower indices, each of which takes the values $1, \ldots, n$, while the index l takes the values $1, \ldots, m$. From this, the notation of Section 4.3, and the continuation formulas of (8.4), the expression for the components of the vector $\underset{k+1}{\eta}$ are found to be

$$\begin{aligned}\eta^l_{i_1 \ldots i_k j} &= \left(\partial_j + \sum_{s=0}^{k} y^r_{j_1 \ldots j_s j} \underset{s}{\partial^{j_1 \ldots j_s}_r}\right)\eta^l_{i_1 \ldots i_k} \\ &\quad - y^l_{i_1 \cdots i_k i}(\partial_j + y^r_j \underset{0}{\partial_r})\xi^i,\end{aligned} \tag{10.2}$$

where the summation over the repeated indices i, r, l and any possible (counting the symmetry) sets $(j_1, \ldots, j_s)$ is assumed.

In applications the values $k = 0$ and $k = 1$ are met especially often. Setting $k = 0$ in (10.2), the *formulas* for the components of the first continuation vector $\underset{1}{\eta}$ are found to be

$$\eta^l_j = \partial_j \eta^l + y^s_j \underset{0}{\partial}_s \eta^l - y^l_s \partial_j \xi^s - y^l_s y^r_j \underset{0}{\partial}_r \xi^s. \tag{10.3}$$

When $k = 1$ the *formulas of the second continuation* are obtained, which consist of (10.3), and from (10.2) the formulas for the components of the vector $\underset{l}{\eta}$ are found to be

$$\eta^l_{ij} = \partial_j \eta^l_i + y^s_j \underset{0}{\partial}_s \eta^l_i + y^s_{rj} \underset{1}{\partial}{}^r_s \eta^l_i - y^l_{is} \partial_j \xi^s - y^l_{is} y^r_j \underset{0}{\partial}_r \xi^s. \tag{10.4}$$

11. Example

For the group of projective transformations of the plane (example in Section 1.10), $X = R(x)$, $Y = R(y)$, $Z = R^2(x, y)$, and equations (7.1) are

$$x' = f(x, y, a) = \frac{x}{1 - ax}, \qquad y' = g(x, y, a) = \frac{y}{1 - ax}.$$

The infinitesimal operator of this group is $\zeta \cdot \partial = x^2 \partial_x + xy\partial_y$. Since all spaces $Y_k = \mathscr{L}^k_s(R; R)$ are one-dimensional here, then all $\underset{k}{y} \in R$, and it is possible to use the simplified notations $\underset{k}{y} = y_k$. The consecutive calculation of the first continuation gives

$$\underset{1}{D} f = (1 - ax)^{-2}, \qquad \underset{1}{D} g = (ay + (1 - ax)y_1)(1 - ax)^{-2}$$

and, according to equation (5.4),

$$y'_1 = ay + (1 - ax)y_1.$$

Consequently, $\eta_1 = \partial_a y'_1(0) = y - xy_1$ and the first continuation of the operator $\zeta \cdot \partial$ is

$$\underset{1}{\zeta} \cdot \partial = x^2 \partial_x + xy\partial_y + (y - xy_1)\partial_{y_1}.$$

To construct the second continuation, it is found that

$$\underset{2\,1}{D v} = (\partial_x + y_1 \partial_y + y_2 \partial_{y_1})(ay + (1 - ax)y_1) = (1 - ax)y_2.$$

By equation (5.4) it is found that $v_2 = (1 - ax)^3 y_2$, and then

$$\eta_2 = \partial_a((1 - ax)^3 y_2)(0) = -3xy_2.$$

The second continuation of the operator $\zeta \cdot \partial$ is therefore

$$\underset{2}{\zeta} \cdot \partial = x^2\partial_x + xy\partial_y + (y - xy_1)\partial_{y_1} - 3xy_2\partial_{y_2}.$$

12. Differential Invariants

Invariants of the continuous group $\underset{k}{G}^1$ are called the *differential invariants of the group* G^1. By themselves they represent mappings $F: Z_k \to P$. An invariant of the group $\underset{k}{G}^1$, actually dependent on the vector $\underset{k}{y}$, is called a *differential invariant* of the group G^1 of the kth order.

Because of the criterion (3.4.1), all differential invariants F of the group $G^1(\zeta)$ of order not more than k are the solutions of the differential equation

$$(\underset{k}{\zeta} \cdot \partial)F(\underset{k}{z}) = 0, \tag{12.1}$$

in which $\zeta \cdot \partial$ is the continuation operator (8.6). The universal differential invariant of kth order generates the general solution of equation (12.1).

For example, for the group of projective transformations of the plane (from Section 4.11), all differential invariants F of order not more than the second must be solutions of the equation

$$x^2\partial_x F + xy\partial_y F + (y - xy_1)\partial_{y_1} F - 3xy_2\partial_{y_2} F = 0.$$

Here the universal invariant, as a mapping $R^4(x, y, y_1, y_2) \to R^3$, is easily constructed by finding three independent first integrals of a characteristic system of ordinary differential equations

$$\frac{dx}{x^2} = \frac{dy}{xy} = \frac{dy_1}{y - xy_1} = \frac{dy_2}{-3xy_2},$$

which can be taken as

$$J(x, y, y_1, y_2) = (y/x, y - xy_1, x^3y^2).$$

Therefore, the general form of the differential invariant of order not more than two is

$$F(x, y, y_1, y_2) = \Phi \circ J(x, y, y_1, y_2) = \Phi(y/x, y - xy_1, x^3y_2),$$

where $\Phi: R^3 \to P$ is an arbitrary function in three variables.

If the manifold $\Psi \subset Z_k$, given with the aid of some mapping $\psi: Z_k \to R^s$ which is an equation of the form

$$\psi(\underset{k}{z}) = 0, \tag{12.2}$$

is an invariant manifold of the group $\underset{k}{G^1}$, then Ψ is also called an *invariant differential manifold of the group* G^1.

Differential invariants and invariant differential manifolds of the groups G^1 are one of the fundamental objects of the group analysis of differential equations. Development of the corresponding theory is connected with the solution of a number of concrete problems, which will be considered in the following chapters.

REFERENCES

The first detailed exposition of the basic elements of the continuous transformation group theory was given in a three-volume monograph by Lie–Engel [1]. For the following one hundred years, this theory has undergone significant changes. Although in relatively recent published books [2, 3] it was still presented in Lie's form, its modern exposition [4] has little in common with the initial one. The necessity for group analysis of differential equations as a continuation theory in its own right, as presented in Section 4, is given here for the first time. The modern conditions and the main results of the group analysis of differential equations have been elucidated in a number of surveys, for example, in the article [5] and in the report [6]. The results and solutions for a number of concrete problems, considered in the present book, are systematized in [7]. In the survey [8] is found information about the results found by methods of Lie group theory before 1973.

[1] Lie S., Engel F. Teorie der Transformationsgruppen, Bd. 1—3. Leipzig, Teubner, 1888, 1890, 1893.
[2] Эйзенхарт Л. П. Непрерывные группы преобразований. М., ИЛ, 1974.
[3] Чеботарев Н. Г. Теория групп Ли. М., Гостехиздат, 1949.
[4] Понтрягин Л. С. Непрерывные группы. М., Гостехиздат, 1954.
[5] Овсянников Л. В. Групповые свойства уравнений механики — В сб. «Механика сплошной среды и родственные проблемы анализа», М., «Наука», (1972), 381—393.
[6] Овсянников Л. В. Some Problems Arising in Group Analysis of Differential Equations. — Proc. of Symposium «Symmetry, Similarity and Group Theoretic Methods in Mechanics», Calgary, Canada, Aug. 19—21 (1974), 181—202.
[7] Ибрагимов Н. Х. Группы Ли в некоторых вопросах математической физики. Новосибирск, НГУ, 1972.
[8] Овсянников Л. В., Ибрагимов Н. Х. Групповой анализ дифференциальных уравнений механики. — Итоги науки и техники. Общая механика, 2, М. (1975), 5—52.

CHAPTER

II

Groups Admitted by Differential Equations

For the aims of group analysis it is found to be convenient and essential to interpret the differential equation E (system of differential equations) as a manifold in a continuous space. A geometrical language permits a natural definition of the concept of invariance of a differential equation, relative to a group of transformations, and the use of an infinitesimal criterion for invariance. According to Lie's suggestion, one says that E admits a group (or operator) if E is contained in an invariant manifold of a continuous group. In the present chapter the problem is posed and solved: Find and study the set of all groups G^1, admitted by the differential equation E.

This problem is a remarkable example in which the infinitesimal approach (that is, using an operator instead of the group) results in a linearization of the problem. The requirement of invariance of E leads to a system of linear and homogeneous differential equations in which the components of an admitted operator are found—the system of determining equations DE. Here the accurate and well-defined algorithm of transition $E \to DE$ operates, which can be implemented on a computer. The theoretical importance of such constructions is that with them the concept of the full group GE of transformations, admitted by E, appears.

The most important property of the full group GE for applications is that it operates on the set of the solutions, that is, any solution of the equation E is transformed again into a solution of the same equation E. The study of this operation and its utilization for the construction of solutions is one of the main problems of the group analysis of differential equations, which will be considered in Chapter V.

Another practical problem considered in this chapter is the problem of group classification of differential equations $E(\theta)$, containing a so-called "arbitrary element" θ. Here all necessary information is developed and algorithms are generated to solve this problem, which permits the possibility of purposeful selection of values and forms of experimentally defined physical quantities and relations θ, contained in the mathematical model $E(\theta)$.

The determining equations uniquely generate a new mathematical object—the Lie algebra of operators. Although a detailed abstract description of this algebra will be given in Chapter IV, there is already sufficient machinery here to become acquainted with it. This permits an understanding of the result of calculations of an admitted group as the common solution of the determining equations DE, which turns out to be a Lie algebra of operators. There arises an inevitable repetition which is justified because a practical solution of many problems of group analysis often deals not with groups of transformations themselves, but with their Lie algebras of operators.

§5. DETERMINING EQUATIONS

1. System of Differential Equations

Below, some of the concepts and notations used in Section 4 are reintroduced. To be considered are the finite-dimensional spaces $X = R^n(x)$, $Y = R^m(y)$, and $Z = X \times Y = R^{n+m}(z)$, where $z = (x, y)$. Let the Z_k be those continuation spaces introduced in Section 4.2.

The equation

$$\varepsilon(\underset{k}{z}) = 0 \tag{1.1}$$

generated with the aid of some mapping $\varepsilon\colon Z_k \to R^s$ is called a *differential equation.*

The natural numbers n, m, k, and s are fundamental characteristics of the differential structure (or "kind") of equation (1.1). Therefore, equation (1.1) will sometimes be denoted by the symbol $E(n, m, k, s)$. If the numbers n, m, k, s have fixed values (or in any problem play a minor role) then equation (1.1) will merely be denoted by E.

From a geometrical point of view, equation (1.1) is given some manifold in a continuous space Z_k; this manifold will also be denoted by the symbol E.

Every equation of type (1.1), written in component form, can be considered as a system with s scalar differential equations for m unknown functions $y = (y^1, \ldots, y^m)$ in n independent variables $x = (x^1, \ldots, x^n)$. That is why equation (1.1) is also called a *system of differential equations*, or, in abbreviated terms, the system E.

The number k is called the *order of the system* (1.1) if any one of its scalar equations can be solved in terms of any component of the vector $\underset{k}{y} \in Y_k$.

Further, it will be assumed that the given system mapping ε of (1.1) satisfies the condition under which the continuous space Z_k can be represented as the product $Z \times \overline{Z}_k$, where $\overline{Z}_k = Y_1 \times \cdots \times Y_k$, $\underset{k}{z} = (z, \underset{k}{\bar{z}})$. Then for any mapping $\varepsilon\colon Z_k \to R^s$ it is possible to generate a particular mapping $\varepsilon_z\colon \overline{Z}_k \to R^s$, operating according to the formula $\varepsilon_z(\underset{k}{\bar{z}}) = \varepsilon(z, \underset{k}{\bar{z}})$.

Assumption. *The mapping ε belongs to the class $C_\infty(Z_k)$ and the manifold E is given by equation* (1.1) *regularly* (see Section 3.11), *and also for any $z \in Z$ the particular mapping ε_0 has a rank in $\overline{Z}_k$ equal to s.*

2. Fundamental Definition

Let a set M, a subset N, and a transformation $T \in \tau(M)$ be given. It is said that N is *invariant relative to T* if $T(N) = N$. Obviously, the union of all transformations T, relative to which the given subset N is invariant, generates a subgroup of the group $\tau(M)$.

In the group analysis of differential equations the continuous space Z_k plays the role of the set M and the system of differential equations of the form (1.1), considered as a manifold $E \in Z_k$, plays the role of the set N. With this all equations, manifolds, different relations, etc., are considered in the framework of the local theory (see Section 1.8), although later the exposition, as a rule, "forgets terms" (see Section 3.3). In terms of these objects the fundamental definition is as follows:

A system of differential equations *E admits a transformation h of the space Z* if the manifold $E \subset Z_k$ is invariant relative to the continuous transformation $\underset{k}{h}$ of the space Z_k. If the system E admits every transformation from some group G^1, then it is said that *E admits the group G^1*. If the system E admits a group $G^1(\zeta)$, characterized by its operator $\zeta \cdot \partial$, then it is also said that the *system E admits the operator $\zeta \cdot \partial$*. With this h, G^1, or $\zeta \cdot \partial$ is accordingly called the *transformation*, the *group*, or the *operator admitted by the system E*. With respect to this definition the following problem arises: Find all groups G^1, of local transformations of the space Z, admitted by the given system of differential equations E.

3. Invariance Condition

In the following, the problem of finding all groups G^1 admitted by the system E is considered under the assumption that the mapping ε, giving the manifold E by means of equation (1.1), satisfies conditions in Section 5.1.

In this case the criterion in Section 3.12 of invariance of a manifold relative to a group $G^1(\zeta)$, with an infinitesimal operator $\zeta \cdot \partial$, can be used. This criterion, in the given situation, has the following form: The system E (1.1) satisfying the condition in Section 5.1 admits the group $G^1(\zeta)$ if and only if the condition of invariance

$$(\underset{k}{\zeta} \cdot \partial)\varepsilon(\underset{k}{z})|_E = 0 \tag{3.1}$$

holds.

The solution of the problem posed in Section 5.2 is composed of the following: The condition of invariance (3.1) is considered as an equation relative to the unknown vector field ζ. The structure of this equation depends only upon the given system E and is completely defined by the algorithm of construction.

The process of forming the conditions of invariance consists of three stages: (a) calculation of the continuous operator $\underset{k}{\zeta} \cdot \partial$, (b) operation by this operator on the mapping E, and (c) transition to the manifold E.

In stage (a) a general form of the operator $\zeta \cdot \partial$ is obtained, that is, the components of the vector $\zeta = (\xi, n)$ are found. Here it is important that ζ is a mapping $Z \to Z$, that is, the components ξ, η depend only upon variables $(x, y) \in X \times Y$. Then using the equations of continuation (4.8.4) or (4.8.5), the components $\eta^l_{i_1 \cdots i_s}$ $(1 \le s \le k)$, of the continuation operator $\underset{k}{\zeta} \cdot \partial$, are calculated.

In stage (b) the operator of (a) is applied to the mapping ε, which usually is studied with the aid of analytic expressions of its coordinates through the components of the vector $\underset{k}{z}$. Since the operator $\underset{k}{\zeta} \cdot \partial$ is a scalar differential operator, it can operate on every component of the vector ε separately. As a result, the new mapping $(\underset{k}{\zeta} \cdot \partial)\varepsilon\colon Z_k \to R^s$ is obtained, where the image $(\underset{k}{\zeta} \cdot \partial)\varepsilon(\underset{k}{z})$ consists of s functions $Z_k \to R$.

In stage (c) for the transition to the manifold E, a set of s such components $r^1, \ldots, r^s$ of the vector $\underset{k}{\bar{z}} = (\underset{1}{y}, \ldots, \underset{k}{y})$ is chosen such that relative to them the equations E can be solved "algebraically." This requirement is always satisfied because of the condition in Section 5.1. As a result of this operation the "dependent" components $r = (r^1, \ldots, r^s)$ are expressed through the remaining, "free" components of the vector $\underset{k}{\bar{z}}$, thereby generating a collection of scalars $t = (t^1, \ldots, t^q)$, where the number q depends on E. Then the obtained expressions $r = r(t)$ are substituted into the results of stage (b), i.e., the functions $(\underset{k}{\zeta} \cdot \partial)\varepsilon(\underset{k}{z})$, and these are equated to zero.

4. Determining Equations

The next step in the formation of the conditions of invariance is a stage (d) in which equations (3.1) are expanded relative to the "free" coordinates of the vector $\underset{k}{\bar{z}}$ introduced in stage (c). To accomplish this goal it is noted that the left-hand side, ψ, of equation (3.1) is a function of the independent variables x, y, t and derivatives of components of the vector ζ, which depend on $z = (x, y)$ but not t. That is why condition (3.1), written in the form $\psi(\zeta, x, y, t) = 0$, is expanded to generate a sequence of equations, which can be obtained, for example, with the aid of the expansion of the mapping ψ by the Taylor formula at any point t_0. For this purpose the point $t = 0$ is usually taken, and as a result of the expansion, the sequence of equations

$$\begin{aligned} \psi(\zeta, x, y, 0) &= 0, \\ \partial_t \psi(\zeta, x, y, 0) &= 0, \\ \partial_t^2 \psi(\zeta, x, y, 0) &= 0, \\ &\vdots \end{aligned} \tag{4.1}$$

is obtained.

Equations (4.1) are called the *determining equations* and will be denoted by the symbol DE. They are considered as equations with respect to the unknown vector field ζ or the operator $\zeta \cdot \partial$. The main property of the determining equations follows from the criterion in Section 3.12: whatever group $G^1(\zeta)$ is admitted by equation E, its tangent vector field ζ satisfies the equations DE; conversely, every solution of the equations DE, $\zeta : Z \to Z$ of the class $C_\infty(Z)$, is such a vector field, whose corresponding group $G^1(\zeta)$ is admitted by the equation E. It can be said that those and only those operators $\zeta \cdot \partial$ that are admitted by the equation E satisfy the equations DE.

The determining equations are a system of differential equations for the unknown vector field ζ. Because of their origin this system possesses a number of properties which are an independent object of analysis.

The first important property of the system DE is that all its equations are *linear* and *homogeneous* relative to the components of the unknown vector ζ. This property follows from the fact that it is possessed by the infinitesimal operator $\underset{k}{\zeta} \cdot \partial$ [see formulas of continuation (4.8.4) or (4.8.5)] and, as is easily verified, this property is not violated in all stages (a)–(d) of the construction of the equations DE. Therefore, a set of solutions of the determining equations DE generates a vector field. This field will be denoted with the symbol L in what follows.

5. Full Group

Let Γ be the union of all transformations h of a space Z that belong to the group $G^1(\zeta)$ whose tangent vector field ζ of class C_∞ is a solution of equations DE. Because of the preceding, all transformations belonging to Γ are admitted by the system E. But from the definition in Section 5.2 and the lemma in Section 4.6 it follows that if the system E is invariant relative to the transformations h and $\bar{h}$, then it is invariant relative to their composition $\bar{h} \circ h$. Consequently, for any two transformations $h, \bar{h} \in \Gamma$, the composition $\bar{h} \circ h$ is also admitted by the system E.

In the general case there is no reason to believe that such a composition $\bar{h} \circ h$ belongs to the group G^1 admitted by the system E, and that is why it is unknown whether $\bar{h} \circ h \in \Gamma$. This problem is not trivial even in the local theory. If the composition $\bar{h} \circ h$ is not contained in Γ, then the union of the transformations Γ is not a group. Nevertheless, the set Γ can be considered as generating (with the aid of composition) a group of transformations admitted by the system E. So a group arises considered as finitely generated by the set Γ (that is, every element is a composition of a finite number of elements from Γ). The group is described by the following definition.

The *full group of the system of differential equations E* is a group of local transformations of the space Z, finitely generated by the set Γ of transformations belonging to the one-parameter Lie group G^1 admitted by the system E.

In the sequel, the full group of the system E will be denoted by the symbol GE. In the important practical case, when the space L of solutions of the system DE is finite dimensional, the coincidence $\Gamma = GE$ occurs. This will be shown in chapter IV. It is worthwhile to note also that the system E can admit transformations which do not belong to the full group GE. Certainly, such transformations are not contained in any group G^1 admitted by the system E.

An example of the last possibility is given by any equation which admits the group of all orthogonal transformations of the space R^3 and only it. In this case, Γ is the union of all rotations (see Sections 1.15 and 2.7) and coincides with GE, and a mapping in any coordinate plane (which is an orthogonal transformation of the space R^3) does not belong to GE and cannot be obtained through any combination of rotations.

6. Actions on Solutions

Using the notation from Section 4.4, it is possible to formulate the following definition:

A mapping $u: X \to Y$ is called a *solution of the differential equation* (1.1) if for all $x \in X$ the equality

$$\varepsilon(\underset{k}{u}(x)) = 0 \tag{6.1}$$

holds.

In a number of problems it is convenient to use a geometric definition of the concept of a solution which is equivalent to this one. To formulate it, it is necessary to consider a manifold U_k in the space Z_k, given parametrically by the mapping $\underset{k}{u}: X \to Z_k$, operating by the formula $\underset{k}{z} = \underset{k}{u}(x)$, that is,

$$U_k = \{\underset{k}{z} \mid \underset{k}{z} = \underset{k}{u}(x),\ x \in X\}. \tag{6.2}$$

Then, directly from definition (6.1), it follows that the mapping $u: X \to Y$ is a solution of equation (1.1), considered as a manifold $E \subset Z_k$, if and only if the manifold U_k satisfies the *formula of inclusion*

$$U_k \subset E. \tag{6.3}$$

This formula of inclusion gives a geometric definition to the concept of a solution of the equation E.

To understand a proof of any fundamental results of transformations of solutions it is necessary to refer to Section 4.5, from which it is clear how the transformation h of the space $Z = X \times Y$ acts on the mappings $u: X \to Y$. In particular, the lemma in Section 4.5 states that if the transformed mapping u' is defined by equations (4.5.5), then for the continuation mapping $\underset{k}{u'}$ the formulas

$$\underset{k}{u'}(x') = \underset{k}{h} \circ \underset{k}{u}(x), \qquad x' = f \circ \underset{0}{u}(x) \tag{6.4}$$

hold. Let U'_k denote a manifold in Z_k given by formula (6.2) using the transformed mapping u'.

Theorem. *If the differential equation E admits a transformation h, then any solution $u: X \to Y$ of the equation E is transformed, with the aid of h, into the solution $u': X \to Y$ of the same equation E.*

Proof. If u is a solution, then for the corresponding manifold U_k the formula of inclusion (6.3) is satisfied. So if $\underset{k}{z} = \underset{k}{u}(x) \in U_k$, then, because of the invariance of the manifold E, the point $\underset{k}{z'} = \underset{k}{h}(\underset{k}{z}) = \underset{k}{h} \circ \underset{k}{u}(x)$ also belongs to E. From (6.4) it follows that the point $\underset{k}{z'} = \underset{k}{u'}(x') \in E$ for any $x' = f \circ \underset{0}{u}(x)$. Since $f \circ \underset{0}{u}: X \to X$ is a homeomorphism, the point $\underset{k}{z'} = \underset{k}{u'}(x) \in E$ for any $x \in X$. Consequently, $U'_k \subset E$. ■

Expressing this result in other words, it is possible to say that the transformation $h: Z \to Z$, admitted by the equation E, acts on the set of solutions of this equation. Let GE be the full group of the equation E. An application of this theorem to any transformation $h \in GE$ leads to the important consequence:

Consequence. *For any differential equations E is defined the action of the full group GE on the set of solutions of this equation. In short, the full group GE acts on the set of solutions of the equation E.*

7. Generation of Solutions

The statement of the previous consequence will be fully employed in Chapters V–VII. Here it is only noted how it can be employed for the generation of a solution from any known ones.

Let $u: X \to Y$ be any solution of the equation E and let the group $G^1(h)$ be admitted by the equation E. As a result of a transformation of the solution u, with the aid of the element $h_a \in G^1(h)$, there is obtained a new solution u_a for any $a \in R$. In this way the admitted group G^1 produces a one-parameter family of solutions $\{u_a\}$ from one solution. If there is another admitted group G^1, then this construction can be applied to every solution u_a so as to obtain a two-parameter family of solutions, and so on.

For the concrete generation of the solution it is necessary to have the equations for the transformations $h_a \in G^1(h)$ [see (4.7.1)]

$$x' = f(x, y, a), \qquad y' = g(x, y, a), \tag{7.1}$$

which will be operated on with the common rules for transformations of mappings discussed in Section 4.5. If h_a, given by the formulas (7.1), transforms the mapping u into the mapping u_a, then the value $u_a(x')$ is found by the elimination of the vector x from the equations

$$x' = f(x, u(x), a), \qquad u_a(x') = g(x, u(x), a). \tag{7.2}$$

In the result the transformed solution can be written in the form

$$u_a(x) = (g_a \circ \underset{0}{u}) \circ (f_a \circ \underset{0}{u})^{-1}(x), \tag{7.3}$$

where the mapping $\underset{0}{u}: X \to X \times Y$ operates according to the formula $\underset{0}{u}(x) = (x, u(x))$.

8. Algorithm

The determination of the full group GE of the given system of differential equations E is the first step in the group analysis of the system. The algorithm for the construction of the full group will be described here, and some of its peculiarities and properties will be noted.

The aforementioned algorithm consists of three parts—forming the determining equations DE, constructing the space of solutions L of the equations DE, and generating the set Γ of the groups $G^1(\zeta)$—for all vector fields $\zeta \in L$.

The first part—the forming of the determining equations or the transition $E \to DE$—needs the application of only one operation of differentiation and elimination. It is subdivided into the four stages (a)–(d), described in Sections 5.3 and 5.4. The accumulated experience in developing this part of the algorithm can be stated in two essential suggestions. First, in those cases where the equation E does not have a very complicated structure relative to the components of the vector $\underset{k}{\bar{z}}$, all calculations can be successfully resolved. Second, this part of the algorithm is very laborious because of the necessity to carry out a large number of monotonous and long calculations and transformations of equations having large numbers of terms.

To have some idea how much work is involved it is useful to calculate the total number of defining equations obtained at the end of stage (d). For example, let the given system of equations, (1.1), be first order, $E(n, m, 1, s)$, and quasi-linear, that is, linear (heterogeneous), with respect to the vector $\underset{1}{y}$. Here the total number of components of the vector $\underset{1}{\bar{z}} = \underset{1}{y}$ is equal to nm and the number of "free" components (see Section 5.3) is $nm - s$. Now it needs to be noted that the result given by the formulas of continuation (4.9.3) have additional components η_i^l of the continuous operation $\zeta \cdot \underset{1}{\partial}$ which are quadratic forms (heterogeneous) of componnets of the vector $\underset{1}{y}$. Consequently, because of the linearity of the equation E relative to $\underset{1}{y}$, the left-hand side of the system of equations (3.1) will also be heterogeneous quadratic forms of $q = nm - s$ "free" coordinates $t^1, \ldots, t^q$ after a transition of E [stage (c)]. Now it is easy to calculate that the number of scalar equations (4.1) is equal, in this case, to $s(q + 1)(q + 2)/2$. So for a quasilinear equation $E(n, m, 1, s)$ the total number N of defining equations is

$$N = \tfrac{1}{2}s(nm - s + 1)(nm - s + 2). \tag{8.1}$$

Some concrete numbers, typical of those for applied problems, are given in the following table, where it is assumed that $s = m$:

n	2	2	3	3	4	4
m	2	3	3	4	4	5
N	12	20	84	180	364	680

The analytical calculation can be made for other types of equations $E(n, m, k, s)$. It shows that, generally speaking, the number of the determining equations is very large.

This led to the idea of entrusting the calculation of the transition $E \to DE$ to a computer. This idea has been realized in the program CINO (from the first letters of the words components of infinitesimal operators) and PASSIV program, elaborated in CS SO Akad. Nauk, Novosibirsk, USSR.

The second part—construction of the space of solutions L of the determining equations DE—can cause some difficulties, which are connected with the large number of equations as well as the fact that they are partial differential equations for $n + m$ functions—components of the unknown vector $\zeta(\xi, \eta)$. But the large number N of equations DE must not be alarming for two reasons. First, among these equations there is met, generally speaking, a number of identities, so that the actual number of equations to be solved will be less than N. Second, when the system is very overdetermined, there are relatively few solutions, and to find them is relatively easy. Of course, the construction of the complete solution is not always successful and in a number of cases has to be limited using some useful subspaces instead of the space L. This must be done considering the additional algebraic structure of the space L, which will be discussed in Section 7.

The third part is to generate the set Γ and thereby complete the process of building the full group GE. The possibilities of its concrete realization are connected with the degree of success achieved in the second part of the algorithm. In Chapter IV the case of a finite-dimensional space L, when the generation of the set $\Gamma = GE$ requires only an integration of a system of ordinary differential equations, will be studied in detail. For this problem the theory of the algorithm described above is satisfactory. As to applications, in a number of problems it is enough to have one (or more) of the admitted groups $G^1(\zeta)$, whose tangent vector fields's (ζ) construction has already been described in Section 2.3.

9. Example

If all numbers n, m, k, s are not large, for example ≤ 2, then it is suitable to label the components of the vectors individually. For example, for an equation of the kind $E(2, 2, 1, 2)$ it is convenient to assume that $X = R^2(x, y)$, $Y = R^2(u, v)$, $Z = R^4(x, y, u, v)$, and $Y_1 = R^4(\underset{1}{y})$, where $\underset{1}{y}$ is the matrix

$$\underset{1}{y} = \begin{bmatrix} u_1 & u_2 \\ v_1 & v_2 \end{bmatrix}.$$

In this notation the system E is considered:

$$u_2 = v_1, \qquad v_2 = -uu_1, \tag{9.1}$$

the content of which is revealed through the definition of the solution: This is a mapping $(u, v)\colon R^2 \to R^2$, operating by the formula $(u, v)(x, y) = (u(x, y), v(x, y))$, and $u_1 = \partial_x u(x, y)$, $u_2 = \partial_y u(x, y)$, $v_1 = \partial_x v(x, y)$, $v_2 = \partial_y v(x, y)$. Now it is seen that (9.1) is well known in gas dynamics as the *equations of transonic steady two-dimensional flow of a gas*. For the system (9.1) the problem of constructing the full group GE is solved below.

In the first part, stage (a), the form of the infinitesimal operator,

$$\zeta \cdot \partial = \xi\partial_x + \eta\partial_y + \sigma\partial_u + \tau\partial_v, \tag{9.2}$$

where every coefficient is an unknown mapping $Z \to R$, is given. Then it is required to construct the first continuation of the operator (9.2). To help achieve this goal, components of the operator $\underset{1}{D} = (D_1, D_2)$ of total differentiation are introduced, where

$$D_1 = \partial_x + u_1\partial_u + v_1\partial_v,$$

$$D_2 = \partial_y + u_2\partial_u + v_2\partial_v,$$

and the continuation operator (9.2) has the form

$$\underset{1}{\zeta} \cdot \partial = \zeta \cdot \partial + \xi_1\partial_{u_1} + \xi_2\partial_{u_2} + \eta_1\partial_{v_1} + \eta_2\partial_{v_2}.$$

The additional coefficients are calculated by means of the formulas of the first continuation [e.g., (4.8.4) with $k = 0$] and turn out to be

$$\xi_1 = D_1\sigma - u_1D_1\xi - u_2D_1\eta, \qquad \xi_2 = D_2\sigma - u_1D_2\xi - u_2D_2\eta,$$

$$\eta_1 = D_1\tau - v_1D_1\xi - v_2D_1\eta, \qquad \eta_2 = D_2\tau - v_1D_2\xi - v_2D_2\eta.$$

Stage (b) consists of an operation by the operator $\underset{1}{\zeta} \cdot \partial$ on every equation from (9.1), which leads to the equations

$$\xi^2 = \eta_1, \qquad \eta_2 = -\sigma u_1 - u\xi_1.$$

To write these equations in expanded form, it is suitable to simplify the notation for partial derivatives, and for any mapping $\phi\colon Z \to R$ to write $\partial_x\phi = \phi_x$, $\partial_y\phi = \phi_y$, $\partial_u\phi = \phi_u$, and $\partial_v\phi = \phi_v$. Then, in expanded form, the previous equations become

$$\begin{aligned}\sigma_y &+ u_2\sigma_u + v_2\sigma_v - u_1(\xi_y + u_2\xi_u + v_2\xi_v) - u_2(\eta_y + u_2\eta_u + v_2\eta_v)\\ &= \tau_x + u_1\tau_u + v_1\tau_v - v_1(\xi_x + u_1\xi_u + v_1\xi_v) - v_2(\eta_x + u_1\eta_u + v_1\eta_v),\end{aligned}$$

$$\begin{aligned}\tau_y &+ u_2\tau_u + v_2\tau_v - v_1(\xi_y + u_2\xi_u + v_2\xi_v) - v_2(\eta_y + u_2\eta_u + v_2\eta_v)\\ &= -u_1\sigma - u[\sigma_x + u_1\sigma_u + v_1\sigma_v - u_1(\xi_x + u_1\xi_u + v_1\xi_v)\\ &\quad - u_2(\eta_x + u_1\eta_u + v_1\eta_v)].\end{aligned}$$

Stage (c)—transition to the manifold (9.1)—is carried out quite simply by the substitution in the previous equations of the expressions (9.1) for u_2 and v_2 in terms of u_1 and v_1. After regrouping terms, the result of this stage is written in the form of equations which express the condition of invariance of equations (9.1) relative to the operator (9.2):

$$\begin{aligned}(\sigma_y - \tau_x) &+ u_1(\tau_u + u\sigma_v + \xi_y + u\eta_x) + v_1(\tau_v - \sigma_u + \eta_y - \xi_x)\\ &+ u_1^2(u\eta_u - u\xi_v) + v_1^2(\eta_u - \xi_v) = 0,\\ (u\sigma_x + \tau_y) &+ u_1(\sigma + u\sigma_u - u\tau_v + u\eta_y - u\xi_x) + v_1(\tau_u + u\sigma_v - \xi_y - u\eta_x)\\ &- u_1^2(u^2\eta_v + u\xi_u) - v_1^2(u\eta_v + \xi_u) = 0.\end{aligned}$$

Here it is clearly seen that the left-hand sides are heterogeneous quadratic forms of the "free" variables u_1 and v_1.

Stage (d)—decomposition of the conditions of invariance—is reduced here to setting the coefficients of u_1, v_1, u_1^2, and v_1^2 equal to zero, which gives the determining equations. According to formula (8.1), 12 equations must be obtained, but in the given case four of them are identically satisfied. There remain only eight nontrivial determining equations, which after some examination can be written in the form of two subsystems:

$$\xi_y = \eta_u, \qquad \xi_u = -u\eta_v, \qquad \xi_y = -u\eta_x, \qquad \xi_x = \eta_y + \sigma/(2u) \tag{9.3}$$

and

$$\tau_u = -u\sigma_v, \qquad \tau_v = \sigma_u + \sigma/(2u), \qquad \tau_x = \sigma_y, \qquad \tau_y = -u\sigma_x. \tag{9.4}$$

The determining equations are formed and the first part of the algorithm in Section 5.8 is complete.

The second part—construction of the space of solutions of the system (9.3), (9.4)—for this example can also be completed. It should be noted that equations (9.4) give an expression for all the derivatives of τ. From them six conditions of consistency follow, all of which are equations of the second order for the functions σ. It is easy to check that the necessary consequences of these equations are the equalities $\sigma_x = \sigma_y = 0$. But then from (9.4) it follows that $\tau_x = \tau_y = 0$ also. Therefore, the components σ and τ are actually functions only of the variables u and v, and as such, they must satisfy the first two equations of (9.4).

By analogy, equations (9.3) give the expressions for all derivatives of ξ, which have six more conditions of consistency. The necessary consequences of these are the equations

$$\eta_x = -\tfrac{1}{2}\sigma_v, \qquad \sigma_u = \sigma/u. \tag{9.5}$$

The general solution of the second of these equations is $\sigma = uf(v)$. Substituting this expression into the first two equations of (9.4) gives the general form of the functions σ and τ, which satisfy (9.4), (9.5):

$$\sigma = 2C^1uv + 2C^2u, \qquad \tau = C^1(-\tfrac{2}{3}u^3 + \tfrac{3}{2}v^2) + 3C^2v + C^3, \tag{9.6}$$

where C^1, C^2, and C^3 are arbitrary constants. Upon substituting (9.6) into (9.5), the equation $\eta_x = -C^1u$ is obtained, which, when adjoined to equations (9.3), permits one to find the form of the components ξ, η:

$$\begin{aligned} \xi &= C^1(-xv + yu^2) + C^2x + C^4x + \xi^0, \\ \eta &= -C^1(xu + 2yv) + C^4y + \eta^0, \end{aligned} \tag{9.7}$$

where C^4 is an additional arbitrary constant and ξ^0 and η^0 are functions mapping $R^2(u, v) \to R$, satisfying the equations

$$\xi^0_v = \eta^0_u, \qquad \xi^0_u = -u\eta^0_v. \tag{9.8}$$

So the general equation for the solution of equations (9.3) and (9.4) is given by the formulas (9.6) and (9.7), which depend on four arbitrary constants C^i and on the arbitrary solutions (ξ^0, η^0) of the system (9.8). Since the linear equations (9.8) have an infinite set of linearly independent solutions, the space of solutions L is *infinite dimensional* here. It is easily seen that L can be represented as the direct sum $L = L^4 \oplus L^\infty$, where L^4 is the four-dimensional subspace of the solutions for which $\xi^0 = \eta^0 = 0$ and the subspace L^∞ is infinite dimensional and consists of all solutions for which all the $C^i = 0$.

The basis in L^4 can be chosen by consecutively setting one of the constants C^i equal to one and the rest equal to zero. This gives rise to the operators

$$\begin{aligned} \zeta_1 \cdot \partial &= (-xv + yu^2)\partial_x - (xu + 2yv)\partial_y + 2uv\partial_u + (-\tfrac{2}{3}u^3 + \tfrac{3}{2}v^2)\partial_v, \\ \zeta_2 \cdot \partial &= x\partial_x + 2u\partial_u + 3v\partial_v, \\ \zeta_3 \cdot \partial &= \partial_v, \\ \zeta_4 \cdot \partial &= x\partial_x + y\partial_y. \end{aligned} \tag{9.9}$$

The general operator $\zeta \cdot \partial \in L^4$ is an arbitrary linear combination of the operators (9.9),

$$\zeta \cdot \partial = C^i\zeta_i \cdot \partial. \tag{9.10}$$

To the subspace L^∞ there corresponds an infinite set of linearly independent operators of the form

$$\zeta^0 \cdot \partial = \xi^0(u, v)\partial_x + \eta^0(u, v)\partial_y. \tag{9.11}$$

With these the second part of the algorithm is completed. Further study of the space L is related to its algebraic structure, with which Section 7 is concerned.

The third part of the algorithm—generation of the set Γ of all groups $G^1(\zeta)$ corresponding to the operators $\zeta \cdot \partial$—in this example can also be completed. As will be shown in Chapter IV, for this it is enough to find the groups $G^1(\zeta_i)$ $(i = 1, 2, 3, 4)$, with basis operators (9.9), and the general form of the group $G^1(\zeta^0)$, with the operator (9.11). For the operator of translation $\zeta_3 \cdot \partial$ and the operators of dilatation $\zeta_2 \cdot \partial$ and $\zeta_4 \cdot \partial$, the construction is obvious (see Section 1.15). For the operator $\zeta_i \cdot \partial$ the solution of the Lie equations are

$$x' = \left[x - 3au^{3/2}\frac{3vx - 2u^2y}{4u^3 + 9v^2}\right]w^{1/6}, \qquad u' = uw^{2/3},$$

$$y' = \left[y - 3au^{3/2}\frac{2ux + 3vy}{4u^3 + 9v^2}\right]w^{-1/6}, \qquad v' = (v - au^{3/2})w; \quad (9.12)$$

$$w = \frac{4u^3 + 9v^2}{4u^3 + 9(v - au^{3/2})^2}.$$

Finally, for the operator (9.11), the transformations of the group $G^1(\zeta^0)$ with any concrete solution (ξ^0, η^0) of the system (9.8) are

$$x' = x + a\xi^0(u, v), \qquad y' = y + a\eta^0(u, v), \qquad u' = u, \qquad v' = v. \quad (9.13)$$

Here the following statement is correct: The set of transformations, generated by the transformation elements of the groups $G^1(\zeta_i)$ and $G^1(\zeta^0)$, is congruent with the set Γ, which in turn is congruent with the full group GE of the system (9.1). The proof is connected with an addition to the theory developed in Chapter IV.

§6. GROUP CLASSIFICATION PROBLEM

1. General Considerations

For any particular set of differential equations or groups of transformations the problem of classifying those objects, which belong to the given set, is resolved by introducing various characterizing criteria. For differential equations the classification criteria can be of the type (n, m, k, s), which represent properties of linearity and quasi-linearity, dynamic and static properties, and many others. For the transformation groups as such, the criteria are isomorphisms and homomorphisms, structural properties, similarity, inclusion, and many others. But in group analysis these two sets are studied together, on the basis of the correspondence $E \to GE$ between the system of the differential equations E and its full group GE, established in Section 5.

This correspondence is a mapping which leads to an expansion of classification possibilities for every one of the aforementioned sets. As the mapping $E \to GE$ is not one-to-one, the influence of classification characteristics of the transformation groups on classification possibilities of differential equations must be, generally speaking, stronger than the reciprocal influence. That is why the problem of group classification of differential equations (rather than the differential equation classification of groups) is first considered.

One of the strong classification criteria is similarity of the groups. If the full groups of any equations are similar, then the group properties of these equations, generally speaking, are the same. For example, if system E transforms by a "change of variables" into the system E', then the full groups GE and GE' turn out to be similar. Consequently, the group analysis of one equation E is actually the group analysis for all equations of the class obtained from E by a "change of variables."

In what follows the problem of such a classification of differential equations induced by a partial order—defined by inclusion—existing in a set of transformation groups of a fixed space Z is raised. In its fullest, this problem is very difficult because of the fact that the same group GE might be the full group of equations E with completely different structures. That is why, here, the problem is restricted to consideration of the class of equations with a fixed structure, varying only because of so-called "arbitrary elements."

It turns out, in spite of this restriction of the group classification problem for differential equations, that it still plays a major role in applied problems, permitting optimal selection for analytic forms of experimental relations included in mathematical models of physical processes.

2. Equation Transformation

Let there be an equation E of the type (5.1.1) with the mapping $\varepsilon\colon Z_k \to R^s$. With the aid of transformations $p \in \tau(Z)$ and $q \in \tau(R^s)$ (diffeomorphisms of the class C_∞, where p satisfies the condition in Section 4.5), a new mapping $\varepsilon' = q \circ \varepsilon \circ \underset{k}{p}^{-1}$ can be constructed, where the kth continuation $\underset{k}{p}$ of the mapping p is defined in Section 4.5. The mapping ε' will be called *similar* to the mapping ε and the pair (p, q) is the *similar transformation*. The equation $E'\colon \varepsilon'(\underset{k}{z}) = 0$, or

$$q \circ \varepsilon \circ \underset{k}{p}^{-1}(\underset{k}{z}) = 0, \tag{2.1}$$

is equivalent to equation (5.1.1), if and only if $q(0) = 0$. When this condition is satisfied, the manifold $E' \subset Z_k$, given by equation (2.1), is said to be similar to the manifold E. Obviously, the transition $E \to E'$ is realized by the transformation $\underset{k}{p}$ of the space Z_k, under which operation the formula $\underset{k}{z}' = \underset{k}{p}(\underset{k}{z})$ maps the point $z_k \in E$ to the point $\underset{k}{z}' \in E'$.

The main classification property of the pair (E, GE) is established by the following simple but important statement.

Theorem. *The full groups GE and GE' of the "equivalent" equations E and E' are similar and the relation of similarity* (1.11.2) *of the transformations $h \in GE$ and $h' \in GE'$ is $h' = p \circ h \circ p^{-1}$.*

Proof. Since E admits the transformation h, it follows that $\underset{k}{\bar{z}} = \underset{k}{h}(\underset{k}{z}) \in E$ for any $\underset{k}{z} \in E$. The transformation $\underset{k}{p}$ maps the point $\underset{k}{z}$ to the point $\underset{k}{z}' = \underset{k}{p}(\underset{k}{z}) \in E'$, where $\underset{k}{z}'$ is a point of the manifold E'. Since

$$q \circ \varepsilon \circ \underset{k}{p}^{-1}(\underset{k}{h}'(\underset{k}{z}')) = (q \circ \varepsilon \circ \underset{k}{p}^{-1}) \circ (\underset{k}{p} \circ \underset{k}{h} \circ \underset{k}{p}^{-1})(\underset{k}{p}(\underset{k}{z})) = q \circ \varepsilon(\underset{k}{\bar{z}}) = 0,$$

$\underset{k}{h}'(\underset{k}{z}') \in E'$. ■

In practice this theorem is broadly used in studying the group properties of any equation E, as a preliminary simplification of this equation without having to construct a suitable transformation of similarity (p, q). This is not different from the aforementioned classical "change of variables," which does not change the group structure of the differential equation.

3. Arbitrary Element

Now consider an auxiliary space $T = R^l$ of vectors $t \in T$, called the *space of parameters*, and the mapping $\varepsilon\colon T \times Z_k \to R^s$. It is supposed that for any $t \in T$ the partial mapping $\varepsilon_t\colon Z_k \to R^s$ satisfies the condition in Section 5.1.

For every mapping $\theta\colon Z \to T$, operating by the formula $t = \theta(z)$, the complex mapping $\varepsilon(\theta)\colon Z_k \to R^s$, operating according to $\varepsilon(\theta)(\underset{k}{z}) = \varepsilon(\theta(z), \underset{k}{z})$, is defined. This construction permits us to define a class of differential equations, representatives of which are given by assignment of the mapping $\theta\colon Z \to T$ and which have the form

$$\varepsilon(\theta)(\underset{k}{z}) = 0. \tag{3.1}$$

Equation (3.1) will be called the *equation with the arbitrary element* θ and will be denoted by $E(\theta)$.

The given definition can be expanded to consider the arbitrary element θ as a mapping $Z_{k'} \to T$ for some $0 < k' < k$. Many properties resulting from such an expanded interpretation of an arbitrary element are analogous to those which are obtained from the given definition.

One can say that all equations of the class (3.1) have the same differential structure (sometimes this is expressed by saying that these equations are of the same "kind"), defined by the partial mapping $\varepsilon_t\colon Z_k \to R^s$, and differ from one another only by the "value" of the arbitrary element θ. This construction permits study, by group analysis, of such differential equations which contain some types of parameters or functions not strictly fixed in a practical investigation.

The concept of an arbitrary element arises from the fact that many concrete problems of mathematical physics require the solution of differential equations which contain experimentally determined parameters or functions. Often these are given in the form of tables or graphs. These parameters or functions play the role of the arbitrary element. The techniques of group classification permit us to choose purposefully the analytic form of such functions. One of our goals is the determination of a *maximal full group.*

4. Equivalent Transformations

Because of the actions of the transformation $p\colon Z \to Z$, from Section 6.2, the arbitrary element θ is transformed into $\theta \circ p^{-1}$. For the entire class of equations $E(\theta)$, the similarity transformation operates on the mapping $\varepsilon(\theta)\colon Z_k \to R^s$ by the formula

$$\varepsilon(\theta) \to \varepsilon'(\theta) = q \circ \varepsilon(\theta \circ p^{-1}) \circ \underset{k}{p}^{-1}.$$

From this the equation $\varepsilon'(\theta)(\underset{k}{z}) = 0$ is obtained, which can be written in the form

$$\varepsilon(\theta \circ p^{-1}) \circ \underset{k}{p}^{-1}(\underset{k}{z}) = 0 \tag{4.1}$$

with the assumption $q(0) = 0$.

For the problem of group classification it is essential to define such transformations, which operate only on the arbitrary element θ, while maintaining the differential structure of the equation $E(\theta)$. With this goal, consider the transformations w of the space $T \times Z$, operating by the formula $w(t, z) = (\psi(t, z), p(z))$, where p is a transformation of the space Z, and the mapping $\psi\colon T \times Z \to T$ possesses the property that for every $z \in Z$ the partial mapping $\psi_z\colon T \to T$ is a transformation of the space T. The continuation $\underset{k}{w}$ of the transformation w is called its continuation only with respect to the second component, that is, the mapping $\underset{k}{w}\colon T \times Z_k \to T \times Z_k$, according to

$$\underset{k}{w}(t, \underset{k}{z}) = (\psi(t, z), \underset{k}{p}(\underset{k}{z})).$$

Definition. The transformation w of the space $T \times Z$ is called an *equivalent transformation* of the mapping $\varepsilon\colon T \times Z_k \to R^s$ if a transformation q of the space R^s exists for which equality of the mappings

$$\varepsilon \circ \underset{k}{w} = q \circ \varepsilon \tag{4.2}$$

holds.

The utility of equivalent transformations for group classification is given by means of two properties of those transformations.

First, *equivalent transformations operate only on an arbitrary element θ, maintaining the structure of the differential equation $E(\theta)$.* To show this it is enough to write the equality (4.2) at the point $(\theta(\bar{z}), \underset{k}{\bar{z}})$, where $\bar{z} = p^{-1}(z)$, in the form

$$q \circ \varepsilon(\theta \circ p^{-1}(z), \underset{k}{p}^{-1}(\underset{k}{z})) = \varepsilon(\bar{\theta}(z), \underset{k}{z}),$$

where $\bar{\theta} = \psi \circ (\theta \circ p^{-1}, p^{-1})$. Because of (4.1), $\varepsilon(\bar{\theta})(\underset{k}{z}) = 0$ is equivalent to equation (3.1) with the same mapping ε. Consequently, equation $E(\bar{\theta})$ is the same as $E(\theta)$ but with another arbitrary element $\bar{\theta}$ instead of θ. That is why, when applied to the equation, the transformation w can also be called an equivalent transformation of an arbitrary element θ. This transformation can be displayed in the explicit form

$$\theta(z) \to \bar{\theta}(z) = \psi(\theta(p^{-1}(z), p^{-1}(z))). \tag{4.3}$$

In addition, since equation $E(\bar{\theta})$ is equivalent to equation $E(\theta)$, it follows from the theorem in Section 6.2 that the groups admitted by these equations are similar.

Second, *equivalent transformations generate a group.* Actually, if w' is another equivalent transformation (together with q' corresponding to it), then

$$\varepsilon \circ \underset{k}{w} \circ \underset{k}{w'} = q \circ \varepsilon \circ w' = q \circ q' \circ \varepsilon,$$

and it is noted that $\underset{k}{w} \circ \underset{k}{w'} = \underset{k}{w \circ w'}$. The last equality follows from the definition of $\underset{k}{w}$ and the fundamental property of the continuation (the lemma in Section 4.6), so if $w' = (\psi', p')$, then

$$\underset{k}{w} \circ \underset{k}{w'} = (\psi_z \circ \psi'_z, \underset{k}{p} \circ p'_k) = (\psi_z \circ \psi'_z, p \circ p') = \underset{k}{w \circ w'}.$$

It is appropriate to note that because of definition (4.2), the continuous equivalent transformation $\underset{k}{w}$ leaves the equation $E(t)$,

$$\varepsilon(t, \underset{k}{z}) = 0, \tag{4.4}$$

invariant. That is why any equivalent transformation w is admitted by equation (4.4) in the sense of Section 5.2.

As a rule, further considerations will be restricted to such equivalent transformations which belong to a one-parameter group G^1 of transformations of the form w of the space $T \times Z$. The group $GE(t)$, generated by such groups G^1 of equivalent transformations, will be called the *equivalence group* of any equation $E(\theta)$, obtained from the equation $E(t)$ by means of the concrete mapping $\theta: Z \to T$.

Determination of the equivalence group is done with the aid of the algorithm in Section 5.8. With this it is required to consider the specific construction of the unknown transformation w, as well as know in advance possible special properties of the arbitrary element (for example, independence of some components of the vector z).

5. Classification Problem

Let $GE(\theta)$ be the full group of the equation $E(\theta)$, given by equation (3.1). As θ runs through the set of mappings $Z \to T$, there is obtained a set of groups $GE(\theta)$. The group GE_0, given by the intersection of all groups $GE(\theta)$, will be called the *kernel of the full groups*. Because of its definition, the group GE_0 is admitted by the equation $E(\theta)$ for any arbitrary element θ. Since $GE_0 \subset GE(\theta)$ for every concrete θ, every group $GE(\theta)$ is an expansion of the group GE_0.

Every concrete mapping $\theta: Z \to T$ can be called a *specialization of the arbitrary element*. In these terms there arises the following problem of group classification for a class of equations.

Problem. *For a class of differential equations find the kernel of the full groups, GE_0, and all specializations of the arbitrary element θ, giving an expansion of the group GE_0.*

The search for the kernel of the full groups is simplified because of the following theorem, whose formulation has the following meaning. According to the definition, every equivalent transformation $w: T \times Z \to T \times Z$ consists of two components $w = (\psi, p)$, the second of which is some transformation of the space Z. Thus it can be said that the equivalent transformation w contains the transformation p.

Theorem. *The kernel of the full groups GE_0 is included in the equivalence group $GE(t)$.*

Proof. Let $\zeta \cdot \partial$ be an operator of the group $G^1 \subset GE(t)$. Because of the special construction of the equivalent transformation w, this operator can be represented in the form $\bar{\zeta} \cdot \partial = \tau \cdot \partial_t + \zeta \cdot \partial_z$, with the tangent vectors $\tau: Z \times T \to T$ and $\zeta: Z \to Z$. Then the condition of invariance for equation (4.4), relative to the group $G^1(\bar{\zeta})$, will be

$$[(\tau \cdot \partial_t)\varepsilon(t, \underset{k}{z}) + (\underset{k}{\zeta} \cdot \partial_z)\varepsilon(t, \underset{k}{z})]\,|_{E(t)} = 0. \tag{5.1}$$

On the other hand, the invariance condition of the equation $E(\theta)$, relative to the group $G^1(\zeta) \subset GE_0$ with the tangent vector $\zeta: Z \to Z$, is

$$[((\zeta \cdot \partial_z)\theta(z)) \cdot \partial_t \varepsilon(\theta(z), \underset{k}{z}) + (\underset{k}{\zeta} \cdot \partial_z)\varepsilon(\theta(z), \underset{k}{z})]\,|_{E(\theta)} = 0. \tag{5.2}$$

For the group GE_0, equation (5.2) must be satisfied whatever the mapping θ is. Thus $\theta(z)$ can be replaced by t everywhere except in the term $(\zeta \cdot \partial_z)\theta(z)$. Of course, it is always possible to choose the mapping θ as a solution of the equation $(\zeta \cdot \partial_z)\theta = \tau$. After these operations the invariance condition (5.2) is completely equivalent to (5.1). Consequently, since the tangent vector ζ of the group $G^1(\zeta) \subset GE_0$ satisfies equation (5.1), $G^1(\zeta)$ is contained in the group $GE(t)$. ■

6. Description of the Solution Procedure

To solve the problem of classification, the invariance condition (5.2) was constructed, which is decomposed relative to "free" components of the vectors $\underset{1}{y}, \ldots, \underset{k}{y}$. In the resulting determining equations $DE(\theta)$, a hypothesis is made that θ (and the derivative $\partial\theta$) is arbitrary, which leads to an additional decomposition of the equations $DE(\theta)$ generating the determining equations DE_0 of the kernel of the full groups. The set of equations DE_0 generates a space of vector fields L_0. In the finite-dimensional case the space L_0 is defined by the stipulation of its basis.

Then the determining equations $DE(\theta)$ are to be solved. Because of the overdefinition of the system $DE(\theta)$, there is a related formation of a congruence condition, as a result of which equations arise which contain only the arbitrary element θ. The latter equations are called *classifying*. For every solution θ there is obtained a subspace $L(\theta)$ of the vector fields tangent to the groups G^1 that are admitted by the equation $E(\theta)$. Because of the theorem in Section 6.5, every space $L(\theta)$ contains L_0 and therefore (in the finite-dimensional case) can be given by indicating the complement of the basis for the space L_0 with respect to the basis for the space $L(\theta)$.

A classification of the resulting equations is obtained according to the criterion of equivalence. For this, as noted previously, the equivalence group $GE(t)$ of the equation $E(\theta)$ is constructed as the full group of the equation $E(t)$. Its determining equations are obtained from the condition of invariance (5.1).

An important property of the classifying equations is that they admit the equivalence group $GE(t)$. From this fact it follows that the classifying equations, according to their construction, depend only on the differential structure of the equation $E(\theta)$, and this structure is invariant relative to any equivalent transformations. Because of the theorem in Section 5.6, this property implies that the transformations of the arbitrary element, in the form (4.3) for $w \in GE(t)$, operate on a set of solutions of the classifying equations. Thus, the classification problem becomes the decomposition of the set of all solutions of the classifying equations into equal classes of solutions, relative to the operation of the group $GE(t)$ on this set.

This operation is defined by formula (4.3), and with its aid, from every class of equivalent solutions, there are chosen typical representatives which have the simplest possible analytic form.

The final result for the solution of the group classification problem for the equation $E(\theta)$ is a table indicating the selected representatives of the arbitrary element θ and associated spaces of vector fields $L(\theta)$ (their bases), arranged in order of increase with respect to inclusion.

7. Example: Nonlinear Heat Conduction

Consider the equation with the form $E(2, 1, 2, 1)$, which in individual coordinate notation $X = R^2(x, t)$, $Y = R(y)$, $Z = R^3(x, t, y)$, $Y_1 = R^2(y_1, y_2)$, and $Y_2 = R^3(y_{11}, y_{12}, y_{22})$ has the form

$$y_2 = \theta(y)y_{11} + \theta'(y)y_1^2. \tag{7.1}$$

Here $\theta\colon R \to R$ is an arbitrary element and $\theta'(y) = \partial\theta(y)$. If $u\colon R^2(x, t) \to R$ is a solution of this equation, given by the formula $y = u(x, t)$, then its continuation is

$$y_1 = \partial_x u(x, t), \qquad y_2 = \partial_t u(x, t), \qquad y_{11} = \partial_x^2 u(x, t), \qquad \ldots .$$

Because of this notation, the solution satisfies the identity

$$\partial_t u = \partial_x[\theta(u)\partial_x u],$$

which shows that (7.1) is a one-dimensional equation of nonlinear heat conduction, where $u(x, t)$ is temperature at the point x at the moment of time t and the coefficient of heat conduction θ is a function of temperature. For equation (7.1) the problem of group classification, with an arbitrary element θ, is solved here. It is assumed that $\theta > 0$.

The operator admitted by equation (7.1) is to be found in the form

$$\zeta \cdot \partial = \xi\partial_x + \eta\partial_t + \sigma\partial_y, \tag{7.2}$$

where every one of the functions ξ, η, σ is an unknown function $Z \to R$. With the aid of the operators of total differentiation D_1, D_2, the components of the operator D are

$$D_1 = \partial_x + y_1\partial_y + y_{11}\partial_{y_1} + y_{12}\partial_{y_2},$$
$$D_2 = \partial_t + y_2\partial_y + y_{12}\partial_{y_1} + y_{22}\partial_{y_2},$$

the operator (7.2) can be calculated as

$$\underset{1}{\zeta} \cdot \partial = \zeta \cdot \partial + \alpha_1\partial_{y_1} + \alpha_2\partial_{y_2},$$
$$\alpha_1 = D_1\sigma - y_1 D_1\xi - y_2 D_1\eta, \qquad \alpha_2 = D_2\sigma - y_1 D_2\xi - y_2 D_2\eta,$$

and its second continuation becomes

$$\underset{2}{\zeta} \cdot \partial = \underset{1}{\zeta} \cdot \partial + \beta_{11}\partial_{y_{11}} + \beta_{12}\partial_{y_{12}} + \beta_{12}\partial_{y_{22}},$$
$$\beta_{11} = D_1\alpha_1 - y_{11}D_1\xi - y_{12}D_1\eta, \qquad \beta_{12} = \cdots, \qquad \beta_{22} = \cdots.$$

The expressions for β_{12} and β_{22} are not contained in the result of the operator $\underset{2}{\zeta} \cdot \partial$ acting on equation (7.1).

Further calculations are carried out according to stages (b)–(d) of the first part of the algorithm in Section 5.8. The transition on the manifold E is accomplished by the substitution of the expression y_2 from (7.1). Here the role of "free" components is played by y_1, y_{11}, y_{12}, and y_{22}. As a result of a decomposition of the invariant condition with respect to these variables, one obtains after some work, the following equations DE in which the appropriate derivatives ∂_x, ∂_t, ∂_y, and $\theta'' = \partial_y^2\theta$ are denoted by the subscripts x, t, and y:

$$\eta_x = 0, \qquad \eta_y = 0, \qquad \xi_y = 0, \qquad \sigma_{yy} = 0, \qquad \theta(2\xi_x - \eta_t) - \theta'\sigma = 0,$$
$$\theta(2\sigma_{xy} - \xi_{xx}) + 2\theta'\sigma_x + \xi_t = 0, \qquad \theta\sigma_{xx} - \sigma_t = 0. \tag{7.3}$$

The determining equations DE_0 (kernel of the full groups) are obtained from (7.3) as a result of the assumption that the quantities θ and θ' are independent "free" parameters. Consequently, the equations DE_0 consist of the equations from the first row of (7.3) and the additional equations

$$\sigma = 0, \qquad \xi_t = 0, \qquad \xi_{xx} = 0, \qquad 2\xi_x - \eta_t = 0.$$

From these it follows that $\xi = \xi(x), \eta = \eta(t)$, and the function ξ is linear. That is why the general solution of the equations DE_0 is

$$\xi = C_3 x + C_1, \qquad \eta = 2C_3 t + C_2, \qquad \sigma = 0, \tag{7.4}$$

where C_1, C_2, C_3 are arbitrary constants. So the group GE_0 is generated by transformations belonging to three groups $G'(\zeta)$, whose tangent vectors $\zeta = (\xi, \eta, \sigma)$ are obtained from (7.4) by setting one of the constants C_i equal to one and the rest equal to zero. Consequently, the basis vectors for the space L_0 are

$$\zeta_1 = (1, 0, 0), \qquad \zeta_2 = (0, 1, 0), \qquad \zeta_3 = (x, 2t, 0). \tag{7.5}$$

The equivalence group of equation (7.1) is sought in the class of these equations together with the condition that the arbitrary element depend only on y and that $\theta'(y) = \partial\theta(y)$. Thus the transformations, operating on θ, must be admitted by the system of equations

$$\partial_x \theta = 0, \qquad \partial_t \theta = 0, \qquad \partial_y \theta = \theta'. \tag{7.6}$$

To find the equivalence group, equation (7.1) is written in the form

$$y_2 = \theta y_{11} + \theta' y_1^2, \tag{7.7}$$

and the problem is to solve (7.6), (7.7), thereby constructing the full group of transformations of the space $Z \times T = R^s(x, t, y, \theta, \theta')$. The infinitesimal operator is sought in the form

$$\bar{\zeta} \cdot \partial = \zeta \cdot \partial + \pi \partial_\theta + \rho \partial_{\theta'}, \tag{7.8}$$

where $\zeta \cdot \partial$ is an operator of the form (7.2) and π, ρ are unknown functions $Z \times T \to R$. Then, because of the independence of the vector ζ on the variables θ, θ', the invariance condition must be decomposed also with respect to those variables.

Here the calculation is quite elementary, and it is easy to find the final result. The space L turns out to be six dimensional and for all admitted vectors $\rho = 0$. The basis vectors (7.8) can be chosen as (zero values of the last component are omitted)

$$\begin{aligned} &\bar{\zeta}_1 = (1, 0, 0), \qquad \bar{\zeta}_2 = (0, 1, 0), \qquad \bar{\zeta}_3 = (0, 0, 1), \\ &\bar{\zeta}_4 = (x, 2t, 0), \qquad \bar{\zeta}_5 = (0, 0, y), \qquad \bar{\zeta}_6 = (0, -t, 0, \theta). \end{aligned} \tag{7.9}$$

Comparison of (7.5) with (7.9) gives an illustration of the theorem in Section 6.5.

As θ is a function only of y, the equivalent transformations operating on θ are generated by the groups $G^1(\bar{\zeta}_3)$, $G^1(\bar{\zeta}_5)$, and $G^1(\bar{\zeta}_6)$. It is then easy to check that the three-dimensional (with independent parameters a, b, c) family of transformations,

$$\bar{y} = ay + b, \qquad \bar{\theta} = c\theta,$$

is obtained. From here it follows that the general formula for equivalent transformations of the arbitrary element (4.3) can be written in the form

$$\bar{\theta}(y) = c\theta(ay + b). \tag{7.10}$$

Now one can begin the classification. The first classification step is to consider two possibilities: $\theta' \neq 0$ or $\theta' = 0$. The equation $\theta' = 0$ is classifying (see Section 6.6), and obviously it is invariant relative to the transformation (7.10). This possibility is related to a linear equation, obtained from (7.1), and will be considered later.

Let $\theta' \neq 0$. Then from (7.3) the following expression is obtained:

$$\sigma = (\theta/\theta')(2\xi_x - \eta_t). \tag{7.11}$$

As $\sigma_{yy} = 0$, and ξ, η do not depend on y,

$$(\theta/\theta')''(2\xi_x - \eta_t) = 0,$$

where derivatives with respect to y are denoted by primes. The assumption $2\xi_x - \eta_t = 0$ gives $\sigma = 0$, and then from (7.3) the equalities $\xi_t = 0$, $\xi_{xx} = 0$ follow, that is, the equations DE_0 are obtained. An alternative to this assumption is the equality

$$(\theta/\theta')'' = 0, \tag{7.12}$$

which is a classifying equation. An expansion of the full group $GE(\theta)$, in comparison with GE_0, is possible only if the function θ satisfies equation (7.12). From (7.12) it follows that

$$\theta/\theta' = ky + l,$$

with some constants k, l. If the transition to the equivalent function $\bar{\theta}$, according to (7.10), is desired, then this equation will take the form (the bar above θ is omitted)

$$\theta/\theta' = ky + (kb + l)/a.$$

Now it can be seen that it is necessary to study two possibilities: $k = 0$ and $k \neq 0$.

If $k = 0$, then with the choice $a = l$ there is obtained the equation $\theta' = \theta$ with the general solution $\theta = ce^y$. Because of (7.10), a representative of this class is the function $\theta = e^y$. In this case, from (7.11), $\sigma = 2\xi_x - \eta_t$ is obtained, and the rest of the determining equations (7.3) (the last row) are found to be

$$\xi_t = 0, \qquad \xi_{xx} = 0, \qquad \eta_{tt} = 0.$$

The general solution of the whole system $DE(e^y)$ is

$$\xi = C_3 x + C_1, \qquad \eta = (2C_3 - C_4)t + C_2, \qquad \sigma = C_4,$$

where the C_i are arbitrary constants. A comparison with (7.4) shows that the space of solutions $L(e^y)$ is found by an expansion (at the expense of the constant C_4) with the aid of the vector $\zeta_4 = (0, -t, 1)$.

If $k \neq 0$, then the choice $b = -l/k$ gives the equation $\theta/\theta' = ky$ or $\theta'/\theta = my^{-1}$, where $m = 1/k$. The general solution is $\theta = cy^m$, a representative of which, because of (7.10), is the function $\theta = y^m$. Here (7.11) gives $\sigma = k(2\xi_x - \eta_t)y$, and the rest of equations (7.3), after the decomposition with respect to the variable y, become

$$\xi_t = 0, \qquad (3m + 4)\xi_{xx} = 0, \qquad 2\xi_{xxx} - 2\xi_{xt} + \eta_{tt} = 0. \tag{7.13}$$

The assumption $3m + 4 = 0$ is classifying. If $3m + 4 \neq 0$, then the expansion L_0 takes place on only one vector $\zeta_4 = (mx, 0, 2y)$. If $3m + 4 = 0$, that is, $m = -4/3$, then the solution of the rest of the determining equations (7.3) and (7.13) shows that in addition to the vector $\zeta_4 = (-2x, 0, 3y)$, one independent vector $\zeta_5 = (-x^2, 0, 3xy)$ is obtained.

So in the case $\theta' \neq 0$, the following classification scheme is found (the upper index in the symbol of a space L^k of the solutions of the determining equations indicates the dimension of the space):

$$L_0^3 \subset \begin{cases} L^4(e^y) \\ L^4(y^m) \subset L^5(y^{-4/3}). \end{cases} \tag{7.14}$$

8. Linear Equation Case

If in the previous example $\theta = \text{const}$, then equation (7.1) becomes linear and equal to the equation

$$y_2 = y_{11}. \tag{8.1}$$

In this case the system of determining equations (7.3) is simplified:

$$\begin{gathered} \eta_x = 0, \qquad \eta_y = 0, \qquad \xi_y = 0, \qquad 2\xi_x = \eta_t, \\ \sigma_{yy} = 0, \qquad 2\sigma_{xy} = -\xi_t, \qquad \sigma_{xx} = \sigma_t. \end{gathered} \tag{8.2}$$

The general solution of system (8.2) is found without difficulty to be

$$\xi = C_1 + C_3 x - 2C_4 t - 4C_5 xt, \qquad \eta = C_2 + 2C_3 t - 4C_5 t^2,$$

$$\sigma = (C_6 + C_4 x + C_5(x^2 + 2t))y + u^0(x, t),$$

where the C_i are arbitrary constants and u^0 is any solution of equation (8.1). Consequently, the space $L(1)$ turns out to be infinite dimensional for equation (8.1). Expressing this fact in another way, one says that the *full group is infinite dimensional.*

The property that the full group is infinite dimensional always holds for any linear equations E. From the viewpoint of the operation of the group on the set of solutions, this property is equivalent to the fact that the set of solutions of a linear equation generates an infinite-dimensional vector space (solutions can be added and multiplied by scalars without going outside the solution set). The analogous fact has already been mentioned in the example in Section 5.9.

If we denote the infinite part of the space $L(1)$, generated by the vectors $\zeta = (0, 0, u^0(x, t))$, by the symbol L^∞, then $L(1)$ can be represented as the direct sum $L(1) = L^6(1) \oplus L^\infty$, where the six-dimensional space $L^6(1)$ is obtained when $u^0 = 0$. The basis of $L^6(1)$ can be chosen to consist of vectors (7.5), arising from the constants C_1, C_2 and C_3, and the additional vectors obtained from the constants C_4, C_5, and C_6,

$$\zeta_4 = (-2t, 0, xy), \qquad \zeta_5 = (-4xt, -4t^2, (x^2 + 2t)y), \qquad \zeta_6 = (0, 0, y). \tag{8.3}$$

With these the problem of group classification for equation (7.1) is completely solved.

§7. LIE ALGEBRA OF OPERATORS

1. Commutator

Let X be a Banach space. Consider the vector fields $\xi: X \to X$ of class $C_\infty(X)$. As the operations of summation of vector fields and multiplication of them by a real number also give vector fields, the set of all vector fields of the space X itself generates a vector space (over the field R). This space will be denoted by the symbol $\Xi(X)$ (or merely Ξ, if it does not cause a misunderstanding) and will be called the *space of vector fields on the space* X. In the local theory the vector fields on the open set $V \subset X$ of the class $C_\infty(V)$ are considered, according to which the space of the vector fields $\Xi(V)$ is generated from the set V.

By the *commutator* of the vector fields $\xi_1, \xi_2 \in \Xi$ is meant a vector field, denoted by the symbol $[\xi_1, \xi_2]$, and defined by the formula

$$[\xi_1, \xi_2] = \partial\xi_2\langle\xi_1\rangle - \partial\xi_1\langle\xi_2\rangle. \tag{1.1}$$

Using the idea of the infinitesimal operator in Section 2.6, this definition can be rewritten in the form

$$[\xi_1, \xi_2] = (\xi_1 \cdot \partial)\xi_2 - (\xi_2 \cdot \partial)\xi_1. \tag{1.2}$$

The idea of the commutator of vector fields is usually transferred to or defined for field operators. In particular, the operator $[\xi_1, \xi_2] \cdot \partial$ is called a commutator of the operators $\xi_1 \cdot \partial$, $\xi_2 \cdot \partial$, which is expressed by the formula

$$[\xi_1 \cdot \partial, \xi_2 \cdot \partial] = [\xi_1, \xi_2] \cdot \partial. \tag{1.3}$$

This formula means that the symbol of a commutator [,] of vector fields is interchangeable with scalar multiplication of a vector field by a differential operator ∂. Because of this it makes no difference whether we talk about a commutator of operators or a commutator of vector fields.

2. Action on Mapping

The original concept of an operator commutator occurs in the definition of the action of operators on mappings when such an action needs to be carried out repeatedly, but with different operators. From this, the following formula holds:

$$(\xi_1 \cdot \partial)(\xi_2 \cdot \partial) - (\xi_2 \cdot \partial)(\xi_1 \cdot \partial) = [\xi_1, \xi_2] \cdot \partial, \tag{2.1}$$

which expresses the fact that the results of action of the left- and right-hand sides on any mapping are equal.

To check equation (2.1), it is necessary to remember that the result of an action of the operator $\xi \cdot \partial$ on any differentiable mapping $F\colon X \to Y$ is defined as the value of the derivative ∂F on the vector ξ, that is, $(\xi \cdot \partial)F = \partial F\langle \xi \rangle$. Consequently, because of this definition and the differentiation rules, it follows that

$$(\xi_1 \cdot \partial)(\xi_2 \cdot \partial)F = \partial(\partial F\langle \xi_2 \rangle)\langle \xi_1 \rangle = \partial^2 F\langle \xi_1, \xi_2 \rangle + \partial F\langle \partial \xi_2 \langle \xi_1 \rangle \rangle.$$

Substituting the vectors ξ_1, ξ_2 and taking into account the symmetric property of the second derivative $\partial^2 F$, after the indicated subtraction in (1.1), results in

$$\begin{aligned}(\xi_1 \cdot \partial)(\xi_2 \cdot \partial)F - (\xi_2 \cdot \partial)(\xi_1 \cdot \partial)F &= \partial F\langle \partial \xi_2 \langle \xi_1 \rangle - \partial \xi_1 \langle \xi_2 \rangle \rangle \\ &= \partial F\langle [\xi_1, \xi_2] \rangle = ([\xi_1, \xi_2] \cdot \partial)F,\end{aligned}$$

that is, equality (2.1).

These calculations show that the operator equality (2.1) is identical to the definition of the commutator $[\xi_1, \xi_2]$.

3. Algebraic Properties

Definition (1.1) introduces a binary operation in the space of vector fields Ξ relating the vector field $[\xi_1, \xi_2] \in \Xi$ to the ordered pair (ξ_1, ξ_2). This operation is called a *commutation operation.*

The introduction of the operation of commutation gives a certain algebraic structure to the field Ξ. In Chapter IV this structure will be described axiomatically and in abstract terms. Here, because of the constructive definition of the commutation operation on a space of vector fields, the following algebraic properties result:

1° *The operation of commutation is bilinear.* This means that for any vector fields ξ_1, ξ_2, ξ_3 and any numbers a, b the following identities hold:

$$[a\xi_1 + b\xi_2, \xi_3] = a[\xi_1, \xi_3] + b[\xi_2, \xi_3],$$
$$[\xi_1, a\xi_2 + b\xi_3] = a[\xi_1, \xi_2] + b[\xi_1, \xi_3].$$

2° *The operation of commutation is antisymmetric.* This means that for any vector fields ξ_1, ξ_2 the identity

$$[\xi_1, \xi_2] = -[\xi_2, \xi_1]$$

is satisfied.

3° *The operation of commutation satisfies the Jacobi relation.* That is to say that for any vector fields ξ_1, ξ_2, ξ_3 the identity

$$[[\xi_1, \xi_2], \xi_3] + [[\xi_2, \xi_3], \xi_1] + [[\xi_3, \xi_1], \xi_2] = 0$$

is satisfied.

Verification of these properties is easily carried out for any of the definitions of the commutator (1.1), (1.2), or (2.1). For example, to check 3° for (2.1) it is enough to note that

$$\begin{aligned}[[\xi_1, \xi_2], \xi_3] \cdot \partial &= ([\xi_1, \xi_2] \cdot \partial)(\xi_3 \cdot \partial) - (\xi_3 \cdot \partial)([\xi_1, \xi_2] \cdot \partial) \\ &= (\xi_1 \cdot \partial)(\xi_2 \cdot \partial)(\xi_3 \cdot \partial) - (\xi_2 \cdot \partial)(\xi_1 \cdot \partial)(\xi_3 \cdot \partial) \\ &\quad - (\xi_3 \cdot \partial)(\xi_1 \cdot \partial)(\xi_2 \cdot \partial) + (\xi_3 \cdot \partial)(\xi_2 \cdot \partial)(\xi_1 \cdot \partial)\end{aligned}$$

and make a cyclic permutation of the indexes $1 \to 2 \to 3 \to 1$. After summation of the resulting three equations all terms on the right-hand side sum to zero and 3° is established.

4. Definitions

Consider a subspace L of the space of vector fields Ξ, that is a subset $L \subset \Xi$, which is itself a vector space.

The space $L \subset \Xi$ is called a *Lie algebra of operators* (or *vector fields*) if it is closed relative to the operation of commutation; that is, for any $\xi_1, \xi_2 \in L$ also $[\xi_1, \xi_2] \in L$.

Examples of Lie algebras are the whole space Ξ and the zero subspace $\{0\} \subset \Xi$, which consists of only a zero vector field. Another example is given by any one-dimensional Lie algebra—a set of all vector fields of the form $a\xi$, obtained by multiplication of a fixed vector field ξ by various numbers $a \in R$.

By the *dimensionality of the Lie algebra of operators L* is meant the dimensionality of L thought of as a vector space. A finite-dimensional Lie algebra, of dimensionality r, is also called an r-dimensional Lie algebra and denoted by the symbol L^r. In particular, any one-dimensional Lie algebra has dimensionality equal to unity and is denoted by the symbol L^1.

The Lie algebra L is called *Abelian* (or *commutative*) if $[\xi_1, \xi_2] = 0$ for any $\xi_1, \xi_2 \in L$. For instance, any one-dimensional Lie algebra L^1 is Abelian.

Let L be a Lie algebra of operators and let the subset $L' \subset L$ also be a Lie algebra. Then L' is called a *subalgebra of the Lie algebra L*. In other words, the subspace $L' \subset L$ is a subalgebra in L if $[\xi_1, \xi_2] \in L'$ for any $\xi_1, \xi_2 \in L'$. In particular, any Lie algebra $L \subset \Xi$ can be considered as a subalgebra of the Lie algebra Ξ.

The Lie subalgebra $L' \subset L$ is called an ideal if for any $\xi' \in L'$ and any $\xi \in L$ the commutator $[\xi', \xi] \in L'$.

5. Structural Tensor

Let the Lie algebra of operators $L^r \subset \Xi(X)$ be given. The finite dimensionality of the vector space L^r means that it has a basis—a collection of r linearly independent vector fields $\{\xi_i\} = (\xi_1, \ldots, \xi_r)$. For any vector field $\xi \in L^r$ there exists a uniquely defined collection of numbers, with respect to ξ, $\{c^i\} = (c^1, \ldots, c^r)$—coefficients of the vector field ξ in the basis $\{\xi_i\}$, with which the equality $\xi = c^i\xi_i$ holds.

For any $i, j = 1, \ldots, r$ the commutator $[\xi_i, \xi_j] \in L^r$, so the decomposition formula with respect to the basis $\{\xi_i\}$ is

$$[\xi_i, \xi_j] = C^k_{ij}\xi_k. \tag{5.1}$$

The numbers C^k_{ij} are called *structural constants of the Lie algebra L^r for the basis $\{\xi_i\}$*.

The following criterion, obvious but of very practical importance, is used in applications to determine whether a vector space $L^r \subset \Xi$ is a Lie algebra or not.

Criterion. *A vector space $L^r \subset \Xi$ is a Lie algebra of vector fields if and only if the vectors of any basis $\{\xi_i\} \subset L^r$ satisfy relations of the form* (5.1).

It is easy to verify, by transformation from one basis in L^r to another, that the numbers C^k_{ij} are components of a third-rank tensor. Thus, the collection of numbers $\{C^k_{ij}\}$ is also called a *tensor of structural constants* or just the *structural tensor of the Lie algebra L^r*. This tensor, of course, is independent of the bases in L^r and consequently will be denoted by one symbol, C. Then the structural constants C^k_{ij}, defined by equation (5.1), are the components of the structural tensor C with the basis $\{\xi_i\}$.

The structural constants C_{ij}^k, for the Lie algebra L^r, have analogous properties to 3.1°–3.3° of the commutation operation. Property 3.1° of bilinearity of the commutator holds for the C_{ij}^k since they are components of the (structural) tensor C. The property 3.2° of antisymmetry is antisymmetry of C_{ij}^k with respect to the lower indices:

$$C_{ij}^k = -C_{ji}^k. \tag{5.2}$$

Finally, property 3.3°—Jacobi's identity—becomes the relation

$$C_{ij}^s C_{sk}^l + C_{jk}^s C_{si}^l + C_{ki}^s C_{sj}^l = 0, \tag{5.3}$$

which is also called Jacobi's relation for the structural constants.

6. Linear Mappings of a Lie Algebra

In the theory of Lie algebras certain classes of linear mappings of Lie algebras play an important role as vector spaces. Let L and K be two Lie algebras of operators with $L \subset \Xi(X)$ and $K \subset \Xi(Y)$, where the Banach spaces X and Y, generally speaking, are different.

The linear mapping $\gamma: L \to K$ is called a *homomorphism* (or *homomorphic mapping*) *of the Lie algebra L into the Lie algebra K* if for any $\xi, \xi' \in L$ the relation

$$\gamma\langle[\xi, \xi']\rangle = [\gamma\langle\xi\rangle, \gamma\langle\xi'\rangle] \tag{6.1}$$

is satisfied. When (6.1) holds for the homomorphism $\gamma: L \to K$, the image $\gamma(L)$ is a subalgebra of the Lie algebra K.

A homomorphism $\gamma: L \to K$ is called an *isomorphism* if the linear mapping γ is one-to-one. The cases of isomorphism when $\gamma(L) = K$ or when $\gamma(L) \subset K$ should be separated. If the isomorphism γ is such that $\gamma(L) = K$, then it is said to be an *isomorphism onto*, or there is an isomorphic correspondence of the Lie algebras L and K. In this case the Lie algebras L and K are called *isomorphic*. If the strict inclusion $\gamma(L) \subset K$ holds, then γ provides an *isomorphic insertion* of the Lie algebra L into the Lie algebra K; in this case the inverse isomorphism γ^{-1} is defined only on the subalgebra $\gamma(L) \subset K$.

If a homomorphism $\gamma: L \to K$ is given, then the set of those $\xi \in L$ for which $\gamma\langle\xi\rangle = 0$ is called *the kernel of the homomorphism* γ. It is obvious that the kernel of any homomorphism consists of at least one zero (zero vector field). The kernel of the homomorphism γ is denoted by the natural symbol $\gamma^{-1}(0)$.

For any homomorphism $\gamma: L \to K$ whatsoever, the kernel $\gamma^{-1}(0)$ is an ideal in L. Actually, if $\xi_0 \in \gamma^{-1}(0)$, then $\gamma\langle\xi_0\rangle = 0$ and for any $\xi \in L$

$$\gamma\langle[\xi, \xi_0]\rangle = [\gamma\langle\xi\rangle, 0] = 0,$$

whence $[\xi, \xi_0] \in \gamma^{-1}(0)$ also.

The isomorphism $\alpha: L \to L$ is called an *automorphism of the Lie algebra L*. In other words, an automorphism of the Lie algebra L is a linear transformation of the vector space L which satisfies the identity (6.1).

7. Isomorphism Criterion

It is clear that isomorphic finite-dimensional Lie algebras have equal dimensionality. To establish isomorphism of the Lie algebras L^r and K^r the following criterion is used:

Lemma. *The Lie algebras L^r and K^r are isomorphic if and only if their structural tensors are congruent. This means that if $\gamma: L^r \to K^r$ is an isomorphism, then the structural constants of the Lie algebra L^r in the basis $\{\xi_i\}$ are equal to the corresponding constants of the Lie algebra K^r in the basis $\{\gamma\langle\xi_i\rangle\}$; conversely, if in the Lie algebras L^r and K^r it is possible to select the bases $\{\xi_i\}$ and $\{\eta_i\}$ such that the structural constants of the given algebras bases are correspondingly equal, then the Lie algebras L^r and K^r are isomorphic.*

Proof. If in L^r the structural constants are defined by (5.1) and $\gamma: L^r \to K^r$ is an isomorphism, then from (6.1) it follows that

$$[\gamma\langle\xi_i\rangle, \gamma\langle\xi_j\rangle] = C_{ij}^k \gamma\langle\xi_k\rangle. \tag{7.1}$$

Conversely, (5.1) applied in the algebra K^r also gives

$$[\eta_i, \eta_j] = C_{ij}^k \eta_k. \tag{7.2}$$

Since the linear transformation $\gamma: L^r \to K^r$ is uniquely defined by the data in its basis vectors, it follows that $\gamma\langle\xi_i\rangle = \eta_i$ $(i = 1, \ldots, r)$, whence both relations (7.1) and (7.2) are satisfied. From this (6.1) follows for any pair of basis vectors (ξ_i, ξ_j). Thus, because of the bilinearity of the commutator, equation (6.1) holds for any $\xi, \xi' \in L^r$. ■

8. Example

If an r-dimensional space of vector fields L^r is given by specification of its basis $\{\xi_i\}$, then whether L^r is a Lie algebra can be established, with the aid of the criterion in Section 7.5, by the construction of a *table of commutators*. This table is an $r \times r$ matrix in which the commutator $[\xi_i, \xi_j]$ is placed at the intersection of the ith row and the jth column. As a result of property 3.3° this matrix is *antisymmetric* and it is actually necessary to calculate only its upper half.

As an example take the space L^6 of operators admitted by the linear equation of heat conduction (6.8.1). The basis in L^6 is formed by the vector fields (6.7.5) and (7.8.3) on the space $Z = R^3(x, t, y)$. This basis is

$$\zeta_1 = (1, 0, 0), \qquad \zeta_2 = (0, 1, 0), \qquad \zeta_3 = (x, 2t, 0),$$
$$\zeta_4 = (-2t, 0, xy), \qquad \zeta_5 = (-4xt, -4t^2, (x^2 + 2t)y), \qquad \zeta_6 = (0, 0, y). \tag{8.1}$$

Calculation by formula (1.2), where it is supposed that $\partial = (\partial_x, \partial_t, \partial_y)$, leads to the following table of commutators:

	ζ_1	ζ_2	ζ_3	ζ_4	ζ_5	ζ_6
ζ_1	0	0	ζ_1	0	$2\zeta_4$	0
ζ_2	0	0	$2\zeta_2$	$-2\zeta_1$	$-4\zeta_3 + 2\zeta_6$	0
ζ_3	$-\zeta_1$	$-2\zeta_2$	0	ζ_4	$2\zeta_5$	0
ζ_4	0	$2\zeta_1$	$-\zeta_4$	0	0	0
ζ_5	$-2\zeta_4$	$4\zeta_3 - 2\zeta_6$	$-2\zeta_5$	0	0	0
ζ_6	0	0	0	0	0	0

(8.2)

From this table it follows that the criterion in Section 7.5 holds. Consequently, L^6 is a Lie algebra. With the aid of the table one can draw a conclusion about the structure of this Lie algebra. For simplification of the notation, below, the subspaces in L^6 are encoded by giving only the numbers i of their basis vectors $\{\zeta_i\}$; for instance, $L^3(2, 3, 6)$ means a subspace with the basis $(\zeta_2, \zeta_3, \zeta_6)$.

In the chain of inclusions

$$L^3(1, 2, 6) \subset L^4(1, 2, 3, 6) \subset L^5(1, 2, 3, 4, 6) \subset L^6$$

every subspace is a subalgebra in the next, and L^3 is an Abelian ideal of the entire Lie algebra L^6. We can also note a series of other structural relations; for example, L^4 is an ideal in L^5, etc.

9. Invariance and Relation to Similarity

According to the definition in Section 1.16, the vector field $\bar{\xi} \in \Xi(X)$ is called *similar* to the vector field $\xi \in \Xi(X)$ if there exists a transformation p of the space X for which the formula of similarity of vector fields

$$\bar{\xi}(\bar{x}) = \partial p(x)\langle\xi(x)\rangle, \qquad \bar{x} = p(x) \tag{9.1}$$

holds. In this notation the property of invariance of the similar transformation (9.1) is formulated below.

Lemma. *A commutator of the vector fields, similar to the given ones, is similar to the commutator of the given vector fields, that is, the equality*

$$[\bar{\xi}_1, \bar{\xi}_2] = \overline{[\xi_1, \xi_2]} \tag{9.2}$$

is satisfied.

Proof. Let ∂ still be the symbol of differentiation of the mappings $x \to f(x)$, and let $\bar{\partial}$ be the symbol of differentiation of the mapping $\bar{x} \to f(\bar{x})$. Then, if $\bar{x} = p(x)$ and $\bar{u}(\bar{x}) = v(x)$, the equation $\bar{\partial}\bar{u} = \partial v \circ (\partial p)^{-1}$ holds. According to definition (1.1), the left-hand side of equation (9.2) in this notation is

$$[\bar{\xi}_1, \bar{\xi}_2] = \bar{\partial}\bar{\xi}_2\langle\bar{\xi}_1\rangle - \bar{\partial}\bar{\xi}_1\langle\bar{\xi}_2\rangle.$$

Using the previous formula of differentiation and the definition of similarity (9.1), the following is obtained:

$$\begin{aligned}\bar{\partial}\bar{\xi}_2\langle\bar{\xi}_1\rangle &= \partial(\partial p\langle\xi_2\rangle) \circ (\partial p)^{-1}\langle\partial p\langle\xi_1\rangle\rangle = \partial(\partial p\langle\xi_2\rangle)\langle\xi_1\rangle \\ &= \partial^2 p\langle\xi_2, \xi_1\rangle + \partial p\langle\partial\xi_2\langle\xi_1\rangle\rangle.\end{aligned}$$

Subtracting the analogous formula with permuted indices and taking into account the symmetry of the second derivative $\partial^2 p$ give

$$[\bar{\xi}_1, \bar{\xi}_2] = \partial p\langle\partial\xi_2\langle\xi_1\rangle - \partial\xi_1\langle\xi_2\rangle\rangle = \partial p\langle[\xi_1, \xi_2]\rangle,$$

which, because of the definition of similarity (9.1), is congruent with $\overline{[\xi_1, \xi_2]}$. ■

Consequence. *If $L \subset \Xi(X)$ is a Lie algebra and all vector fields $\xi \in L$ are subjected to the transformation of similarity p by the formula* (9.1), *then the union $\bar{L}$ of the transformed vector $\bar{\xi}$ is also a Lie algebra.*

This follows directly from the equality (9.2).

The Lie algebra $\bar{L}$ obtained by this construction *is said to be related to the Lie algebra L.* It is clear that related Lie algebras are similar. Actually, the mapping $\gamma\colon L \to \bar{L}$, defined by the formula of similarity of vector fields (9.1), is one-to-one and linear because the lemma retains the commutator.

In particular, the result stated in Section 3.8 can be restated to read as follows: The one-dimensional Lie algebra $L^1(\xi)$ is similar to the Lie algebra $L^1(\lambda)$, where λ is a constant, $\lambda \in X$.

10. Group Commutator

If $G^1(f_1)$ and $G^1(f_2)$ are Lie groups of transformations of the space X generated by the mappings f_1 and f_2, then for any $a \in R$ it is possible to form the *group commutator*

$$q_a = f_{2a} \circ f_{1a} \circ f_{2a}^{-1} \circ f_{1a}^{-1}. \tag{10.1}$$

For every $a \in R$ the group commutator is a transformation of the space X. That is why the mapping $q\colon X \times R \to X$, operating by the formula $q(x, a) = q_a(x)$, generates a one-parameter family of transformations of the space X, where $q(x, 0) = x$. Although this family, generally speaking, is not the group G^1, to it one can associate a completely defined vector field on X by analogy with what has been done for the groups G^1 (see Section 1.13). This field is constructed as a field of tangent vectors to the "orbits" $q_x(R)$ of the points x when $a = 0$. But here it is impossible to calculate such tangent vectors by the formula $\partial_a q_x(0)$ because this derivative turns out to be identically equal to zero. This results because of the "bad" parameterization of the curves $q_x(R)$ by the parameter a at the point $a = 0$, since $\partial_a q(x, 0) = 0$. As it turns out, a "good" parameterization is achieved if we assume $a = \sqrt{t}$. Then the unknown vector field is given by the formula

$$\partial_t q(x, \sqrt{t})|_{t=0} = \tfrac{1}{2}\partial_a^2 q(x, 0).$$

Lemma. *The tangent vector field of the group commutator* (10.1) *exists for any* $G^1(f_1)$ *and* $G^1(f_2)$, *and if* ξ_1 *and* ξ_2 *are tangent vector fields for these groups, then*

$$\partial_t q(x, \sqrt{t})|_{t=0} = [\xi_1, \xi_2]. \tag{10.2}$$

Proof. For any group $G^1(f)$, with tangent vector field ξ, the Lie equation (1.14.1) yields the Taylor formula for expansion of the mapping f_a:

$$f_a(x) = x + a\xi(x) + \tfrac{1}{2}a^2\partial\xi(x)\langle\xi(x)\rangle + a^3F(x, a).$$

Here the norm $|F(x, a)|$ is limited to a neighborhood of the point $a = 0$. [In the future this fact will be expressed by writing $|F(x, a)| < \infty$.] For the groups $G^1(f_1)$ and $G^1(f_2)$, having tangent vector fields ξ_1 and ξ_2, use of the previous formula gives

$$\begin{aligned} f_{2a} \circ f_{1a} = I_X &+ a(\xi_1 + \xi_2) + \tfrac{1}{2}a^2(\partial\xi_1\langle\xi_1\rangle + \partial\xi_2\langle\xi_2\rangle) \\ &+ a^2\partial\xi_2\langle\xi_1\rangle + a^3F'_a, \end{aligned} \tag{10.3}$$

where the mapping $F'_a\colon X \to X$ is such that $|F'_a(x)| < \infty$. In addition, if a mapping $h_a\colon X \to X$ has the form

$$h_a(x) = x + a\lambda(x) + a^2\mu(x) + a^3\nu(x, a),$$

where $|\nu(x, a)| < \infty$, then the composition $q_a \circ h_a$ by the Taylor formula is

$$\begin{aligned} q_a \circ h_a(x) = x &+ a(\partial_a q(x, 0) + \lambda(x)) + a^2(\tfrac{1}{2}\partial_a^2 q(x, 0) \\ &+ \partial_x(\partial_a q(x, 0))\langle\lambda(x)\rangle + \mu(x)) + a^3\nu_1(x, a), \end{aligned}$$

where $|\nu_1(x, a)| < \infty$. Because of (10.3) the composition $f_{1a} \circ f_{2a}$ has the form of the mapping h_a, with $\lambda = \xi_1 + \xi_2$, $\mu = \frac{1}{2}(\partial\xi_1\langle\xi_1\rangle + \partial\xi_2\langle\xi_2\rangle) + \partial\xi_1\langle\xi_2\rangle$. Thus definition (10.1), taken in the form $q_a \circ f_{1a} \circ f_{2a} = f_{2a} \circ f_{1a}$

with (10.3) and the previous formula of expansion of the composition $q \circ (f_{1a} \circ f_{2a})$, leads to the relation

$$a\partial_a q(x, 0) + a^2(\tfrac{1}{2}\partial_a^2 q(x, 0) + \partial_x(\partial_a q(x, 0))\langle \xi_1(x) + \xi_2(x)\rangle$$
$$+ a^2 \partial\xi_1(x)\langle \xi_2(x)\rangle = a^2 \partial\xi_2 \langle \xi_1(x)\rangle + a^3 \omega(x, a),$$

where $|\omega(x, a)| < \infty$. Upon division by a^2 and using the fact that $\partial_a q(x, 0) = 0$, passage to the limit $a \to 0$ gives

$$\tfrac{1}{2}\partial_a^2 q(x, 0) + \partial\xi_1(x)\langle \xi_2(x)\rangle = \partial\xi_2(x)\langle \xi_1(x)\rangle.$$

Therefore, it follows from definition (1.1) that (10.2) is correct. ■

11. Admitted Operators

Here and in the following, the facts necessary to understand the group properties of differential equations are established.

According to Section 3.3, the mapping $F: X \to Y$ is invariant relative to the group $G^1(\xi)$ of transformations of the space X, characterized by its operator $\xi \cdot \partial$, if for every $x \in X$ the equation $(\xi \cdot \partial)F(x) = 0$ is satisfied. In this case it is said that *the mapping F admits the operator $\xi \cdot \partial$.*

Obviously, if the mapping F admits the operators $\xi_1 \cdot \partial$ and $\xi_2 \cdot \partial$, then it also admits their commutator $[\xi_1, \xi_2]$. Indeed, because of (7.2),

$$([\xi_1, \xi_2] \cdot \partial)F(x) = (\xi_1 \cdot \partial)(\xi_2 \cdot \partial)F(x) - (\xi_2 \cdot \partial)(\xi_1 \cdot \partial)F(x) = 0.$$

The analogous fact for manifolds is not so trivial. According to Section 3.12, the manifold Ψ, defined with the aid of the mapping $\psi: X \to R^s$, is invariant relative to the group $G^1(\xi)$ if and only if the equation $(\xi \cdot \partial)\psi(x)|_\Psi = 0$ holds. In this case it is also said that *the manifold Ψ admits the operator $\xi \cdot \partial$.*

Lemma. *If the manifold Ψ admits the operators $\xi_1 \cdot \partial$ and $\xi_2 \cdot \partial$, then it also admits their commutator $[\xi_1, \xi_2] \cdot \partial$.*

Proof. Let $G^1(\xi_1)$ and $G^1(\xi_2)$ be the Lie groups of transformations of the space X generated by the mappings f_1 and f_2, respectively. As the manifold Ψ is invariant relative to any transformations $f_{1a} \in G^1(\xi_1)$, $f_{2a} \in G^1(\xi_2)$, Ψ also is invariant relative to the group commutator q_a. This means that for any point $x \in \Psi$, $\psi \circ q(x, \sqrt{t}) = 0$ holds. But then, because of the lemma in Section 7.10,

$$0 = \partial_t(\psi \circ q(x, \sqrt{t}))|_{t=0} = \partial\psi(q)\langle \partial_t q(x, \sqrt{t})\rangle|_{t=0}$$
$$= \partial\psi(x)\langle [\xi_1, \xi_2]\rangle = ([\xi_1, \xi_2] \cdot \partial)\psi(x). \quad ■$$

12. Continuation of Commutator

Here the important property of permutability of the commutation and continuation operations is established. The notation and facts from Section 4 are used below. In addition, the Banach spaces X, Y, $Z = X \times Y$ and the corresponding continuation spaces Z_k will be used.

In Section 4.8 the kth continuation of any vector field $\zeta \in \Xi(Z)$ is defined as a tangent vector field of the continuation group $\underset{k}{G}^1(\zeta)$ and, corresponding to it, the continuation operator $\underset{k}{\zeta} \cdot \partial$. If there are two vector fields $\zeta, \zeta' \in \Xi(Z)$, then with the aid of the operations of continuation and commutation it is possible to generate two vector fields (or two operators) on the continuous space Z_k: the commutator $[\underset{k}{\zeta}, \underset{k}{\zeta'}]$ of the kth continuations of the vector fields ζ and ζ' and the kth continuation of their commutator $\underset{k}{[\zeta, \zeta']}$. The following lemma is established.

Lemma. *The commutator of the kth continuations is equal to the kth continuation of the commutator of these operators, that is,*

$$[\underset{k}{\zeta}, \underset{k}{\zeta'}] = \underset{k}{[\zeta, \zeta']}. \tag{12.1}$$

Proof. An initial point is the fundamental property of the continuous transformations as formulated in the lemma in Section 4.6. It is possible to give two versions of the proof, in each of which the almost obvious property of permutability of a continuation operation with one of the operations is used: differentiation with respect to a parameter or the transformation of similarity. Both versions use the groups $G^1(\zeta)$, generated by the mappings $h: Z \times R \to Z$, with corresponding continuation groups $\underset{k}{G}^1(\zeta)$ from mappings $\underset{k}{h}: Z_k \times R \to Z_k$.

First version. Because of the definition in Section 4.8 and the equality $\zeta = \partial_a h(0)$, the formula of permutability of the continuation operation with the operation of differentiation with respect to the parameter a,

$$\underset{k}{\zeta} = \underbrace{\partial_a h(0)}_{k} = \partial_a \underset{k}{h}(0),$$

holds. Let a group $G^1(\zeta')$ be generated by the mapping h'. From the lemma in Section 7.10 it follows that the commutator $[\underset{k}{\zeta}, \underset{k}{\zeta'}]$ of continuation vector fields can be calculated, using the group commutator $\underset{k}{h'_a} \circ \underset{k}{h_a} \circ \underset{k}{h'^{-1}_a} \circ \underset{k}{h^{-1}_a}$, by the formula

$$[\underset{k}{\zeta}, \underset{k}{\zeta'}] = \partial_t(\underset{k}{h'_{\sqrt{t}}} \circ \underset{k}{h_{\sqrt{t}}} \circ \underset{k}{h'^{-1}_{\sqrt{t}}} \circ \underset{k}{h^{-1}_{\sqrt{t}}})|_{t=0}.$$

Using the lemma in Section 4.6, the just noted property, and, once again, the lemma in Section 7.10, this expression is equal to

$$\underbrace{\partial_t(h'_{\sqrt{t}} \circ h_{\sqrt{t}} \circ h'^{-1}_{\sqrt{t}} \circ h^{-1}_{\sqrt{t}})}_{k}|_{t=0} = \underbrace{\partial_t(h'_{\sqrt{t}} \circ h_{\sqrt{t}} \circ h'^{-1}_{\sqrt{t}} \circ h^{-1}_{\sqrt{t}})|_{t=0}}_{k} = [\underset{k}{\zeta}, \underset{k}{\zeta'}],$$

which was to be established.

Second version. Let p be a transformation of the space Z. Considered as a transformation of the similarity group, according to Section 1.11, it transforms the mapping h_a into the mapping $\bar{h}_a = p \circ h_a \circ p^{-1}$, from which the analogous formula of similarity (1.16.1) of the corresponding vector fields $\bar{\zeta}(\bar{z}) = \partial p(z)\langle\zeta(z)\rangle$ follows, where $\bar{z} = p(z)$. From the lemma in Section 4.6 the equality

$$\overline{\boxed{\underset{k}{h_a}}} = \underset{k}{\boxed{\bar{h}_a}}$$

is correct, where the left-hand side is a result of the transformation $\underset{k}{h_a}$, using the similarity continuation p_k. From this follows the formula of permutability of the vector field ζ with the transformation of similarity

$$\overline{\boxed{\underset{k}{\zeta}}} = \underset{k}{\boxed{\bar{\zeta}}}.$$

The combination of this fact with the lemma in Section 7.9 shows that equation (12.1) is invariant relative to the transformation of similarity.

Now it must be noted that according to the theorem of similarity in Section 3.8, for the given vector field ζ' such a transformation of similarity p can always be found such that the similar vector field $\bar{\zeta}'$ will be constant on the space Z. Thus it is enough to prove equality (12.1) for the case when the vector field $\zeta' = \lambda$ is constant. In this case equality (12.1) is equal to

$$(\lambda \cdot \partial)\underset{k}{\zeta} = \underbrace{(\lambda \cdot \partial)\zeta}_{k}. \tag{12.2}$$

As $\underset{k}{\zeta} = (\xi, \eta, \underset{1}{\eta}, \ldots, \underset{k}{\eta})$,

$$(\lambda \cdot \partial)\underset{k}{\zeta} = ((\lambda \cdot \partial)\xi, (\lambda \cdot \partial)\eta, (\lambda \cdot \partial)\underset{1}{\eta}, \ldots, (\lambda \cdot \partial)\underset{k}{\eta}).$$

On the other hand, since λ is constant, the operator $\lambda \cdot \partial$ is permutable with the operator of total differentiation D, so from formula (4.8.5) the equality

$$(\lambda \cdot \partial)\underset{s}{\eta} = D^s((\lambda \cdot \partial)\eta - \underset{1}{y} \cdot (\lambda \cdot \partial)\xi) + \underset{s+1}{y} \cdot (\lambda \cdot \partial)\xi = \underbrace{(\lambda \cdot \partial)\eta}_{s} \qquad (s = 1, \ldots, k)$$

follows, whence (12.2) is obvious. ■

13. Admitted Lie Algebra

Application of the previous results to differential equations leads to the following theorem, in whose formulation the definition in Section 5.2 is used.

Theorem. *If a differential equation admits the operators $\zeta \cdot \partial$ and $\zeta' \cdot \partial$, then it also admits their commutator $[\zeta, \zeta'] \cdot \partial$.*

Proof. According to the definition in Section 5.2, the differential equation E admits the operator $\zeta \cdot \partial$ if the manifold $E \subset Z_k$ is invariant relative to the operator $\underset{k}{\zeta} \cdot \partial$. As a result of the lemma in Section 7.11, a manifold E, invariant relative to operators $\underset{k}{\zeta} \cdot \partial$ and $\underset{k}{\zeta'} \cdot \partial$, is also invariant relative to their commutator $[\underset{k}{\zeta}, \underset{k}{\zeta'}] \cdot \partial = \underset{k}{[\zeta, \zeta']} \cdot \partial$, where the last equality holds because of the lemma in Section 7.12. Thus, according to the definition in Section 5.2, the equation E admits the operator $[\zeta, \zeta']$. ■

From this theorem and the definition in Section 7.4 the next result follows, which has considerable value.

Consequence. *The set of all operators admitted by the given differential equation generates a Lie algebra.*

If there is a Lie algebra L of operators admitted by the given differential equation, then it is said that this equation admits a Lie algebra; in this case L is called the *admitted Lie algebra of the given equation.*

The largest admitted Lie algebra of equation E, containing all operators admitted by the equation, will be called the *full Lie algebra of the equation E* and is denoted with the symbol LE.

In these terms it can be said that any admitted Lie algebra of the equation E is a subalgebra of the full Lie algebra LE.

Now it is clear that in the earlier examples of calculations of the full groups what were actually calculated were some admitted Lie algebras of the corresponding equations. For example, the system of equations in Section 5.9 for the transonic flow of a gas admits the full infinite-dimensional Lie algebra LE, which can be represented as the direct sum of two subspaces $LE = L^4 \oplus L^\infty$, where the basis in L^4 is generated by the operators (5.9.9) and the general form of the operator $\zeta \cdot \partial \in L^\infty$ is given by formula (5.9.11). Through the construction of a table of commutators of (5.9.9), one can be convinced that L^4 is a Lie algebra. The operators (5.9.11) also generate a Lie algebra L^∞ which is Abelian. It is obvious that the commutators $[\zeta^0, \zeta] \cdot \partial$, for any operator $\zeta \cdot \partial \in LE$, have a form of the operators $\zeta^0 \cdot \partial$, which implies that the subalgebra $L^\infty \subset LE$ is an ideal in LE.

The equation $E(\theta)$ of nonlinear heat conduction, with $\theta \neq 1$, admits (depending on the value of the arbitrary element θ) the full Lie algebra presented in (6.7.14). It is easy to check that the Lie algebra L_0, admitted by equation $E(\theta)$, independent of the value of the arbitrary element θ, is an ideal in every one of the algebras L^4, L^5 given in (6.7.14).

14. Linear Equations

A differential equation E is called *linear* (and *homogeneous*) if the mapping defined in the form (5.1.1) $\varepsilon: Z_k \to R^s$ is such that for every $x \in X$ the particular mapping $\varepsilon_x: Y \times Y_1 \times \cdots \times Y_k \to R^s$ is a linear mapping.

The full Lie algebra of operators

$$\zeta(z) \cdot \partial = \xi(x, y) \cdot \partial_x + \eta(x, y) \cdot \partial_y, \tag{14.1}$$

admitted by the linear equation E, always contains an Abelian subalgebra of the form

$$\zeta^0(z) \cdot \partial = v(x) \cdot \partial_y, \tag{14.2}$$

where the mapping $v: X \to Y$. An essential property of such operators is that they are defined by the solutions of the equation E.

Theorem. *The linear equation E admits an operator of the form* (14.2) *if and only if the mapping v is a solution of the equation E.*

Proof. Let G^1 be the group of transformations of the space $Z = X \times Y$ corresponding to the operator (14.2) and generated by the mapping

$$x' = x, \qquad y' = y + av(x). \tag{14.3}$$

If $u: X \to Y$ is a solution of the equation E, then the transformations of this group generate from it, according to (5.7.3), a one-parameter family of solutions

$$u'(x) = u(x) + av(x).$$

From this it follows that if the equation E admits the operator (14.2), then the mapping v is a solution of the equation E. Conversely, if v is a solution, then the group G^1, generated by the mapping (14.3), is admitted by the equation E. Actually, here $f = I_X$, $g = I_Y + av$, and equations (4.5.3), (4.5.4) give $\underset{k}{y'} = \underset{k}{y} + a\underset{k}{v}$. Thus, because of the linearity of the mapping ε, the equality $\varepsilon(\underset{k}{z'}) = \varepsilon(\underset{k}{z}) + a\varepsilon(\underset{k}{v})$, in which $\varepsilon(\underset{k}{v}) = 0$ from the definition of the solution in Section 5.6, is true. This means that the mapping ε is an invariant of the group (14.3) and so admits this group, as well as the operator (14.2). ■

Since the set of all solutions of a linear equation generates a vector space, this property belongs to the union of operators of the form (14.2). It is obvious that the commutator of any two such operators is equal to zero. Consequently, their union is an Abelian Lie algebra. Generally speaking, this Lie algebra is infinite dimensional and thus will be denoted L^∞.

Another typical Lie algebra, admitted by a linear equation, is the set of operators of the form

$$\zeta(z)\cdot\partial = \xi(x)\cdot\partial_x + \beta(x)\langle y\rangle\cdot\partial_y, \tag{14.4}$$

where the component of the vector field $\xi\colon X \times Y \to Y$ does not depend on $y \in Y$ and p is the mapping $X \to \mathscr{L}(Y)$, so the component of the vector field η is linear relative to y. Of course, here the mappings ξ and β are not so simply defined as in the previous case; they must be solutions of the corresponding determining equations *DE*.

As the commutator of operators of the form (14.4),

$$\begin{aligned}[\zeta_1, \zeta_2]\cdot\partial = [\xi_1, \xi_2]\cdot\partial_x &+ ((\xi_2\cdot\partial_x)\beta_2 \\ &- (\xi_2\cdot\partial_x)\beta_1 + \beta_2\circ\beta_1 - \beta_1\circ\beta_2)\langle y\rangle\cdot\partial_y,\end{aligned}$$

is again an operator of the form (14.4), the union of all such operators, admitted by the equation E, also generates a Lie algebra. This Lie algebra usually turns out to be finite dimensional and is an L^r. It is easy to verify that a direct sum of the vector spaces $L^r \oplus L^\infty$ is also a Lie algebra. Using this, the commutator of any operators $\zeta^0\cdot\partial \in L^\infty$ and $\zeta\cdot\partial \in L^r$,

$$[\zeta^0\cdot\partial, \zeta\cdot\partial) = (\beta\langle v\rangle) - (\xi\cdot\partial_x)v)\cdot\partial_y, \tag{14.5}$$

also belongs to the Lie algebra L^∞, because of the previous theorem. Thus the following statement is established.

Consequence. *The subalgebra L^∞ is an ideal in $L^r \oplus L^\infty$.*

The importance of the two previously defined typical Lie algebras is that for the majority of linear equations their full Lie algebra of operators is $L^r \oplus L^\infty$. In other words, for linear equations, as a rule, the nontrivial admitted operators have the form (14.4). An obvious criterion for the full Lie algebra of the linear equation E to have the structure $L^r \oplus L^\infty$, where L^r consist of the operators of the form (14.4), is that the following must be satisfied. *The equations*

$$\partial_y\xi = 0, \qquad \partial_y^r\eta = 0 \tag{14.6}$$

must be part of the system of determining equations DE.

Equation (14.5) is interesting since it can be considered an *infinitesimal formula for producing solutions.* Actually, using the established theorem together with the mapping v, the component of the operator (14.5) is also a solution. That is why the infinitesimal formula to produce the solutions u', derived from the solution u with the aid of operators of the form (14.4), is such that

$$u'(x) = \beta\langle u(x)\rangle - (\xi \cdot \partial_x)u(x). \tag{14.7}$$

Actually, this formula can be obtained from the general formula for the production of solutions (5.7.5) with the group of transformations G^1, the operator of which is (14.4). For that it is only necessary to note that if $\{u_a\}$ is a one-parameter family of solutions produced by this group, then the derivative $\partial_a u_a|_{a=0}$ is also a solution because of the linearity of the equation E. A calculation of this derivative of the right-hand side of equation (5.7.3) gives exactly the expression (14.7).

For example, as is shown in Section 6.8, the full Lie algebra of the linear equation of heat conduction (6.8.1) is $L^6 \oplus L^\infty$. Here the infinitesimal formulas of the form (14.7) for the production of solutions with the aid of the operators $\zeta_1, \ldots, \zeta_5$ are

$$u' = u_x,\ u' = u_t,\ u' = xu_x + 2tu_t,$$

$$u' = xu + 2tu_x,\ u' = (x^2 + 2t)u + 4xtu_x + 4t^2u_t,$$

where $u_x = \partial_x u, u_t = \partial_t u$. The operator ζ_6 doesn't produce any new solutions.

15. Abstract Determining Equations

Because of the theorem of Section 7.13, for any differential equation E, the determining equations DE have the property that if ζ and ζ' are two solutions, then the commutator $[\zeta, \zeta']$ is also a solution of the equations DE. This property can be used as the basis for an axiomatic definition of a special class of differential equations.

It is said that the *system of differential equations, with unknown vector field* $\xi\colon X \to X$, *is an abstract system of determining equations (briefly an AS-system) if this system is linear and homogeneous and if for any two solutions* ξ, ξ' *the commutator* $[\xi, \xi']$ *is also a solution.*

Because of this definition, the set of all solutions of an AS-system generates a Lie algebra of vector fields on the space X. The previous property means that these equations are always an AS-system.

It is natural for the following question to be asked: For every Lie algebra of vector fields (operators) $L \subset \Xi(X)$ can an AS-system be found such that the space of its solutions is congruent with L? In Chapter VII it will be shown, based on the theory of differential invariants, that the answer is yes for finite-dimensional Lie algebras.

REFERENCES

The first applications of Lie theory devoted to these problems belong to Lie in a number of works [1–3]. He introduced the idea of an admitted group and also gave the first examples of group classification of differential equations. These problems were stated in modern language only in the past 20 years [4–7], in connection with the group analysis of concrete equations of mechanics and physics. We also note those investigations in which attempts have been made to modify the Lie theory in its applications to differential equations [8, 9]. The problem of "mechanization" of the calculations, connected with the construction of the determining equations, was solved in papers [10] and [11]. The algorithms given in Section 6 for solving the problem of group classification are stated here for the first time. The example from Section 6.7 is presented in the paper [12]. Recently this result was unexpectedly used in the theory of difference schemes [13]. The qualitative properties of abstract determining equations are developed in [14].

[1] L i e S. Über die Integration durch bestimmte Integrale von einer Klasse lineare partiellen Differentialgleichungen. — Arch. Math., **6**, Helf 3 (1881), 328—368.
[2] L i e S. Classification und Integration von gewöhnlichen Differentialgleichungen zwischen x, y, die eine Gruppe von Transformationen gestatten. — Arch. Math. Naturv., Christiania **9** (1883), 371—393.
[3] L i e S. Allgemeine Untersuchungen über Differentialgleichungen, die eine continuirliche, endliche Gruppe gestatten. — Math. Ann. **25**, № 1, (1885), 71—151.
[4] О в с я н н и к о в Л. В. Группы и инвариантно-групповые решения дифференциальных уравнений. — ДАН СССР **118**, № 3, (1958), 439—442.
[5] О в с я н н и к о в Л. В. Групповые свойства дифференциальных уравнений. Новосибирск, Изд. СО АН СССР, 1962.
[6] О в с я н н и к о в Л. В. Лекции по теории групповых свойств дифференциальных уравнений. Новосибирск, НГУ, 1966.
[7] K u c h a r c z y k P. Teoria grup Liego w zastosowaniu do równán róznizkowych czastkowych. Warszawa, PAN, 1967.
[8] H a r r i s o n B. K., E s t a b r o o k F. S., Geometric approach to invariance groups and solution of partial differential systems. — J. Math. Phys. **12**, № 4 (1971), 653—666.
[9] A n d e r s o n R., K u m e i S., W u l f m a n C. E. Generalisation of the concept of invariance of differential equations. Results of applications to some Schrödinger equations. — Phys. Rev. Lett. **28**, № 15 (1972), 988—991.
[10] К а т к о в В. Л., К о с т ю к о в а Н. И. Процессор КИНО. — В сб. «Динамика сплошной среды», вып. 1. Новосибирск (1969), 48—60.

[11] Костюкова Н. И. Процессор ПАССИВ. — Вычисл. мат. и вычисл. техн., вып. 3. Харьков, Физ.-техн. ин-т низких температур АН УССР, (1972), 38—41.
[12] Овсянников Л. В. Групповые свойства уравнения нелинейной теплопроводности. — ДАН СССР, **125**, № 3 (1959), 492—495.
[13] Яненко Н. Н., Шокин Ю. И. О групповой классификации разностных схем для системы одномерных уравнений газовой динамики. — В сб. «Некоторые проблемы математики и механики». Л., «Наука», (1970), 277—283.
[14] Овсянников Л. В. Групповое свойство определяющих уравнений. — В сб. «Динамика сплошной среды», вып. **7**. Новосибирск, (1971), 5—11.

CHAPTER

III

Full Groups of Concrete Systems of Equations

The effectiveness of a theory must be confirmed by nontrivial examples of its application. To achieve this goal, this chapter is devoted to the calculation of the full group and the group classification of equations having practical importance.

The initiation of research on the group properties of ordinary differential equations was due to S. Lie, who had in mind the development of a well-founded theory for integrating equations such as the one created by Abel for solving algebraic equations. For equations of the first order such a theory can be considered to be known at least as far as nontrivial operators admit such equations (see Sections 8.5 and 8.10). But as it turns out, the problem of determining nontrivial solutions is equivalent to the problem of integration of the original equations. The explanation why the dissatisfaction caused by this fact is replaced by optimism in the case of ordinary differential equations of higher order will be given in this chapter.

In mathematical physics linear equations containing partial derivatives of the second order with respect to two independent variables are met especially often. All calculations necessary for an exhaustive classification are given for such equations. The complete formulas are calculated for the hyperbolic (and also parabolic) normal form. It is hoped that the interested reader will use the appropriate formulas for the elliptic case himself. The equations of the two-dimensional boundary layer provide an example of a degenerate system with a nontrivial classification, and the equations of gas dynamics are considered as an example of a special system of equations.

The example of equations for gas dynamics is interesting because the values of the physical constant (index of the adiabatic curve) permitting an expansion of the full group are revealed by group analysis.

§8. ORDINARY DIFFERENTIAL EQUATIONS

1. Systems of First-Order Equations

Differential equation of the type $E(1, m, 1, m)$ are considered here. In this case $X = R(x)$, $Y = R^m(y)$, $Z = R^{m+1}(z)$, where $z = (x, y)$ and the first continuation space is $Y_1 = R^m(\underset{1}{y})$. If it is assumed that the mapping ε, which is used to describe the equation $\varepsilon(x, y, \underset{1}{y}) = 0$, satisfies the conditions of the implicit function theorem relative to $\underset{1}{y}$, then it can be solved for $\underset{1}{y}$. Further, it is supposed that this is done and the equation E is assigned the normal form

$$\underset{1}{y} = f(x, y), \tag{1.1}$$

with a mapping $f\colon Z \to R^m$ of the class $C_\infty(Z)$.

In component form it is usually supposed that $y = (y_1, \ldots, y_m)$, $\underset{1}{y} = (y'_1, \ldots, y'_m)$, $f = (f_1, \ldots, f_m)$ and equation (1.1) is written as the system

$$y'_k = f_k(x, y_1, \ldots, y_m) \qquad (k = 1, \ldots, m),$$

which is called a *system of ordinary differential equations of the first order*.

The problem of finding and studying the structure of the full Lie algebra of equation (1.1) is solved in the sequel.

2. Determining Equations

The algorithm in Section 5.8 is used below. The operator, admitted by equation (1.1), is to be found in the form

$$\zeta \cdot \partial = \xi \partial_x + \eta \cdot \partial_y, \tag{2.1}$$

with the components $\xi\colon Z \to X$ and $\eta\colon Z \to Y$. Using the total differentiation operator $\underset{1}{D} = \partial_x + \underset{1}{y} \cdot \partial_y$, the first continuation of operator (2.1) is found to be

$$\underset{1}{\zeta} \cdot \partial = \xi \partial_x + \eta \cdot \partial_y + \underset{1}{\eta} \cdot \underset{1}{\partial},$$

$$\underset{1}{\eta} = \underset{1}{D}\eta - \underset{1}{y}\underset{1}{D}\xi = \partial_x \eta + \partial_y \eta \langle \underset{1}{y} \rangle - \underset{1}{y} \partial_x \xi - (\underset{1}{y} \cdot \partial_y \xi)\underset{1}{y}.$$

The result of the operation of operator $\underset{1}{\zeta} \cdot \partial$ on equation (1,1),

$$\underset{1}{\eta} = \xi \partial_x f + \eta \cdot \partial_y f,$$

gives the invariance condition

$$\partial_x \eta + \partial_y \eta \langle f \rangle - f \partial_x \xi - (f \cdot \partial_y \xi) f = \xi \partial_x f + \partial_y f \langle \eta \rangle.$$

After transition to the manifold (1.1), stage (d) in Section 5.4 of decomposition of the invariance condition is absent here since in this situation there are no "free" components of the vector $\underset{1}{y}$. That is why in this case the invariance condition is congruent to the determining equation *DE* for equation (1.1).

3. General Solution Analysis

In component form the determining equation *DE* is a system with m scalar equations for $m + 1$ unknown functions of the vector $\zeta = (\xi, \eta^1, \ldots, \eta^m)$. Thus one of these functions can be arbitrarily selected, and because of that the full Lie algebra of equation (1.1) will surely be infinite dimensional.

This can be clearly seen by introducing a new unknown mapping $\beta: Z \to Y$, connected with the vector $\zeta = (\xi, \eta)$ by the formula

$$\eta = \xi f + \beta. \tag{3.1}$$

Using expressions for the derivatives of the vector

$$\partial_x \eta = \xi \partial_x f + f \partial_x \xi + \partial_x \beta,$$

$$\partial_y \eta \langle f \rangle = (f \cdot \partial_y \xi) f + \xi \partial_y f \langle f \rangle + \partial_y \beta \langle f \rangle,$$

the determining equation is simplified to

$$\partial_x \beta + \partial_y \beta \langle f \rangle = \partial_y f \langle \beta \rangle.$$

To introduce vector fields on the space Z set

$$\phi = (1, f), \qquad \beta_1 = (0, \beta).$$

Then the final determining equation is written, using the symbol of the commutator, in this elegant form:

$$[\phi, \beta_1] = 0. \tag{3.2}$$

The mapping $\beta_1 = 0$ is a solution of equation (3.2). With this solution, equation (3.1) becomes $\eta = \xi f$. From this the following conclusion results: *Equation* (1.1) *admits the operator*

$$\phi \cdot \partial = \partial_x + f \cdot \partial_y, \tag{3.3}$$

and all operators of the form $\zeta_0 \cdot \partial = \xi\phi \cdot \partial$ with any mapping $\xi: Z \to R$. Since for any ξ, ξ', the commutator

$$[\xi\phi, \xi'\phi] = (\xi\phi \cdot \partial\xi' - \xi'\phi \cdot \partial\xi)\phi$$

is again a vector of the form ζ_0, then the operators $\zeta_0 \cdot \partial$ generate a Lie algebra. This infinite-dimensional Lie algebra will be denoted by L^∞.

If we subtract $\xi\phi \cdot \partial$ from the operator (2.1), then we obtain an operator for which the component $\xi = 0$. It is clear that the set L^0 of operators with $\xi = 0$, that is, the set of operators of the form $\beta_1 \cdot \partial = \beta \cdot \partial_y$, is the direct complement of L^∞ in the full Lie algebra LE, so $LE = L^0 \oplus L^\infty$.

To obtain a representation of a "general" solution defining equation (3.2), the group $G^1(\phi)$, the operator of which is $\phi \cdot \partial$, is brought into consideration. Let $J: Z \to Y$ be a universal operator (see Section 3.5) of this group. As the vector field $\phi = (1, f) \in \Xi(Z)$ has a nonzero component in X, so the linear mapping $\partial_y J(x, y): Y \to Y$ is a linear homeomorphism of the space Y.

Deduction of the unknown representation is based on the following result.

Lemma. *The mapping $\beta: Z \to Y$ is a solution of equation* (3.2) *if and only if the mapping $\Phi: Z \to Y$, related to β through the formula*

$$\Phi = \partial_y J\langle\beta\rangle,$$

is an invariant of the group $G^1(\phi)$.

Proof. According to the definition of the commutator (7.1.3), the identity

$$([\phi, \beta_1] \cdot \partial)J = (\phi \cdot \partial)(\beta_1 \cdot \partial)J - (\beta_1 \cdot \partial)(\phi \cdot \partial)J.$$

holds. But $(\phi \cdot \partial)J = 0$ because of the criterion in Section 3.4 of the group invariant for $G^1(\phi)$. Consequently, applying this to the lemma formula, this identity becomes

$$\partial_y J\langle[\phi, \beta_1]\rangle = (\phi \cdot \partial)\Phi.$$

After this the two statements follow from the criterion in Section 3.4 and from the fact that $\partial_y J \in \mathscr{L}(Y)$ is a homeomorphism. ■

A representation of a general solution of equation (3.2) follows from the lemma as

$$\beta = (\partial_y J)^{-1}\langle\Phi\rangle, \tag{3.4}$$

where $\Phi: Z \to Y$ is an arbitrary invariant of the group $G^1(\phi)$. This is why the space L^0 is infinite dimensional.

4. Structure of the Full Lie Algebra

If β and β' are two solutions of equation (3.2), then the commutator $[\beta, \beta']$ is also a solution of this equation. Now $[\beta, \beta']_1 = [\beta_1, \beta'_1]$ and Jacobi's identity 3° in Section 7.3 gives

$$\begin{aligned}[\phi, [\beta_1, \beta'_1]] &= [\beta_1, [\phi, \beta'_1]] - [\beta'_1, [\phi, \beta_1]] \\ &= [\beta_1, 0] - [\beta'_1, 0] = 0.\end{aligned}$$

Consequently, the space L^0 is a Lie algebra.

Further, the subalgebra L^∞ is an ideal in LE. It follows from this that the commutator of any solution β_1 of the equation (3.2) and any vector $\xi\phi \in L^\infty$,

$$[\beta_1, \xi\phi] = (\beta \cdot \partial_y \xi)\phi,$$

also belongs to the Lie algebra L^∞.

From that the structure of the full Lie algebra of equation (1.1) is described by its decomposition into the direct sum $LE = L^0 \oplus L^\infty$ of the infinite-dimensional subalgebra L^0 and the infinite-dimensional ideal L^∞. Here L^0 consists of all operators of the form $\beta \cdot \partial_y$, with any solution β of equation (3.2) [all these solutions are given by equation (3.4)], and the ideal L^∞ consists of the operator $\xi\phi \cdot \partial$, obtained by a multiplication of the operator (3.3) with the arbitrary mapping $\xi: Z \to R$.

5. Equation Dimensionality Reduction

The operators $\zeta_0 \cdot \partial \in L^\infty$ are obtained "free of charge" without any essential analytic construction. But from the viewpoint of the problem of integrating the system of ordinary differential equations, knowledge of these operators gives no essential information. Obviously, the problem of constructing the groups G^1 corresponding to these operators is actually equivalent to the problem of integration of the original equation (1.1).

Clearly, the additional operators $\beta \cdot \partial_y \in L^0$, as the formula (3.4) shows, are expressed by means of solutions of the original equation (1.1). Thus knowledge of any operator $\beta \cdot \partial_y$ must be connected with any realization of the integration process. This can be a basis for the fact that the operators $\beta \cdot \partial_y$ can be used effectively for integration (even if partial) of equation (1.1). With this there must be taken into consideration that the problem of constructing a universal invariant for the groups of the form $G^1(\beta)$, with a tangent vector field β, satisfying equation (3.2), essentially differs from the same problem for the group $G^1(\phi)$. And it may turn out (as it often does in practice) that this problem may be significantly simpler. It can be said that success in solving the system of ordinary differential equations in analytic form is significantly enhanced by the knowledge of the Lie algebra of the operators $L \subset L^0$. In Chapter V this fact will be considered in depth. One particular but, in practice, very useful result in which a reduction of the dimensionality is well understood is a contraction of equation (1.1) to the equation of the kind $E(1, m-1, 1, m-1)$, discussed in the theorem below.

Theorem. *Knowledge of the universal invariant of the operator $\beta \cdot \partial_y$ ($\beta \neq 0$), satisfying equation* (3.2), *permits a reduction of the dimensionality of equation* (1.1).

Proof. The group $G^1(\beta)$ of transformations of the space Y is considered for every $x \in X$. Let $K'\colon Y \to R^{m-1}$ be its universal invariant. The function $H\colon Y \to R$, satisfying the equation $(\beta \cdot \partial_y)H = 1$, is constructed, and the mapping K' is supplemented by the mapping

$$K = (K', H)\colon Y \to R^m.$$

As will be shown below, the mapping $\bar{y} = K(x, y)$, when x is fixed, is a local transformation of the space Y. Thus equation (1.1), transformed with the aid of this mapping according to the relations

$$\underset{1}{\bar{y}} = \underset{1}{D}K = \partial_x K + (\underset{1}{y} \cdot \partial_y)K = (\partial_x + \mathrm{f} \cdot \partial_y)\mathrm{K} = (\phi \cdot \partial)\mathrm{K},$$

becomes the equation

$$\underset{1}{\bar{y}} = (\phi \cdot \partial)K(x, y), \qquad \bar{y} = K(x, y). \tag{5.1}$$

In the identity, found from the definition of the commutator (7.1.3),

$$([\phi, \beta_1] \cdot \partial)K = (\phi \cdot \partial)(\beta \cdot \partial_y)K - (\beta \cdot \partial_y)(\phi \cdot \partial)K;$$

the left-hand side and the first term on the right-hand side are equal to zero because of the definition of β and the construction of K. Consequently, this identity leads to the equation $(\beta \cdot \partial_y)((\phi \cdot \partial)K) = 0$, which shows that $(\phi \cdot \partial)K$ is an invariant of the group $G^1(\beta)$. From the universality of the invariant K it follows that such a mapping $\Phi\colon R^{m-1} \to R^m$, for which $(\phi \cdot \partial)K(x, y) = \Phi \circ K'(x, y) = \Phi(\bar{y}')$, exists, where $\bar{y}'$ is a component of the representation $\bar{y} = (K', H)$. This is why, if we assume $\Phi = (\Phi', P)$, $\bar{y} = (\bar{y}', p)$ and remember that Φ can depend on x, equation (5.1) can be written as a system of two equations

$$\underset{1}{\bar{y}'} = \Phi'(x, \bar{y}'), \qquad \partial_x p = P(x, \bar{y}'). \tag{5.2}$$

The first of these has the type $E(1, m-1, 1, m-1)$, and the second is integrable with one quadrature if a solution of the first one is known.

It is left to show that $\bar{y} = K(x, y)$ is a local transformation of the space X. To do this it is enough to prove the existence of a point (x_0, y_0) such that the linear mapping $\partial_y K(x_0, y_0): Y \to X$ is a linear homeomorphism of the space Y. If such a point does not exist, then there is a vector space $\alpha: Z \to Y$, for which $\partial_y K\langle\alpha(x, y)\rangle = 0$, or, in the components of $K = (K', H)$,

$$(\alpha \cdot \partial_y)K' = 0, \qquad (\alpha \cdot \partial_y)H = 0.$$

The first of these identities means that, for a fixed x, the mapping $K': Y \to R^{m-1}$ is a universal invariant for the group $G^1(\alpha)$, and the second that H is an invariant of this group. Because of the universality of K', it follows that there exists a mapping $\Psi: R \to R$ such that $H = \Psi \circ K'$. But then $(\beta \cdot \partial_y)H = \partial\Psi\langle(\beta \cdot \partial_y)K'\rangle = 0$, which contradicts the definition of the mapping H. ■

6. Examples

In the case of an equation of the form $E(1, 1, 1, 1)$, that is, one ordinary differential equation of the first order, the vector $\underset{1}{y}$ is one dimensional and denoted here by a more descriptive symbol y_x. Let this equation have the form

$$y_x = f(x, y). \tag{6.1}$$

It can be said that all classical specializations of this equation, when it is integrated in quadratures, are based on a knowledge of some group of transformation G^1 of the space $R^2(x, y)$ admitted by equation (6.1). In these cases the transformation of the unknown y by the formula $\bar{y} = J(x, y)$, where $J: R^2(x, y) \to R$, is a universal invariant of the admitted group G^1. This changes (6.1) to an equation with separated variables.

For example, a "generalized homogeneous" equation $y_x = x^{k-1}F(yx^{-k})$ admits the operator $x\partial_x + ky\partial_y$ and the universal invariant of the group G^1 is $J = yx^{-k}$. A transformation to the variable $\bar{y} = yx^{-k}$ changes this equation to

$$x\bar{y}_x = F(\bar{y}) - k\bar{y},$$

which is integrated with one quadrature.

Another method of integrating equation (6.1) in quadratures is connected with the concept of an integrating factor, which is the function $\mu: R^2(x, y) \to R$ satisfying the equation

$$\partial_x \mu + \partial_y(f\mu) = 0. \tag{6.2}$$

It is easy to verify that if we set $\mu = 1/\beta$, then condition (6.2) becomes (3.2). That is why, because of equation (3.1), a knowledge of the operator $\xi\partial_x + \eta\partial_y$ admitted by equation (6.1) gives the integrating factor $\mu = 1/(\eta - \xi f)$.

For example, the equation

$$y_x = y/(x(y + \ln x))$$

admits the operator $xy\partial_x$. Thus the integrating factor is $\mu = (y + \ln x)y^{-2}$. Upon multiplying by this, the equation is written with the operator of total differentiation $D_1 = \partial_x + y_x\partial_y$ as

$$D_1(\ln y - (\ln x)/y) = 0.$$

From this the general integral $\ln y - (\ln x)/y = c$ is found, where c is an arbitrary constant.

7. Higher Order Equations

Essentially new facts appear for scalar ordinary differential equations of order $k > 1$, that is, for equations of the type $E(1, 1, k, 1)$. In this case $X = R(x,)$, $Y = R(y)$, $Z = R^2(x, y)$ and all spaces Y_s are one dimensional, in connection with which the symbol $\underset{s}{y}$ can be changed to the simpler form y_s; hence $\underset{s}{z} = (x, y, y_1, \ldots, y_s)$. The equation E is reduced to the normal form

$$y_k = f(x, y, y_1, \ldots, y_{k-1}), \tag{7.1}$$

with the given mapping $f: Z_{k-1} \to R$.

The operator admitted by equation (7.1) is to be found in the form (2.1). To write the condition of invariance, the kth continuation of the operator (7.1)

$$\underset{k}{\zeta} \cdot \partial = \xi\partial_x + \eta\partial_y + \eta_1\partial_{y_1} + \cdots + \eta_k\partial_{y_k}$$

is needed. According to formula (4.8.5),

$$\eta_l = D^l\eta - (D^l(y_1\xi) - y_{l+1}\xi),\ D = \partial_x + y_1\partial_y + y_2\partial_{y_1} + \cdots.$$

By the method of induction with respect to l the truth of the formulas:

$$D^l\eta = \sum_{\alpha, r, s} A_\alpha^{lrs} y_1^{\alpha_1} \cdots y^{\alpha_l} \partial_x^r \partial_y^s \eta \tag{7.2}$$

is established, where A_α^{lrs} are positive integers, $\alpha = (\alpha_1, \ldots, \alpha_l)$, and the sum is taken with respect to all indices α, r, s, which satisfy the relations

$$\alpha_1 + 2\alpha_2 + \cdots + l\alpha_l = l - r,$$

$$\alpha_1 + \alpha_2 + \cdots + \alpha_l = s, \qquad 0 \le s \le l.$$

In a completely analogous fashion,

$$D^l(y_1\xi) - y_{l+1}\xi = \sum_{\alpha, r, s} B_\alpha^{lrs} y_1^{\alpha_1} \cdots y_1^{\alpha_l} \partial_x^r \partial_y^s \xi \tag{7.3}$$

is found, where B_α^{lrs} are positive integers and the indices α, r, s satisfy the relations

$$\alpha_1 + 2\alpha_2 + \cdots + l\alpha_l = l - r + 1,$$

$$\alpha_1 + \alpha_2 + \cdots + \alpha_l = s + 1, \qquad 0 \leq s \leq l.$$

The condition of invariance of equation (7.1) is

$$(\underset{k}{\eta} - (\underset{k-1}{\zeta} \cdot \partial)f)|_{y_k = f(\underset{k-1}{z})} = 0. \tag{7.4}$$

When $k > 1$, this condition contains as "free" parameters the variables $y_1, \ldots, y_{k-1}$, with respect to which it must be decomposed. This leads, generally speaking, to an overdefined system DE for two functions ξ, η and causes a finite dimensionality of the full Lie Algebra LE admitted by equation (7.1). There arises the problem of group classification of equation (7.1) with respect to the arbitrary element—the right-hand side f, which has not yet been completely solved. One of the known results for equations of the second order is given below.

8. Second-Order Equation

For equations of the type $E(1, 1, 2, 1)$

$$y_2 = f(x, y, y_1), \tag{8.1}$$

the second continuation of the operator (2.1),

$$\underset{2}{\zeta} \cdot \partial = \xi\partial_x + \eta\partial_y + \eta_1\partial_{y_1} + \eta_2\partial_{y_2},$$

is needed. Formulas (7.2) and (7.3) give the expressions below, in which the short notation for derivatives $\eta_x = \partial_x\eta$, $\eta_y = \partial_y\eta$, $\xi_{xy} = \partial_x\partial_y\xi$, etc., are used:

$$\eta_1 = \eta_x + y_1(\eta_y - \xi_x) - y_1^2\xi_y,$$

$$\begin{aligned}\eta_2 = \eta_{xx} &+ y_1(2\eta_{xy} - \xi_{xx}) + y_1^2(\eta_{yy} - 2\xi_{xy}) \\ &- y_1^3\xi_{yy} + y_2(\eta_y - 2\xi_x - 3y_1\xi_y).\end{aligned}$$

In this notation the condition of invariance (7.4) is

$$\begin{aligned}\eta_{xx} &+ y_1(2\eta_{xy} - \xi_{xx}) + y_1^2(\eta_{yy} - 2\xi_{xy}) - y_1^3\xi_{yy} + (\eta_y - 2\xi_x - 3y_1\xi_y)f \\ &= \xi\partial_x f + \eta\partial_y f + (\eta_x + y_1(\eta_y - \xi_x) - y_1^2\xi_y)\partial_{y_1} f.\end{aligned} \tag{8.2}$$

Equation (8.2) must be decomposed with respect to the "free" parameter y_1. The easiest way to do this is to assume that the function f is given in the form of a power series

$$f(x, y, y_1) = \sum_{n=0}^{\infty} f_n(x, y)y_1^n.$$

With this the operation of decomposition is reduced to the substitution of this series into (8.2) and then comparing terms with like powers of the parameter y_1. There results, generally speaking, an infinite system of equations *DE* in which the first four determining equations, corresponding to the powers y_1^0, y_1^1, y_1^2, and y_1^3, are the most essential:

$$\begin{aligned}
\eta_{xx} &= f_{0x}\xi + f_{0y}\eta - f_0\eta_y + 2f_0\xi_x + f_1\eta_x,\\
2\eta_{xy} - \xi_{xx} &= f_{1x}\xi + f_{1y}\eta + 2f_2\eta_x + f_1\xi_x + 3f_0\xi_y,\\
\eta_{yy} - 2\xi_{xy} &= f_{2x}\xi + f_{2y}\eta + 3f_3\eta_x + f_2\eta_y + 2f_2\xi_y,\\
-\xi_{yy} &= f_{3x}\xi + f_{3y}\eta + 4f_4\eta_x + 2f_3\eta_y - f_3\xi_x + f_2\xi_y,\\
&\vdots
\end{aligned}$$

The finite-dimensionality property of the full admitted Lie algebra L^r and the maximal possible value of its dimensionality r are established by the following theorem, proved for the first time by Sophus Lie.

Theorem. *Equation* (8.1) *admits the space L^r of operators of type* (2.1) *only for values $r \leq 8$. The maximum possible value $r = 8$ is achieved for the equation* $y_2 = 0$.

Proof. The first four determining equations are rewritten in the abbreviated form

$$\begin{aligned}
\eta_{xx} &= \phi_1, & 2\eta_{xy} - \xi_{xx} &= \phi_2,\\
\eta_{yy} - 2\xi_{xy} &= \phi_3, & \xi_{yy} &= \phi_4,
\end{aligned} \tag{8.3}$$

where the ϕ_i are some linear functions of the values $\xi, \eta, \xi_x, \xi_y, \eta_x, \eta_y$. By the differentiation of equations (8.3) and algebraic combination certain expressions for all derivatives of the third order of the unknown functions are found. These expressions can be written in abbreviated form as one vector relation:

$$\partial^3\zeta = \psi(x, y, \zeta, \partial\zeta, \partial^2\zeta), \tag{8.4}$$

where $\zeta = (\xi, \eta)$ and the right-hand side ψ are linear relative to $\zeta, \partial\zeta, \partial^2\zeta$. The general solution of such an equation may depend on not more than 12 arbitrary constants, on which equation (8.3) superimposes four independent relations. Thus not more than eight arbitrary constants remain, which implies that the dimensionality of the solution space $r \leq 8$. The actual number of arbitrary constants which remain in the solution of the system *DE* may be even less than eight, if in addition to equations (8.3) there are some other non-trivial equations in *DE*.

In the case of the equation $y_2 = 0$, all $\phi_i = 0$ in (8.3) and in (8.4) $\psi = 0$ also. The general solution of the equation $\partial^3 \zeta = 0$ has the form

$$\zeta = (A_1, B_1) + (A_2, B_2)x + (A_3, B_3)y + (A_4, B_4)xy + (A_5, B_5)x^2 + (A_6, B_6)y^2,$$

with arbitrary constants A_j, B_j. Equations (8.3) subordinate these constants to the relations

$$B_5 = 0, \qquad B_4 - A_5 = 0, \qquad B_6 - A_4 = 0, \qquad A_6 = 0.$$

Finally, the general solution of the determining equations is

$$\xi = A_1 + A_2 x + A_3 y + A_4 xy + B_4 x^2, \qquad \eta = B_1 + B_2 x + B_3 y + A_4 y^2 + B_4 xy, \tag{8.5}$$

where A_j, B_j are independent. A set of operators $\zeta \cdot \partial$ with the components (8.5) generates the Lie algebra L^8. ■

As one of the consequences of this theorem the classification problem for a linear equation of the second order can be formulated as the study of

$$y_2 = a(x)y_1 + b(x)y. \tag{8.6}$$

Consequence. *Any equation* (8.6) *admits the Lie algebra* L^8.

Proof. As is well known, a transformation of the unknown y by $\bar{y} = p(x)y$ reduces equation (8.6) to the same equation, but with $a = 0$. For this $p = \exp(-\frac{1}{2}\int a\,dx)$ must be taken. In addition, the equation $y_2 = b(x)y$ becomes the equation $\bar{y}_2 = 0$ after transformation using a solution $u \neq 0$:

$$\bar{y} = y/u(x), \qquad \bar{x} = \int (u(x))^{-2}\,dx.$$

Consequently, any equation (8.6) is identical to the equation $y_2 = 0$ in the sense of Section 6.2, and the result follows from the theorem in Section 6.2. ■

9. Complete Systems

A linear equation of the type $E(n, 1, 1, s)$, for which $X = R^n(x)$, $Y = R(y)$, $Y_1 = R^n(\underset{1}{y})$, where $\underset{1}{y} = (y_1, \ldots, y_n)$, is considered. This equation is defined with the aid of the mapping $\psi: X \to \mathscr{L}(Y_1; R^s)$,

$$\psi(x)\langle \underset{1}{y} \rangle = 0.$$

The mapping ψ can be represented by a matrix (ψ_j^i) of size $s \times n$. Then in component form this equation has the form of the system E

$$\psi_j^i(x)y_i = 0 \qquad (j = 1, \ldots, s). \tag{9.1}$$

For the solution $u: X \to R$, operating by the formula $y = u(x)$, this will be $y_i = \partial_i u(x)$, where $\partial_i = \partial/\partial_x^i$ and $\psi_j y_i = (\psi_j \cdot \partial)u(x)$ with the operator $\partial = (\partial_1, \ldots, \partial_n)$. This is why the solution u of the system (9.1) must satisfy a system of identities $(\psi_j \cdot \partial)u(x) = 0$ for every operator

$$\psi_1 \cdot \partial, \qquad \ldots, \qquad \psi_s \cdot \partial, \tag{9.2}$$

and because of the criterion in Section 3.4 this can be interpreted as a compatible invariant of the operator set (9.2). But, as is shown in Section 7.11, if this is so, then the solution u is an invariant of any commutator $[\psi_j, \psi_l] \cdot \partial$. This fact leads to two possibilities, depending upon whether new operators appear from this or not. The situation in which new operators appear will often be met later, so it is to the point to introduce a special concept.

Let $\xi_1, \ldots, \xi_m$ be vector fields on the space X. It is said that these fields (or the corresponding operators $\xi_j \cdot \partial$) are *linearly dependent* if there exist functions $\omega^j: X \to R$ $(j = 1, \ldots, m)$, not all of them equal to zero, such that the identity

$$\omega^j \xi_j(x) = 0$$

is satisfied; in the opposite case it is said that the vector fields ξ_j (operators $\xi_j \cdot \partial$) are *not linearly dependent* (*linearly independent*).

It is clear that if operators are linearly dependent, then and only then can one of them be linearly expressed in terms of others, that is, can be represented in the form of a linear combination of the others, with variable coefficients.

The system (5.1) is called *complete* if the commutator $[\psi_j, \psi_l] \cdot \partial$ of any two operators from (9.2) is expressible as a linear combination of the operators in (9.2).

As is known, the determination of solutions of a complete system is reduced to the integration of a system of ordinary differential equations. In this theory it is established that any complete system of equations (9.1) is equivalent to a Jacobi system in which $s \leq n$ and $[\psi_j, \psi_l] = 0$ for any j, $l = 1, \ldots, s$. Reduction of an arbitrary complete system to a Jacobi one needs only satisfaction of algebraic operations, which is realized with the aid of a mapping $v: X \to \mathscr{L}(R^s)$, where $v(x)$ is a transformation of the space R^s, for every $x \in X$, and the matrix $(\bar{\psi}_j^i)$ of the mapping $\bar{\psi} = v\langle x \rangle: X \to R^s$ contains a unit matrix I_{R^s}, for example, $\bar{\psi}_j^i = \delta_j^i$ for $i, j = 1, \ldots, s$.

In what follows a Jacobi system (9.1) is considered whose operators are not linearly dependent; in particular, their number $s \leq n$. The limited problem, to find the Lie algebra of operators

$$\xi(x) \cdot \partial = \xi^i(x)\partial_i \tag{9.3}$$

admitted by the system, is solved below.

An application of the algorithm, described in Section 5.3, leads to the condition of invariance of the system (9.1), denoted by E:

$$[\psi_j, \xi]|_E = 0 \qquad (j = 1, \ldots, s).$$

The decomposition of this condition relative to $n - s$ "free" parameters, which are those components of the vector $\underset{1}{y} = (y_1, \ldots, y_n)$ which can be expressed in terms of s coordinates, using equation (9.1), leads to determining equations of the form

$$[\psi_j, \xi] = \lambda_j^l \psi_l \qquad (j = 1, \ldots, s), \tag{9.4}$$

where (λ_j^l) is the matrix of some mapping $\lambda: X \to \mathscr{L}(R^s)$.

From (9.4) it follows at once that the set of operators (9.3), admitted by the system (9.1), generates a Lie algebra L. Because of completeness of the system (9.1), any operator from (9.2) satisfies equations (9.4); moreover, any operator

$$\xi \cdot \partial = v^j \psi_j \cdot \partial \tag{9.5}$$

is expressible linearly by means of the operators (9.2), with arbitrary mappings $v^j: X \to R$. It turns out that the Lie algebra L is completely determined by these operators.

Theorem. *The Lie algebra of operators* (9.3), *admitted by the Jacobi system* (9.1), *in which operators* (9.2) *are not linearly dependent, is the set of all operators of the form* (9.5).

Proof. It is enough to prove that the functions v^j, with which the operator (9.5) satisfies the equations (9.4) for given λ_j^l, exist. Substituting (9.5) into (9.4), and using the property that (9.1) is a Jacobi system, leads to the equation $(\psi_j \cdot \partial v^l)\psi_l = \lambda_j^l \psi_l$. Because of the linear independence of the operators (9.2), this equation is decomposed into separate equations for every coefficient v^l:

$$(\psi_j \cdot \partial)v^l = \lambda_j^l \qquad (j = 1, \ldots, s). \tag{9.6}$$

As is known from the theory of differential equations, because of the permutability of equations (9.2) [system (9.1) is a Jacobi system], the conditions of integrability for equations (9.6) are that for all $k, j = 1, \ldots, s$ the identities

$$(\psi_k \cdot \partial)\lambda_j^l = (\psi_j \cdot \partial)\lambda_k^l \tag{9.7}$$

must be satisfied. These identities are verified with the aid of the Jacobi relation 3° in Section 7.3:

$$[\psi_k, [\psi_j, \xi]] + [\psi_j, [\xi, \psi_k]] + [\xi, [\psi_k, \psi_j]] = 0.$$

After the expression (9.4) is substituted, the fact that (9.1) is a Jacobi system is used, and the linear independence of the operator (9.2) is considered, this relation takes exactly the same form as equation (9.7). ■

10. Conclusions

The considerations of this section demonstrate in the case of ordinary differential equations of the first order and in the case of complete systems, whose integration problem leads to the solution of systems of ordinary differential equations, that the admitted full Lie algebra of operators turns out to be infinite dimensional. But the problem of construction of even some groups G^1, which can be used to generate solutions of the corresponding system E (to say nothing about the full group GE), is not simpler than the integration of the original equations. Thus knowledge of the infinite-dimensional component of the full Lie algebra actually does not provide the means for forming effective algorithms for solving the aforementioned differential equations.

The possibility of a direct search for solutions is found only when it is possible to find the admitted operators, which are essentially simpler than those operators included in the system E. Examples of these are the ordinary differential equations of higher orders.

Although this aforementioned fact is a bit disappointing in relation to the perspective of using group analysis for solving ordinary differential equations of the first order, nevertheless knowledge of the structure of the full admitted Lie algebra of operators can be useful in general theoretical investigations, which are not directly connected with the problem of searching for concrete solutions. As will be shown in Chapter V, the situation is significantly different for the case of partial differential equations. Here the algorithm of group analysis is effective.

The results obtained for ordinary differential equations of the first order differ from those obtained for equations of higher order. This fact, in the light of the fact that one equation $E(1, 1, k, 1)$ of order k is equivalent to the system $E(1, k, 1, k)$, with k equations of the first order, leads to a question about the relation between the full groups of such equations. This problem has not been completely studied. The difficulty here is that the continuous spaces on which the corresponding manifolds lie have different dimensionalities. The manifold $E(1, 1, k, 1)$ lies in the space $Z_k = R^{k+1}$, while the manifold $E(1, k, 1, k)$ is in the space $Z_1 = R^{2k+1}$. With this the transformation $E(1, k, 1, k) \to E(1, 1, k, 1)$ can be considered as the result of a mapping $Z_1 \to Z_k$, while the inverse transformation $E(1, 1, k, 1) \to E(1, k, 1, k)$ is not a mapping.

§9. SECOND-ORDER LINEAR EQUATIONS WITH TWO INDEPENDENT VARIABLES

1. Statement of the Problem

The specific coordinate notation, namely, $X = R^2(x, y)$, $Y = R(z)$, $Z = R^3(x, y, z)$, $Y_1 = R^2(z_1, z_2)$, $Y_2 = R^3(z_{11}, z_{12}, z_{22})$, is used for equations of the type $E(2, 1, 2, 1)$ throughout Section 9. The general linear equation of this type has the form

$$a^{11}z_{11} + 2a^{12}z_{12} + a^{22}z_{22} + b^1z_1 + b^2z_2 + cz = 0,$$

with coefficients which are *analytic functions*, that is, mapping $a^{ik}, b^i, c: X \to R$ of class $C_A(X)$.

The problem of group classification (see Section 6) in relation to the arbitrary element—the set of coefficients $\{a^{ik}, b^j, c\}$—is considered for this equation. The most general transformations of equivalence, which retain the differential structure of this equation, are

$$\bar{x} = u(x, y), \qquad \bar{y} = v(x, y), \qquad z = \omega(x, y)\bar{z}. \tag{1.1}$$

From the theory of second-order differential equations it is known that this equation is changed by a suitable transformation of equivalence (1.1) to one of the following *normal forms*: to the *hyperbolic normal* form

$$z_{12} + Az_1 + Bz_2 + Cz = 0, \tag{1.2}$$

to the *elliptic normal* form

$$z_{11} + z_{22} + Az_1 + Bz_2 + Cz = 0, \tag{1.3}$$

or to the *parabolic normal* form

$$z_{11} + Az_1 + Bz_2 + Cz = 0, \tag{1.4}$$

with known coefficients $A, B, C: X \to R$, of the class $C_A(X)$.

From the viewpoint of the local analytic theory, the elliptic form is equivalent to the hyperbolic form in the (considered) case of analytic coefficients. The corresponding transformations of equivalence (1.1) are obtained by using variables in the complex field as independent variables via the formulas $\bar{x} = x + iy, \bar{y} = x - iy, \bar{z} = z$. As a result of this transformation, equation (1.3) becomes the equivalent equation

$$4\bar{z}_{12} + (A + iB)\bar{z}_1 + (A - iB)\bar{z}_2 + C\bar{z} = 0$$

and any classification analytic results for equations (1.2) and (1.3) can be put into one-to-one correspondence.

The problem of group classification of equations of the form (1.2), with relation to the arbitrary element $\theta = (A, B, C)$, is considered in Section 2.6.

2. Laplace Invariants

For equations of the form (1.2) the set of equivalence transformations (1.1) is restricted to the following:

$$\bar{x} = \alpha(x), \qquad \bar{y} = \beta(y), \qquad z = \omega(x, y)\bar{z}. \tag{2.1}$$

The equipollent equations (1.2) are called *equivalent with respect to the function* if they transform into each other when the transformations (2.1) take the form $\alpha(x) = x$, $\beta(y) = y$, that is, only at the expense of the transformation $z = \omega\bar{z}$.

It is convenient to distinguish classes of equipollent equations (1.2), relative to the equivalence transformations (2.1), in terms of so-called Laplace invariants of those equations. The Laplace invariants of equation (1.2) are the functions

$$h = \partial_x A + AB - C, \qquad k = \partial_y B + AB - C. \tag{2.2}$$

The justification for the name "invariants" for these functions is given in the following two lemmas, in which equation (1.2) is denoted by the symbol (h, k)—a pair of its Laplace invariants.

Lemma 1. *The equation $(\bar{h}, \bar{k})$ is equivalent—with respect to the function—to the equation (h, k) if and only if $\bar{h} = h$, $\bar{k} = k$.*

Proof. As a result of the substitution of $z = \omega\bar{z}$ into (1.2), an equation of the same structure, with the coefficients

$$\bar{A} = A + \frac{\omega_2}{\omega}, \qquad \bar{B} = B + \frac{\omega_1}{\omega}, \qquad \bar{C} = C + A\frac{\omega_1}{\omega} + B\frac{\omega_2}{\omega} + \frac{\omega_{12}}{\omega}, \tag{2.3}$$

is found, where $\omega_1 = \partial_x\omega$, $\omega_2 = \partial_y\omega$, $\omega_{12} = \partial_x\partial_y\omega$. From this

$$\begin{aligned}\bar{h} &= \partial_x\bar{A} + \bar{A}\bar{B} - \bar{C}\\ &= \partial_x A + \frac{\omega_{12}}{\omega} - \frac{\omega_1\omega_2}{\omega^2} + AB + A\frac{\omega_1}{\omega} + B\frac{\omega_2}{\omega} + \frac{\omega_1\omega_2}{\omega^2} - \bar{C} = h\end{aligned}$$

is found, and similarly for k. On the contrary, if $\bar{h} = h$ and $\bar{k} = k$ and $\bar{A}, \bar{B}, \bar{C}$ are coefficients of the equation $(\bar{h}, \bar{k})$, then from the definition (2.2) the equality

$$\partial_x(\bar{A} - A) = \partial_y(\bar{B} - B)$$

follows. Because of this a function ω satisfying the first two equations (2.3) exists. With this function the equality $\bar{h} = h$ turns out to be equipollent to the third equation of (2.3). Thus the substitution $z = \omega\bar{z}$ transforms the equation (h, k) into the equation $(\bar{h}, \bar{k})$. ■

Consequence. *Equation* (1.2) *is equivalent—with respect to the function—to the equation* $z_{12} = 0$ *if and only if* $h = k = 0$.

Lemma 2. *The equation* $(\bar{h}, \bar{k})$ *is equipollent to the equation* (h, k) *if and only if there exist functions* $\bar{x} = \alpha(x)$ *and* $\bar{y} = \beta(y)$ *such that for all* $(x, y) \in X$ *the relations*

$$h(x, y) = \alpha'(x)\beta'(y)\bar{h}(\bar{x}, \bar{y}), \qquad k(x, y) = \alpha'(x)\beta'(y)\bar{k}(\bar{x}, \bar{y}), \tag{2.4}$$

where $\alpha' = \partial_x\alpha$, $\beta' = \partial_y\beta$, *are satisfied.*

Proof. The transformation (2.1) can be generated in two parts: in part (a); $\bar{x} = x$, $\bar{y} = y$, $z = \omega\bar{z}$; and in part (b), $\bar{x} = \alpha(x)$, $\bar{y} = \beta(y)$, $z = \bar{z}$. For part (a) the Laplace invariants are retained because of Lemma 1. In the study of part (b) they are transformed, as an easy calculation shows, precisely by the formulas (2.4). That is why, if the equation $(\bar{h}, \bar{k})$ is equipollent to the equation (h, k), the relations (2.4) are satisfied. On the contrary, if these equations hold, then in part (b) the functions α, β from (2.4) transform equation (h, k) into some equation $(\tilde{h}, \tilde{k})$, and $\tilde{h} = \bar{h}, \tilde{k} = \bar{k}$. Because of Lemma 1, the equation $(\tilde{h}, \tilde{k})$ is equivalent—with respect to the function—to the equation $(\bar{h}, \bar{k})$, that is, it transforms into it in the case of part (a). ■

3. Laplace Series

Besides (2.1) there are other transformations of equation (1.2), the so-called *Laplace series*, which also maintain the differential structure of this equation.

An auxiliary function z^1, connected with z by the equations

$$z_2 + Az = z^1, \qquad z_1 + Bz^1 = hz, \tag{3.1}$$

is introduced. It is easy to check that the elimination of z^1 from these equations gives exactly equation (1.2). Conversely, when $h \neq 0$, elimination of z from system (3.1) leads to an equation for z^1, which also has the form of (1.2). This new equation will be called the *x-Laplace transformation of equation* (1.2). Its Laplace invariants h^1, k^1 are found by a simple calculation and turn out to be

$$h^1 = 2h - k - \partial_x\partial_y \ln h, \qquad k^1 = h. \tag{3.2}$$

Another auxiliary function, z^2, connected with z by the equations

$$z_1 + Bz = z^2, \qquad z_2^2 + Az^2 = kz, \tag{3.3}$$

is also introduced in an analogous way. It is easy to verify that elimination of z^2 from these equations gives exactly equation (1.2). Conversely, when $k \neq 0$, elimination of z from the system (3.3) leads to an equation for z^2, which also has the form (1.2). This new equation will be called the *y-Laplace transformation of equation* (1.2). Its Laplace invariants h^2, k^2 are

$$h^2 = k, \qquad k^2 = 2k - h - \partial_x\partial_y \ln k. \tag{3.4}$$

The equalities $(h^1)^2 = (h^2)^1 = h$, $(k^1)^2 = (k^2)^1 = k$ are established by direct calculation. They show by Lemma 1 that the y-Laplace transformation of the equation (h^1, k^1) is equivalent—with respect to the function—to the equation (h, k) and the x-Laplace transformation of the equation (h^2, k^2) is equivalent—with respect to the function—to (h, k). Therefore, the set of nonequivalent equations derived from the given equation (h, k) is "one dimensional." To illustrate this property the assumption will be made that $h = h^0$ and $k = h^{-1}$ so that the initial equation has the notation (h^0, h^{-1}). Let h^n, for any integer number n, be defined by the recurrence relation

$$h^{n+1} + h^{n-1} = 2h^n - \partial_x\partial_y \ln(h^n). \tag{3.5}$$

Then (3.2) shows that when $n = 0$, this formula gives the invariant h^1 of the equation (h^1, k^1) and when $n = -1$, the invariant $k^2 = h^{-2}$ of the equation (h^2, k^2). It is easy to check that the x-Laplace transformation changes (h^n, h^{n-1}) into (h^{n+1}, h^n) and the y-Laplace transformation converts (h^{n+1}, h^n) into (h^n, h^{n-1}) for any integer number n. So the *Laplace series*

$$\ldots; (h^{-2}, h^{-3}); (h^{-1}, h^{-2}); (h^0, h^{-1}); (h^1, h^0); (h^2, h^1); \ldots$$

is obtained, where the terms to the right are realized by x-transformations, and those to the left by y-Laplace transformations.

The Laplace series can be continued as long as any of the invariants h^n is not identically zero. A remarkable property of this series is that if for any n $h^n = 0$, then for the original equation it is possible to construct the *formula for the general solution* by quadratures, with two arbitrary functions. If the Laplace series "interrupts" from both ends, then it is possible to construct the general solution of the original equation without quadratures. These elegant analytic facts are not proved because they are not used in what follows.†

† One can find these facts in Ref. [4], given at the end of this chapter.

The case when the invariants h^n are in a constant ratio is important and is discussed below.

Lemma. *If the equation* (h, k) *is such that the ratios*

$$k/h = p, (\partial_x \partial_y \ln h)/h = q \tag{3.6}$$

are constant, then all invariants of the Laplace series for this equation have a constant ratio.

Proof. In the Laplace series notation the equality (3.6) is rewritten in the form

$$h^{-1} = ph^0, \qquad \partial_x \partial_y \ln h^0 = qh^0.$$

From equation (3.5), with $n = 0$, the ratio $h^1/h^0 = 2 - p - q$ is constant. On this basis, the constancy of the ratio h^n/h^0 is obtained from equation (3.5) by induction with respect to n. ■

In the case in which p, q are constants, for the determination of the ratio h^n/h^0, with n any number, it is necessary to consider equation (3.5) as an equation in the finite-dimensional differences

$$h^{n+1} - 2h^n + h^{n-1} = -qh^0,$$

with the initial conditions $h^{-1} = ph^0$, $h^0 = h^0$. The solution of this problem (clearly unique) is

$$h^n/h^0 = 1 + (1 - p)n - \tfrac{1}{2}qn(n + 1). \tag{3.7}$$

4. Determining Equations

When searching for the operator in the form $\alpha\partial_x + \beta\partial_y + \gamma\partial_z$ with components $\alpha, \beta, \gamma: Z \to R$, admitted by equation (1.2), it turns out that a portion of the determining equation system will be equations of the form (7.14.6); that is, the equations $\partial_z \alpha = \partial_z \beta = 0$, $\partial_z^2 \gamma = 0$ hold. From this fact and from the results of Section 7.14 it follows that for equation (1.2) its full Lie algebra is $LE = L^r \oplus L^\infty$, where L^r is generated by the operators

$$\zeta \cdot \partial = \xi(x, y)\partial_x + \eta(x, y)\partial_y + \sigma(x, y)z\partial_z. \tag{4.1}$$

Obviously, the transformations of dilatation with the operator $z\partial_z$ are admitted by any equation (1.2), that is, they are contained in the kernel of the full groups of the class to which equation (1.2) belongs. Thus the operator $z\partial_z$ can be included in L^∞ and we can assume that the function σ, in the operator (4.1), is correctly defined up to a constant multiple.

The determining equations are calculated, following the algorithm in Section 5.8, applying the continuous operator (4.1) twice:

$$\underset{2}{\zeta} \cdot \partial = \zeta \cdot \partial + \tau_1 \partial_{z_1} + \tau_2 \partial_{z_2} + \tau_{11} \partial_{z_{11}} + \tau_{12} \partial_{z_{12}} + \tau_{22} \partial_{z_{22}},$$

$$\tau_1 = \sigma_x z + (\sigma - \xi_x) z_1 - \eta_x z_2, \quad \tau_2 = \sigma_y z - \xi_y z_1 + (\sigma - \eta_y) z_2,$$

$$\tau_{11} = \sigma_{xx} z + (2\sigma_x - \xi_{xx}) z_1 - \eta_{xx} z_2 + (\sigma - 2\xi_x) z_{11} - 2\eta_x z_{12},$$

$$\begin{aligned} \tau_{12} = {} & \sigma_{xy} z + (\sigma_y - \xi_{xy}) z_1 + (\sigma_x - \eta_{xy}) z_2 - \xi_y z_{11} \\ & + (\sigma - \xi_x - \eta_y) z_{12} - \eta_x z_{22}, \end{aligned}$$

$$\tau_{22} = \sigma_{yy} z - \xi_{yy} z_1 + (2\sigma_y - \eta_{yy}) z_2 - 2\xi_y z_{12} + (\sigma - 2\eta_y) z_{22},$$

and for equation (1.2) we get the condition of invariance

$$\begin{aligned} & \sigma_{xy} z + (\sigma_y - \xi_{xy}) z_1 + (\sigma_x - \eta_{xy}) z_2 - \xi_y z_{11} \\ & \quad -(\sigma - \xi_x - \eta_y)(A z_1 + B z_2 + C z) - \eta_x z_{22} \\ & \quad + A(\sigma_x z + (\sigma - \xi_x) z_1 - \eta_x z_2) + B(\sigma_y z - \xi_y z_1 + (\sigma - \eta_y) z_2) \\ & \quad + C\sigma z + (\xi \partial_x A + \eta \partial_y A) z_1 + (\xi \partial_x B + \eta \partial_y B) z_2 + (\xi \partial_x C + \eta \partial_y C) z = 0 \end{aligned}$$

where $\sigma_x = \partial_x \sigma$, $\sigma_{xy} = \partial_x \partial_y \sigma$, and so on.

The decomposition, relative to the "free" parameters z, z_1, z_2, z_{11}, z_{22}, leads to the determining equations, which after a little manipulation and using Laplace invariants (2.2) are shown to be equipollent to the following system:

$$\xi_y = 0, \qquad \eta_x = 0,$$

$$\partial_x(\sigma + B\xi + A\eta) = (h - k)\eta, \qquad \partial_y(\sigma + B\xi + A\eta) = (k - h)\xi, \tag{4.2}$$

$$\partial_x(k\xi) + \partial_y(k\eta) = 0, \qquad \partial_x(h\xi) + \partial_y(h\eta) = 0. \tag{4.3}$$

The first two equations show that ξ depends only on x and η only on y. The two equations (4.2) define the function σ after ξ and η are found. The compatibility condition of these equations follows from (4.3). With these the function σ is uniquely defined by equations (4.2), up to a constant term which corresponds to an assumption about the Lie algebra L^r of operators (4.1) admitted by equation (1.2). In this way equations (4.3) are responsible for the general solution of the determining equations and for results of group classification. Indeed, using Lemma 2 in Section 9, it is easy to verify that equations (4.3) are invariant relative to any equivalence transformation (2.1). Because of that, (4.3) are the classifying equations in the sense of Section 6.6.

5. General Solution Analysis

We now restrict our attention to the case when at least one of the invariants h, k is different from zero. If this classification assumption is not satisfied, that is, $h = k = 0$, then according to the consequence in Section 9.2, equation (1.2) is equipollent to the equation $z_{12} = 0$ and admits an infinite-dimensional Lie algebra of operators

$$\zeta^0 \cdot \partial = \xi(x)\partial_x + \eta(y)\partial_y, \tag{5.1}$$

with arbitrary functions ξ and η.

To be definite, let $h \neq 0$. In the notation of (3.6), equations (4.3) become

$$\xi p_x + \eta p_y = 0,$$

which shows that either $p = k/h$ is an invariant of the group G^1, with the operator (4.1), or $p = \text{const}$.

If the ratio p of the Laplace invariants of equations (1.2) *is not a constant, then the Lie algebra of operators* (4.1) *admitted by this equation has dimension not higher than one.* Actually, if (ξ, η) and (ξ_1, η_1) satisfy the previous equation, then, when $p \neq \text{const}$, $\xi_1 = Q\xi, \eta_1 = Q\eta$, where Q is a constant, as it must be a function only of x and only of y at the same time. But then from (4.2) it also follows that $\sigma_1 = Q\sigma$, whence the operator $\zeta_1 \cdot \partial = Q\zeta \cdot \partial$.

Let $p = \text{const}$. Then of the two equations (4.3) only one remains, namely, the second. It can be rewritten in the form

$$\xi\partial_x \ln h + \eta\partial_y \ln h + \xi' + \eta' = 0,$$

where $\xi' = \partial_x \xi$, $\eta' = \partial_y \eta$. Applying the operator $\partial_x \partial_y$ to this equation, we get

$$\xi\partial_x \ln(\partial_x\partial_y \ln h) + \eta\partial_y \ln(\partial_x\partial_y \ln h) + \xi' + \eta' = 0.$$

If we subtract the previous equation from this one and use the notation of (3.6), then the new equation

$$\xi q_x + \eta q_y = 0$$

appears. Thus it is again possible to conclude that either $q = (\partial_x\partial_y \ln h)/h$ is an invariant of the group G^1, with the operator (4.1), or $q = \text{const}$. From this a conclusion analogous to the previous one follows. From these arguments the following classifying result is obtained:

If for equation (1.2) *even one of the functions p, q is not identically constant, then the Lie algebra of the operators* (4.1) *admitted by this equation is not more than one dimensional.*

In the case when q is a constant, the invariant h cannot be arbitrary, but must be a solution of the Liouville equation

$$\partial_x\partial_y \ln h = qh. \tag{5.2}$$

Lemma. *The general solution of equation* (5.2), *in the case when q is a constant, has the form*

$$\begin{aligned} h(x, y) &= \frac{2}{q}\frac{\alpha'(x)\beta'(y)}{(\alpha(x) + \beta(y))^2} \qquad (q \neq 0), \\ h(x, y) &= \alpha'(x)\beta'(y) \qquad (q = 0), \end{aligned} \tag{5.3}$$

where α, β are arbitrary functions and $\alpha' = \partial_x \alpha$, $\beta' = \partial_y \beta$.

Proof. The case when $q = 0$ is trivial. If $q \neq 0$, then (5.3) is a solution of equation (5.2), for any α, β, which is easily checked by direct substitution. Let $h = \chi^{-2}$; then (5.3) becomes the equation

$$\chi\chi_{xy} - \chi_x\chi_y = -q/2,$$

whose differentiation ∂_x leads to an equipollent consequence $\partial_y \ln(\chi_{xx}/\chi) = 0$, from which it follows, using a function $r: x \to r(x)$, that

$$\chi_{xx} = r\chi.$$

If χ_1 and χ_2 are two linearly independent solutions of this linear ordinary differential equation of the second order, then its general solution is $\chi = c_1(y)\chi_1(x) + c_2(y)\chi_2(x)$, with arbitrary functions c_1, c_2. This expression can be written in the form

$$\chi = m(x)n(y)(\alpha(x) + \beta(y)).$$

Substitution of this into the equation for χ leads to the result

$$2\alpha'\beta'(mn)^2 = q,$$

because of which $h = \chi^{-2}$ has the form (5.3). ■

6. Classification Theorem

The most essential result of the group classification of equation (1.2) is formulated here.

Theorem. *Equation* (1.2) *admits more than a one-dimensional Lie algebra of operators* (4.1) *if and only if the functions*

$$p = k/h, \qquad q = (\partial_x\partial_y \ln h)/h$$

are identically constant. If p and q are constants, then equation (1.2) *is equipollent to the Euler–Poisson equation* ($q \neq 0$)

$$z_{12} - \frac{2/q}{x + y} z_1 - \frac{2p/q}{x + y} z_2 + \frac{4p/q^2}{(x + y)^2} z = 0 \tag{6.1}$$

or to the equation ($q = 0$)

$$z_{12} + xz_1 + pyz_2 + pxyz = 0 \tag{6.2}$$

and its Lie algebra of operators (4.1) *is three dimensional.*

Proof. The necessity for the functions p and q to be constant is established in the previous discussion. If p and q are constant, then because of the lemma in Section 9.5 the Laplace invariants of equation (1.2) are (case $q \neq 0$)

$$h(x, y) = \frac{2}{q} \cdot \frac{\alpha'(x)\beta'(y)}{(\alpha(x) + \beta(y))^2}, \qquad k(x, y) = \frac{2p}{q} \cdot \frac{\alpha'(x)\beta'(y)}{(\alpha(x) + \beta(y))^2}.$$

On the other hand, the Laplace invariants $\bar{h}$, $\bar{k}$ of equation (6.1) are

$$\bar{h}(x, y) = \frac{2}{q} \cdot \frac{1}{(x + y)^2}, \qquad \bar{k}(x, y) = \frac{2p}{q} \cdot \frac{1}{(x + y)^2}.$$

Comparison of these expressions with the previous ones shows that conditions (2.4) of Lemma 2 in Section 9.2 are satisfied, whence equation (1.2) is equipollent to equation (6.1). In the case $q = 0$, the lemma in Section 9.5 gives for the invariants of equation (1.2)

$$h = \alpha'(x)\beta'(y), \qquad k = p\alpha'(x)\beta'(y)$$

in the situation where the invariants of equation (6.2) are $\bar{h} = 1$, $\bar{k} = p$. That is why, again because of Lemma 2 in Section 9.2, equation (1.2) is equipollent to equation (6.2).

Since equipollent equations admit similar Lie algebras (see Section 7.9), to conclude the proof it is left to establish that equations (6.1) and (6.2) admit a three-dimensional Lie algebra of operators (4.1). In the case when $q \neq 0$, the second equation from (4.3), with the invariant $\bar{h}$ of the equation (6.1), becomes

$$x\xi' - 2\xi + y\eta' - 2\eta + y\xi' + x\eta' = 0.$$

Application of the operation $\partial_x\partial_y$ gives $\xi'' + \eta'' = 0$, so that $\xi'' = -\eta'' = 2C_0 = \text{const}$. Two integration give the expressions

$$\xi = C_0 x^2 + C_1 x + C_2, \qquad \eta = -C_0 y^2 + C_3 y + C_4,$$

whose substitution into the previous equation shows that $C_3 = C_1, C_4 = -C_2$. So the general form of the components ξ, η satisfying equations (4.3) is

$$\xi = C_0 x^2 + C_1 x + C_2, \qquad \eta = -C_0 y^2 + C_1 y - C_2.$$

Integrating equations (4.2), we get the result

$$\sigma = C_0(2/q)(px - y).$$

Finally, the three-dimensional Lie algebra L^3 of operators (4.1), by equation (6.1), is obtained, and the basis in L^3 is generated with the operators

$$\begin{aligned} \zeta_1 \cdot \partial &= \partial_x - \partial_y, \qquad \zeta_2 \cdot \partial = x\partial_x + y\partial_y, \\ \zeta_3 \cdot \partial &= x^2\partial_x - y^2\partial_y + (2/q)(px - y)z\partial_z. \end{aligned} \tag{6.3}$$

When $q = 0$, the calculations are as simple and give the Lie algebra of operators (4.1), admitted by equation (6.2), with the basis

$$\zeta_1 \cdot \partial = x\partial_x - y\partial_y, \qquad \zeta_2 \cdot \partial = \partial_x - yz\partial_z, \qquad \zeta_3 \cdot \partial = \partial_y - pxz\partial_z. \tag{6.4}$$

Consequence. *Two equations of the form* (1.2) *admitting the three-dimensional Lie algebra of operators* (4.1) *are equipollent if and only if they have the same values of the parameters p and q.*

Proof. If the considered equations have the same p and q, then according to the established theorem, they both are equipollent to one of the equations (6.1) or (6.2). On the contrary, if the given equations are equipollent, then equality of the p values follows from (2.4). Equality of the q values also follows from the relations (2.4), because of which

$$(\partial_x\partial_y \ln h)/h = (\partial_x\partial_y \ln(\alpha'\beta'\bar{h})/(\alpha'\beta'\bar{h}) = (\partial_{\bar{x}}\partial_{\bar{y}} \ln \bar{h})/\bar{h}. \qquad \blacksquare$$

7. Parabolic Normal Form

The differential structure of equation (1.4) is retained under all transformations (1.1) if $\partial_x v = 0$. With the aid of a suitable transformation (1.1) it is easily verified that equation (1.4) is equipollent to such an equation in which $A = 0$ and $B = -1$. Thus the case of a parabolic normal form can be limited to the class of equations

$$z_{11} = z_2 + Hz, \tag{7.1}$$

with an arbitrary element $H: X \to R$. In what follows this equation is denoted by $E(H)$.

The most general equivalence transformation maintaining the differential structure of equation (7.1) depends on three arbitrary functions $y \to a(y), b(y), c(y)$, which have the form

$$\bar{x} = ax + b, \qquad \bar{y} = \int a^2\, dy, \qquad z = \bar{z} \exp\left(\frac{a'}{4a}x^2 + \frac{b'}{2a}x + c\right). \tag{7.2}$$

Under the operation of this transformation the coefficient H becomes the coefficient $\bar{H}$, defined by the formula

$$a^2\bar{H}(\bar{x}, \bar{y}) = H(x, y) + \frac{aa'' - 2a'^2}{4a^2}x^2 + \frac{ab'' - 3a'b'}{4a^2}x + \left(c' - \frac{a'}{2a} - \frac{b'^2}{4a^2}\right), \tag{7.3}$$

where $a' = \partial_y a$, $a'' = \partial_y^2 a$, and so on.

The equation $E(\bar{H})$ is equipollent to the equation $E(H)$, relative to the equivalence transformation (7.2), *if and only if the coefficients $\bar{H}$ and H are related by equation* (7.3).

The determining equations $DE(H)$ for the operators (4.1), admitted by equation (7.1), are obtained by applying the algorithm in Section 5.8 and are found to be

$$\begin{gathered} \eta_x = 0, \qquad 2\xi_x - \eta_y = 0, \qquad 2\sigma_x + \xi_y = 0, \\ \xi\partial_x H + \eta\partial_y H + \eta_y H = \sigma_{xx} - \sigma_y. \end{gathered} \tag{7.4}$$

The general solution of the system, consisting of the first three equations, is found without any difficulty and is expressible by means of three arbitrary functions $y \to \phi(y), \psi(y), \chi(y)$ by the equations

$$\xi = 2x\phi' + \psi, \qquad \eta = 4\phi, \qquad 2\sigma = -x^2\phi'' - x\psi' - \chi, \tag{7.5}$$

where the derivatives ∂_y are denoted by a prime. Substitution of these expressions into the last equation in (7.4) leads to the equation, relating the function ϕ, ψ, χ, and H:

$$2(2x\phi' + \psi)\partial_x H + 8\phi\partial_y H + 8\phi' H = x^2\phi''' + x\psi'' + \chi' - 2\phi''. \tag{7.6}$$

Let $h = \partial_x^3 H$. Applying this operator to equation (7.6) gives

$$(2x\phi' + \psi)\partial_x h + 4\phi\partial_y h + 10\phi' h = 0.$$

Clearly, the assumption $h = 0$ is classifying. If $h \neq 0$, then for $k = \partial_x \ln h$ in the previous equation, the result $(2x\phi' + \psi)\partial_x k + 4\phi\partial_y k + 2\phi' k = 0$ is obtained. With the aid of these two equations it follows that

$$(2x\phi' + \psi)\partial_x p + 4\phi\partial_y p = 0, \tag{7.7}$$

for the quantity $p = kh^{-1/5}$. From here, in a manner analogous to the result of Section 9.5, it can be concluded that if $p \neq \text{const}$, then equation (7.1) admits not more than one operator (4.1). Thus, to have a Lie algebra L^r which is more than one dimensional, it is necessary that the function h, with constant p, must satisfy the classifying equation $h^{-1/5}\partial_x \ln h = p$, which is equipollent to the equation

$$\partial_x h = ph^{6/5}. \tag{7.8}$$

8. Classification of Parabolic Forms

The classification equation (7.8) is satisfied by the function $h = 0$. In this case $\partial_x^3 H = 0$ and the arbitrary element H has the form

$$H = \beta x^2 + \gamma x + \delta$$

for some functions $y \to \beta(y)$, $\gamma(y)$, $\delta(y)$. Obviously, a suitable equivalence transformation (7.2) can always be found under which the function H transforms into $\bar{H} = 0$. Equation (7.1), when $H = 0$, has already been considered in Section 6.8, where its full Lie algebra L^5 was found.

In what follows it is assumed that $h \neq 0$. If in equation (7.8) the constant $p = 0$, then it becomes $\partial_x^4 H = 0$, so that the arbitrary element has the form

$$H = \alpha x^3 + \beta x^2 + \gamma x + \delta,$$

where α, β, γ, δ are some functions of the variable y with $\alpha \neq 0$. It is easy to verify that this function H is equivalent to

$$H = x^3 + \gamma(y)x.$$

Substitution into the determining equation (7.6) shows that in this case $\phi' = \psi = \chi' = 0$ and $\phi\gamma' = 0$. This means that either the Lie algebra L^r is the zero algebra (if $\phi = 0$), or it is L^1 (if $\phi = C_0$ and $\gamma' = 0$). But the last possibility gives a dependence of the function H only on x. This case can be considered as trivial, so if $\partial_y H = 0$, equation (7.1) always admits the operator ∂_y.

So nontrivial Lie algebras L^r are possible only if it is assumed that $p \neq 0$. In this case, the integration of equation (7.8) gives

$$h = \partial_x^3 H = -24m(x + \lambda)^{-5}, \qquad 24m = (5/p)^5,$$

with a function $y \to \lambda(y)$. Three integrations lead to the general expression for the arbitrary element

$$H = m(x + \lambda)^{-2} + \beta x^2 + \gamma x + \delta,$$

which by a suitable equivalence transformation [(7.2), (7.3)] becomes a simpler form, namely,

$$H(x, y) = m(x + \mu(y))^{-2}.$$

In its turn, this function is equivalent to

$$H(x, y) = mx^{-2} + \theta(y)x \tag{8.1}$$

for some function $\theta: y \to \theta(y)$.

Substitution of the function (8.1) into the determining equation (7.6), and its decomposition relative to the variable x, gives the equations

$$\phi''' = 0, \qquad \psi = 0, \qquad \chi' = 2\phi'', \qquad 3\phi'\theta + 2\phi\theta' = 0. \tag{8.2}$$

From the last equation in (8.2) the function ϕ is uniquely defined, up to a constant multiplier, so in the case $\theta \neq 0$ equation (7.1) admits not more than the Lie algebra L^1.

Finally, if $\theta = 0$, then the first three equations in (8.2) remain. Their solution is [because of (7.5) the function χ must be defined up to a constant term]

$$\phi = C_1 + C_2 y + C_3 y^2, \qquad \psi = 0, \qquad \chi = 4C_3 y.$$

By equations (7.5) this solution gives the Lie algebra L^3 of operators (4.1), admitted in this case by equation (7.1), and its basis is

$$\zeta_1 = (0, 1, 0), \qquad \zeta_2 = (x, 2y, 0), \qquad \zeta_3 = (-4xy, -4y^2, (x^2 + 2y)z). \tag{8.3}$$

9. Classification Result

The final result for the group classification (7.1) of second-order equations with parabolic normal forms is given in the following theorem.

Theorem. *Equation* (7.1) *admits any operator* (4.1) *if and only if it is equipollent to an equation of this type with the function H that does not depend on one of the coordinates x or y. Expansion of the full Lie group is possible only if equation* (7.1) *is equipollent to the same equation with the function* $H = mx^{-2}$ (m = const); *the last equation, when* $m \neq 0$, *admits the algebra* L^3 *of operators* (8.3), *and when* $m = 0$ *the Lie algebra found in Section* 6.8.

Proof. Because of the foregoing results it is only necessary to prove the first part of the statement. It is easy to verify, using the formula for the similarity of vector fields in Section 1.16, that under the transformations defined by equations (7.2), the vector $(\xi, \eta, \sigma z)$ becomes the vector $(\bar{\xi}, \bar{\eta}, \bar{\sigma}\bar{z})$, with the components

$$\bar{\xi} = a\xi + (a'x + b')\eta, \qquad \bar{\eta} = a^2\eta, \qquad \bar{\sigma} = \sigma + \rho_x\xi + \rho_y\eta,$$

where

$$\rho = \frac{a'}{4a}x^2 + \frac{b'}{2a}x + c.$$

Because of equations (7.5), these imply that transformations of the functions ϕ, ψ, χ are given by the equations

$$\bar{\phi} = a^2\phi, \qquad \bar{\psi} = a\psi + 4b'\phi, \qquad \bar{\chi} = \chi - \frac{b'}{2a}\psi - 4c'\phi. \tag{9.1}$$

Let equation (7.1) admit the nonzero operator (4.1) with the functions ϕ, ψ, χ. If $\phi \neq 0$, then it is possible to choose functions a, b, c such that equations (9.1) become $\bar{\phi} = \text{const} \neq 0, \bar{\psi} = \bar{\chi} = 0$. In this case the determining equation (7.6), after the transformations (7.2) and (7.3), will take the form $4\bar{\phi}\partial_{\bar{y}} - \bar{H} = 0$, from which it follows that $\bar{H}$ does not depend on $\bar{y}$. But if $\phi = 0$ and $\psi \neq 0$, then with the aid of (9.1) it is possible to make $\bar{\psi} = \text{const} \neq 0$ and $\bar{\chi} = 0$, and the determining equation (7.6), after transformations (7.2) and (7.3), will be $\bar{\psi}\partial_{\bar{x}}\bar{H} = 0$, that is, $\bar{H}$ does not depend on $\bar{x}$. Finally, from (7.6), it follows that if $\phi = \psi = 0$, then the operator (4.1) is the zero operator. ■

Consequence. *If equation* (7.1) *admits two linearly independent operators of the form* (4.1), *then it is equipollent to the equation*

$$z_{11} = z_2 + mx^{-2}z.$$

§10. BOUNDARY LAYER EQUATIONS

1. Description of the System of Equations

The well-known equations from hydrodynamics of the two-dimensional nonstationary boundary layer (Prandtl equations) are considered. They describe the motion of a viscous incompressible fluid near an impervious solid surface for large Reynolds numbers. This system has the form $E(3, 3, 2, 3)$ and will be described in the notation used in hydrodynamics. Here $X = R^3(t, x, y)$, $Y = R^3(u, v, p)$, where t is time; x, y are space coordinates; u, v are components of the velocity vector; and p is pressure. The operator of differentiation of the mapping $Z = X \times Y \to P$ is written in the form $\partial = (\partial_t, \partial_x, \partial_y, \partial_u, \partial_v, \partial_p)$, and for the partial derivatives the abbreviated notation with indices is used; for example, $\partial_t u = u_t$, $\partial_x u = u_x$, $\partial_y^2 u = u_{yy}$, and so on.

Without loss of generality it is possible to assume that the fluid density and the viscous coefficient are equal to one. With these assumptions the system studied is

$$\begin{aligned} u_t + uu_x + vu_y + p_x &= u_{yy}, \\ p_y &= 0, \\ u_x + v_y &= 0. \end{aligned} \tag{1.1}$$

The pressure p plays a special role in this system. Because of the second equation, p is a function only of (t, x). In hydrodynamics problems this function is often assumed to be given. In this case, if it assumed that p is given, then the system (1.1) will become a system of the form $E(3, 2, 2, 2)$:

$$\begin{aligned} u_t + uu_x + vu_y + \theta &= u_{yy}, \\ u_x + v_y &= 0. \end{aligned} \tag{1.2}$$

Unlike system (1.1), here the function θ can be considered as an arbitrary element, while system (1.1) does not contain arbitrary elements. Thus, for these two systems of equations, the problems of group analysis are different: for system (1.1) it is merely to calculate its full group (the full Lie algebra of operators); for system (1.2) it is group classification relative to the arbitrary element $\theta\colon (t, x) \to \theta(t, x)$. The following is devoted to the solution of these problems.

2. Preliminary Information on the Operator

For the components of the unknown operator it is expedient to use the individual coordinate notation, in which a component is denoted by the same letter that a scalar variable has, but with a star above it. With this notation the operator admitted by system (1.1) has the form

$$\zeta \cdot \partial = \overset{*}{t} \cdot \partial_t + \overset{*}{x}\partial_x + \overset{*}{y}\partial_y + \overset{*}{u}\partial_u + \overset{*}{v}\partial_v + \overset{*}{p}\partial_p, \tag{2.1}$$

the components of which, $\overset{*}{t}, \ldots, \overset{*}{p}$, are the unknown mappings $Z \to R$. The continuation of the operator formulas (2.1) is constructed with the aid of the operator of total differentiation $D = (D_t, D_x, D_y)$, with the components

$$\begin{aligned} D_t &= \partial_t + u_t\partial_u + v_t\partial_v + p_t\partial_p + \cdots, \\ D_x &= \partial_x + u_x\partial_u + v_x\partial_v + p_x\partial_p + \cdots, \\ D_y &= \partial_y + u_y\partial_u + v_y\partial_v + p_y\partial_p + \cdots + u_{yy}\partial_u \ + \cdots, \end{aligned}$$

where the displayed terms actually take part in the construction of the determining equations.

In the beginning it is worthwhile to restrict the functional form of the components of the operator (2.1) by noting that it must be admitted by $u_x + v_y = 0$, which is the last equation of system (1.1), (1.2). The algorithm in Section 5.8, in the process of whose realization the condition of invariance is taken on in the entire manifold (1.1) of the continuation space Z_2, leads to the representation

$$\overset{*}{t} = 2\alpha(t), \qquad \overset{*}{x} = \beta(t, x), \qquad \overset{*}{y} = \gamma(t, y),$$
$$\overset{*}{u} = (\sigma + \beta_x)u + \lambda, \qquad \overset{*}{v} = (\sigma + \gamma_y)v + \mu, \tag{2.2}$$

in which the functions $\sigma, \lambda, \mu: X \to R$ satisfy the equations

$$\sigma_x + \beta_{xx} = 0, \qquad \sigma_y + \gamma_{yy} = 0, \qquad \lambda_x + \mu_y = 0, \tag{2.3}$$

which are components of the determining equations of any system (1.1), (1.2).

3. Determining Equations

Application of the algorithm in Section 5.8 to the second equation in (1.1) gives the representation of the last component of the operator (2.1) as

$$\overset{*}{p} = \psi(t, x, p). \tag{3.1}$$

Finally, the first equation, with the aid of the algorithm in Section 5.8, generates the following determining equations:

$$\beta_{tx} = \alpha_{tt}, \qquad \beta_{xx} = 0, \qquad \gamma_y = \alpha_t;$$
$$\sigma = -2\alpha_t, \qquad \lambda = \beta_t, \qquad \mu = \gamma_t; \qquad \psi_x = -\lambda_t, \qquad \psi_p = 2\beta_x - 4\alpha_t. \tag{3.2}$$

The complete system of determining equations DE for equation (1.1) consists of equations (2.3) and (3.2).

For the system $E(\theta)$ of equations (1.2), with the arbitrary element θ, the algorithm in Section 5.8 gives the same determining equations (2.3), (3.2) with the following changes: in (3.2) the last two equations are excluded and their place is taken by the determining equation

$$2\alpha\partial_t\theta + \beta\partial_x\theta = (\beta_x - 4\alpha_t)\theta - \lambda_t. \tag{3.3}$$

4. General Solution

Substituting the expressions (3.2) for λ, μ into the last equation of (2.3) and using the first three equations in (3.2) shows that $\alpha_{tt} = 0$. Because of this, the general solution of equations (2.3)–(3.2) is found to be

$$2\alpha = C_1 + 2C_2 t, \qquad \beta = C_3 x + \delta(t), \qquad \gamma = C_2 y + \varepsilon(t), \qquad \sigma = -2C_2,$$
$$\lambda = \delta'(t), \qquad \mu = \varepsilon'(t), \qquad \psi = (2C_3 - 4C_2)p - \delta''(t)x + \pi(t), \tag{4.1}$$

where C_1, C_2, C_3 are arbitrary constants, $t \to \delta(t)$, $\varepsilon(t)$, $\pi(t)$ are arbitrary functions, and derivatives with respect to t are denoted by primes. On the basis of this solution, the full Lie algebra of operators (2.1), admitted by the system (1.1), is obtained from equations (2.2) and (3.1) and has the components

$$\begin{gathered}\overset{*}{t} = C_1 + 2C_2 t, \qquad \overset{*}{x} = C_3 x + \delta(t), \qquad \overset{*}{y} = C_2 y + \varepsilon(t),\\ \overset{*}{u} = (C_3 - 2C_2)u + \delta'(t), \qquad \overset{*}{v} = -C_2 v + \varepsilon'(t), \\ \overset{*}{p} = (2C_3 - 4C_2)p - \delta''(t)x + \pi(t).\end{gathered} \tag{4.2}$$

Here the constant C_1 gives the operator ∂_t of translation with respect to time and the constants C_2 and C_3 give the operators of dilatation

$$2t\partial_t + 2x\partial_x + y\partial_y - v\partial_v, \qquad x\partial_x + u\partial_u + 2p\partial_p.$$

The arbitrary function δ, ε give the operator

$$\delta(t)\partial_x + \varepsilon(t)\partial_y + \delta'(t)\partial_u + \varepsilon'(t)\partial_v - \delta''(t)x\partial_p,$$

which corresponds to the transformation of translation of the coordinate system in the plane $R^2(x, y)$, which is moving progressively in time by an arbitrary law. The arbitrary function π gives the operator $\pi(t)\partial_p$, corresponding to the fact that the pressure p is defined by equation (1.1) correctly up to a term which is equal to an arbitrary function of time.

The remarkable peculiarity of the resulting full Lie algebra LE of the system E, assigned by equations (1.1), is that it is infinite dimensional: its operators depend on three arbitrary functions δ, ε, π, of the time t.

5. The Given Pressure Case

With a change to the system $E(\theta)$, given by equations (1.2), the operator (2.1) is "shorter" because of the discarded term $\overset{*}{p}\partial_p$, and to the determining equations (2.3) and (3.2) equation (3.3) is added. For the group classification of equations (1.2) it is first necessary to find the kernel of the full Lie algebra of the system $E(\theta)$.

Obviously, equation (3.3) with the arbitrary function θ can only be satisfied if $\alpha = \beta = \lambda_t = 0$. The addition of these equations to system (2.3), (3.2) shows that the general solution (4.1) must have $C_1 = C_2 = C_3 = 0$ and $\delta = 0$. Thus the kernel of the full Lie algebra L_0, of the equation $E(\theta)$, is generated by the operator

$$\zeta \cdot \partial = \varepsilon(t)\partial_y + \varepsilon'(t)\partial_v, \tag{5.1}$$

with one arbitrary function $t \to \varepsilon(t)$.

Next the equivalence transformations for the class of equations $E(\theta)$, which must be transformations of the space $R^5(t, x, y, u, v)$, must be calculated. Application of the algorithm from Section 6 gives the following result. The general transformation of equivalence consists of all transformations corresponding to the kernel of the full Lie algebra (5.1), and of transformations, depending on three constants k, l, m and one arbitrary function $t \to \omega(t)$, which are given by the equations

$$\bar{t} = k^2 t + l, \qquad \bar{x} = mk^4(x + \omega), \qquad \bar{y} = ky,$$
$$\bar{u} = mk^2(u + \omega'), \qquad \bar{v} = k^{-1}v, \tag{5.2}$$

where $\omega' = \partial_t \omega$. With these, the arbitrary element θ is transformed into

$$\bar{\theta}(\bar{t}, \bar{x}) = m(\theta(t, x) - \omega''(t)). \tag{5.3}$$

6. Group Classification

The determining equation (3.3), using the already determined expressions of the functions (4.1), assumes the form

$$(C_1 + 2C_2 t)\theta_t + (C_3 x + \delta(t))\theta_x = (C_3 - 4C_2)\theta - \delta''(t), \tag{6.1}$$

where the notations $\theta_t = \partial_t \theta$, $\theta_x = \partial_x \theta$ are used. This makes clear the general structure of the classification equation which is responsible for the group classification of the system $E(\theta)$: Any concrete function, with which it is possible to extend the kernel of the full Lie algebra (5.1), must be a solution of the equation

$$(A + Bt)\theta_t + (C_x + v(t))\theta_x = (C - 2B)\theta - v''(t), \tag{6.2}$$

with some constants A, B, C and a function $t \to v(t)$.

It is easy to verify that equation (6.2) is invariant relative to the equivalence transformations (5.2), (5.3), supplemented with the following transformations of the constants A, B, C and the function v:

$$\bar{A} = k^2 A - lB, \qquad \bar{B} = B, \qquad \bar{C} = C, \qquad \bar{v} = mk^4(v + (A + Bt)\omega' - C\omega). \tag{6.3}$$

In the first classification possibility, the function $\theta = 0$ or, more generally, any function θ depending only on t, is always equal, because of (5.3), to the function $\theta = 0$. In this case the determining equation will assume the form $\delta'' = 0$ and the general solution will be $\delta = C_4 + C_5 t$. According to equations (4.2), this gives a five-dimensional expression L^5 for the kernel of the full Lie algebra. The kernel has the basis

$$\zeta_1 \cdot \partial = \partial_t, \qquad \zeta_2 \cdot \partial = 2t\partial_t + y\partial_y - 2u\partial_u - v\partial_v,$$
$$\zeta_3 \cdot \partial = x\partial_x + u\partial_u, \qquad \zeta_4 \cdot \partial = \partial_x, \qquad \zeta_5 \cdot \partial = t\partial_x + \partial_u. \tag{6.4}$$

Then some alternative possibilities are considered: (a) all constants A, B, C are equal to zero, and (b) one of them is not zero.

***Case* (a).** Here the classification equation (6.2) has the form $\nu\theta_x = -\nu''$. The solution of this equation, correct up to the equivalence transformations (5.2), (5.3), is $\theta = \phi(t)x$ with some completely determined function $\phi: t \to \phi(t)$. The substitution of this function into the determining equation (6.1) and the subsequent decomposition relative to the "free" variable x gives the equations

$$C_1\phi' + 2C_2(t\phi' + 2\phi) = 0, \qquad \delta'' + \phi\delta = 0. \tag{6.5}$$

If $C_1 = C_2 = 0$, then the first equation does not impose any limitations on the function ϕ. The second one always has two linearly independent solutions δ_1 and δ_2 and its general solution is $\delta = C_4\delta_1 + C_5\delta_2$. Consequently, in this case, the kernel of the full Lie algebra L_0 is expanded by the subalgebra L^3 with the basis given by the operator $\zeta_3 \cdot \partial$, from (6.4), and the operators

$$\begin{aligned} \zeta_4^0 \cdot \partial &= \delta_1(t)\partial_x + \delta_1'(t)\partial_u, \\ \zeta_5^0 \cdot \partial &= \delta_2(t)\partial_x + \delta_2'(t)\partial_u. \end{aligned} \tag{6.6}$$

If $C_2 \neq 0$, then equation (6.5) is equivalent to the equation $t\phi' + 2\phi = 0$, with the general solution $\phi = qt^{-2}$, where $q = \text{const}$. Here the second equation (6.5) is $t^2\delta'' + q\delta = 0$. When $q = n(1 - n) \neq \frac{1}{4}$, it has two linearly independent solutions:

$$\delta_1 = t^n, \qquad \delta_2 = t^{1-n},$$

and when $q = \frac{1}{4}$, the two solutions are

$$\delta_1 = t^{1/2}, \qquad \delta_2 = t^{1/2}\ln t.$$

So, if $\theta = qt^{-2}x$, then the subalgebra of the Lie algebra L_0 is L^4, with a basis consisting of the operators $\zeta_2 \cdot \partial$ and $\zeta_3 \cdot \partial$ from (6.4) and the operators (6.6), with the previously determined functions δ_1 and δ_2.

Finally, if $C_2 = 0$ but $C_1 \neq 0$, then from (6.5) it follows that $\phi' = 0$, so that $\phi = \text{const}$. At the expense of the equivalence transformation, this constant can be made equal to ± 1, whence the second equation in (6.5) becomes $\delta'' \pm \delta = 0$. So for the function $\theta = \pm x$, the extending Lie algebra is again L^4, but with a basis consisting of the operators $\zeta_1 \cdot \partial$ and $\zeta_3 \cdot \partial$ from (6.4) and the operators (6.6), where either $\delta_1 = \sin t$, $\delta_2 = \cos t$ (for $\theta = x$), or $\delta_1 = \sinh t$, $\delta_2 = \cosh t$ (for $\theta = -x$).

***Case* (b).** If even one of the constant A, B, C is not equal to zero, then from (6.3) it follows that for any function ν it is possible to select a function ω from the equivalence transformation (5.2), (5.3) for which $\bar{\nu} = 0$. Consequently, in this case the arbitrary element is equivalent to the function θ, satisfying the equation

$$A\theta_t + B(t\theta_t + 2\theta) + C(x\theta_x - \theta) = 0. \tag{6.7}$$

Here various possibilities appear, depending on whether the constants B and C are equal to zero or not. There are four subcases:

$$\begin{array}{llll} (\mathrm{b}_1) & C = B = 0, & (\mathrm{b}_2) & C = 0, \quad B \neq 0, \\ (\mathrm{b}_3) & C \neq 0, \quad B = 0, & (\mathrm{b}_4) & C \neq 0, \quad B \neq 0. \end{array}$$

(b_1) Equation (6.7) is $\theta_t = 0$, with the general solution $\theta = \psi(x)$. The determining equation (6.1) assumes the form

$$C_3(x\psi' - \psi) + \delta\psi' + 4C_2\psi = -\delta''.$$

Since ψ is an arbitrary function of x it must be that $C_2 = C_3 = 0$ and $\delta = 0$, because of which the one-dimensional Lie algebra L^1 remains, with the operators $\zeta_1 \cdot \partial$ from (6.4), where the operator $\zeta_1 \cdot \partial$ corresponds to the constant C_1 in (4.2). Differentiation with respect to x gives the equation

$$C_3 x\psi'' + \delta(t)\psi'' + 4C_2\psi' = 0.$$

Obviously, the assumption $\psi'' = 0$ brings back the cases $\theta = 0$ or (a). Thus it is assumed that $\psi'' \neq 0$, and then the function δ must be a constant $\delta = C_4$. Then the determining equation will be

$$(C_3 x + C_4)\psi' = (C_3 - 4C_2)\psi.$$

When $C_3 \neq 0$, the solution of this equation is equivalent to a power function $\psi = x^n$, where $n \neq 0$ and $n \neq 1$. With this function the relations $C_4 = 0$, $(n-1)C_3 = 4nC_2$ are obtained in the equation, which gives a Lie subalgebra $L^2 \subset L^5$ with the basis of operators

$$\zeta_1 \cdot \partial, \qquad \zeta_2 \cdot \partial + \frac{4n}{n-1}\zeta_3 \cdot \partial.$$

If $C_3 = 0$, then the solution is equivalent to the function e^x and $C_4 = C_3 - 4C_2$. So when $\theta = e^x$, the subalgebra $L^3 \subset L^5$, with the basis

$$\zeta_1 \cdot \partial, \qquad \zeta_2 \cdot \partial - 4\zeta_4 \cdot \partial, \qquad \zeta_3 \cdot \partial + \zeta_4 \cdot \partial,$$

is obtained.

(b_2) As in (6.3) it follows that when $B \neq 0$ we can assume $A = 0$, so here the function θ is equivalent to the solution of the equation $t\theta_t + 2\theta = 0$, that is, the function $\theta = \psi(x)t^{-2}$. With this function the determining equation (6.1) assumes the form

$$-2C_1 t^{-1}\psi + C_3(x\psi' - \psi) + \delta\psi' = -t^2\delta'',$$

differentiation ∂_x of which gives the equation

$$-2C_1 t^{-1}\psi' + C_3 x\psi'' + \delta\psi'' = 0.$$

From this, with the arbitrary function ψ, the relations $C_1 = C_3 = 0$ and $\delta = 0$ follow, because of which only L^1 and $\zeta_2 \cdot \partial$ remain from (6.4). The assumption $\psi'' = 0$ brings back the cases $\theta = 0$ or (a). If $\psi'' \neq 0$, then it is necessary to have $C_3 = 0$ and $(t\delta)' = 0$, that is, $\delta = C_4 t^{-1}$. Thus the previous equation shows that the function ψ must be a solution of the equation $\psi'' = q\psi'$, where $q \neq 0$, or it must be the function $\psi = e^x$ up to equivalence. With this, the determining equation is reduced to the equality $C_4 = 2C_1$. Consequently, for $\theta = e^x t^{-2}$ the Lie algebra L^2 with the operators

$$\zeta_2 \cdot \partial, \qquad \zeta_1 \cdot \partial + 2t^{-1}\partial_x - 2t^{-2}\partial_u$$

is found.

(b$_3$) Obviously, here the assumption $A = 0$ brings us back to the cases $\theta = 0$ or (a). If $A \neq 0$, then at the expense of the equivalence transformation (6.3) we can assume $A = \pm C$, because of which equation (6.7) becomes $\pm\theta_t + x\theta_x = \theta$, with the general solution $\theta = e^{\mp t} f(xe^{\pm t})$, where f is an arbitrary function. Substitution of the function $\theta = e^{-t} f(z)$, $z = xe^t$, into the determining equation (6.1) gives

$$(C_1 + C_3 + 2C_2 t)(zf' - f) + e^t\delta f' = -4C_2 f - e^t\delta'',$$

from which, for the arbitrary f, the Lie algebra L^1 is obtained, the operator of which is defined by the relation $C_1 + C_3 = 0$, that is $\zeta_1 \cdot \partial - \zeta_3 \cdot \partial$. The differentiation ∂_z leads to the equation

$$(C_1 + C_3 + 2C_2 t)zf'' + e^t\delta f'' + 4C_2 f' = 0.$$

The assumption $f'' = 0$ brings us back to the cases $\theta = 0$ or (a). If $f'' \neq 0$, then from this equation it follows that $C_2 = 0$ and $(e^t\delta)' = 0$, from which $\delta = C_4 e^{-t}$. But then it is also necessary that $C_4 = 0$ and $C_1 + C_3 = 0$, and there are no new operators, but $\zeta_1 \cdot \partial - \zeta_3 \cdot \partial$ remains.

(b$_4$) Here as in (b$_2$) it is possible to assume that $A = 0$, which leads equation (6.1) to

$$t\theta_t - nx\theta_x = -(n + 2)\theta,$$

where $n \neq 0$. The general solution of this equation is $\theta = t^{-n-2} f(z)$, where $z = xt^n$. With this function θ the determining equation (6.1) becomes

$$(nC_1 + (2nC_2 + C_3)t)(zf' - f) + t^{n+1}\delta f' - 2C_1 f = -t^{n+3}\delta''.$$

With an arbitrary function f, this equation requires the relations $C_1 = 0$, $2nC_2 + C_3 = 0$, $\delta = 0$, which gives the Lie algebra L' with the operator $\zeta_2 \cdot \partial - 2n\zeta_3 \cdot \partial$. By the differentiation ∂_z the equation

$$(nC_1 + (2nC_2 + C_3)t)zf'' + t^{n+1}\delta f'' - 2C_1 f' = 0$$

is found, and here the assumption $f'' = 0$ brings us back to the cases $\theta = 0$ or (a). If $f'' \neq 0$, then from this equation with the aid of differentiation ∂_t

the relations $2nC_2 + C_3 = 0$ and $(t^{n+1}\delta)' = 0$ follow, from which $\delta = C_4 t^{-n-1}$. With these results the determining equation assumes the form

$$C_4(f' + (n+1)(n+2)) + C_1(nzf' - (n+2)f) = 0.$$

Here $C_1 \neq 0$, so it is possible to assume $C_4 = nqC_1$, with a new constant q. Because of this relation the previous equation is classifying and

$$n(z+q)f' = (n+2)f - n(n+1)(n+2)q.$$

The general solution of this equation is $f = N(z+q)^{(n+2)/n} + n(n+1)q$ ($N = \text{const}$), because of which the function $\theta = t^{-n-2}f(z)$ becomes

$$\theta = N(x + qt^{-n})^{(n+2)/n} + n(n+1)qt^{-n-2}.$$

But if in (5.2) it is assumed that $m = k = 1$ and $\omega = qt^{-n}$, then from equation (5.3) there is obtained $\bar{\theta} = N\bar{x}^{(n+2)/n}$. Consequently, the function θ is found to be equivalent to the function $\theta = x^{n_1}$, already considered in (b_1).

A summary of the foregoing results for the group classification of system (1.2) is given in Table 6.1, where the first column gives the indices of the cases considered; the second, the kind of function; the third, the specializations contained in the arbitrary function θ; the fourth, the dimensionality r of the resulting Lie algebra L^r; the fifth, a list of the operators (vector fields) and the sixth, some additional information. All operators with the double index have the form $\delta(t)\partial_x + \delta'(t)\partial_u$.

TABLE 6.1

	θ	S^a	r	Operators	Notes
	0	$\sim$	5	$\zeta_1, \zeta_2, \zeta_3, \zeta_4, \zeta_5$	(6.4)
(a)	$\phi(t)x$	$\sim$	3	$\zeta_3, \zeta_4^0, \zeta_5^0$	$\delta'' + \phi\delta = 0$
		$n(1-n)t^{-2}$	4	$\zeta_2, \zeta_3, \zeta_4^1, \zeta_5^1$	$\delta_1 = t^n$, $\delta_2 = t^{1-n}$
		$(\frac{1}{4})t^{-2}$	4	$\zeta_2, \zeta_3, \zeta_4^2, \zeta_5^2$	$\delta_1 = t^{1/2}$, $\delta_2 = t^{1/2}\ln t$
		1	4	$\zeta_2, \zeta_3, \zeta_4^3, \zeta_5^3$	$\delta_1 = \sin t$, $\delta_2 = \cos t$
		-1	4	$\zeta_2, \zeta_3, \zeta_4^4, \zeta_5^4$	$\delta_1 = \sinh t$, $\delta_2 = \cosh t$
(b_1)	$\psi(x)$	$\sim$	1	ζ_1	
		x^n	2	$\zeta_1, \zeta_2 + \dfrac{4n}{n-1}\zeta_3$	$n(n-1) \neq 0$
		e^x	3	$\zeta_1, \zeta_2 - 4\zeta_4, \zeta_3 + \zeta_4$	
(b_2)	$\psi(x)t^{-2}$	$\sim$	1	ζ_2	
		e^x	2	$\zeta_2, \zeta_1 + \zeta_4^5$	$\delta = 2t^{-1}$
(b_2)	$e^{\pm t}f(xe^{\mp t})$	$\sim$	1	$\zeta_1 \pm \zeta_3$	
(b_4)	$t^{-n-2}f(xt^n)$	$\sim$	1	$\zeta_2 - 2n\zeta_3$	

[a] S is specialization.

7. Steady-State Boundary Layer

Side by side with (1.1), the analogous system in which the unknowns u, v, p do not depend on time is interesting. This system of the steady-state boundary layer equations, having the form $E(2, 3, 2, 3)$, where $X = R^2(x, y)$, $Y = R^3(u, v, p)$, $Z = R^5\,(x, y, u, v, p)$, is

$$\begin{aligned} uu_x + vu_y + p_x &= u_{yy}, \\ p_y &= 0, \\ u_x + v_y &= 0. \end{aligned} \tag{7.1}$$

If the pressure p is a given function of x, then with the function $\theta = \partial_x p$, the analog of system (1.2) is obtained:

$$\begin{aligned} uu_x + vu_y + \theta &= u_{yy}, \\ u_x + v_y &= 0. \end{aligned} \tag{7.2}$$

For the systems (7.1) and (7.2) the problems of group analysis, which were solved above for the systems (1.1) and (1.2), are solved here. The admitted operator is again sought in the form (2.1) with $\overset{*}{t} = 0$ and components $\overset{*}{x}, \dots, \overset{*}{p}: Z \to R$.

Let E denote the system (7.1) and $E(\theta)$ the system (7.2). Determination of the determining equations DE and their general solution is accomplished without any difficulties by

$$\begin{aligned} \overset{*}{x} &= C_1 + (C_2 + 2C_3)x, \qquad \overset{*}{y} = C_3 y + \phi, \\ \overset{*}{u} &= C_2 u, \qquad \overset{*}{v} = -C_3 v + \phi' u, \end{aligned} \tag{7.3}$$

$$\overset{*}{p} = 2C_2 p + C_4, \tag{7.4}$$

where $C_1, \dots, C_4$ are arbitrary constants, $\phi: x \to \phi(x)$ is an arbitrary function, and $\phi' = \partial_x \phi$. As this general solution contains an arbitrary function, so the full Lie algebra LE turns out to be infinite dimensional.

The infinite-dimensional part L^∞ of the Lie algebra LE consists of the operators $\phi\partial_y + \phi' u \partial_v$, with an arbitrary function $\phi: x \to \phi(x)$. An easy calculation shows that L^∞ is an ideal in LE. The direct complement of this ideal is the subspace $L^4 \subset LE$, generated by the operators corresponding to the constants $C_1, \dots, C_4$ in (7.3) and (7.4):

$$\begin{aligned} \zeta_1 \cdot \partial &= \partial_x, \qquad \zeta_2 \cdot \partial = x\partial_x + u\partial_u + 2p\partial_p, \\ \zeta_3 \cdot \partial &= 2x\partial_x + y\partial_y - v\partial_v, \qquad \zeta_4 \cdot \partial = \partial_p. \end{aligned} \tag{7.5}$$

These operators form a basis in L^4. It is easy to verify that L^4 is a Lie algebra; consequently, it is a subalgebra in LE. So the full Lie algebra of equations (7.1) is given as the direct sum $LE = L^\infty \oplus L^4$ of its infinite-dimensional ideal L^∞ and the four-dimensional subalgebra L^4 with the basis (7.5).

In studying $E(\theta)$ the variable p is not included in the components of the space Z, so $Z = R^4(x, y, u, v)$. The construction of the determining equations $DE(\theta)$ and their solution gives the following result.

The representation of the general solution (7.3) also holds, but in it the constants C_1, C_2, C_3 are connected with one more determining equation, for the arbitrary element θ,

$$(C_1 + (C_2 + 2C_3)x)\theta' = (C_2 - 2C_3)\theta, \tag{7.6}$$

where $\theta' = \partial_x \theta$. The infinite-dimensional ideal L^∞ is contained in the Lie algebra $LE(\theta)$ for any concrete arbitrary element θ. Because of the omission of the variable p, the decomposition of the full Lie algebra $LE(\theta)$, as a direct sum, has the form $LE(\theta) = L^\infty \oplus L^r$, where L^r is some subalgebra of the three-dimensional Lie algebra L^3 with the basis consisting of the three first operators (7.5). The determination of the Lie algebra L^r is realized in the process of a solution of the group classification for the system $E(\theta)$ based on the determining equation (7.6).

It is obvious that the complete Lie algebra L^3 is obtained for the case $\theta = 0$. On the other hand, as with an arbitrary θ, the equality (7.6) is possible only if $C_1 = C_2 = C_3 = 0$, so the kernel of the full Lie algebra L^r is null. Thus the kernel of the full Lie algebra $LE(\theta)$ coincides with L^∞,

8. Group Classification

Equation (7.6) shows that the classifying equation of the system $E(\theta)$ must have the form

$$(A + Bx)\theta' = C\theta, \tag{8.1}$$

with some constants A, B, C. The solution of the group classification problem is based on application of the following equivalence transformations of the system $E(\theta)$:

$$\bar{x} = kx + l, \qquad \bar{y} = m^{-1}y, \qquad \bar{u} = km^2u, \qquad \bar{v} = mv, \qquad \bar{\theta} = km^4\theta, \tag{8.2}$$

where k, l, m are real parameters and $km \neq 0$.

As a result of transformations (8.2) the constants in equation (8.1) are changed into $\bar{A} = kA - lB$, $\bar{B} = B$, $\bar{C} = C$, in connection with which it is convenient to separate two possibilities: (a) $B = 0$ and (b) $B \neq 0$.

***Case* (a).** If $B = 0$, then it must be assumed that $A \neq 0$, for otherwise it will turn out that either $\theta = 0$ or $C = 0$, which situations have already been considered. If $A \neq 0$, then the solution of equation (8.1) is $\theta = qe^{nx}$, with some constants q, n. Because of (8.2) this function is equivalent to either $\pm e^x$ (if $n \neq 0$) or 1 (if $n = 0$).

With the function $\theta = \pm e^x$ equation (7.6) leads to the relations $C_1 = -4C_3$, $C_2 = -2C_3$, which give the Lie subalgebra L^1 with the operator

$$4\partial_x + 2u\partial_u - y\partial_y + v\partial_v.$$

For the function $\theta = 1$ the relation $C_2 = 2C_3$ is found, which gives the Lie subalgebra L^2 with the basis

$$\zeta_1 \cdot \partial, \qquad 2\zeta_2 \cdot \partial + \zeta_3 \cdot \partial.$$

***Case* (b).** If $B \neq 0$, then with the aid of a suitable equivalence transformation, it is possible to take $A = 0$. But if $A = 0$ and $B \neq 0$, then the solution of equation (8.1) is a power function $\theta = qx^n$, and it can be assumed that $q \neq 0$ and $n \neq 0$. This function is equivalent to the function $\theta = \pm x^n$. The substitution of it into the determining equation (7.6) leads to the relations

$$C_1 = 0, \qquad (n - 1)C_2 + 2(n + 1)C_3 = 0.$$

As a consequence, the Lie algebra L^1 with the operator $2(n + 1)\zeta_2 - (n - 1)\zeta_3$ is found.

A summary of the results for the group classification problem of equations (7.2) is found in Table 8.1. In the first column are the indices of the cases considered; in the second, specializations of the function θ; in the third, the dimensionality r of the calculated Lie algebra L^r; in the fourth, the basis of the operators (vector fields); and in the fifth, some additional information. The operators in this table are those of (7.5), except the term $2p\partial_p$ is dropped in the operator $\zeta_2 \cdot \partial$.

TABLE 8.1

	θ	r	Operators	Notes
	0	3	$\zeta_1, \zeta_2, \zeta_3$	(7.5)
(a)	$\pm e^x$	1	$4\zeta_1 + 2\zeta_2 - \zeta_3$	
	1	2	$\zeta_1, 2\zeta_2 + \zeta_3$	
(b)	$\pm x^n$	1	$2(n + 1)\zeta_2 - (n - 1)\zeta_3$	$n \neq 0$

§11. GAS DYNAMICS EQUATIONS

1. Description of the System of Equations

The equations for the adiabatic motion of a gas in the space R^n $(n=1,2,3)$, connecting components of the velocity vector $u \in R^n$, density ρ, and pressure p, as functions of the coordinates of a point $x \in R^n$ and time t are

$$\rho\, d_t u + \operatorname{grad} p = 0, \qquad d_t \rho + \rho \operatorname{div} u = 0, \qquad d_t S(p, \rho) = 0.$$

Here the symbol d_t denotes a "material derivative," namely, the differential operator $\partial_t + u \cdot \partial_x$, and $S: R^2 \to R$ is a given function, called the entropy. It is assumed that the function S is sufficiently smooth and satisfies the condition $\partial_p S \neq 0$, because of which it is possible to introduce another given function $A: R^2 \to R$, defined by the formula

$$A = -\rho\, \partial_\rho S / \partial_p S.$$

In the case of a polytropic gas, this function is $A = \gamma p$, where the constant γ is called the index of the adiabatic curve. Using the function A, it is possible to express the last equation in an equipollent form:

$$d_t p + A \operatorname{div} u = 0.$$

The considered system has the type $E(n+1, n+2, 1, n+2)$. In what follows it is studied for any positive integers n; this does not create any additional difficulties compared with the "physical" cases $n = 1, 2, 3$. The calculations utilize the notation

$$X = R^{n+1}(t, x^1, \ldots, x^n) \qquad \text{and} \qquad Y = R^{n+2}(u^1, \ldots, u^n, p, \rho).$$

With these, the operator of differentiation of a mapping $X \to Y$ is written as $\partial_x = (\partial_0, \partial_1, \ldots, \partial_n)$, where $\partial_0 = \partial_t$, and the "material derivative" is given by the formula $\partial_0 + u^i \partial_i$. In addition, the first derivatives are denoted by a shorthand in accordance with relations

$$\partial_\alpha u^i = u^i_\alpha, \qquad \partial_\alpha p = p_\alpha, \qquad \partial_\alpha \rho = \rho_\alpha \qquad (\alpha = 0, 1, \ldots, n).$$

Below, the agreement is used that Latin indices $i, j, k, \ldots$ assume values $1, \ldots, n$ and Greek indices $\alpha, \beta, \ldots$ take values $0, 1, \ldots, n$. In this notation the system studied has the form

$$\begin{aligned} u^i_0 + u^k u^i_k + \rho^{-1} p_i &= 0, \\ \rho_0 + u^k \rho_k + \rho u^k_k &= 0, \\ p_0 + u^k p_k + A u^k_k &= 0. \end{aligned} \tag{1.1}$$

For the system (1.1), denoted as well by $E(A)$, the problem of group classification relative to the arbitrary element A as a function $(p, \rho) \to A(p, \rho)$ under the assumption that $A \neq 0$ is solved below.

2. Determining Equations

For the components of the operator admitted by the system (1.1), some additional notation is also introduced, namely,

$$\zeta \cdot \partial = \xi^\alpha \partial_\alpha + \eta^i \partial_{u^i} + \sigma \partial_p + \tau \partial_\rho, \tag{2.1}$$

where ξ^α, η^i, σ, τ are unknown mappings $X \times Y \to R$.

Application of the algorithm given in Section 5.8 for the transition $E \to DE$ can be carried out without any difficulty. As (1.1) is a system of first order, only the first continuation of the operator (2.1) is needed. The transition on the manifold E is realized merely by excluding the values u_0^i, p_0, ρ_0 using (1.1), and in the resulting conditions of invariance all remaining components u_k^i, p_k, ρ_k are "free" parameters. As the equations (1.1) are quasi-linear, the total number N_n of determining equations, obtained after decomposition, can be calculated by formula (5.8.1). That calculation gives $N_1 = 30$, $N_2 = 180$, and $N_3 = 680$.

The decomposition of the invariance conditions relative to the "free" parameters leads to the following results. Elimination of the product of quadratic terms, by setting them equal to zero, is equivalent to satisfaction of the equalities

$$\partial_\rho \xi^\alpha = \partial_p \xi^\alpha = \partial_{u^i} \xi^\alpha = 0 \qquad (\alpha = 0, 1, \ldots, n; i = 1, \ldots, n). \tag{2.2}$$

Elimination of the product of the first-power terms is equipollent to satisfaction of the equations

$$\partial_i \xi^0 = 0, \qquad \partial_1 \xi^1 = \partial_2 \xi^2 = \cdots = \partial_n \xi^n = \lambda, \qquad \partial_k \xi^i + \partial_i \xi^k = 0, \tag{2.3}$$

$$\eta^i = u^k \partial_k \xi^i - u^i \partial_0 \xi^0 + \partial_0 \xi^i, \tag{2.4}$$

$$\partial_\rho \sigma = \partial_{u^i} \sigma = 0, \qquad \partial_p \tau = 0, \qquad \tau = \rho(\partial_p \sigma + 2\partial_0 \xi^0 - 2\lambda), \tag{2.5}$$

$$\sigma \partial_p A + \tau \partial_\rho A = A \partial_p \sigma, \tag{2.6}$$

where the auxiliary function $\lambda: X \to R$ is introduced. Finally, the remaining terms, with the null power of the "free" parameters, form the system

$$\begin{aligned} &\partial_0 \eta^i + u^k \partial_k \eta^i + \rho^{-1} \partial_i \sigma = 0, \\ &\partial_0 \tau + u^k \partial_k \tau + \rho \partial_k \eta^k = 0, \\ &\partial_0 \sigma + u^k \partial_k \sigma + A \partial_k \eta^k = 0. \end{aligned} \tag{2.7}$$

The equations (2.2)–(2.7) generate the complete system of the determining equations $DE(A)$.

3. Kernel of Full Lie Algebra

If the function A is completely arbitrary, then equation (2.6) can be satisfied only with the values $\sigma = \tau = 0$. In this case, from (2.5), it follows that $\lambda = \partial_0 \xi^0$ and equations (2.7) are reduced to

$$\partial_0 \eta^i + u^k \partial_k \eta^i = 0, \qquad \partial_k \eta^k = 0. \tag{3.1}$$

The substitution of the expression (2.4) into these equations and the resulting decomposition of the resulting equations relative to the "free" parameters, using (2.2) and (2.4), leads to a system of equations containing only the components ξ^i:

$$\partial_k \partial_j \xi^i = 0, \qquad 2\partial_k \partial_0 \xi^i = \delta^i_k \partial_0^2 \xi^0, \qquad \partial_0^2 \xi^i = 0. \tag{3.2}$$

These equations show that all of the second derivatives of the components ξ^i are equal to zero and, in addition, $\partial_0^2 \xi^0 = 0$. Thus the general solution is found without any difficulty and turns out to be

$$\begin{aligned} &\xi^i = C_2 x^i + C^i_k x^k + B^i t + B^i_0 \qquad (C^i_k = -C^k_i), \\ &\xi^0 = C_1 + C_2 t, \qquad \eta^i = C^i_k u^k + B^i, \end{aligned} \tag{3.3}$$

where C_1, C_2, C^i_k, B^i, B^i_0 are arbitrary constants and the matrix (C^i_k) is antisymmetric.

Consequently, the kernel of the full Lie algebra L_0 of the system of equations $E(A)$ is finite dimensional and its dimensionality is equal to the number of independent arbitrary constants in the general solution (3.3), which is found to be

$$2 + \tfrac{1}{2}n(n-1) + 2n = 2 + \tfrac{1}{2}n(n+3).$$

The basis of operators (2.1) of the Lie algebra L_0 corresponding to this solution is

$$\begin{aligned} &\zeta_1 \cdot \partial = \partial_t, \qquad \zeta_2 \cdot \partial = t\partial_t + x^i \partial_i, \\ &\zeta^1_i \cdot \partial = \partial_i, \qquad \zeta^2_i \cdot \partial = t\partial_i + \partial_{u^i}, \\ &\zeta_{ik} \cdot \partial = x^k \partial_i - x^i \partial_k + u^k \partial_{u^i} - u^i \partial_{u^k} \qquad (i < k). \end{aligned} \tag{3.4}$$

4. Preliminary Analysis

From equations (2.5) it follows that $\partial_p^2 \sigma = 0$ and for the function σ the representation

$$\sigma = rp + s \tag{4.1}$$

is obtained, with new unknown functions $x \to r(x)$ and $x \to s(x)$. The substitution of this expression for σ and the corresponding expression for τ, according to (2.5), into (2.6) gives the equation

$$(p\partial_p A + \rho\partial_\rho A - A)r + (\partial_p A)s + 2(\rho\partial_\rho A)(\partial_0 \xi^0 - \lambda) = 0. \qquad (4.2)$$

Further, if the expressions (2.4) and (4.1) are substituted into the first equation in (2.7), then because of (2.2), (2.3), and (2.5) equations (3.2) are obtained and, in addition,

$$\partial_i r = \partial_i s = 0 \qquad (i = 1, \ldots, n),$$

showing that the functions r and s can only depend on t. Besides, from (2.3) and (3.2) it follows that this is also correct for the function λ, as well as for the function ξ^0, because of which the following relation is obtained from (3.2):

$$2\partial_0 \lambda = \partial_0^2 \xi^0. \qquad (4.3)$$

Thus, eliminating τ and η^k from the second equation of (2.7), and considering the previous relations, gives

$$\partial_0 r + (n + 2)\partial_0 \lambda = 0. \qquad (4.4)$$

Finally, eliminating σ and η^k from the third equation of (2.7), and considering the previous equations, leads to the equation

$$\partial_0 s + (nA - (n + 2)p)\partial_0 \lambda = 0. \qquad (4.5)$$

The resulting relations (4.2)–(4.5) provide a foundation for the group classification of the system $E(A)$. The solution of this problem is connected with the application of an easily verified equivalence transformation, in which the values t, x^i, u^i are not changed. However, p, ρ, A are transformed by the formulas

$$\bar{p} = ap + b, \qquad \bar{\rho} = a\rho, \qquad \bar{A}(\bar{p}, \bar{\rho}) = aA(p, \rho), \qquad (4.6)$$

with the parameters $a, b, a \neq 0$.

5. Group Classification

The characteristic classification value is $nA - (n + 2)p$. It is worth noting that if this value is constant, then, at the expense of a suitable equivalence transformation (4.6), it can be set equal to zero. Thus two main cases must be considered:

$$\text{(a)} \quad nA - (n + 2)p \neq \text{const} \qquad \text{and} \qquad \text{(b)} \quad nA = (n + 2)p.$$

***Case* (a).** Here, from (4.5), the equations $\partial_0 s = \partial_0 \lambda = 0$ follow, and from (4.3) and (4.4) also $\partial_0^2 \xi^0 = 0$ and $\partial_0 r = 0$. Consequently, the quantities

r, s, λ, and $\partial_0 \xi^0$ are constants. But now it is obligatory that $\partial_0 \xi^0 = \lambda$, so that in the analogous representation (3.3) of the components ξ^0, ξ^i, η^i, one more constant $q = \partial_0 \xi^0 - \lambda$ will appear, generally speaking. If it is assumed that $\lambda = C_2$, then this representation, including the expressions for σ and τ, becomes

$$\begin{gathered} \xi^i = C_2 x^i + C^i_k x^k + B^i t + B^i_0 \qquad (C^i_k = -C^k_i), \\ \xi^0 = C_1 + (C_2 + q)t, \qquad \eta^i = C^i_k u^k - qu^i + B^i, \\ \sigma = rp + s, \qquad \tau = (r + 2q)\rho. \end{gathered} \tag{5.1}$$

These formulas give the general solution for the entire system of determining equations, excluding equation (4.2), which now has the form (in the simplified notation $\partial_p A = A_p$, $\partial_\rho A = A_\rho$)

$$(pA_p + \rho A_\rho - A)r + A_p s + 2\rho A_\rho q = 0. \tag{5.2}$$

Now it is clear that an expansion of the kernel of the full Lie algebra L_0 is possible, in the considered case, only at the expense of three new constants q, r, s. However, it is possible only on the Lie algebra L^1 or L^2. From (5.1) it is seen that the operators

$$\zeta'_1 \cdot \partial = t\partial_t - u^i \partial_{u^i} + 2\rho\partial_p, \qquad \zeta'_2 \cdot \partial = p\partial_p + \rho\partial_\rho, \qquad \zeta'_3 \cdot \partial = \partial_p \tag{5.3}$$

correspond to the constants q, r, s and the general form of the expanded vector algebra is

$$\zeta' = q\zeta'_1 + r\zeta'_2 + s\zeta'_3. \tag{5.4}$$

To classify the possibilities included in equation (5.2) it is convenient to interpret the coefficients of r, s, q as components of the three-dimensional vector

$$v = (pA_p + \rho A_\rho - A, A_p, 2\rho A_\rho).$$

If the function A is such that three non*copolanar* vectors v can be found (with p and ρ changing), then (5.2) can be satisfied only if $q = r = s = 0$ and the expansion of the kernel L_0 cannot be made. Thus, it is necessary to consider only three subcases, depending on the dimensionality of the affine manifold V, containing all possible vectors v, namely, (a_1) V is two dimensional, (a_2) V is one dimensional, and (a_3) V is null dimensional.

(a_1) Let the constant vector $v' = (a', b', c')$ be orthogonal to V. Then $v' \cdot v = 0$; that is, the function A must be a solution of the equation

$$(a'p + b')A_p + (a' + 2c')\rho A_\rho - a'A = 0. \tag{5.5}$$

If $a' \neq 0$, then a suitable equivalence transformation, (4.6) and (5.5), can be chosen to give $pA_p + (1 + 2m)\rho A_\rho = A$ by selecting $a' = 1$, $b' = 0$, $c' = m$. The general solution of this equation is

$$A = pf(p^{1+2m}\rho^{-1}). \tag{5.6}$$

If $a' = 0$ but $b' \neq 0$, then in the case $c' \neq 0$, $b' = c'$ can be selected to provide a suitable transformation, (4.6) and (5.5), which gives the equation $A_p + 2\rho A_\rho = 0$ with the general solution

$$A = f(\rho e^{-2p}). \tag{5.7}$$

In the case when $c' = 0$ in (5.5), the equation $A_p = 0$ is obtained, with the general solution

$$A = f(\rho). \tag{5.8}$$

Finally, if $a' = b' = 0$ but $c' \neq 0$, then (5.5) becomes $A_\rho = 0$ with the general solution

$$A = f(p). \tag{5.9}$$

f is an arbitrary function $R \to R$ in equations (5.6)–(5.9).

(a_2) Let the constant vector $v' = (a', b', 2c')$ generate V, and suppose the vector v is not constant. Then, with some function $B: R^2 \to R$, the equations

$$pA_p + \rho A_\rho - A = a'B, \qquad A_p = b'B, \qquad \rho A_\rho = c'B$$

will be satisfied. Therefore it follows that

$$A = (b'p + c' - a')B$$

and, in addition,

$$(b'p + c' - a')B_p = 0, \qquad (b'p + c' - a')\rho B_\rho = c'B. \tag{5.10}$$

Now it is seen that $b' = 0$ is necessary because alternatively it would follow from (5.10) that $B = \text{const}$. Further, if $b' = 0$, then it is also necessary that $c' \neq a'$ and $c' \neq 0$. That is why the general solution of system (5.10) has the form $B = B_0\rho^m$, whence

$$A = A_0\rho^m \qquad (m \neq 0). \tag{5.11}$$

With this it can be assumed that $a' = m - 1$ and $c' = m$.

(a_3) If the vector $v = \text{const}$, then it is obvious that either

$$A = \gamma p \qquad (\gamma = \text{const} \neq 0) \tag{5.12}$$

or

$$A = 1,$$

correct up to an equivalence transformation.

In every subcase, additional operators of the form (5.4) are obtained as a result of substitution into (5.2) of one of the determined functions A and concrete values of the constants q, r, s are chosen.

***Case* (b).** Here

$$A = \frac{n+2}{n} p \tag{5.13}$$

and the equality $\partial_0 \lambda = 0$ does not follow from (4.5). But it is easy to see that the quantity $\partial_0 \lambda$ must be constant. Indeed, the only requirements for equations (3.2) to be identically satisfied are that equations

$$\partial_k(\partial_0^2 \xi^i) = \partial_0(\partial_0 \partial_k \xi^i)$$

hold. In view of (3.2) and (2.3), these lead to $\partial_0^2 \lambda = 0$, from which the constancy of $\partial_0 \lambda$ follows.

Equation (4.2) is satisfied by the function (5.13) if and only if $s = 0$. Consequently, the general solution of equations (4.2)–(4.5) has the form

$$\lambda = C_2 + \lambda_0 t, \qquad \xi^0 = C_1 + (C_2 + q_0)t + \lambda_0 t^2,$$
$$r = r_0 - \lambda_0(n+2)t, \qquad s = 0.$$

This solution extends the Lie algebra L_0 to the Lie algebra L^3 at the expense of adding three new operators corresponding to three constants q_0, r_0, λ_0. The constants q_0, r_0 give the operators $\zeta'_1 \cdot \partial$ and $\zeta'_2 \cdot \partial$ already known from (5.3). An essentially new operator, corresponding to the constant λ_0, is from the system (2.3)–(2.5):

$$\zeta'_4 = t^2 \partial_t + tx^i \partial_i + (x^i - tu^i)\partial_{u^i} - (n+2)tp\partial_p - nt\rho\partial_\rho. \tag{5.14}$$

6. Summary of Results

The calculation of the operators in every case considered leads to the following classification result, using equation (5.2).

When there is an arbitrary function A, the system of equations (1.1) admits the full Lie algebra L_0 with dimensionality $2 + \frac{1}{2}n(n+3)$ and with the operator basis (3.4). L_0 is an ideal (see Section 7.4) in any of the admitted Lie algebras $LE(A)$. The representation $LE(A) = L_0 \oplus L^r$ holds, where such expansions are only possible for $r = 1, 2, 3$,

All cases for the expansion of the Lie algebra L_0 are enumerated in Table 6.1. Here the first column gives the indices of the cases classified. In the second column specializations of the functions A are found; in the third, the dimensionality r of the extending Lie algebra L^r; in the fourth, the basis

TABLE 6.1

	A	r	Operators	Notes
	$\sim$	0		
(a$_1$)	$pf(p^{1+2m}\rho^{-1})$	1	$m\zeta_1' + \zeta_2'$	
	$f(\rho e^{-2p})$	1	$\zeta_1' + \zeta_3'$	$f \sim$
	$f(\rho)$	1	ζ_3'	
	$f(p)$	1	ζ_1'	
(a$_2$)	$A_0 \rho^m$	2	$\zeta_3', (m-1)\zeta_1' - 2m\zeta_2'$	$A_0 = \text{const}, \quad m \neq 0$
(a$_3$)	γp	2	ζ_1', ζ_2'	$\gamma = \text{const}$
	1	2	ζ_1', ζ_3'	
(b)	$\frac{n+2}{n} p$	3	$\zeta_1', \zeta_2', \zeta_4'$	$n = 1, 2, \ldots$

of L^r, where the notation of (5.3) and (5.14) is used; and in the fifth, some additional information are found.

Case (b), wherein the maximal expansion of the Lie algebra L^3 is achieved, is especially interesting. The resulting operator (5.14) is nontrivial in the sense that it cannot be obtained, as can some of the other operators, on the basis of so-called "physical considerations."

REFERENCES

Many papers connected with the calculation of full groups and the group classification of differential equations for concrete systems have been published. For ordinary differential equations the first results were obtained by S. Lie.† Classification of second order equations was carried out by Dickson [1]. In Ref. [2] the methods of group analysis are used in the control theory.

The theorem that a linear second-order equation, with two independent variables, admits not more than a three-parameter group of nontrivial transformations belongs to S. Lie [3]. The idea of classifying hyperbolic equations by means of Laplace invariants appeared when the memoir [4] was published; in the present form it first appeared in Ref. [5], where its application to S. A. Chapligin gas dynamics equations is also given. The paper [6] is also related to this topic.

The group analysis of the equations for the two-dimensional boundary layer first appeared in Ref. [7]. The exhaustive analysis of these equations appearing in Section 10 is presented here for the first time. Investigations [8] and [9], devoted to the three-dimensional boundary layer, are interesting.

† See Ref. [2] in Chapter II.

The equations of gas dynamics were studied in Refs. [10] and [11].† Reference [2], in which applications to the exterior Cartan forms are studied is of some interest.

There are also a number of publications [13–23] (the claim is not made that the list is complete) on the group analysis of other concrete equations in mechanics and physics.

[1] Dickson L. E. Differential equations from the group standpoint. — Ann. Math. 25 (1924), 287—378.

[2] Павловский Ю. Н. Групповые свойства управляемых динамических систем и фазовые организационные структуры. — Ж. вычисл. мат. и мат. физ. 14, № 4 (1974), 862—872 и № 5 (1974), 1093—1103.

[3] Lie S. Über die Integration durch bestimmte Integrale von einer Klasse linearer partiellen Differentialgleichungen. — Arch. Math. 6, Heft 3 (1881), 328—368.

[4] Darboux G. Leçons sur la thèorie générale des surfaces 3 (1889).

[5] Овсянников Л. В. Групповые свойства уравнений С. А. Чаплыгина. — Ж. прикл. мех. и техн. физ. № 3 (1960), 126—145.

[6] Hainzi J. Zur Gruppeninvarianz lineare partieller Differentialgleichungen mit konstanten Koeffizienten. — Arch. Rat. Mech. Anal. 59, № 2 (1975), 149—158.

[7] Павловский Ю. Н. Исследование некоторых инвариантных решений уравнений пограничного слоя. — Ж. вычисл. мат. и мат. физ. 1, № 2 (1961), 280—294.

[8] Ланкерович М. Я. Групповые свойства уравнений трехмерного пограничного слоя на произвольной поверхности. — В сб. «Динамика сплошной среды», вып. 7. Новосибирск (1971), 12—24.

[9] Каплан В. С. Условия существования инвариантных решений уравнений трехмерного ламинарного пограничного слоя на развертывающихся поверхностях. — Уч. зап. ЦАГИ 3, № 3 (1972), 36—44.

[10] Гладышев М. Т. Групповая классификация дифференциальных уравнений, описывающих одномерное неустановившееся движение жидкости. — Дифференц. уравнения 2, № 5 (1966), 695—700.

[11] Суровихин К. П. Групповая классификация уравнений, описывающих одномерные нестационарные течения газа. — ДАН СССР 156, № 3 (1964), 533—536.

[12] Суровихин К. П. О групповой классификации методом Картана уравнений одномерного течения газа. — ДАН СССР 172, № 1 (1966), 55—58.

[13] Ибрагимов Н. Х. Групповые свойства некоторых дифференциальных уравнений. Новосибирск, «Наука», 1967.

[14] Ибрагимов Н. Х. Групповые свойства волновых уравнений для частиц с нулевой массой. — ДАН СССР 178, № 3 (1968), 566—568.

[15] Ибрагимов Н. Х. Об инвариантности уравнений Дирака. — ДАН СССР, 185, № 6 (1969), 1226—1228.

[16] Бучнев А. А. Группа Ли, допускаемая уравнениями движения идеальной несжимаемой жидкости. — В сб. «Динамика сплошной среды», вып. 7. Новосибирск (1971), 212—214.

[17] Ленский Э. В. О групповых свойствах уравнений движения нелинейной вязко-пластической среды. — Вестник МГУ, мат., мех., № 5 (1966), 116—125.

† See also Ref. [4] in Chapter II. The complete classification was first published in Ref. [5] in Chapter II.

[18] Л е о н о в а М. А. Групповая классификация и инвариантные решения уравнений течения и теплообмена вязко-пластической среды. — Ж. прикл. мех. и техн. физ. № 4 (1966), 3—18.
[19] М а м о н т о в Е. В. К теории нестационарных околозвуковых течений. — ДАН СССР **185**, № 3 (1969), 538—540.
[20] Н г у е н В а н Д ь е п. Об уравнениях пограничного слоя жидкости с моментными напряжениями. — Прикл. мат. и мех. **32**, № 4 (1968), 748—753.
[21] О в с я н н и к о в Л. В. Уравнения динамической конвекции моря. Новосибирск, Ин-т гидродинамики СО АН СССР, 1967.
[22] Ч и р к у н о в Ю. А. Групповое свойство уравнений Ламе. — В сб. «Динамика сплошной среды», вып. 14. Новосибирск (1973), 138—140.
[23] M ü l l e r E. A., M a t s c h a t K. Über das Auffinden von Ähnlichkeitslösungen partieller Differentialgleichungssysteme unter Benutzung von Transformationsgruppen, mit Anwendung auf Probleme der Strömungsphysik. — Miszellaneen angew. Mech. Berlin (1962), 190—205.

CHAPTER

IV

The Lie Theory

Further development of group analysis methods for differential equations is based on the complete theory of local Lie groups and Lie algebras. This chapter is devoted to the exposition of that theory. It starts by emphasizing a series of fundamental ideas.

In the Lie theory an essential role is played by the so-called three main Lie theorems, which in totality lead to the determination of a unique correspondence between the (local) Lie groups and Lie algebras. These theorems are established here for arbitrary Banach groups and Lie algebras, for which the parametric space can also be infinite dimensional. An important fact is that in every class of isomorphic groups there is a canonical representative, that is, a group all of whose one-parameter subgroups are straight lines. This result is obtained in the third "inverse" Lie theorem using a differential equation whose solution is constructed from composite operators in the form of a known SCH-series. The domain of convergence of the SCH-series, for an arbitrary Banach space, is evaluated. Together with this one more fundamental result of the Lie theory about the analyticity of the multiplication law is obtained.

The main facts of the Lie algebra theory are expounded recapitulatively, and some important theorems (for example, the Levi theorem about separation of a radical) are presented without proof. Applications of group analysis methods to the study of operations on the set of solutions, admitted by the equation's group, are connected with the classification of the subalgebras of the given Lie algebra (up to similarity). Unfortunately, this algebra problem has not yet been solved. Thus the concept introduced here of an algorithm for an optimal system of subalgebras (or subgroups) remains to be completely established. The Lie algebras of operators, admitted by differential equations, often have nontrivial kernels, and sometimes they are also difficult to solve. Optimal systems of subalgebras are needed which can be constructed by "rule-of-thumb" methods in every concrete case, perhaps leaning on the Levi and Maltsev–Harish–Chandra theorems, as well as on other general facts of the theory of Lie algebras.

The determination of the correspondence of Lie groups and algebras, subgroups and subalgebras, and also the groups of internal automorphisms is accomplished in the more or less traditional way. Use of the SCH-series, for these purposes, is very helpful.

The Lie groups of transformations are treated here too as representations of Lie groups of local transformations on an arbitrary Banach space. Once more three main Lie theorems are demonstrated, but they are concerned particularly with groups of transformations. To obtain constructively the Lie group of the finite-dimensional space transformations from its Lie operator algebra, it is recommended to construct the canonical group of the second kind, which is useful in some theoretical problems.

Elements of the similarity theory of transformation groups are given at the end of the chapter. The new idea (compared to abstract Lie groups) of transitivity (intransitivity) of a transformation group is introduced, and infinitesimal criteria for the properties of similarity and transitivity are discussed.

The results of this chapter make it possible to describe adequately the contents of the first three chapters. But, together with the Lie groups theory, the group analysis of differential equations must deal with Lie "infinite groups," whose theory cannot be considered complete.

§12. THE LOCAL LIE GROUP

1. Definitions

Let B be a Banach space. Zero neighborhoods of the space B will be denoted with the symbol O. In particular, O_ρ is an open sphere of radius ρ whose center is at the zero element.

The pair (O, ϕ), consisting of the zero neighborhood $O \subset B$ and the mapping $\phi: O \times O \to B$, is called a *group kernel* if the following axioms hold:

1°. $\phi(a, 0) = \phi(0, a) = a$ for every $a \in O$.

2°. $\phi(\phi(a, b), c) = \phi(a, \phi(b, c))$ for every such $a, b, c \in O$ for which $\phi(a, b) \in O$ and $\phi(b, c) \in O$.

3°. $\phi \in C_\infty(O \times O)$.

Two group kernels (O, ϕ) and (O', ϕ') are regarded as indistinguishable if $\phi = \phi'$ on the intersection $O \cap O'$. The following definition is justified by the observation that the property of indistinguishability is a set-theoretical indication of equivalence.

A nonempty class of indistinguishable group kernels (O, ϕ) is called a *local* (*Banach*) *Lie group*.

A local (Banach) Lie group is completely characterized by defining the mapping ϕ, which satisfies the axioms 1°–3° for any neighborhood of zero $O \subset B$, and that is why it is denoted (B, ϕ). In what follows, when there is no possibility of misunderstanding, the local (Banach) Lie group (B, ϕ) will merely be called a *group* (B, ϕ).

The mapping ϕ is called a *multiplication law* (or simply *multiplication*) in the group (B, ϕ), and the value $\phi(a, b)$ is the *product* of a by b, which for short, will also be denoted by $a \cdot b$. In particular, axiom 2° can by written as $(a \cdot b) \cdot c = a \cdot (b \cdot c)$, which is an expression of the associativity of multiplication in the group (B, ϕ).

The concept "product is defined" is introduced inductively: the product $a \cdot b$ is defined if $a, b \in O$; the product $a \cdot b \cdot c$ is defined if the products $a \cdot b$ and $b \cdot c$ are defined and belong to O, and so on.

2. Properties of Multiplication

Two important properties of the multiplication law ϕ, which justify the utilization of the term "group," follow from axioms 1°–3° of Section 12.1. Let the group kernel (O, ϕ) be given.

(α) There exists a number $\rho > 0$ and a mapping $i: O_\rho \to O$ of the class $C_\infty(O_\rho)$, for which

$$\phi(i(a), a) = \phi(a, i(a)) = 0$$

is satisfied for every $a \in O_\rho$.

Proof. Let ∂_1 and ∂_2 denote differentiation of the mapping ϕ with respect to the first and second arguments, respectively. It is obvious that

$\partial_1\phi(0, 0) = \partial_2\phi(0, 0) = I_B$ follows from 1° and 3° in Section 12.1. Consequently, the mapping ϕ satisfies all the required conditions of the well-known implicit function theorem of analysis. Therefore there exist $\rho > 0$ and mappings $i_1, i_2: O_\rho \to O$ of the class $C_\infty(O_\rho)$, such that for $a \in O_\rho$ the equalities $\phi(i_1(a), a) = 0$, $\phi(a, i_2(a)) = 0$ hold, or briefly, $i_1(a) \cdot a = 0$, $a \cdot i_2(a) = 0$. But because of 2° the relations $i_1(a) = i_1(a) \cdot a \cdot i_2(a) = i_2(a)$ are correct and it is possible to assume $i = i_1 = i_2$. ■

In view of this property for the multiplication ϕ, the element $i(a)$ is the inverse for the element a and is therefore denoted by the usual symbol a^{-1}. With this, the role of *unit element* is played by the zero element of the space B. Thus i is an *inverse mapping*, operating by the formula $a^{-1} = i(a)$.

(β) *For any positive integers n a real numbers $\rho(n) > 0$ can be found such that for all $a_1, \ldots, a_n \in O_{\rho(n)}$ all products $b_1 \cdot b_2 \cdot b_3 \cdots b_n$ have the property that every b_k is either a_k or a_k^{-1}.*

Proof. For $n = 2$ the correctness of the statement follows from property (α), and for $n > 2$ it is established by induction. ■

(γ) *Let the number ρ be chosen in accordance with property* (α). *Then the mapping $\psi: O_\rho \times O_\rho \to B$, operating by the formula $\psi(a, b) = a \cdot b^{-1}$, is continuous and, moreover, it is of class $C_\infty(O_\rho \times O_\rho)$.*

Proof. As $\psi(a, b) = \phi(a, i(b))$, the correctness of the statement follows from axiom 3° in Section 12.1, property (α), and known theorems of analysis about continuation of composition differentiability. ■

Properties (α)–(γ) show that the only difference between the group kernel (O, ϕ) and a group, in its precise definition, is that in (O, ϕ) the operations of multiplication and inversion of the elements from O are defined and continuous not for all of the elements from O, but only for those elements from a sufficiently small neighborhood O_ρ of the unit element. Thus the group kernel [or the group (B, ϕ)] is a topological group kernel, or in the present sense, a local topological group.

If the space B is finite dimensional, the group (B, ϕ) is called a *local Lie group*.

Example. Let l be a linear continuous functional on the space B, that is, the linear mapping $B \to R$, operating by the formula $a \to l\langle a \rangle$. Then the mapping ϕ, given by

$$\phi(a, b) = a + b + l\langle a \rangle b,$$

defines the group (B, ϕ). Satisfaction of the axioms 1°–3° in Section 12.1 is easily verified. Here the operation of multiplication is defined and continuous on the entire space B, but for the operation of inversion this is not so. Indeed, a calculation of the inverse mapping gives the expression

$$a^{-1} = -(1 + l\langle a \rangle)^{-1} a,$$

and the guaranteed properties of the definition and smoothness of the mapping $a \to a^{-1}$ are correct only on the neighborhood of zero

$$O = \{a \mid |l\langle a \rangle| < 1\}.$$

This example is the simplest among all possible examples of non-Abelian groups (B, ϕ) for the case of an arbitrary Banach space B. In what follows, some elements of the theory of the groups (B, ϕ) will be illustrated using this example.

3. Local Isomorphism

Let B and B' be Banach spaces and O a neighborhood of zero of the space B. The union $\Gamma(B, B')$ of the mapping $\gamma: B \to B'$, of the class $C_\infty(O)$, for which $\gamma(0) = 0$, is considered. The mapping $\gamma \in \Gamma(B, B')$ will be called *locally invertible* if there exist neighborhoods of zero $O \subset B$ and $O' \subset B'$ with $O' \subset \gamma(O)$, $O \to \gamma(O)$ is a one-to-one correspondence, and $\gamma^{-1}: O' \to B$ is a mapping of the class $C_\infty(O')$. It is known from analysis that the mapping $\gamma \in \Gamma(B, B')$ is locally invertible if the derivative $\partial\gamma(0)$ is a linear isomorphism of the spaces B and B', that is, the linear mapping $\partial\gamma(0): B \to B'$ is a one-to-one mapping of B onto B' and the inverse mapping $(\partial\gamma(0))^{-1}: B' \to B$ is continuous.

Let the groups (B, ϕ) and (B', ϕ') be given. The mapping $\gamma \in \Gamma(B, B')$ is called a *local homomorphism of the group* (B, ϕ) *into* (or *onto*) *the group* (B', ϕ') if $\gamma \circ \phi = \phi' \circ \gamma$, or more accurately, if there exists a number $\rho > 0$ such that for all $a, b \in O_\rho \subset B$

$$\gamma(\phi(a, b)) = \phi'(\gamma(a), \gamma(b)). \tag{3.1}$$

Equation (3.1) means that a homomorphism of the group is a mapping which commutes with multiplication.

A local homomorphism γ of the group (B, ϕ) onto the group (B', ϕ') is called a *local isomorphism of the group* (B, ϕ) *onto the group* (B', ϕ') if the mapping γ is locally invertible. In this case it is also said that the *group* (B', ϕ') *is locally isomorphic to the group* (B, ϕ).

Note. If the group (B, ϕ) and the invertible mapping $\gamma \in \Gamma(B, B')$ are given, then it is possible to determine the new mapping ϕ', operating by the formula

$$\phi'(a, b) = \gamma^{-1}(\phi(\gamma(a), \gamma(b))).$$

It is easy to check that ϕ' satisfies axioms 1°–3° in Section 12.1; that is, it is a new multiplication law. Comparison of this formula with (3.1) shows that the resulting group (B, ϕ') is locally isomorphic to the group (B, ϕ).

A local isomorphism of the group (B, ϕ) onto (B, ϕ) is called a *local automorphism* of this group. Obviously, the set of all local automorphisms, denoted by $\mathrm{Aut}(B, \phi)$, generates some group of local transformations of the space B which retains the null element of that space.

To shorten the statements, in what follows, the adjective "local" is dropped if it will not create misunderstanding. Then, the terminology homomorphism, isomorphism, or automorphism of the groups (B, ϕ) is employed.

It is obvious that every group (B, ϕ) is isomorphic to itself and that if the group (B', ϕ') is isomorphic to the group (B, ϕ), then the group (B, ϕ) is also isomorphic to the group (B', ϕ'). The last statement follows from the fact that the equality $\gamma \circ \phi = \phi' \circ \gamma$, in the case of isomorphism, is equipollent to $\gamma^{-1} \circ \phi' = \phi \circ \gamma^{-1}$. Further, if γ is an isomorphism of the group (B, ϕ) onto the group (B', ϕ') and γ' is an isomorphism of the group (B', ϕ') onto the group (B'', ϕ''), then $\gamma' \circ \gamma$ is an isomorphism of the group (B, ϕ) onto the group (B'', ϕ''). This follows from the relations

$$\gamma' \circ \gamma \circ \phi = \gamma' \circ \phi' \circ \gamma = \phi'' \circ \gamma' \circ \gamma.$$

So an isomorphic relation of groups (B, ϕ) is a set-theoretical criterion of equivalence, under which the collection of all groups (B, ϕ) is separated into nonintersecting classes of isomorphic groups. A class of groups isomorphic to the group (B, ϕ) is denoted by $Q(B, \phi)$.

The main purpose of Lie theory, presented below, is to provide an exhaustive characterization of the class $Q(B, \phi)$ in terms of the wider concept of normalized (Banach) Lie algebras and to separate every given class into the simplest canonical representative. This will be especially convenient when subgroups, homomorphisms, and automorphisms of the group (B, ϕ) are studied.

4. Equation and First Theorem of Lie

For the given group (B, ϕ) an *auxiliary* mapping $A: B \to \mathscr{L}(B)$, defined by

$$A(a) = \partial_2 \phi(a, 0), \tag{4.1}$$

is introduced. From axiom 1° in Section 12.1 it follows that $A(0) = I_B$ is the identity mapping of the space B. Consequently, for sufficiently small ρ and all $a \in O_\rho$, the linear mapping $A(a) \in \mathscr{L}(B)$ is a linear homeomorphism of the space B. Consequently, for $a \in O_\rho$ there is an inverse mapping $V(a) = (A(a))^{-1} \in \mathscr{L}(B)$, which also can be defined directly by

$$V(a) = \partial_2 \phi(a^{-1}, a). \tag{4.2}$$

Indeed, because of 1.2° in Section 12.1, the equality $\phi(a^{-1}, \phi(a, b)) = b$ results, whose differentiation (with respect to the argument b at the point $b = 0$) gives

$$\partial_2 \phi(a^{-1}, \phi(a, 0)) \circ \partial_2 \phi(a, 0) = \partial_2 \phi(a^{-1}, a) \circ A(a) = I_B.$$

Differentiation of the associative relation 2° in Section 12.1 with respect to the argument c at the point $c = 0$ gives

$$\partial_2 \phi(\phi(a, b), 0) = \partial_2 \phi(a, b) \circ \partial_2 \phi(b, 0),$$

which, because of the definitions of the auxiliary mappings A and V, can be rewritten in the form

$$\partial_2 \phi(a, b) = A(\phi(a, b)) \circ V(b).$$

This equation, holding for all $a, b \in O_\rho$, expresses the fact that the partial mapping $\phi_a : O_\rho \to B$, defined by $\phi_a(b) = \phi(a, b)$, is a solution of the differential equation (with an initial condition)

$$\partial \phi(b) = A(\phi) \circ V(b), \qquad \phi(0) = a. \tag{4.3}$$

Equation (4.3) is called the *Lie equation.* The first *Lie theorem*, which follows, determines the connection between the group (B, ϕ) and the property of complete integrability of equation (4.3).

Theorem. *If A and V are auxiliary mappings of the group (B, ϕ), then the mapping $A(\phi) \circ V(b) \in \mathscr{L}(B)$ is completely integrable in the null neighborhood of the space B; that is, there exists $\rho > 0$ such that equation* (4.3) *has the solution ψ satisfying the condition $\psi(c) = a$, for any $c, a \in O_\rho$.*

On the contrary, let the mapping A: $O_\rho \to \mathscr{L}(B)$ of the class $C_\infty(O_\rho)$ be given, where $A(0) = I_B$ and for any $a \in O_\rho$ the mapping $A(a)$ is a linear homeomorphism of the space B onto B, and let $V(a) = (A(a))^{-1}$. If the mapping $A(\phi) \circ V(b)$ is completely integrable in the null neighborhood of the space B, then the solution $\phi = \phi_a$ of problem (4.3) *defines the group (B, ϕ) with the multiplication law $\phi(a, b) = \phi_a(b)$ and for this group A is its auxiliary mapping* (4.1).

Proof. (Direct statement) The mapping $\psi : O_{\rho(3)} \to B$, defined by $\psi(b) = \phi(a, \phi(c^{-1}, b))$ [the symbol $\rho(n)$ is defined by (β) in Section 12.2], with $a, c \in O_{\rho(3)}$ fixed, is considered. Then $\psi(c) = a$ and, because of (4.3) and associativity of the composition of linear mappings,

$$\begin{aligned}\partial \psi(b) &= \partial_2 \phi(a, \phi(c^{-1}, b)) \circ \partial_2 \phi(c^{-1}, b)\\ &= A(\psi) \circ V(\phi(c^{-1}, b)) \circ A(\phi(c^{-1}, b)) \circ V(b) = A(\psi) \circ V(b).\end{aligned}$$

Consequently, ψ is a solution of equation (4.3) satisfying the initial condition $\psi(c) = a$ for any $c, a \in O_{\rho(3)}$, which means that the inverse mapping $A(\phi) \circ V(b)$ is completely integrable.

(Converse statement) Let ϕ_a be a solution of the problem (4.3) and $\phi(a, b) = \phi_a(b)$. Then $\phi(a, 0) = a$ because of the initial condition in (4.3). In addition, when $a = 0$ the only solution of problem (4.3) is $\phi_0(b) = b$, so that $\phi(0, b) = b$. From this, satisfaction of axiom 1° in Section 12.1 is established. Further, let $x = \phi(a, b)$, $y = \phi(b, c)$, $u = \phi(x, c)$, $v = \phi(a, y)$. From the construction of the mappings $c \to u(c) = \phi_x(c)$, $c \to v(c) = \phi_a \circ \phi_b(c)$, the following equations hold:

$$\partial u(c) = \partial_2 \phi(x, c) = A(u) \circ V(c), \qquad u(0) = x,$$
$$\partial v(c) = \partial_2 \phi(a, y) \circ \partial_2 \phi(b, c) = A(v) \circ V(y) \circ A(y) \circ V(c) = A(v) \circ V(c),$$
$$v(0) = \phi(a, b) = x.$$

Consequently, both mappings u and v are solutions of the same problem (4.3) (with initial value x instead of a), and from its uniqueness both solutions coincide. But the equality $u(c) = v(c)$ is equipollent to $\phi(\phi(a, b), c) = \phi(a, \phi(b, c))$, so axiom 2° in Section 12.1 is established. Finally, axiom 3° in Section 12.1 is also satisfied because, from the assumptions of the theorem, it follows that the right-hand side of equation (4.3) is infinitely differentiable and therefore the solution of problem (4.3) is also infinitely differentiable in the null neighborhood of the space B. This fact is known from analysis. Correctness of the last statement of the theorem follows directly from equation (4.3) when $b = 0$, since $V(0) = I_B$. ■

Consequence. *The group (B, ϕ) is uniquely defined by the assignment of an auxiliary mapping A satisfying the conditions given in the theorem.*

Note. The results presented here and below, for the theory of Lie groups (B, ϕ), are based on application of the auxiliary mapping $A(a) = \partial_2 \phi(a, 0)$, describe the so-called "right structure" of the group (B, ϕ). In an analogous fashion a theory of the "left structure" can be developed, based upon the auxiliary mapping $A^*(a) = \partial_1 \phi(0, a)$. These two theories lead to the same result.

Example. For the group (B, ϕ), given in the example of Section 12.2, the auxiliary mappings are

$$A(a) = (1 + l\langle a \rangle) I_B, \qquad V(a) = (1 + l\langle a \rangle)^{-1} I_B.$$

The Lie equation (4.3) has the form

$$\partial \phi(b) = (1 + l\langle \phi \rangle)(1 + l\langle b \rangle)^{-1} I_B.$$

5. Canonical Multiplication

Functions $g: \Delta \to B$, of the class $C_\infty(\Delta)$, defined on intervals of the real axis $\Delta = (-\alpha, \alpha)$ and operating by $t \to g(t)$, where $g(0) = 0$, are considered. The image $g(\Delta) \subset B$ is a smooth curve in B, in particular there is a vector $p = \partial g(0) \in B$ which is tangent to this curve at the zero of the space B.

The curve $g(\Delta) \subset B$ is called a *one-parameter subgroup of the group* (B, ϕ) if $\partial g(0) \neq 0$ and if for any $t, s, t + s \in \Delta$ the equality

$$\phi(g(t), g(s)) = g(t + s) \tag{5.1}$$

holds. The vector $p = \partial g(0)$ is called the *directing vector* of the one-parameter subgroup $g(\Delta)$.

The problem of determining one-parameter subgroups of the given group (B, ϕ) is resolved by finding the solutions g of the functional equation (5.1) when the mapping ϕ is given. It turns out that this equation is equipollent to some differential equation.

Theorem. *If A is an auxiliary mapping* (4.1) *of the group* (B, ϕ), *then any one-parameter subgroup $g(\Delta)$, with the directing vector p, satisfies the differential equation*

$$\partial g = A(g)\langle p\rangle, \qquad g(0) = 0, \tag{5.2}$$

with the given initial condition. Conversely, no matter what the nonzero vector $p \in B$ is, the solution $t \to g(t)$ of problem (5.2) *gives a one-parameter subgroup of the group* (B, ϕ) *with the directing vector p.*

Proof. Differentiation of equation (5.1) with respect to s, at the point $s = 0$, gives (5.2) because of definition (4.1). Conversely, if g is a solution of problem (5.2), then because of the Lie equation (4.3) and equations (5.2) the functions $s \to u(s) = \phi(g(t), g(s))$ and $s \to v(s) = g(t + s)$ satisfy the equations

$$\begin{aligned} \partial u(s) &= \partial_2 \phi(g(t), g(s))\langle \partial g(s)\rangle = A(u) \circ V(g(s)) \circ A(g(s))\langle p\rangle \\ &= A(u)\langle p\rangle, \qquad u(0) = g(t), \\ \partial v(s) &= A(v)\langle p\rangle, \qquad v(0) = g(t). \end{aligned}$$

Consequently, for every fixed t the functions u and v are solutions of the same differential equation (5.2) with identical initial values $g(t)$. From the theorem about uniqueness of the solution to the Cauchy problem for ordinary differential equations, these equations must coincide, which gives equation (5.1). Finally, as $A(0) = I_B$, then from (5.2) with $t = 0$ the equation $\partial g(0) = p$ follows. ■

The simplest structure is in those one-parameter subgroups for which $g(\Delta)$ is a straight line in B, that is, when $g(t) = tp$. In the general case there are no one-parameter subgroups for the group (B, ϕ) which are straight lines of the kind tp. They are all essentially "curves." The opposite case is distinguished by the following definition.

The group (B, ω) is called *canonical* if a straight line $g(t) = tp$ is its one-parameter subgroup. The multiplication law ω in the canonical group is called *canonical multiplication.*

The property for multiplication to be canonical is expressed by the relation

$$\omega(tp, sp) = (t + s)p. \tag{5.3}$$

Criterion of Canonicality. *The group (B, ω) is canonical if and only if its auxiliary mappings A and V satisfy the relations*

$$A(a)\langle a\rangle = a, \qquad V(a)\langle a\rangle = a \tag{5.4}$$

for all $a \in O_\rho$.

Proof. If the group (B, ω) is canonical, then from (5.3) it follows that $\partial_s \omega(ta, sa)|_{s=0} = a$. On the other hand, from the Lie equation (4.3),

$$\partial_s \omega(ta, sa)|_{s=0} = \partial_2 \omega(ta, 0)\langle a\rangle = A(ta)\langle a\rangle.$$

Comparison of these two expressions when $t = 1$ gives the relation (5.4) for A and by a similar argument the expression for V in (5.4). Conversely, let (5.4) be satisfied. Then for any $p \in B$, the mapping $(t, p) \to tp$ is such that $(tp)|_{t=0} = 0$ and

$$\partial_t(tp) = p = A(tp)\langle p\rangle.$$

Therefore, the mapping $t \to tp$ is a solution of problem (5.2). Because of Theorem 12.5, this mapping gives a one-parameter subgroup. Consequently, the conditions for a group to be canonical are satisfied. ■

6. Canonical Isomorphism

One of the fundamental results of Lie theory is that in every class of isomorphic groups $Q(B, \phi)$ there is a canonical representative (B, ω).

Theorem. *For any group (B, ϕ) there is an isomorphism χ onto some canonical group (B, ω) which satisfies the condition $\partial\chi(0) = I_B$.*

Proof. To construct the unknown isomorphism, the dependence of the solution of problem (5.2) on the vector $p \in B$, as a parameter, is employed.

This gives the mapping $(t, p) \to g(t, p)$ of a null neighborhood $U \subset R \times B \to B$. Here it will be shown that in this mapping the image really depends only on the product tp. Indeed, for sufficiently small $|\lambda - 1|$, where $\lambda \in R$, the transformation $t \to t/\lambda$, $p \to \lambda p$, $g \to g$ does not change the equations of problem (5.2). [Problem (5.2) admits this mapping!] Thus the mapping $(t, p) \to g(t/\lambda, \lambda p)$ is a solution, for any λ, and because of the solution's uniqueness it coincides with the mapping $(t, p) \to g(t, p)$. From this it follows that there exists $\rho > 0$ such that the image $g(t, p)$ is defined for all $p \in O_\rho$ and $t \in (-\alpha, \alpha)$, where $\alpha > 1$. Then in the relation $g(t, p) = g(t/\lambda, \lambda p)$ it is possible to assume $\lambda = t$, which gives a representation of the solution in the form $g(t, p) = g(1, tp)$. So the mapping $\chi\colon O_\rho \to B$ defined by

$$\chi(p) = g(1, p) \tag{6.1}$$

appears.

Now it is confirmed that χ possesses the properties in the definition in Section 12.3 necessary for an isomorphism, that is, $\chi \in \Gamma(B, B)$ and is locally invertible. First of all, if $p = 0$, then the unique solution of problem (5.2) is $g = 0$, and then from (6.1) it follows that $\chi(0) = 0$. Further, the right-hand side of equation (5.2) is a mapping of the class $C_\infty(O_\rho \times O_\rho)$, because of which the solution $(t, p) \to g(t, p)$, as is well known from the theory of differential equations, is a mapping of the class $C_\infty(\Delta \times O_\rho)$. Thus $\chi \in \Gamma(B, B)$. To establish the invertibility of χ the derivative $\partial_\chi(tp)\langle h\rangle = \partial_2 g(t, p)\langle h\rangle$, with the arbitrary vector $h \in B$, is considered. The solution of the "variation equations," namely, the equations

$$\partial_t(\partial_2 g(t, p)\langle h\rangle) = \partial A(g)\langle \partial_2 g(t, p)\langle h\rangle, p\rangle + A(g)\langle h\rangle,$$
$$\partial_2 g(0, p)\langle h\rangle = 0,$$

are sought for problem (5.2).

At the point $p = 0$ these equations will be

$$\partial_t(\partial_2 g(t, 0)) = I_B, \qquad \partial_2 g(0, 0) = 0,$$

which have the unique solution $\partial_2 g(t, 0) = tI_B$. Thus equation (6.1) gives $\partial_\chi(0) = I_B$. As was noted at the beginning of Section 3, invertibility of the mapping χ follows from these results.

From the note in Section 12.3 the group (B, ω), with multiplication law defined by

$$\omega(p, q) = \chi^{-1}(\phi(\chi(p), \chi(q))), \tag{6.2}$$

is isomorphic to the initial group (B, ϕ). It must be shown that (B, ω) is a canonical group. Actually, according to the construction for $p \in O_\rho$ as the image of the interval Δ, the mapping $\chi(tp) = g(t, p)$ is a one-parameter subgroup in the group (B, ϕ), with directing vector p. Consequently, because of the definition in Section 12.5, the equalities

$$\begin{aligned}\omega(tp, sp) &= \chi^{-1}(\phi(g(t, p), g(s, p))) = \chi^{-1}(g(t + s, p)) \\ &= \chi^{-1}(\chi((t + s)p)) = (t + s)p\end{aligned}$$

hold, that is, the multiplication ω is canonical. ■

The isomorphism χ of the group (B, ϕ) onto the canonical group (B, ω), satisfying the condition $\partial\chi(0) = I_B$, is called a *canonical isomorphism of the group* (B, ϕ).

Consequence. *For any group* (B, ϕ) *there exists a neighborhood* $O_\rho \subset B$ *such that for every* $a \in O_\rho$ *there is one and only one one-parameter subgroup included in* O_ρ *which contains the element* a.

Proof. For a canonical group this statement is obvious. For any group (B, ϕ) it follows from the established theorem and from the fact that under an isomorphism one-parameter subgroups transform into one-parameter subgroups. ■

This fact means that the elements of all possible one-parameter subgroups of the group (B, ϕ) fill up some null neighborhood of the space B. Sometimes this is expressed by saying that the group (B, ϕ) is "interwoven" with its one-parameter groups.

Example. The group given in the example in Section 12.2 is not canonical. To construct its canonical isomorphism it is necessary to solve the corresponding equation (5.2), which, according to the example in Section 12.4, has the form

$$\partial_t g = (1 + l\langle g\rangle)p, \qquad g(0) = 0.$$

The solution of this problem is

$$g(t, p) = (l\langle p\rangle)^{-1}(\exp(l\langle tp\rangle) - 1)p.$$

The unknown canonical isomorphism is found from (6.1) to be

$$\chi(p) = (l\langle p\rangle)^{-1}(\exp(l\langle p\rangle) - 1)p.$$

To determine the multiplication law in the corresponding canonical group it is also necessary to find the inverse mapping χ^{-1}, that is, to solve the equation $\chi(p) = a$ relative to the vector $p \in B$. This solution is

$$p = \chi^{-1}(a) = (l\langle a\rangle)^{-1}(\ln(1 + l\langle a\rangle))a.$$

Now the canonical multiplication law is obtained from (6.1), and after some work it is found to be

$$\omega(p, q) = \frac{\exp(l\langle p\rangle) - 1}{\exp(l\langle p + q\rangle) - 1} \cdot \frac{l\langle p + q\rangle}{l\langle p\rangle l\langle q\rangle}(pl\langle q\rangle - ql\langle p\rangle) + \frac{l\langle p + q\rangle}{l\langle q\rangle} q.$$

It is easy to verify the property of canonicality for this multiplication ω.

7. Homomorphisms of Canonical Groups

The homomorphisms of groups (B, ϕ), defined in Section 3, are local mappings. The essential property of canonical groups is that for them any homomorphism can be continued until a global mapping of the corresponding Banach spaces results.

Theorem. *If both groups (B, ω) and (B', ω') are canonical, then any local homomorphism γ of the group (B, ω) into the group (B', ω') can be continued until the linear mapping $\gamma\colon B \to B'$ results.*

Proof. Because of equation (3.1) and the canonicality of the groups, for any $a \in B$ with sufficiently small norm

$$\gamma(\omega(a, a)) = \gamma(2a) = \omega'(\gamma(a), \gamma(a)) = 2\gamma(a),$$

that is, $\gamma(2a) = 2\gamma(a)$. Analogously, it is established that for any integer n the equality $\gamma(na) = n\gamma(a)$ holds (induction with respect to n). Further, if $mb = a$, then $m\gamma(b) = \gamma(a)$ or $\gamma((1/m)a) = (1/m)\gamma(a)$. Therefore, for any integers m, n

$$\gamma\left(\frac{n}{m} a\right) = \frac{n}{m} \gamma(a).$$

So the equality $\gamma(\lambda a) = \lambda\gamma(a)$ is established for $a \in O_\rho$ and any rational number λ, $|\lambda| < 1$. As a result of continuity of the mapping γ on O_ρ this result holds for any $\lambda \in (-1, 1)$. As γ is continuously differential on O_ρ, and $\gamma(0) = 0$,

$$\partial\gamma(0)\langle a\rangle = \lim_{t\to 0} t^{-1}(\gamma(ta) - \gamma(0)) = \lim_{t\to 0} t^{-1}(t\gamma(a)) = \gamma(a).$$

So on O_ρ the homomorphism γ coincides with the linear mapping $\partial\gamma(0)\colon B \to B'$, which is limited to some null neighborhood of the space B. Thus $\gamma(a) = \partial\gamma(0)\langle a\rangle$ holds for all $a \in B$. ■

Important consequences follow from this theorem. First of all, it gives an essential addition to the theorem in Section 12.6.

Consequence 1. *There is only one canonical isomorphism χ of the group (B, ϕ).*

Proof. Let $\chi: B \to B$ be a canonical isomorphism of the group (B, ϕ) onto the canonical group (B, ω), constructed as in the theorem in Section 12.6. If $\chi': B \to B$ is another canonical isomorphism of the group (B, ϕ) onto the canonical group (B, ω'), then $\gamma = \chi' \circ \chi^{-1}$ is an isomorphism of the group (B, ω) onto the group (B, ω'). Because of the theorem just established, γ is a linear mapping of B to B. Thus from the equation $\chi' = \gamma \circ \chi$ the relation $\partial\chi'(0) = \gamma \circ \partial\chi(0)$ follows, from which $\gamma = I_B$ and $\chi' = \chi$ are obtained. ■

In addition, it is possible to give a description of all canonical representatives for a given class of isomorphic groups.

Consequence 2. *In every class of isomorphic groups $Q(B, \phi)$ there is one and only one canonical representative, up to a linear isomorphism.*

Proof. Any two canonical representatives are related by an isomorphism, which, because of the theorem just established, is linear. ■

In addition to this consequence it is useful to note that *any linear isomorphism of a canonical group always maps it onto a canonical group.*

8. Second Theorem of Lie

This theorem is actually merely a specialization of the well-known Frobenius theorem from analysis concerning complete integrability of a mapping $f: X \times Y \to \mathscr{L}(X; Y)$, that is, concerning the condition of existence of a solution of the problem $\partial u(x) = f(x, y)$, $u(x_0) = y_0$, for any initial data (x_0, y_0), so as to recover the mapping $u: X \to Y$. This condition states that for any point $(x, y) \in X \times Y$ the bilinear mapping

$$(h, k) \to \partial_1 f(x, y)\langle h, k\rangle + \partial_2 f(x, y)\langle f(x, y)\langle h\rangle, k\rangle \tag{8.1}$$

is symmetric. Here $h, k \in X$ are arbitrary vectors.

Theorem. *If A and V are auxiliary functions of the group (B, ϕ) and $O_\rho \subset B$, then the linear mapping $A(\phi) \circ V(b): B \to B$ is completely integrable on $O_\rho \times O_\rho$ if and only if there is a bilinear mapping $C \in \mathscr{L}^2(B; B)$ such that, for any $a \in O_\rho$ and $h, k \in B$, the relation*

$$\partial V(a)\langle h, k\rangle - \partial V(a)\langle k, h\rangle = C\langle V(a)\langle k\rangle, V(a)\langle h\rangle\rangle \tag{8.2}$$

or an equipollent relation

$$\partial A(a)\langle A(a)\langle k\rangle, h\rangle - \partial A(a)\langle A(a)\langle h\rangle, k\rangle = A(a)\langle C\langle k, h\rangle\rangle \tag{8.3}$$

is satisfied.

Proof. Because of the Frobenius theorem, the problem is reduced to the transformation of condition (8.1) by applying the mapping $(b, \phi) \to f(b, \phi)$, operating by the definition $f(b, \phi) = A(\phi) \circ V(b)$. With this, it is useful to establish that

$$\partial_1 f\langle h, k\rangle = A\langle \partial V\langle h, k\rangle\rangle, \qquad \partial_2 f\langle h, k\rangle = \partial A\langle h, V\langle k\rangle\rangle.$$

Thus the bilinear mapping (8.1) here is

$$(h, k) \to A(\phi)\langle \partial V(b)\langle h, k\rangle\rangle + \partial A(\phi)\langle A(\phi)\langle h\rangle, V(b)\langle k\rangle\rangle.$$

Differentiation of the identity $A(\phi) \circ V(\phi)\langle k\rangle = k$ gives the relation

$$\partial A(\phi)\langle h, k\rangle = -A(\phi)\langle \partial V(\phi)\langle h, A(\phi)\langle k\rangle\rangle\rangle,$$

because of which the symmetry of the first bilinear mapping is equipollent to the symmetry of the following:

$$(h, k) \to \partial V(b)\langle h, k\rangle - \partial V(\phi)\langle A(\phi) \circ V(b)\langle h\rangle, A(\phi) \circ V(b)\langle k\rangle\rangle.$$

Since the vector $(A(b)\langle h\rangle, A(b)\langle k\rangle)$, together with the vector (h, k), is arbitrary, the last bilinear equation can be changed to

$$(h, k) \to \partial V(b)\langle A(b)\langle h\rangle, A(b)\langle k\rangle\rangle - \partial V(\phi)\langle A(\phi)\langle h\rangle, A(\phi)\langle k\rangle\rangle.$$

The symmetry of this bilinear mapping is equipollent to the relation

$$\begin{aligned}\partial V(b)\langle A(b)\langle h\rangle, A(b)\langle k\rangle\rangle - \partial V(b)\langle A(b)\langle k\rangle, A(b)\langle h\rangle\rangle \\ = \partial V(\phi)\langle A(\phi)\langle h\rangle, A(\phi)\langle k\rangle\rangle - \partial V(\phi)\langle A(\phi)\langle k\rangle, A(\phi)\langle h\rangle\rangle,\end{aligned}$$

which must be satisfied for any point $(b, \phi) \in O_\rho \times O_\rho$. In particular, here it can be assumed that $\phi = 0$. Then, on the right-hand side the bilinear mapping $C\colon B \times B \to B$ given by

$$C\langle k, h\rangle = \partial V(0)\langle h, k\rangle - \partial V(0)\langle k, h\rangle \tag{8.4}$$

appears, from which the previous equality, after the exchange of (h, k) for $(V(b)\langle h\rangle, V(b)\langle k\rangle)$ and b for a, will be exactly the form (8.2). Relation (8.3) is obtained from (8.2) with the aid of

$$\partial V(a)\langle h, k\rangle = -V(a)\langle \partial A(a)\langle h, V(a)\langle k\rangle\rangle\rangle,$$

calculated by differentiation of the identity $A(a) \circ V(a) = I_B$. ∎

Either equation (8.2) or (8.3) is called the *equation of Maurer–Cartan.*

9. Structural Operator

The bilinear mapping C, defined by equation (8.4), is called a *structural operator of the group* (B, ϕ).

Since $\partial V(0) = -\partial A(0)$, equation (8.4) is equipollent to

$$C\langle k, h\rangle = \partial A(0)\langle k, h\rangle - \partial A(0)\langle h, k\rangle. \tag{9.1}$$

According to the theorem in Section 12.8, for every group (B, ϕ) there is a unique structural operator C corresponding to it. It turns out that the converse statement is also correct, that is, for every given bilinear mapping $C: B \times B \to B$ satisfying certain conditions, it is possible to find a unique group (B, ϕ) for which C is the structural operator. This fact is contained in the third Lie theorem, which is divided into "direct" and "converse" theorems. The third theorem of Lie clarifies the necessary properties of the structural operator.

Theorem. *The structural operator C of the group (B, ϕ) possesses the following properties (here h, k, l are arbitrary elements from B):*

1° *C is a bilinear mapping $B \times B \to B$.*

2° *The mapping C is antisymmetric:*

$$C\langle h, k\rangle = -C\langle k, h\rangle.$$

3° *The mapping C satisfies the Jacobi relation*

$$C\langle C\langle h, k\rangle, l\rangle + C\langle C\langle k, l\rangle, h\rangle + C\langle C\langle l, h\rangle, k\rangle = 0.$$

4° *The mapping C is bounded, that is, there exists a number $\delta > 0$ such that*

$$|C\langle h, k\rangle| \leq \delta |h| \, |k|.$$

Proof. Properties 1° and 2° follow from the definition (8.4). Relation 3° is obtained by differentiation with respect to a at the point $a = 0$, which results, taking into consideration property 2°, in the equation

$$\partial^2 V\langle l, h, k\rangle - \partial^2 V\langle l, k, h\rangle = C\langle \partial V\langle l, k\rangle, h\rangle - C\langle \partial V\langle l, h\rangle, k\rangle.$$

Now it is easy to make the circular permutation $h \to k \to l \to h$ twice and then add all three equations. The property of symmetry of the second derivative, which for the linear mapping $\partial^2 V: B^3 \to B$ is written

$$\partial^2 V(l, h, k) = \partial^2 V(h, l, k),$$

indicates that the sum on the left-hand side will be equal to zero. On the right-hand side, after applying definition (8.4), the exact expression written in 3°

will be obtained. Finally, property 4° follows from the fact that the derivative $\partial V \in C_\infty(O_\rho)$, and therefore $\partial V(0)$ is a continuous bilinear mapping $B \times B \to B$. In 4° it can be assumed that $\delta = 2|\partial V(0)|$. ∎

Example. The structural operator of the group (B, ϕ) from the examples in Sections 12.2 and 12.4, calculated by (9.1), is

$$C\langle h, k\rangle = l\langle k\rangle h - l\langle h\rangle k.$$

10. Banach Algebra of Lie

The mapping $C: B \times B \to B$ with the properties 1°–3° of Section 12.9 endows the space B with a structural Lie algebra, the operator of which is defined by

$$[h, k] = C\langle h, k\rangle. \tag{10.1}$$

Although this fact will be strictly determined only in the following section, it is already possible to note here the complete agreement of properties 1°–3° in Section 12.9 of the mapping C with the properties, emphasized in Section 7.3, of the commutator of vector fields. Moreover, on the space B it is possible to consider a family of vector fields of $\{\xi_h\}$, where every vector field ξ_h is defined by assignment of the vector $h \in B$ and which operates by $\xi_h(a) = A(a)\langle h\rangle$. Then the family $\{\xi_h\}$ generates a Lie algebra of vector fields on the space B of the exact definition in Section 7.4, as equation (8.3) is equipollent to the equality $[\xi_h, \xi_k] = \xi_{C\langle h, k\rangle}$.

The additional property 4° in Section 12.9 means that this Lie algebra is normalized. Later, in connection with these remarks, the Banach space B with the mapping $C: B \times B \to B$, possessing the properties 1°–4° in Section 12.9, will be called a *Banach algebra of Lie* and denoted by (B, C).

It can be said that because of the (direct) third theorem, for every group (B, ϕ) there corresponds a Banach algebra of Lie (B, C) resulting from equations (8.4) or (9.1). Naturally the question arises; How does this correspondence operate on the class of isomorphic groups $Q(B, \phi)$? To answer this question, it is necessary to find out how the structural operator is transformed under group isomorphism.

Lemma. *Under the isomorphism γ of the group (B, ϕ) onto the group (B', ϕ), the structural operators of these groups are related by the equation*

$$C'\langle \partial\gamma(0)\langle h\rangle, \partial\gamma(0)\langle k\rangle\rangle = \partial\gamma(0)\langle C\langle h, k\rangle\rangle. \tag{10.2}$$

Proof. Differentiation of equation (3.1) with respect to the argument b, at $b = 0$, and application of definition (4.1) give

$$A'(\gamma(a))\langle\partial\gamma(0)\langle k\rangle\rangle = \partial\gamma(a)\langle A(a)\langle k\rangle\rangle$$

for any vector $k \in B$. Differentiation of this equation with respect to a, at $a = 0$, for any vector $h \in B$ leads to the relation

$$\partial A'(0)\langle\partial\gamma(0)\langle h\rangle, \partial\gamma(0)\langle k\rangle\rangle = \partial^2\gamma(0)\langle h, k\rangle + \partial\gamma(0)\langle\partial A(0)\langle h, k\rangle\rangle.$$

From this equation, as the result of permutation of the vectors h, k, subtraction term by term, application of symmetry of the second derivative, and use of the definition (9.1), equation (10.2) appears. ■

Consequence 1. *The canonical isomorphism of the group (B, ϕ) has the same structural operator.*

Proof. This fact follows from the definition in Section 12.6 and equation (10.2). ■

The Banach algebra of Lie (B', C') is said to be *isomorphic* to the Banach algebra of Lie (B, C) if there exists an inverse linear mapping u of B onto B' such that $C' \circ u = u \circ C$, or, more precisely, for any vectors $h, k \in B$

$$C'\langle u\langle h\rangle, u\langle k\rangle\rangle = u\langle C\langle h, k\rangle\rangle. \tag{10.3}$$

It is obvious that the isomorphic relation separates the set of all Banach algebras of Lie into nonintersecting classes of isomorphic Banach algebras of Lie. The class containing the given Banach algebra of Lie (B, C) will be denoted by $Q(B, C)$.

Consequence 2. *There is a mapping of the set $\{Q(B, \phi)\}$ of the class $Q(B, \phi)$ of isomorphic groups into the set $\{Q(B, C)\}$ of the class $Q(B, C)$ of the Banach algebras of Lie.*

Proof. This fact follows from the lemma and definition (10.3). ■

In the converse of the third theorem of Lie it will be established that the mapping mentioned in Consequence 2 is a one-to-one onto mapping.

Consequence 3. *The isomorphism γ of the canonical group (B, ω) onto the canonical group (B', ω') is, at the same time, an isomorphism of the corresponding Banach algebras of Lie (B, C) and (B', C').*

Proof. According to the theorem in Section 12.7, the isomorphism γ is an invertible linear mapping $B \to B'$. Consequently, the statement is correct because of the lemma and definition (10.3). ■

11. Group Determination by Means of Its Structural Operator

If a structural operator, that is, the mapping $C: B \times B \to B$, is given and possesses the properties 1°–4° in Section 12.9, then to reconstruct the group (B, ϕ) it is enough to find the auxiliary mapping A as a solution of the Maurer–Cartan equation (8.3), for a given C. Indeed, with the resulting A and its corresponding V, the mapping $A(\phi) \circ V(b)$ will be completely integrable (the theorem in Section 12.8) and the multiplication law ϕ will be found as the unique solution of the Lie equation (4.3). From this, in general, the solution of equation (8.3) is not unique. But for canonical groups it is unique.

Lemma. *In the case of the canonical group (B, ϕ), the auxiliary mapping $A: O_\rho \to \mathscr{L}(B)$ is uniquely determined by the structural operator of this group.*

Proof. Consider mapping $(t, a) \to tV(ta)$. For arbitrary fixed $h \in B$ it is assumed that $w(t) = tV(ta)\langle h\rangle$. Differentiation of w with respect to t gives the expression

$$\partial_t w = t\partial V(ta)\langle a, h\rangle + V(ta)\langle h\rangle.$$

On the other hand, from the canonical criterion in Section 12.5, the equation $a = V(ta)\langle a\rangle$ holds, and its differentiation leads to the equation

$$h = t\partial V(ta)\langle h, a\rangle + V(ta)\langle h\rangle.$$

By subtracting this equation term by term from the previous one and applying equation (8.2), one obtain the equation

$$\partial_t w - h = tC\langle V(ta)\langle h\rangle, V(ta)\langle a\rangle\rangle,$$

which, because of the bilinearity of C, the definition of w, and the relation $V(ta)\langle a\rangle = a$, is equipollent to

$$\partial_t w = h + C\langle w, a\rangle, \qquad w(0) = 0. \tag{11.1}$$

For fixed $a, h \in B$ this is a linear ordinary differential equation (with an initial condition), whose solution is the mapping $w: R \to B$. As the right-hand side is a mapping of the class $C_\infty(R \times B)$, so is the mapping $(t, w): R \times B \to B$. Therefore, because of the well-known existence and uniqueness theorem for ordinary differential equations, the mapping $w: R \to B$ is uniquely determined by assignment of its structural operator C. Consequently, $V(a)\langle h\rangle = w(1)$, $V(a)$, and $A(a) = (V(a))^{-1}$ are uniquely determined, also. ■

The differential equation (11.1) is of major importance in Lie theory because it forms the basis for constructing the algorithm generated by the structural operator.

12. Third Theorem of Lie

The direct theorem has already been established as the theorem in Section 12.9. That is why only the converse theorem of Lie is stated and proved here.

Theorem. *For every mapping $C: B \times B \to B$ satisfying conditions* 1°–4° *in Section* 12.9, *there is a corresponding group (B, ϕ) for which the given mapping C is its structural operator.*

Proof. *The program*: The group (B, ϕ) is constructed in three stages. The first consists of the solution of problem (11.1). Its solution is a function $w = W(t, a, h)$ which depends on the parameters a, h, included in equation (11.1). It is established that the mapping $(a, h) \to W(1, a, h)$ is linear with respect to h and thus generates the mapping $V: B \to \mathscr{L}(B)$. It is established that the mapping V constructed in this way satisfies the equation of Maurer–Cartan (8.2) with the given C and $V(0) = I_B$. The second stage consists of determining the mapping A, where $A(a) = (V(a))^{-1}$. The third stage is the solution of the Lie equation (4.3) with the calculated A and V. The results of the theorems in Sections 12.4 and 12.8 then provide the unknown group (B, ϕ).

Details: Because of the well-known theorem about the continuous dependence of an ordinary differential equation's solution on the parameters of the right-hand side, the solution of problem (11.1) as a mapping $(t, a, h) \to W(t, a, h)$, in any case, belongs to the class $C_\infty(R \times B \times B)$. When t and a are fixed, the mapping W is linear with respect to h: additivity and homogeneity follow from the linearity of problem (11.1) with respect to w and h, and continuity follows from the continuous dependence of the solution on h.

From the properties of the mapping W, it is clear that the vector $t^{-1}W(t, a, h)$ actually depends only on two "variables," the product ta and h. Since the transformation $t \to t/\lambda$, $a \to \lambda a$, $w \to w/\lambda$ does not change the equations of problem (11.1), both $W(t, a, h)$ and $\lambda W(t/\lambda, \lambda a, h)$ are solutions for any $\lambda \in R$. But because of uniqueness of the solution, the relation $W(t, a, h) = \lambda W(t/\lambda, \lambda a, h)$ must hold, and from this, when $\lambda = t$, the equation $W(t, a, h) = tW(1, ta, h)$ is obtained. Thus, it is possible to assume

$$W(t, a, h) = tV(ta)\langle h\rangle. \tag{12.1}$$

The mapping $V: B \to \mathscr{L}(B)$ of the class $C_\infty(B)$ is defined by this equation.

To prove that the constructed mapping V satisfies the Maurer–Cartan equation (8.2), the function $z: R \to B$, depending on the pair of vectors $h, k \in B$ and given by the formula

$$z(t) = \partial V(ta)\langle h, k\rangle - \partial V(ta)\langle k, h\rangle - C\langle V(ta)\langle k\rangle, V(ta)\langle h\rangle\rangle,$$

is introduced into these considerations. The derivative of this function with respect to t is

$$\partial_t z = \partial^2 V(ta)\langle a, h, k\rangle - \partial^2 V(ta)\langle a, k, h\rangle - C\langle \partial V(ta)\langle a, k\rangle, V(ta\langle h\rangle\rangle \\ - C\langle V(ta)\langle k\rangle, \partial V(ta)\langle a, h\rangle\rangle.$$

The values of the derivatives ∂V and $\partial^2 V$ contained here are found with the aid of certain relations. The first is obtained by substitution of expression (12.1) into equation (11.1):

$$t\partial V(ta)\langle a, h\rangle = h - V(ta)\langle h\rangle + tC\langle V(ta)\langle h\rangle, a\rangle;$$

and the second, by differentiation of the first with respect to a:

$$t\partial^2 V(ta)\langle k, a, h\rangle = -2\partial V(ta)\langle k, h\rangle + C\langle V(ta)\langle h\rangle, k\rangle \\ + tC\langle \partial V(ta)\langle k, h\rangle, a\rangle.$$

As a result of these substitutions, an application of the definition of the function z and property 2° in Section 12.9 of the mapping C, the expression $t\partial_t z + 2z - tC\langle z, a\rangle$ appears, and the other part of the right-hand side is equal to zero because of property 3° in Section 12.9 of the mapping C. Therefore, for z the ordinary differential equation

$$t\partial_t z + 2z = tC\langle z, a\rangle$$

is obtained. After simplification by the substitution $z' = t^2 z$ the problem becomes

$$\partial_t z' = C\langle z', a\rangle, \qquad z'(0) = 0.$$

The unique solution of this problem is $z' = 0$, so $z = 0$ also. But the equality $z(t) = 0$, when $t = 1$, is equipollent to equation (8.2), so, indeed, the constructed mapping V satisfies the equation of Maurer–Cartan.

The relation $V(0) = I_B$ follows from equation (11.1), since when $a = 0$ that equation becomes $\partial_t \omega = h$, which has the function $w = th$ as its solution. With this solution equation (12.1) gives $V(0) = I_B$. From the continuity of the mapping V there exists a null neighborhood $O_\rho \subset B$ such that for every $a \in O_\rho$ the linear mapping $V(a)\colon B \to B$ will be a linear homeomorphism of the space B. That is why the mapping $A\colon O_\rho \to \mathscr{L}(B)$, defined by $A(a) = (V(a))^{-1}$, is now determined, and A is of class $C_\infty(O_\rho)$.

With the A and V determined the linear mapping $A(\phi) \circ V(b)$ is completely integrable because of the theorem in Section 12.8. Thus the Lie equation (4.3) has the unique solution ϕ, and an application of the theorem in Section 12.4 gives the group (B, ϕ).

Finally, from the equation of Maurer–Cartan (8.2), it follows that the given mapping C is the structural operator of the constructed group (B, ϕ).

■

The group (B, ϕ) obtained from the given mapping C with properties 1°–4° in Section 12.9 [or from the given Banach algebra of Lie (B, C)] by the process described in the proof of this theorem is said to be *generated by the mapping C*.

Note. *The group (B, ϕ), generated by the mapping C, is canonical.* Indeed, if in (11.1) it is assumed that $h = a$, then since $C\langle a, a\rangle = 0$ the solution of problem (11.1) is $w = ta$. Thus formula (12.1) determining the mapping V will be $a = V(ta)\langle a\rangle$, and this, according to the canonical criterion in Section 12.6, is enough to draw the necessary conclusion.

Consequence 1. *If C is a structural operator of the group (B, ϕ) and if C generates the group (B, ϕ'), then the group (B, ϕ') is isomorphic to the group (B, ϕ).*

Proof. The group (B, ϕ) is isomorphic to the canonical representative (B, ω) of the class $Q(B, \phi)$ and the groups (B, ϕ) and (B, ω) have the same structural operator (Consequence 1 of the lemma in Section 12.10). Therefore, by the lemma in Section 12.11 and the previous note, $\phi' = \omega$. So the group $(B, \phi') = (B, \omega)$ and it is isomorphic to (B, ϕ). ■

Consequence 2. *Isomorphic Banach algebras of Lie (B, C) and (B', C') generate isomorphic groups (B, ω) and (B', ω').*

Proof. Let γ be an isomorphism of (B, C) onto (B', C'). Since γ is an invertible linear mapping of B onto B', γ can be considered as an isomorphism of the group (B, ω) onto some canonical group (B', ψ), to which the Banach algebra of Lie (B', C') corresponds (Consequence 3 of the lemma in Section 12.10). So both groups (B', ω') and (B', ψ) are canonical and have the same structural operator C'. From the lemma in Section 12.11 it follows that $\psi = \omega'$. ■

Consequence 3. *The set $\{Q(B, \phi)\}$ of classes of isomorphic groups $Q(B, \phi)$ is in one-to-one correspondence with the set $\{Q(B, C)\}$ of isomorphic Banach Lie algebras.*

Proof. The existence of the mapping $\{Q(B, \phi)\} \to \{Q(B, C)\}$ was determined in Consequence 2 of the lemma in Section 12.10. The inverse mapping is obtained in the converse third theorem of Lie and its uniqueness is guaranteed by Consequence 2. ■

The result given in Consequence 3 about the correspondence between Lie groups and algebras is one of the fundamental facts of Lie theory.

13. Example

Using the theorem in Section 12.12, the construction of the group (B, ω) from its structural operator C, for the example of the group in Section 12.2, can be followed all the way. With the structural operator found in Section 12.9, equation (11.1) is

$$\partial_t w = h + l\langle w\rangle a - l\langle a\rangle w, \qquad w(0) = 0.$$

The solution of this problem is found, without any difficulty, to be

$$\begin{aligned} w = W(t, a, h) &= (l\langle a\rangle)^{-1}(1 - \exp(-l\langle ta\rangle))h \\ &\quad + (l\langle a\rangle)^{-2}(l\langle ta\rangle - 1 + \exp(-l\langle ta\rangle))l\langle h\rangle a. \end{aligned}$$

The mapping V is found from (12.1) to be

$$\begin{aligned} V(a)\langle h\rangle &= (l\langle a\rangle)^{-1}(1 - \exp(-l\langle a\rangle))h \\ &\quad + (l\langle a\rangle)^{-2}(l\langle a\rangle - 1 + \exp(-l\langle a\rangle))l\langle h\rangle a. \end{aligned}$$

To construct the Lie equation it is also necessary to find the mapping $A(a)$, which requires the inverse to $V(a)$, that is, for the given vector $k \in B$, to solve the equation $V(a)\langle h\rangle = k$ relative to the vector $h \in B$. The solution is given by

$$\begin{aligned} h = A(a)\langle k\rangle &= (1 - \exp(-l\langle a\rangle))^{-1}[l\langle a\rangle k \\ &\quad - (l\langle a\rangle)^{-1}(l\langle a\rangle - 1 + \exp(-l\langle a\rangle))l\langle k\rangle a]. \end{aligned}$$

With these A and V the Lie equation (4.3) is completely integrable, so the solution of problem (4.3) can be found by "integration along a ray." This means that it is enough to find the mapping $\psi: R \to B$ operating by $\psi(t) = \omega(tb)$. Because of the canonicality of the unknown multiplication law ω (note in Section 12.12), a simpler problem is obtained for ψ, namely,

$$\partial_t \psi = A(\psi)\langle b\rangle, \qquad \psi(0) = a,$$

whose solution is $\omega(a, b) = \psi(1)$. In the present example this equation has the form

$$\begin{aligned} \partial_t \psi &= (1 - \exp(-l\langle\psi\rangle))^{-1}[l\langle\psi\rangle b - (l\langle\psi\rangle)^{-1}(l\langle\psi\rangle \\ &\quad - 1 + \exp(-l\langle\psi\rangle))l\langle b\rangle\psi], \qquad \psi(0) = a. \end{aligned}$$

In spite of the apparent complexity of this equation, it is easily solved. First, it is found that $l\langle\psi\rangle = l\langle a + tb\rangle$, after which it is left to integrate the linear equation. An integration gives

$$\begin{aligned} \psi(t) &= l\langle a + tb\rangle(\exp(l\langle a + tb\rangle) - 1)^{-1}[(l\langle b\rangle)^{-1}(\exp(l\langle a + tb\rangle))b \\ &\quad + (l\langle a\rangle)^{-1}(\exp(l\langle a\rangle) - 1)a - (l\langle b\rangle)^{-1}(\exp(l\langle a\rangle))b]. \end{aligned}$$

From this the multiplication law $\omega(a, b) = \psi(1)$ is found, which after transformation is

$$\omega(a, b) = \frac{\exp(l\langle a\rangle) - 1}{\exp(l\langle a + b\rangle) - 1} \cdot \frac{l\langle a + b\rangle}{l\langle a\rangle l\langle b\rangle}(al\langle b\rangle - bl\langle a\rangle) + \frac{l\langle a + b\rangle}{l\langle b\rangle} b.$$

As was expected, this result coincides with that found in the example in Section 12.6.

14. Canonical Multiplication Analyticity

Let X and Y be Banach spaces. The mapping $f: X \to Y$ is called *analytic in a neighborhood of the point* $x_0 \in X$ if it can be expressed in the form of a series

$$f(x) = f(x_0) + \sum_{n=1}^{\infty} u_n(x - x_0)^n, \tag{14.1}$$

where $u_n \in \mathscr{L}_s^n(X; Y)$ is a symmetric n-linear mapping $X^n \to Y$,

$$u_n(\xi) = u_n\langle \xi, \ldots, \xi\rangle \qquad (\xi \in X),$$

and for some $\rho > 0$ the series (14.1) converges in norm if $|x - x_0| < \rho$. The coefficients of the series (14.1), that is, the n-linear mappings u_n, are calculated by means of the formula

$$u_n = \frac{1}{n!}\partial^n f(x_0) \qquad (n = 1, 2, \ldots),$$

because of which (14.1) is the Taylor series of the mapping f near the point x_0.

Analysis of the algorithm in Section 12.12 for the construction of the multiplication law, using the structural operator, gives the following result.

Theorem. *The multiplication law of a canonical group is analytic in the null neighborhood of the space $B \times B$.*

Proof. Because of Consequence 2 of the theorem in Section 12.7, it is sufficient to prove the result for the group (B, ω) generated by the structural operator C. The right-hand side of the linear equation (11.1) is analytic on the whole space B as a function of the vector a. Thus the solution $W(t, a, h)$ of problem (11.1) is analytic for all $t \in R$ and $a, h \in B$. So the mapping V, defined by this solution, is analytic on the whole space B. Because of $A(a) = (V(a))^{-1}$, $V(0) = I_B$, the mapping A is also analytic, but only in some null neighborhood $O \subset B$. Consequently, the right-hand side of the Lie equation (4.3) is analytic on $O \times O$. Then from the theorem of analysis about analyticity of the solutions of analytic differential equations, the multiplication law will also be an analytic mapping $\omega: O_\rho \times O_\rho \to B$, for some $\rho > 0$. ■

One more fundamental result of the Lie theory is determined by this theorem: the analyticity of local Lie groups.

It is clear that series of the type (14.1), giving analytic mappings V, A, and ω in the null neighborhood of the space B, must be *standard* in the sense that their coefficients are uniquely determined by the structural operator C and they can be expressed by means of C. The goal of what follows below is to obtain the exact formulas for these coefficients.

Here the following simplifying notations are convenient. The commutator $[a, b] = C\langle a, b\rangle$, and for $n > 2$ recurrently defined "long" commutators

$$[a_1, \ldots, a_{n-1}, a_n] = [[a_1, \ldots, a_{n-1}], a_n] \qquad (n = 3, 4, \ldots)$$

are introduced. Those "long" commutators in which $a_2 = \cdots = a_n$ are denoted by an abbreviated symbol, namely,

$$a = [a, b^0], \qquad [a, \underbrace{b, \ldots, b}_{m}] = [a, b^m] \qquad (m = 1, 2, \ldots).$$

The solution of equation (11.1) is sought in the form of a series in powers of t,

$$w(t) = \sum_{n=1}^{\infty} \frac{1}{n!} t^n w_n,$$

whose substitution into the equation gives the recurrence formulas

$$w_1 = h, \qquad w_n = [w_{n-1}, a] \qquad (n = 2, 3, \ldots).$$

Comparison of these with the "long" commutator definition shows that for any $n = 1, 2, \ldots$, the coefficients take the form $w_n = [h, a^{n-1}]$. From equation (12.1) for the mapping V the representation

$$V(a)\langle h\rangle = \sum_{n=1}^{\infty} \frac{1}{n!} [h, a^{n-1}] \tag{14.2}$$

is obtained. Convergence of the series (14.2) follows from 4° in Section 12.9, from which it follows that

$$|[h, a^{n-1}]| \leq \delta^{n-1} |a|^{n-1} |h|.$$

This bound shows that series (14.2) converges in norm on the entire space B.

Construction of the mapping A is equipollent to solution of the equation $V(a)\langle h\rangle = k$, relative to h. It is more convenient to find $A(ta) = (V(ta))^{-1}$ than to solve the equation $V(ta)\langle h\rangle = k$. This solution is to be found in the form of the series

$$V(ta)\langle h\rangle = \sum_{n=1}^{\infty} \beta_n t^{n-1} [h, a^{n-1}],$$

the coefficients β_n (numbers) of which are calculated, by the method of undetermined coefficients, from the identity $A(ta) \circ V(ta)\langle h \rangle = h$, namely from

$$\sum_{n=1}^{\infty} \beta_n t^{n-1} \left[\sum_{m=1}^{\infty} \frac{1}{m!} t^{m-1}[h, a^{m-1}], a^{n-1} \right] = h.$$

Since $[[h, a^r], a^s] = [h, a^{r+s}]$, comparison of the coefficients of like powers of t gives rise to the recurrence formulas

$$\beta_1 = 1, \qquad \beta_n = -\sum_{s=1}^{n-1} \beta_s/(n - s + 1)! \qquad (n = 2, 3, \ldots).$$

It is easy to verify that with these numbers β_n the following identity, with respect to z, is obtained:

$$(e^z - 1) \sum_{n=1}^{\infty} \beta_n z^n = z^2.$$

Because of this the β_n are equal to the coefficients of the power series for the analytic function $z^2/(e^z - 1)$. The radius of convergence of this power series is equal to 2π, whence the estimate

$$|\beta_n| < 4(2\pi)^{1-n} \qquad (n = 1, 2, \ldots) \tag{14.3}$$

is found. So the mapping A is realized in the form of a series

$$A(a)\langle h \rangle = \sum_{n=1}^{\infty} \beta_n [h, a^{n-1}], \tag{14.4}$$

which, by (14.3), converges in norm in the null neighborhood $O_\rho \subset B$, where $\rho = 2\pi/\delta$.

The solution ω of the Lie equation (4.3) is found by "integration along the rays," where the function $\psi: R \to B$ defined by $\psi(t) = \omega(tb)$ is sought. Because of the canonicity of the group, that is, $V(a)\langle a \rangle = a$ [obvious from (14.2)], the equation

$$\partial_t \psi = \partial\omega(tb)\langle b \rangle = A(\omega) \circ V(tb)\langle b \rangle = A(\psi)\langle b \rangle$$

holds, whence the Lie equation becomes

$$\partial_t \psi = A(\psi)\langle b \rangle, \qquad \psi(0) = a, \tag{14.5}$$

where the linear mapping $A(\psi)$ is given by series (14.4). And here, in general, it is possible to construct the solution in the form of a power series in powers of t. But such a result would be difficult to utilize, so it is more convenient to use another method, which is presented in Section 15.

15. The Shur–Campbell–Hausdorff Series

An effective method for solving problem (14.5) is to decompose the solution according to increasing powers of homogeneous expressions, which are homogeneous relative to the vectors a and b. To help achieve this goal, problem (14.5) is first transformed by the change $a \to xa$, $b \to xb$, with a real parameter x. This leads to the problem for the mapping $(t, x) \to \psi(t, x)$,

$$\partial_t \psi = xA(\psi)\langle b\rangle, \qquad \psi(0, x) = xa. \tag{15.1}$$

The solution of problem (15.1) is sought in the form of a series in powers of x:

$$\psi(t, x) = \sum_{m=1}^{\infty} \psi_m(t)x^m. \tag{15.2}$$

Here the coefficients $\psi_m(t)$ are proportional to the derivatives with respect to x at the point $x = 0$, namely, $m!\psi_m(t) = \partial_x^m \psi(t, 0)$. The functions ψ_m satisfy the "equations in variations," obtained by differentiation of equation (15.1) m times with respect to x, at the point $x = 0$,

$$\begin{aligned} \partial_t \psi_1 &= b, & \psi_1(0) &= a, \\ \partial_t \psi_{m+1} &= \rho_m, & \psi_{m+1}(0) &= 0 \qquad (m = 1, 2, \ldots), \end{aligned} \tag{15.3}$$

where the right-hand sides ρ_m are

$$\rho_m = \frac{1}{m!} \partial_x^m (A(\psi)\langle b\rangle)|_{x=0}.$$

They are calculated by differentiating the series, obtained from (14.4),

$$A(\psi)\langle b\rangle = \sum_{n=0}^{\infty} \beta_{n+1}[b, \underbrace{\psi, \ldots, \psi}_{n}]$$

m times. Because of the multilinear nature of the "long" commutator, the coefficient p_m differs from zero in the decomposition

$$[b, \underbrace{\psi, \ldots, \psi}_{n}] = \sum_{m=n}^{\infty} p_m x^m$$

only if $n \leq m$, and it is given by

$$p_m = \sum_{|\alpha|=m} [b, \psi_{\alpha_1}, \ldots, \psi_{\alpha_n}],$$

where $|\alpha| = \alpha_1 + \cdots + \alpha_n$. The sum is taken over all permutations of the number m in the form of a sum of n integers, $\alpha_i > 0$. Finally, for the right-hand sides of equations (15.3) the expressions

$$\rho_m = \sum_{n=1}^{m} \sum_{|\alpha|=m} \beta_{n+1}[b, \psi_{\alpha_1}, \ldots, \psi_{\alpha_n}] \qquad (m = 1, 2, \ldots) \tag{15.4}$$

result.

Equations (15.3) and (15.4) permit the calculation of all coefficients, one after the other, of the series (15.2) and, consequently, determine the multiplication law $\omega(a, b) = \psi(1, 1)$. The resulting representation is obtained in the series form

$$\begin{aligned} \omega(a, b) &= \sum_{m=1}^{\infty} q_n(a, b) \\ &= (a + b) + \tfrac{1}{2}[a, b] + \tfrac{1}{12}([a, [a, b]] + [[a, b], b]) + \cdots, \end{aligned} \tag{15.5}$$

where all the mappings $q_m: (a, b) \to q_m(a, b) = \psi_m(1)$ are homogeneous relative to the power m and are expressed in terms of the structural operator $C\langle a, b\rangle = [a, b]$. Series (15.5) is the famous *series of Shur–Campbell–Hausdorff*, or, for short, the SCH-series. Finally, convergence of this series (15.5) must be investigated.

Theorem. *If the structural operator C satisfies the inequality $|C\langle a, b\rangle| \leq \delta|a|\,|b|$, then the series* (15.5) *converges uniformly in norm in the null neighborhood $U \subset B \times B$, where*

$$U = \{(a, b) \,|\, |a| + 4|b| < \pi/\delta\}. \tag{15.6}$$

Proof. The method of majorants is used. Using (15.4), bounds on the solution of problem (15.3) take the form

$$\begin{aligned} |\psi_1(t)| &\leq |a| + t|b|, \\ |\psi_{m+1}(t)| &\leq \int_0^t \sum_{n=1}^{m} \sum_{|\alpha|=m} |b|\,|\beta_{n+1}|\delta^n |\psi_{\alpha_1}(\tau)| \cdots |\psi_{\alpha_n}(\tau)|\, d\tau, \end{aligned} \tag{15.7}$$

for $t \geq 0$.

As one majorant, the following problem is considered for the mapping $\xi: R \to R$, in which p, q, σ are real numbers and $p > 0$, $q > 0$:

$$\partial_t \xi = xq\sigma/(\sigma - \xi), \qquad \xi(0) = xp. \tag{15.8}$$

Problem (15.8) will be solved by the same method used to solve (15.1), that is, the solution is sought in the form of the series

$$\xi(t, x) = \sum_{m=1}^{\infty} \xi_m(t)x^m. \tag{15.9}$$

Consequently, the system

$$\partial_t \xi_1 = q, \qquad \xi_1(0) = p,$$
$$\partial_t \xi_{m+1} = \hat{\rho}_m, \qquad \xi_{m+1}(0) = 0 \qquad (m = 1, 2, \ldots),$$

will be analogous to the system (15.3), and $\hat{\rho}_m$ calculated by

$$\hat{\rho}_m = \frac{1}{m!} \partial_x^m (q\sigma/(\sigma - \xi))|_{x=0}$$

is given by the expression, analogous to (15.4),

$$\hat{\rho}_m = \sum_{n=1}^{m} \sum_{|\alpha|=m} q\sigma^{-n} \xi_{\alpha_1} \cdots \xi_{\alpha_n}.$$

Thus the coefficients of series (15.9) can be found from the recurrence relations

$$\xi_1 = p + tq,$$
$$\xi_{m+1}(t) = \int_0^t \sum_{n=1}^{m} \sum_{|\alpha|=m} q\sigma^{-n} \xi_{\alpha_1}(\tau) \cdots \xi_{\alpha_n}(\tau)\, d\tau. \tag{15.10}$$

Comparison of inequalities (15.7) with equalities (15.10) shows that if $|\psi_i(t)| \le \xi_i(t)$ for $i = 1, \ldots, m$, then also $|\psi_{m+1}(t)| \le \xi_{m+1}(t)$, if for all n the inequality $|b||\beta_{n+1}|\delta^n \le q\sigma^{-n}$ is satisfied. Taking into consideration the estimate (14.3), it is possible to achieve the desired result by assuming $\sigma = 2\pi/\delta$ and $q \ge 4|b|$. Besides, the inequality $|\psi_1(t)| \le \xi_1(t)$ is guaranteed if $p = |a|$ and $q \ge |b|$. So with the values

$$\sigma = 2\pi/\delta, \qquad p = |a|, \qquad q = 4|b| \tag{15.11}$$

the inequalities $|\psi_m(t)| \le \xi_m(t)$ hold, with $t \ge 0$ and for all $m = 1, 2, \ldots$. Thus, from the convergence of series (15.9), with $t = x = 1$, in the neighborhood $U_0 \subset R^2(p, q)$, the uniform convergence in norm of the series (15.5) follows in the null neighborhood $U = \{(a, b)|(|a|, 4|b|) \in U_0\} \subset B \times B$. The region U_0 is easily found from the exact solution of problem (15.8). The calculation gives

$$\xi(1, x) = \sigma - [\sigma^2 - 2\sigma(p + q)x + p^2x^2]^{1/2},$$

from which it follows that $U_0 = \{(p, q)|2(p + q) < \sigma, p \ge 0, q \ge 0\}$. From (15.11) the estimate (15.6) is immediate. ∎

16. The Finite-Dimensional Case

In the classical theory of local Lie groups the Banach space B is finite dimensional. Usually it is assumed that $B = R^r$ and the group (B, ϕ) is denoted by G^r or, if necessary, $G^r(\phi)$. In this case all the relations are written in component form. The components of the vector $a \in R^r$ are denoted by a^α, so that $a = (a^1, \ldots, a^r)$. The multiplication law of vectors $a, b \in R^r$ is given by means of r of the functions $\phi^\alpha \colon R^r \times R^r \to R$:

$$(a \cdot b)^\alpha = \phi^\alpha(a^1, \ldots, a^r, b^1, \ldots, b^r) \qquad (\alpha = 1, \ldots, r).$$

The transformation of the space R^r can be interpreted as a locally invertible transformation to another, generally speaking, "curvilinear" system of coordinates in R^r, or, more simply, as a *transformation of coordinates*. With this the group G^r is itself considered as an object which does not depend on the choice of the coordinate system, and only those properties are to be studied which also do not depend on the choice of the coordinate system.

The auxiliary mappings $A(a)$ and $V(a)$ are described by their matrix elements A^α_β and V^α_β, which satisfy the relations

$$A^\alpha_\beta(a) = \partial_b \phi^\alpha_\beta(a, 0), \qquad V^\alpha_\sigma(a) A^\sigma_\beta(a) = \delta^\alpha_\beta \qquad (\alpha, \beta = 1, \ldots, r).$$

The canonical isomorphism of the group G^r is interpreted as a transformation in this group to the *canonical coordinates of the first kind*. With this, the assignment of the canonical group G^r is its assignment in the canonical coordinates of the first kind. The criterion of canonicity of the group G^r requires the system of canonical coordinates of the first kind to have the form

$$a^\beta A^\alpha_\beta(a) = a^\alpha, \qquad a^\beta V^\alpha_\beta(a) = a^\alpha \qquad (\alpha = 1, \ldots, r).$$

The component form of the action of the structural operator C is

$$(C\langle h, k\rangle)^\alpha = C^\alpha_{\beta\gamma} h^\beta k^\gamma \qquad (\alpha = 1, \ldots, r),$$

where the $C^\alpha_{\beta\gamma}$ are real numbers, called the *structural constants of the group* G^r. Property 1° in Section 12.9 of the structural operator means that the collection of structural constants $\{C^\alpha_{\beta\gamma}\}$ generates a *tensor of the third rank*, twice covariant (indices β, γ) and once contravariant (index α). Property 2° in Section 12.9 is equipollent to antisymmetry of the structural constants relative to the lower indices

$$C^\alpha_{\beta\gamma} = -C^\alpha_{\gamma\beta},$$

and property 3° in Section 12.9 is equipollent to the *Jacobi relation* for the structural constants

$$C^\alpha_{\beta\sigma} C^\sigma_{\gamma\delta} + C^\alpha_{\gamma\sigma} C^\sigma_{\delta\beta} + C^\alpha_{\delta\sigma} C^\sigma_{\beta\gamma} = 0.$$

Finally, property 4° in Section 12.9 is always satisfied and is not a limitation. A direct definition of the structural constants is obtained from equations (8.4) and (9.1), which in component form are

$$C^\alpha_{\beta\gamma} = \partial_\gamma V^\alpha_\beta(0) - \partial_\beta V^\alpha_\gamma(0) = \partial_\beta A^\alpha_\gamma(0) - \partial_\gamma A^\alpha_\beta(0).$$

The criterion of isomorphism of groups G'^r and G^r with structural constants $C'^{\alpha}_{\beta\gamma}$ and $C^{\alpha}_{\beta\gamma}$, respectively, is stated as follows:

If the structural constants of these groups are respectively equal, that is, $C'^{\alpha}_{\beta\gamma} = C^{\alpha}_{\beta\gamma}$, then the group G'^r is isomorphic to the group G^r; conversely, if the group G'^r is isomorphic to the group G^r, then there exists a system of coordinates in R^r such that the structural constants $\bar{C}^{\alpha}_{\beta\gamma}$ of the group G'^r, in this system of coordinates, will be equal to the respective structural constants of the group G^r in the original coordinate system, that is $\bar{C}^{\alpha}_{\beta\gamma} = C^{\alpha}_{\beta\gamma}$.

§13. THE LIE ALGEBRA

1. Definitions and Examples

The vector space L over the field of real numbers is called a *Lie algebra* if on it there is defined the binary operation of commutation $(u, v) \to [u, v]$, corresponding to the vectors $u, v \in L$, whose commutator $[u, v]$ satisfies the axioms:

1° *bilinearity*: for any $u, v, w \in L$ and $a, b \in R$

$$[au + bv, w] = a[u, w] + b[v, w],$$

$$[u, av + bw] = a[u, v] + b[u, w];$$

2° *antisymmetry*: for any $u, v \in L$

$$[u, v] = -[v, u];$$

3° *the Jacobi identity*: for any $u, v, w \in L$

$$[[u, v], w] + [[v, w], u] + [[w, u], v] = 0.$$

If the space L is infinite dimensional, but is a Banach space with the norm $|u|$ of the vector $u \in L$, then the Lie algebra L is called a *Banach algebra of Lie* if one more axiom is satisfied:

4° *continuity*: there is a number $\delta > 0$ such that for any $u, v \in L$

$$|[u, v]| \leq \delta |u| |v|.$$

Some examples of Lie algebras are

a. the set of vector fields $\Xi(X)$ on the Banach space X, the commutator being that determined in Section 7.1;

b. the set of operators admitted by the given differential equation E, considered in Section 7.13;

c. the Banach algebra of Lie (B, C), given by the structural operator C of the group (B, ϕ), which was considered in Section 12.10;

d. the set of vectors a of the three-dimensional Euclidean space R^3 where the operation of commutation is taken as vector multiplication $[a, b] = a \times b$;

e. the set $\mathscr{L}(X)$ of linear mappings $A: X \to X$ of some Banach space X into itself with the operation of commutation $[A, B] = A \circ B - B \circ A$.

The dimensionality of the Lie algebra L is the dimensionality of its vector space L; if it has finite dimensionality r, then the Lie algebra L is called finite dimensional and denoted by L^r; if it is infinite dimensional, then the Lie algebra L is also called infinite dimensional (sometimes denoted by L^∞).

2. Structural Constants

The finite-dimensional Lie algebra L^r is usually given by indication of the *vector basis* $\{u_\alpha\}$ *in the space* L^r. In this case, because of axiom 1° in Section 13.1, the operation of commutation in L^r is fully defined by the table of commutators, that is, by an $r \times r$ square matrix in which the commutator $[u_\beta, u_\gamma]$ is placed at the intersection of the βth row and γth column. Such tables have already been constructed in Section 7.8 for a Lie algebra of operators. With this, the equation for decomposition of the commutator of the basis vectors with respect to the basis is

$$[u_\beta, u_\gamma] = C^\alpha_{\beta\gamma} u_\alpha \qquad (\beta, \gamma = 1, \ldots, r). \tag{2.1}$$

The numbers contained in (2.1), $C^\alpha_{\beta\gamma}$, are called the *structural constants of the Lie algebra* L^r *with respect to basis* $\{u_\alpha\}$. As a result of axioms 1°–3° in Section 13.1, these possess the following easily verified properties, namely, they

1° generate a tensor of the third rank $(C^\alpha_{\beta\gamma})$ over the space L^r,
2° are antisymmetric relative to the lower indexes $C^\alpha_{\beta\gamma} = -C^\alpha_{\gamma\beta}$,
3° satisfy the Jacobi identity

$$C^\alpha_{\beta\sigma} C^\sigma_{\gamma\delta} + C^\alpha_{\gamma\sigma} C^\sigma_{\delta\beta} + C^\alpha_{\delta\sigma} C^\sigma_{\beta\gamma} = 0.$$

These properties can be used as a foundation to construct a Lie algebra L^r. For this it is necessary to choose the basis $\{u_\alpha\}$ in L^r, to give the system of numbers $(C^\alpha_{\beta\gamma})$ with properties 2° and 3°, to define the operation of commutation on the basis vectors by formula (2.1), and to continue the operation on the entire space L^r. Then the commutator of the two vectors $u = a^\beta u_\beta$ and $v = b^\gamma u_\gamma$ will be given by

$$[u, v] = C^\alpha_{\beta\gamma} a^\beta b^\gamma u_\alpha. \tag{2.2}$$

3. Homomorphisms

Let [,] be the symbol of a commutator in the Lie algebra L and $[\ ,\ ]'$ the symbol of a commutator in the Lie algebra L'.

The linear mapping $\Gamma: L \to L'$ is called a *homomorphism* of the Lie algebra L into the Lie algebra L' if for any $u, v \in L$ the equality

$$\Gamma\langle[u, v]\rangle = [\Gamma\langle u\rangle, \Gamma\langle v\rangle]' \tag{3.1}$$

holds. The set of all vectors $u \in L$ transforming into the null vector (zero) of the space L' under the homomorphism $\Gamma: L \to L'$ is called the *kernel of the homomorphism.*

The homomorphism $\Gamma: L \to L'$ is called an *isomorphism* if the mapping $\Gamma: L \to \Gamma(L) \subset L'$ is invertible. One should distinguish the *isomorphism into*, when $\Gamma(L)$ is a subspace in L', and the *isomorphism onto*, when $\Gamma(L) = L'$.

The Lie algebras L and L' are called *isomorphic* if there is an isomorphism L onto L'.

An isomorphism of the Lie algebra L onto L (onto "itself") is called an *automorphism of the Lie algebra L.*

For an automorphism A of the Lie algebra L the determining equality (3.1) will assume the form

$$A\langle[u, v]\rangle = [A\langle u\rangle, A\langle v\rangle]. \tag{3.2}$$

If A_1 and A_2 are automorphisms of the Lie algebra L, then their composition $A_1 \circ A_2$ is also an automorphism because

$$(A_1 \circ A_2)\langle[u, v]\rangle = A_1\langle[A_2\langle u\rangle, A_2\langle v\rangle]\rangle = [(A_1 \circ A_2)\langle u\rangle, (A_1 \circ A_2)\langle v\rangle].$$

Thus the set of all automorphisms of the given Lie algebra is a group, namely, a subgroup of the group of linear homeomorphisms of the space L. The group of all automorphisms of the Lie algebra L is called the *holomorph* of the Lie algebra L and will be denoted by Aut L.

In the finite-dimensional case, isomorphic Lie algebras have the same dimensionality. The criterion of isomorphism for finite-dimensional Lie algebras is stated in terms of their structural constants.

Lemma. *If the structural constants of the Lie algebra L^r are equal to the corresponding structural constants of the Lie algebra L'^r, then these Lie algebras are isomorphic. Conversely, if the Lie algebras L^r and L'^r are isomorphic, then there exists bases in them for which their structural constants are correspondingly equal.*

Proof. Let $C^{\alpha}_{\beta\gamma}$ be the structural constants of the Lie algebra L^r with respect to the basis $\{u_\alpha\}$ and $C'^{\alpha}_{\beta\gamma}$ be the structural constants of the Lie algebra

L'^r with respect to the basis $\{u'_\alpha\}$. If $C^\alpha_{\beta\gamma} = C'^\alpha_{\beta\gamma}$, and for $u \subset L^r$ it is assumed that $u = a^\alpha u_\alpha$, then the linear mapping $\Gamma\colon L^r \to L'^r$, operating by

$$\Gamma\langle u\rangle = a^\alpha \Gamma\langle u_\alpha\rangle = a^\alpha u'_\alpha = u' \subset L'^r,$$

is an isomorphism of L^r onto L'^r. Indeed, if $v = b^\alpha u_\alpha$ also, then because of equation (2.2)

$$\begin{aligned}\Gamma\langle[u, v]\rangle &= a^\beta b^\gamma C^\alpha_{\beta\gamma}\Gamma\langle u_\alpha\rangle = a^\beta b^\gamma C'^\alpha_{\beta\gamma} u'_\alpha \\ &= [u', v'] = [\Gamma\langle u\rangle, \Gamma\langle v\rangle]'.\end{aligned}$$

Conversely, let Γ be an isomorphism of L^r onto L'^r and $\{u_\alpha\}$ be the basis in L^r. Then $\{u'_\alpha\}$, where $u'_\alpha = \Gamma\langle u_\alpha\rangle$, is the basis in L'^r and, because of (2.2) and (3.1),

$$[u'_\beta, u'_\gamma]' = [\Gamma\langle u_\beta\rangle, \Gamma\langle u_\gamma\rangle]' = \Gamma\langle[u_\beta, u_\gamma]\rangle = C^\alpha_{\beta\gamma}\Gamma\langle u_\alpha\rangle = C^\alpha_{\beta\gamma} u'_\alpha,$$

from which it follows that $C'^\alpha_{\beta\gamma} = C^\alpha_{\beta\gamma}$. ■

4. Subalgebras

The vector space $N \subset L$ is called a *subalgebra of the Lie algebra L* if for any $u, v \in N$ it is also true that $[u, v] \in N$.

In the case of an infinite-dimensional Banach algebra of Lie one distinguishes the *closed subalgebra* $N \subset L$, where N is a closed subspace of the Banach space L.

The subalgebra $J \subset L$ is called an *ideal of the Lie algebra L* if for any $u \in L$, $v \in J$ it is also true that $[u, v] \in J$.

In any Lie algebra L one distinguishes two subalgebras (they are also ideals in L): *the null subalgebra*, consisting of one null element of the space L; and the *Lie algebra L itself.* All remaining subalgebras in the Lie algebra L are called its *proper subalgebras and ideals.*

Under the homomorphism of the Lie algebra $\Gamma\colon L \to L'$ the kernel of the homomorphism J is an ideal in L. Since $\Gamma\langle v\rangle = 0$ for $v \in J$, because of (3.1) it follows that for any $u \in L$

$$\Gamma\langle[u, v]\rangle = [\Gamma\langle u\rangle, 0] = 0,$$

which means that $[u, v] \in J$.

The element $z \in L$ is called *central* if $[z, u] = 0$ for any $u \in L$. The union Z of all central elements is called the *center of the Lie algebra L.* The center is a subalgebra in L because for any $z_1, z_2 \in Z$, it is obvious that $az_1 + bz_2 \in Z$ and $[z_1, z_2] = 0$. It is clear that the center Z is an ideal in L.

Let N and M be subspaces of the Lie algebra L. The subspace in L generated by all possible commutators $[u, v]$, where $u \in N$, $v \in M$, is called *Lie's product of the spaces N and M* and is denoted by $[N, M]$.

Lie's product of any two ideals of the Lie algebra L is an ideal in L. This fact easily follows since if N, M are ideals in L, then for any elements $u \in N$, $v \in M$, $w \in L$ the Jacobi identity 3° in Section 13.1 gives $[v, w] \in M$ and $[w, u] \in N$, so that $[[u, v], w] \in [N, M]$.

In particular, the ideal $L^{(1)} = [L, L]$ is called the *commutant (or derived algebra) of the Lie algebra L*. The commutant of the commutant $L^{(2)} = [L^{(1)}, L^{(1)}]$ is called the *second commutant* of the Lie algebra L. By induction one determines the *$(k+1)$th commutant* as a commutant of the kth commutant: $L^{(k+1)} = [L^{(k)}, L^{(k)}]$. In this way a *series of commutants of the Lie algebra L*, appears:

$$L \supset L^{(1)} \supset L^{(2)} \supset \cdots \supset L^{(n)} \supset \cdots. \tag{4.1}$$

For a finite-dimensional Lie algebra L, its commutant series terminates either with a null ideal $L^{(n)} = \{0\}$ or with an $L^{(n)}$ such that $L^{(n-1)} \neq L^{(n)} = L^{(n+1)}$.

By analogy, *Lie's powers of the Lie algebra L* are defined by

$$L^{[2]} = [L, L],\ L^{[3]} = [L^{[2]}, L], \ldots, L^{[k+1]} = [L^{[k]}, L], \ldots,$$

which are the ideals in L. From this, of course, $L^{[2]} = L^{(1)}$. In this way, the *lower central series of the Lie algebra L* appears as

$$L \supset L^{[2]} \supset L^{[3]} \supset \cdots \supset L^{[n]} \supset \cdots. \tag{4.2}$$

For a finite-dimensional Lie algebra L, its lower central series terminates either with the null ideal $L^{[n]} = \{0\}$ or with an $L^{[n]}$ for which $L^{[n-1]} \neq L^{[n]} = L^{[n+1]}$.

Under a homomorphism of the Lie algebra, subalgebras transform into subalgebras and ideals into ideals. Under an isomorphism, a center transforms into a center and a commutant into a commutant.

5. Factor Algebra

Let L be a Lie algebra and J its ideal. For the vectors of the space L, the relation $\sim$ of equivalence relative to the ideal J is introduced: $u \sim v$ if $u - v \in J$. Obviously, this relation is a set-theoretical definition of equivalence, according to which L is decomposed into nonintersecting equivalent classes $U, V, \ldots$.

It is easy to verify that the set $\{U\}$ of these classes generates a Lie algebra if vector operations and the operation of commutation are introduced into it by the following rules: If $u \in U$, $v \in V$, then $aU + bV$ $(a, b \in R)$ is a class containing the element $au + bv$, and $[U, V]$ is the class containing the element $[u, v]$. The correctness of this commutator $[U, V]$ definition follows from the fact that if $u' \sim u$, $v' \sim v$, then

$$[u', v'] = [u' - u, v' - v] + [u' - u, v] + [u, v' - v] + [u, v] \sim [u, v],$$

because the three first terms belong to the ideal J.

The Lie algebra of the equivalence classes, relative to the ideal $J \subset L$ of the Lie algebra L, is called the *factor algebra* (or *algebra of residues*) *of the Lie algebra L with respect to its ideal J* and is denoted by L/J.

There is a *natural homomorphism* Γ of the Lie algebra L on the factor algebra L/J, defined by $\Gamma\langle u\rangle = U$, if $u \in U$. The ideal J is the kernel of this homomorphism.

Lemma 1. *If Γ is a homorphism of the Lie algebra L on the Lie algebra L′ and J is its kernel, then the factor algebra L/J is isomorphic to L′.*

Proof. The mapping $\Gamma': L/J \to L'$, defined on the class U with the aid of the element $u \in U$ by

$$\Gamma'\langle U\rangle = \Gamma\langle u\rangle,$$

does not depend on the selection of the representative $u \in U$ and linearly maps L/J on L' in a one-to-one correspondence. The last statement follows from the fact that the equality $\Gamma'\langle U\rangle = \Gamma'\langle V\rangle$ is equipollent to the equality $\Gamma\langle u\rangle = \Gamma\langle v\rangle$, from which $\Gamma\langle u - v\rangle = 0$, that is, $u \sim v$ and $U = V$. ■

The sum $J_1 + J_2$ of the ideals J_1 and J_2, from the Lie algebra L, is a vector subspace in L, equal to the vector sum of subspaces J_1 and J_2. It is obvious that $J_1 + J_2$ is an ideal in L. In the sequel, the following lemma about isomorphism is used.

Lemma 2. *If J_1 and J_2 are ideals of the Lie algebra L, then the factor algebra $(J_1 + J_2)/J_2$ is isomorphic to the factor algebra $J_1/(J_1 \cap J_2)$.*

Proof. If the element $u \in J_1$ is fixed, then for any $v \in J_2$ the elements $u + v$ are representatives of the same class $U \subset J_1 + J_2$ of equivalent elements in $J_1 + J_2$, relative to the ideal J_2. Thus the class U is uniquely determined by the element u. From this the homomorphism $\Gamma: (J_1 + J_2)/J_2 \to J_1/(J_1 \cap J_2)$ is defined and operates by $\Gamma\langle U\rangle = U'$, where $U' \subset J_1$ is the class of equivalent elements in J_1, relative to the ideal $J_1 \cap J_2$, containing u. If $\Gamma\langle U_1\rangle = \Gamma\langle U_2\rangle$, then $u_1 - u_2 \in J_1 \cap J_2 \subset J_2$, and because of the previous result $U_1 = U_2$. Thus Γ is an isomorphism. ■

6. Structural Features of Subalgebras

Subalgebras, ideals, central ideals, and factor algebras of finite-dimensional Lie algebras L^r can be characterized in terms of the structural constants.

Let the dimensionality of the space N be equal to s and let the basis $\{u_\alpha\}$ in L^r be chosen such that the first s basis vectors $u_1, \ldots, u_s$ give a basis in N. Then the basis $\{u_\alpha\}$ is the union

$$\{u_\alpha\} = \{u_{\alpha'}\} \cup \{u_{\alpha''}\},$$

where $\{u_{\alpha'}\}$ is the basis in N and $\{u_{\alpha''}\}$ the additional elements of the basis in L^r. It is supposed that the index with a prime, of the kind α', assumes the values $1, \ldots, s$ and the index with double primes, of the kind α'', takes values $s+1, \ldots, r$. Let $C^\alpha_{\beta\gamma}$ be the structural constants of the Lie algebra L^r in the basis $\{u_\alpha\}$.

Lemma. *The subspace N of dimensionality s is a subalgebra $N \subset L^r$ if and only if*

$$C^{\alpha''}_{\beta'\gamma'} = 0 \qquad (\beta', \gamma' = 1, \ldots, s;\ \alpha'' = s+1, \ldots, r); \tag{6.1}$$

is an ideal $J \subset L^r$ if and only if

$$C^{\alpha''}_{\beta'\gamma} = 0 \qquad (\gamma = 1, \ldots, r;\ \beta' = 1, \ldots, s;\ \alpha'' = s+1, \ldots, r); \tag{6.2}$$

and is a central ideal $J \subset Z \subset L^r$ if and only if

$$C^{\alpha}_{\beta'\gamma} = 0 \qquad (\alpha, \gamma = 1, \ldots, r;\ \beta' = 1, \ldots, s). \tag{6.3}$$

Also, if $N = J$ is an ideal in L^r, then the factor algebra L^r/J is the Lie algebra L^{r-s} and in the basis $\{U_{\alpha''}\}$, where the class $U_{\alpha''}$ contains $u_{\alpha''}$, it has the structural constants $C^{\alpha''}_{\beta''\gamma''}$.

Proof. The arbitrary vector $u' \in N$ is prescribed in the form $u' = a^{\beta'} u_{\beta'}$. If $v = b^\gamma u_\gamma$ is a vector from L^r, then, according to equation (2.2),

$$[u', v] = C^\alpha_{\beta'\gamma} a^{\beta'} b^\gamma u_\alpha.$$

From this and the definitions in Section 13.4 the statements expressed by equations (6.1)–(6.3) follow. Finally, if $N = J$ is an ideal in L^r, then the dimensionality of the factor algebra L^r/J is equal to $r - s$. Let $U_{\alpha''}$ be an equivalence class containing $u_{\alpha''}$. The $\{U_{\alpha''}\}$ is a basis in L^r/J, and $[U_{\beta''}, U_{\gamma''}]$ is a class containing the element

$$[u_{\beta''}, u_{\gamma''}] = C^\alpha_{\beta''\gamma''} u_\alpha \sim C^{\alpha''}_{\beta''\gamma''} u_{\alpha''},$$

which belongs to the class $C^{\alpha''}_{\beta''\gamma''} U_{\alpha''}$. Consequently, $[U_{\beta''}, U_{\gamma''}] = C^{\alpha''}_{\beta''\gamma''} U_{\alpha''}$. ∎

7. Certain Classes of Lie Algebras

The following classification of special Lie algebras is important in theory and applications.

A Lie algebra L is called *Abelian* (or *commutative*) if $[u, v] = 0$ for all $u, v \in L$.

In an Abelian Lie algebra L, any subspace of the vector space L is an ideal. The factor algebra $L/L^{(1)}$ of any Lie algebra L, relative to its commutant, is Abelian, for if u, v are representatives of the classes U, V, then $[U, V]$ is the class containing $[u, v] \in L^{(1)}$, and therefore $[U, V] = 0$.

A Lie algebra L is called *simple* if in L there are no proper ideals and if its commutant $L^{(1)} \neq \{0\}$.

It is obvious that the Lie algebras L^1 and L^2, with the dimensionalities one and two, are not simple. Simple finite-dimensional Lie algebras exist beginning with dimensionality three. For example, the three-dimensional Lie algebra (d) in Section 13.1 is simple.

A Lie algebra L is called *solvable* if its series of commutants (4.1) terminates with a null ideal.

It is obvious that any Abelian Lie algebra is solvable. This is also true for Lie algebras of dimensionality one and two.

Lemma 1. *Any subalgebra and any homomorphic image of a solvable Lie algebra L are solvable. If L contains a solvable ideal J, such that the factor algebra L/J is solvable, then L is also solvable.*

Proof. Solvability of the subalgebra follows from the fact that every commutant of it is contained in the corresponding commutant of the algebra. Further, under a homomorphism Lie's product (see Section 13.4) always transforms into Lie's product, that is, $\Gamma\langle[N, M]\rangle = [\Gamma\langle N\rangle, \Gamma\langle M\rangle]'$, from which the solvability of the homomorphic image follows. Finally, let Γ be the natural homomorphism of L on L/J, defined in Section 13.5. Then

$$\Gamma\langle L^{(n)}\rangle = (\Gamma\langle L\rangle)^{(n)} = (L/J)^{(n)} = \{0\}$$

for some n. Thus $L^{(n)} \subset J$, and if $J^{(m)} = \{0\}$, then as $L^{(n+m)} \subset J^{(m)}$, it follows that $L^{(n+m)} = \{0\}$. ■

A Lie algebra L is called *semi-simple* if L does not contain solvable ideals different from zero.

Any simple Lie algebra L is semisimple. If this were not the case, then there would be a solvable ideal $J \subset L$, which means $J = L$, and the ideal $J^{(1)}$ would have to be zero, from which it follows that $L^{(1)} = \{0\}$. This is a contradiction of the definition of a simple Lie algebra.

The Lie algebra L is semisimple if it has no nonzero Abelian ideals. Actually, if J is a solvable ideal in L and $J \neq \{0\}$, then for some $n \geq 1$, $J^{(n)} = \{0\}$ and $J^{(n-1)} \neq \{0\}$, where $J^{(n-1)}$ is an Abelian ideal in L.

The Lie algebra L is called *nilpotent* if its lower central series (4.2) ends with the null ideal.

Any one-dimensional (and generally any Abelian) Lie algebra is nilpotent. It is easy to verify that this is not so for a two-dimensional non-Abelian Lie algebra. Thus, a solvable Lie algebra is not always nilpotent.

Lemma 2. *Any nilpotent Lie algebra is solvable.*

Proof. Using the Jacobi identity 3° in Section 13.1, it is easy to prove by induction with respect to l that the commutator of two "long" commutators (the definition is given in Section 12.14) $[[a_1, \ldots, a_k], [b_1, \ldots, b_l]]$ is a linear combination of "long" commutators with length $k + l$. That is why $[L^{[k]}, L^{[l]}] \subset L^{[k+1]}$ for any k, l. Consequently, $L^{(k)} \subset L^{[2^k]}$ for any $k = 1, 2, \ldots$. Thus, if L is nilpotent, that is, $L^{[2^n]} = \{0\}$, for some n, then $L^{(n)} = \{0\}$. ∎

The Lie algebra L is called a *direct sum* of subalgebras N and M, written $L = N \oplus M$, if the space L is the direct sum of the subspaces N and M, that is, if $N \cap M = \{0\}$ and any vector $u \in L$ can be written in the form $u = v + w$, where $v \in N$, $w \in M$.

The subalgebras N and M, as components of the direct sum $L = N \oplus M$, are ideals in L if and only if $[N, M] = \{0\}$.

If in the direct sum $L = J \oplus N$, J an ideal, then the subalgebra N is isomorphic to the factor algebra L/J.

8. The Radical

In the classification of different kinds of entities connected with Lie algebras, one special ideal, contained in any Lie algebra, plays a special role.

Theorem. *Among the solvable ideals of a given finite-dimensional Lie algebra L there is a maximal solvable ideal R, which contains any solvable ideal $J \subset L$. Further, the factor algebra L/R is semisimple.*

Proof. The sum of any two solvable ideals is solvable. Indeed, if J_1 and J_2 are solvable ideals, then the factor algebra $J_1/(J_1 \cap J_2)$ is solvable as the homomorphic image of the solvable ideal J_1 (Lemma 1 in Section 13.7), from which the solvability of the isomorphic factor algebra $(J_1 + J_2)/J_2$ follows (Lemma 2 in Section 13.5), as well as the solvability of the ideal $J_1 + J_2$ (again Lemma 1 in Section 13.7). Consequently, if R is a solvable ideal of a finite-dimensional Lie algebra with maximal dimensionality in

L and J is a solvable ideal, then $R + J = R$. Therefore, $J \subset R$. Further, any solvable ideal of the factor algebra L/R has a form N/R, where N is an ideal in L. The ideal N is solvable because of Lemma 1 in Section 13.7, from which it follows that $N \subset R$ and $N/R = \{0\}$. Thus, the factor algebra L/R is semisimple. ∎

The maximal solvable ideal $R \subset L$ is called the *radical of the Lie algebra L*.

9. The Levi Theorem

The following theorem of Levi gives an important structural characteristic of Lie algebras. It shows that any Lie algebra can be obtained by the direct sum of some semisimple Lie algebra with the maximal solvable ideal.

Theorem. *If R is the radical of a finite-dimensional Lie algebra L, then there is a semisimple subalgebra $N \subset L$ such that $L = N \oplus R$.*†

The semisimple subalgebra N, contained in the direct sum $L = N \oplus R$, is called the *Levi factor* of the Lie algebra L. As opposed to the radical R, the Levi factor, generally speaking, is not uniquely defined.

§14. ASSOCIATED ALGEBRA

1. Algebra of Differentiation

The linear mapping $d: L \to L$, satisfying the condition

$$d\langle[u, v]\rangle = [d\langle u\rangle, v] + [u, d\langle v\rangle], \tag{1.1}$$

is called a *differentiation* of the Lie algebra L.

It is easy to verify that if d_1 and d_2 are differentiations of a Lie algebra, then for any $a, b \in R$ the linear mapping $ad_1 + bd_2$ is a differentiation also. Thus the union of all differentiations of a given Lie algebra L generates a vector space, namely, a subspace in $\mathscr{L}(L)$, which will be denoted by dL.

In addition, if $d_1, d_2 \in dL$, then

$$d_1 \circ d_2\langle[u, v]\rangle = [d_1 \circ d_2\langle u\rangle, v] + [d_2\langle u\rangle, d_1\langle v\rangle] + [d_1\langle u\rangle, d_2\langle v\rangle] + [u, d_1 \circ d_2\langle v\rangle].$$

†For a proof see N. Jacobson, "Lie Algebras." Interscience, New York, 1962.

Permutation of d_1 with d_2 and subtraction of the resulting equation from the previous one gives the relation

$$(d_1 \circ d_2 - d_2 \circ d_1)\langle [u, v] \rangle = [(d_1 \circ d_2 - d_2 \circ d_1)\langle u \rangle, v] + [u, (d_1 \circ d_2 - d_2 \circ d_1)\langle v \rangle].$$

Consequently, with the commutator $[d_1, d_2] = d_1 \circ d_2 - d_2 \circ d_1$, the space dL is a Lie algebra—it is a subalgebra in the Lie algebra, considered in (e) in Section 13.1, of all linear mappings $L \to L$.

The Lie algebra dL is called a *Lie algebra of differentiations of the Lie algebra L.*

The linear mapping $L \to L$, denoted by $\mathrm{ad}(x)$ and operating by

$$\mathrm{ad}(x)\langle u \rangle = [u, x], \tag{1.2}$$

exists for every mapping $x \in L$.

Comparison of the Jacobi identity 3° in Section 13.1, rewritten in the form

$$[[u, v], x] = [[u, x], v] + [u, [v, x]],$$

with definitions (1.1) and (1.2) shows that $\mathrm{ad}(x)$ is a differentiation of the Lie algebra L.

The differentiation $\mathrm{ad}(x)$, defined by formula (1.2), is called an *inner differentiation of the Lie algebra L.*

Lemma. *The union $\mathscr{D}$ of all the inner differentiations of the Lie algebra L is an ideal in the Lie algebra dL.*

Proof. For any $x, y \in L$ and $a, b \in R$ the equation

$$(a\,\mathrm{ad}(x) + b\,\mathrm{ad}(y))\langle u \rangle = [u, ax + by] = \mathrm{ad}(ax + by)\langle u \rangle \tag{1.3}$$

holds. Because of this $\mathscr{D}$ is a vector space in dL. Further, if $x \in L$ and $d \in dL$, then for any $u \in L$

$$(d \circ \mathrm{ad}(x) - \mathrm{ad}(x) \circ d)\langle u \rangle = d\langle [u, x] \rangle - [d\langle u \rangle, x] = [u, d\langle x \rangle] = \mathrm{ad}(d\langle x \rangle)\langle u \rangle.$$

Hence $d \circ \mathrm{ad}(x) - \mathrm{ad}(x) \circ d = \mathrm{ad}(d\langle x \rangle) \in \mathscr{D}$. ■

The Lie algebra $\mathscr{D}$ of all inner differentiations is called an *associated algebra of the Lie algebra L.*

2. Natural Homomorphism

The definition of the associated mapping (1.2) shows that there is a mapping $\mathrm{ad}\colon L \to \mathscr{D}$ of the Lie algebra L on its associated algebra $\mathscr{D}$, operating by $x \to \mathrm{ad}(x)$. The following lemma fixes the algebraic properties of this mapping.

Lemma. *The mapping* $\mathrm{ad}: L \to \mathscr{D}$ *is a homomorphism of the Lie algebra* L *on its associated algebra* $\mathscr{D}$. *The kernel of this homomorphism is the center* $Z \subset L$.

Proof. The linearity of the mapping ad follows from equation (1.3). Further, because of the definition (1.2) and the Jacobi identity 3° in Section 13.1, the equation

$$\begin{aligned}\mathrm{ad}([x, y])\langle u\rangle &= [u, [x, y]] = [[u, x], y] - [[u, y], x] \\ &= (\mathrm{ad}(y) \circ \mathrm{ad}(x) - \mathrm{ad}(x) \circ \mathrm{ad}(y))\langle u\rangle\end{aligned}$$

is obtained. Then, because of the definition of the commutator in $\mathscr{D}$, it follows that $-\mathrm{ad}([x, y]) = [-\mathrm{ad}(x), -\mathrm{ad}(y)]$, that is, the condition of homomorphism (13.3.1) is satisfied. Finally, the equality $\mathrm{ad}(x)\langle u\rangle = 0$, equipollent to the equality $[u, x] = 0$, is obtained for all $u \in L$ if and only if $x \in Z$. ■

The homomorphism $\mathrm{ad}: L \to \mathscr{D}$ is called a *natural homomorphism* of the Lie algebra L on its associated Lie algebra $\mathscr{D}$.

Consequence. *The associated Lie algebra* $\mathscr{D}$ *is isomorphic to the factor algebra* L/Z *of the Lie algebra* L *relative to its center* Z.

This follows from the previous lemma and Lemma 1 in Section 13.5.

3. Representation of Lie Algebra of Operators

Let L be a finite-dimensional Lie algebra and its inner vector space be considered as a Banach space. The mapping $\mathrm{ad}: L \to \mathscr{L}(L)$ gives, on the space L, a family $\mathscr{E}$ of vector fields $e: L \to L$ according to the following rule: Every vector field $e \in \mathscr{E}$, defined by the assignment of a vector $u \in L$, is denoted by the symbol e_u, which acts by

$$e_u(x) = \mathrm{ad}(u)\langle x\rangle = [x, u]. \tag{3.1}$$

The equality $e_{au+bv} = ae_u + be_v$ follows from (3.1) and holds for any $u, v \in L$ and $a, b \in R$. Therefore, family $\mathscr{E}$ is a vector space, namely, a subspace of the space $\Xi(L)$ of all vector fields on the space L.

From definition (3.1) it follows that every vector field e_u is a linear mapping $L \to L$. Thus $e_u \in C_\infty(L)$ for every $u \in L$. In Section 7 the commutator of vector fields has already been determined. As it is obvious that $\partial e_u(x)\langle h\rangle = [h, u]$, a calculation by equation (7.1.1) gives

$$\begin{aligned}[e_u(x), e_v(x)] &= [e_u(x), v] - [e_v(x), u] \\ &= [[x, u], v] - [[x, v], u] = [x, [u, v]] = e_{[u,v]}(x),\end{aligned}$$

where the equality before the last follows from the Jacobi identity. Finally, the important relation

$$[e_u, e_v] = e_{[u,v]} \tag{3.2}$$

is obtained.

Using the definition in Section 7.4, equality (3.2) shows that the space $\mathscr{E}$ is a Lie algebra of vector fields. Also from (3.2) it follows that the mapping $\Gamma: L \to \mathscr{E}$, operating by $\Gamma\langle u\rangle = e_u$, is a homomorphism of the Lie algebra L on the Lie algebra $\mathscr{E}$ of vector fields (3.1).

As was discussed in Section 7, instead of the vector fields e_u one can talk about the corresponding operators $e_u \cdot \partial$, where ∂ is a symbol of differentiation of the mappings $L \to P$. Thus the Lie algebra $\mathscr{E}$ can be considered as a Lie algebra of the operators

$$e_u(x) \cdot \partial = [x, u] \cdot \partial. \tag{3.3}$$

Note. *The Lie algebra $\mathscr{E}$ is isomorphic to the associated algebra $\mathscr{D}$.* The isomorphism $\mathscr{D} \to \mathscr{E}$ is given by equation (3.1). Actually, this equality can be considered merely as a definition of ad(u) using the new notation e_u.

4. Inner Automorphisms

According to the Lie theorem in Section 2.3, in every Banach space L it is possible to construct the group $G^1(e_u)$ of transformations of this space, corresponding to every vector field $e_u \in \mathscr{E}$. The generating mapping of this group must be a solution of the equation $\partial_t x' = e_u(x')$, with the initial condition $x'(0) = x$. Because of definition (3.1) this becomes the problem

$$\partial_t x' = [x', u], \qquad x'(0) = x. \tag{4.1}$$

Since equation (4.1) is linear relative to the unknown x', the solution of problem (4.1) must be the result of operation on the vector $x \in L$ by some linear invertible mapping $L \to L$. This mapping will be denoted by the symbol $A_u(t)$, so the solution of problem (4.1) is $x' = A_u(t)\langle x\rangle$.

It is useful to note that this solution can be deduced by rewriting the equations of problem (4.1) directly for the transformation $A_u(t)$ as

$$\partial_t A_u = \mathrm{ad}(u) \circ A_u, \qquad A_u(0) = I_L,$$

using (3.1). Here the coefficient $\mathrm{ad}(u)$ is a constant mapping (independent of t), and the exact solution is found to be

$$A_u(t) = \exp(t\,\mathrm{ad}(u)). \tag{4.2}$$

From here, the exact expression for the solution of problem (4.1) follows from the generating mapping of the group $G^1(e_u)$, with the condition that $e_u = \mathrm{ad}(u) \neq 0$,

$$x' = A_u(t)\langle x\rangle = \exp(t\,\mathrm{ad}(u))\langle x\rangle. \tag{4.3}$$

Equation (4.2) shows that the resulting transformations $A_u(t)$ of the space L are defined and invertible without any limitations on the values of the group parameter t and the determining vector $u \in L$.

Theorem. *The transformations* (4.2) *of the space L are automorphisms of the Lie algebra L.*

Proof. Under the operation of transformation (4.2) let the vectors $x, y \in L$ transform into the vectors $x', y' \in L$, according to (4.3). Then equation (4.1) is satisfied for x' as well as for y'. Then, from the Jacobi identity, the equations

$$\begin{aligned}\partial_t[x', y'] &= [[x', u], y'] + [x', [y', u]]\\ &= [[x', y'], u], \qquad [x', y'](0) = [x, y]\end{aligned}$$

are found. These show that the commutator $[x', y']$ is a solution of the same problem (4.1) as the function $R \to L$. Therefore, the solution is given by equation (4.3), that is, $[x', y'] = A_u(t)\langle[x, y]\rangle$. This equation, upon rewriting in the form

$$A_u(t)\langle[x, y]\rangle = [A_u(t)\langle x\rangle, A_u(t)\langle y\rangle],$$

coincides with equation (13.3.2), the determining automorphism of the Lie algebra L. ■

The automorphisms $A_u(t)$ found by the foregoing method are called *inner automorphisms of the Lie algebra L.*

The group of transformations of the space L, generated by the transformations $A_u(t)$ which belong to all possible groups $G^1(e_u)$, is called the *group of inner automorphisms of the Lie algebra L.* This group will be denoted by Int L.

To aid in the construction of an isomorphisms group, it should be noted that the Lie algebra $\mathscr{E}$ of operators $e_u \cdot \partial$ is isomorphic to the factor algebra L/Z of the Lie algebra L, relative to its center Z. This fact is obvious because of the lemma in Section 14.2 and the note in Section 14.3. Besides, as will be determined in Section 16, to construct a group of transformations it is enough to find only the groups G^1, generated by the basis vectors of the corresponding Lie algebra.

Let $\{u_\alpha\}$ be a basis in L, chosen in the same way as in Section 13.6, for $N = Z$, and let the one parameter group $G^1(e_{u_\gamma})$, generated by the operator $e_{u_\gamma} \cdot \partial$, be denoted by G_γ^1. Since $e_{u_{\gamma'}}(x) = [x, u_{\gamma'}] = 0$ because $u_{\gamma'} \in Z$ for $\gamma' = 1, \ldots, s$, so the groups G_γ^1 consist of only one identity transformation. Consequently, the groups $G_{\gamma''}^1$ will be nontrivial only for $\gamma'' = s + 1, \ldots, r$.

Assume $x = x^\alpha u_\alpha$ and introduce the structural constants $C_{\beta\gamma}^\alpha$ of the Lie algebra L; then, because of property (13.6.3), $[x, u_{\gamma''}]^\alpha = C_{\beta''\gamma''}^\alpha x^{\beta''}$. Consequently, for the group $G_{\gamma''}^1$, equation (4.1) in the basis $\{u_\alpha\}$ will assume the form of a linear system of scalar equations with constant coefficients

$$\partial_t x'^\alpha = C_{\beta''\gamma''}^\alpha x'^{\beta''}, \qquad x'^\alpha(0) = x^\alpha \qquad (\alpha = 1, \ldots, r). \tag{4.4}$$

From the solution of system (4.4) automorphisms $A_{\gamma''}(t) \in G_{\gamma''}^1$ are given, in the same basis, by the matrix $(A_\sigma^\alpha(t))$, which operates by the equation, equipollent to (4.3),

$$x'^\alpha = A_\sigma^\alpha(t) x^\sigma \qquad (\alpha = 1, \ldots, r). \tag{4.5}$$

The common inner automorphism $A \in \operatorname{Int} L$ is obtained by the following rule: For every $\gamma'' = s + i$ $(i = 1, \ldots, r - s)$, the parameter of the group G_{s+i}^1 is denoted by t^i and the composition of $r - s$ factors

$$A(t^1, t^2, \ldots, t^{r-s}) = A_{s+1}(t^1) \circ A_{s+2}(t^2) \circ \cdots \circ A_r(t^{r-s}) \tag{4.6}$$

is generated. The proof that (4.6) is a common element of the group $\operatorname{Int} L$ will be given in Section 16.

5. Killing's Form

The symmetric bilinear form Φ on the space L, that is, the bilinear mapping $\Phi: L \times L \to R$, is called *invariant* (relative to the group $\operatorname{Int} L$) if for any inner automorphism $A \in \operatorname{Int} L$ and for any $x, y \in L$

$$\Phi\langle A\langle x\rangle, A\langle y\rangle\rangle = \Phi\langle x, y\rangle. \tag{5.1}$$

If in (5.1) we assume $A = A_u(t)$ and differentiate it with respect to t at the point $t = 0$, then, because of the bilinearity of Φ and equations (4.1), an infinitesimal criterion of invariance of the mapping Φ,

$$\Phi\langle [x, u], y\rangle + \Phi\langle x, [y, u]\rangle = 0, \tag{5.2}$$

is found.

A remarkable property of invariant forms is the fact that with their help it is possible to give ideals of a Lie algebra: The set J of elements $y \in L$ for which $\Phi\langle x, y\rangle = 0$ for any element $x \in L$ is an ideal in L. This follows from the fact that J is a subspace in L because of the bilinearity of Φ and, also, for any $y \in J$, $z \in L$ the equality $\Phi\langle x, [y, z]\rangle = -\Phi\langle [x, z], y\rangle = 0$ follows from (5.2), that is, $[y, z] \in J$.

In the structural theory of Lie algebras an important role is played by a special bilinear form K, whose definition is connected with the concept of the trace of a linear mapping. The trace of the linear mapping $A: L \to L$ is defined from the trace of the matrix of this mapping in any basis $\{u_\alpha\} \subset L$. If this matrix is (A^α_β), then the trace, denoted by tr, is given by the formula $\operatorname{tr} A = A^\alpha_\alpha$. Since the quantity $\operatorname{tr} A$ does not depend on the basis in L, this formula determines the mapping $\operatorname{tr}: \mathscr{L}(L) \to R$. The essential property of the mapping tr is its symmetry relative to composition: For every $A, B \in \mathscr{L}(L)$ the property

$$\operatorname{tr}(A \circ B) = \operatorname{tr}(B \circ A)$$

holds.

The symmetric bilinear form K, defined by

$$K\langle x, y\rangle = \operatorname{tr}(\operatorname{ad}(x) \circ \operatorname{ad}(y)) \tag{5.3}$$

is called *Killing's form*. This form is invariant, which is established using the easily verified identity

$$\operatorname{ad}([x, u]) \circ \operatorname{ad}(y) + \operatorname{ad}(x) \circ \operatorname{ad}([y, u]) = A \circ \operatorname{ad}(u) - \operatorname{ad}(u) \circ A,$$

where $A = \operatorname{ad}(x) \circ \operatorname{ad}(y)$, and the property of symmetry of the mapping tr.

With the aid of Killing's form the radical R of the Lie algebra L can be characterized.

Theorem. *The radical R of the finite-dimensional Lie algebra L is a set of elements $x \in L$ for which $K\langle x, y\rangle = 0$ for any element $y \in L^{(1)}$.*†

Let $C^\alpha_{\beta\gamma}$ be the structural constants of the r-dimensional Lie algebra L in the basis $\{u_\alpha\}$. If $x = x^\alpha u_\alpha$ in this basis, then the matrix of the linear mapping $\operatorname{ad}(x): L \to L$, calculated by equations (1.2) and (13.2.2), is

$$(\operatorname{ad}(x)) = (C^\alpha_{\beta\gamma} x^\gamma). \tag{5.4}$$

If $y = y^\alpha u_\alpha$ in the same basis, then the value of Killing's form in the basis $\{u_\alpha\}$ with the vectors $x, y \in L$, calculated according to (5.3), is

$$K\langle x, y\rangle = K_{\alpha\beta} x^\alpha y^\beta, \qquad K_{\alpha\beta} = C^\tau_{\sigma\alpha} C^\sigma_{\tau\beta} \qquad (\alpha, \beta = 1, \ldots, r). \tag{5.5}$$

†For a proof see Jacobson, "Lie Algebras."

6. Structural Properties

Here are collected the facts characterizing the structural properties of solvability and semisimplicity of the Lie algebra, which are useful in classification problems for the group analysis of differential equations.[†]

The following theorem holds if the Lie algebra L is considered over the field of real numbers.

Lie Theorem. *For any r-dimensional solvable Lie algebra L (over the field of real numbers) there exists a chain of ideals*

$$L = L_r \supset L_{r-1} \supset \cdots \supset L_1 \supset \{0\}$$

in which the dimensionality of the ideal L_s is equal to s ($s = 1, 2, \ldots, r$).

With the aid of the Killing's form K the following two criteria of Cartan are stated. The form K is said to be *degenerate* if $\det(K_{\alpha\beta} = 0)$. In the opposite case the form is called *nondegenerate.*

Criterion of Solvability. *The Lie algebra L is solvable if and only if* $K\langle x, x\rangle = 0$ *for any* $x \in L^{(1)}$.

Criterion of Semisimplicity. *The Lie algebra L is semisimple if and only if the Killing's form is not degenerate.*

The classification of a semisimple Lie algebra is based on the following theorem.

Structural Theorem *The finite-dimensional Lie algebra L is semisimple if and only if there is the decomposition*

$$L = L_1 \oplus L_2 \oplus \cdots \oplus L_k,$$

where the L_i are ideals, which are simple Lie algebras.

With the aid of this theorem it is established that every ideal of a semisimple Lie algebra is semisimple and for a semisimple Lie algebra $L^{(1)} = L$.

7. Optimal Subalgebraic Systems

The problem of enumeration of all subalgebras of a given finite-dimensional Lie algebra L is essential for the group analysis of differential equations. For this the group Int L of the inner automorphisms can be considered as known; it is found without any difficulty according to the algorithm presented in Section 14.4. Since, under the operation of an automorphism,

[†] For proofs of theorems see Jacobson, "Lie Algebras."

every subalgebra transforms into a subalgebra of the same dimensionality, so the present problem can be solved correctly, up to transformations determined by the inner automorphisms.

The subalgebras N and M of the Lie algebra L are called *similar* if there is an inner automorphism $A \in \operatorname{Int} L$ such that $A(N) = M$.

By this relation of similarity, which is obviously a set theoretical indicator of equivalence, all subalgebras of the given Lie algebra L are decomposed into classes of similar algebras.

The union of the representatives of similar algebra classes of given dimensionality s (one from every class) will be called the *optimal system* (*of order* s) and denoted by the symbol Θ_s.

So the solution of the stated problem for the finite-dimensional Lie algebra L^r must be tables of optimal systems for every $s = 1, 2, \ldots, r - 1$.

The search for subalgebras of the given dimensionality $s > 1$ is reduced to an algebraic problem. Let $\{u_\alpha\}$ be a basis of the Lie algebra L^r and $C^\alpha_{\beta\gamma}$ be its structural constants in this basis. It is obvious that the vectors $x_i = x_i^\alpha u_\alpha$ $(i = 1, \ldots, s < r)$ generate a basis of the subalgebra N of dimensionality s if and only if the rank of the matrix (x_i^α) is equal to s and the relations $[x_i, x_j] = N^k_{ij} x_k$, or

$$\operatorname{rank}(x_i^\alpha) = s, \qquad C^\alpha_{\beta\gamma} x_i^\beta x_j^\gamma = N^k_{ij} x_k^\alpha \qquad (i, j = 1, \ldots, s;\ \alpha = 1, \ldots, r), \tag{7.1}$$

are satisfied, where N^k_{ij} are the structural constants of the subalgebra N. Consequently, the problem is reduced to the solution of the algebraic equations (7.1) with the unknowns x_i^α, N^k_{ij}.

This difficult algebraic problem has not yet been completely solved, in the sense of the formation of a definite algorithm according to which the optimal system Θ_s can be uniquely constructed. The determination of optimal systems can be done, relatively easily, only for small dimensionality $s \leq 3$. Even in those cases the solution is accomplished by a "semiprimitive" method, which requires choices from a few possibilities at certain stages of the work, and the "simplicity" of the results obtained is not clear.

Thus the theory for constructing optimal systems of subalgebras needs to be developed further. There is a general result for the given problem that gives an important addition to the Levi theorem in Sections 13.9. For the sake of simplicity it is stated in a weak form.

The Maltsev–Harish-Chandra Theorem. *If $L = R \oplus N$, where R is the radical and N the Levi factor of the finite-dimensional Lie algebra L, then for any semisimple subalgebra $N_1 \subset L$ there exists an automorphism $A \in \operatorname{Int} L$ such that $A(N_1) \subset N$.*†

† For a proof see Jacobson, "Lie Algebras."

8. Small Dimensionality

During the construction of optimal systems for a given Lie algebra L a special role is played by the center Z. The point is that every vector $z \in Z$ is invariant relative to any automorphism $A \in \operatorname{Int} L$. Indeed, as $[z, u] = 0$ for any $u \in L$, so the unique solution of the problem (4.1) for the element $z' = A_u(t)\langle z \rangle$, having the form $\partial_t z' = [z', u]$, $z'(0) = z$, is $z' = z$. Thus the central elements generate subalgebras (ideals), which cannot be changed by any automorphisms. They can be included as a direct term in any subalgebra of the Lie algebra L. This means that if the optimal systems are known for the factor algebra L/Z, then they can be considered as known for the entire Lie algebra L. Consequently, it is sufficient to describe methods for constructing optimal systems only for Lie algebras with null centers.

Let the basis $\{u_\alpha\}$ be fixed in a Lie algebra L^r with a null center. The common inner automorphism of this Lie algebra depends on r parameters and is given by equation (4.6), in which it is necessary to assume $s = 0$.

The construction of system Θ_1 begins with the selection of a nonnull vector $x = x^\alpha u_\alpha$ and its image $x' = A\langle x \rangle$, obtained by the operation of the common automorphism $A = A(t^1, \ldots, t^r)$. If (A^α_β) is the matrix of the automorphism A in the basis $\{u^\alpha\}$, then the components of the vectors x' in the basis are given by an equation of the form (4.5)

$$x'^\alpha = A^\alpha_\sigma x^\sigma \qquad (\alpha = 1, \ldots, r).$$

The next step is the selection of values of the parameters $t^1, \ldots, t^r$, on which the automorphism A depends, to achieve the maximum possible simplification of the set $(x'^1, \ldots, x'^r)$ of components of the vector x'. This permits the choice of the simplest representative of each class of similar algebras to which the element x belongs. Usually, this means choosing the maximum possible number of null values for these components. Appearing here are the various alternative cases that give the classes of one-parameter subalgebras and, from them, the optimal system Θ_1.

During the construction of the optimal system Θ_2 it is possible to assume, in advance, that one of the basis vectors of the two-dimensional subalgebra $N \subset L^r$ is taken from the system Θ_1. Let this vector be $x = x^\alpha u_\alpha$. Corresponding to it, a vector $y = y^\alpha u_\alpha$ is chosen, so that the system of equations (7.1) is satisfied. Here that system is

$$C^\alpha_{\beta\gamma} x^\beta y^\gamma = \lambda y^\alpha + \mu x^\alpha \qquad (\alpha = 1, \ldots, r), \tag{8.1}$$

and it contains $r + 2$ unknowns y^α, λ, and μ. The set of solutions for system (8.1) gives a two-dimensional algebra which contains the vector x. As the vectors x and y can be scaled by any nonzero multipliers, instead of the two-parameter set of systems (8.1), with parameters λ and μ, it is enough to consider such systems in which $(\lambda, \mu) = (0, 0), (1, 0), (0, 1)$, and $(1, 1)$. Obviously, the case $(\lambda, \mu) = (1, 1)$ is reducible to the case $(\lambda, \mu) = (1, 0)$ by the change $y \to y + x$. In the case $(\lambda, \mu) = (0, 1)$ there are no solutions. If $\mu = 0$, then (8.1) has the form of an eigenvalue problem for the linear mapping ad(x), with the matrix $(C^{\alpha}_{\beta\gamma} x^{\beta})$, and can be written in the form $\mathrm{ad}(x)\langle y \rangle = \lambda y$. Its solution is reduced to finding the roots of the characteristic polynomial $\det(\lambda I_L - \mathrm{ad}(x))$, which in the expanded form is

$$\det(\lambda I_L - \mathrm{ad}(x)) = \lambda^r - \tau_1(x)\lambda^{r-1} + \tau_2(x)\lambda^{r-2} - \cdots + (-1)^l \tau_{r-l}(x)\lambda^l, \tag{8.2}$$

where $\tau_{r-l}(x) \neq 0$ and $l \geq 0$. As $\mathrm{ad}(x)\langle x \rangle = 0$ and $x = 0$, it follows that $l > 0$.

Corresponding to every real nonzero root of the polynomial (8.2) is one or more linearly independent solution of equation (8.1). [This is when $\mu = 0$. In the case $l > 1$ there are also one or more solutions corresponding to the root $\lambda = 0$. If $l = 1$, and there are no nonzero real roots, then the element x does not belong to any two-dimensional algebra of this kind—with $\mu = 0$ in (8.1).]

Consequently, it follows that such vectors $x \in L$ can exist that do not belong to any two-dimensional Lie algebra L.

The polynomial (8.2), considered for the variable $x \in L^r$, is called *Killing's polynomial* (or the *characteristic polynomial*) *of the Lie algebra* L^r. The maximal value of the number l in the expression of Killing's polynomial (8.2), obtained when the vector x ranges over the whole space L^r, is called the *rank of the Lie algebra* L^r. That element $x \in L^r$ for which the value l in (8.2) is equal to the rank is called a *regular element of the Lie algebra* L^r.

After the construction of all two-dimensional subalgebras, for every $x \in \Theta_1$, they need to be simplified by the action of automorphisms in a manner analogous to the way it was done for the one-dimensional algebra. As the result, the optimal system Θ_2 is obtained.

The system Θ_3 can be developed by the *method of expansion* of two-dimensional subalgebras. For this take any two-dimensional subalgebra from the system Θ_2, with the basis $\{x_1, x_2\}$, and find a vector $y = y^{\alpha} u_{\alpha}$ such that the triple (x_1, x_2, y) generates a basis of a three-dimensional algebra. For that it is necessary and sufficient that the vector y satisfy the equations, following from system (7.1),

$$\left.\begin{aligned} C^{\alpha}_{\beta\gamma} x_1^{\beta} y^{\gamma} &= \lambda_1 y^{\alpha} + \mu_1 x_1^{\alpha} + \nu_1 x_2^{\alpha} \\ C^{\alpha}_{\beta\gamma} x_2^{\beta} y^{\gamma} &= \lambda_2 y^{\alpha} + \mu_2 x_1^{\alpha} + \nu_2 x_2^{\alpha} \end{aligned}\right\} \quad (\alpha = 1, \ldots, r) \tag{8.3}$$

with $r + 6$ unknowns $y^{\alpha}, \lambda_i, \mu_i, \nu_i$. Every solution of system (8.3) is linearly independent of $\{x_1, x_2\}$ and gives a three-dimensional subalgebra of the Lie algebra L^r.

The question arises: Are all three-dimensional subalgebras obtained by this process? Obviously, this question is equipollent to the following: Do all three-dimensional subalgebras contain a two-dimensional subalgebra? More generally, the problem can be reduced to the question, Are there any three-dimensional real Lie algebras that do not contain two-dimensional subalgebras?

The answer to the last question is in the affirmative, and the Lie algebra with the basis $\{u_1, u_2, u_3\}$ and the structure

$$[u_1, u_2] = u_3, \qquad [u_2, u_3] = u_1, \qquad [u_3, u_1] = u_2 \tag{8.4}$$

is an example. This Lie algebra corresponds to the rotation group for the three-dimensional space R^3 in the sense of Section 12.10 and thus will be called the *Lie algebra of rotations*. This example is the only possible one, as will shortly be demonstrated.

Lemma. *Any three-dimensional real Lie algebra not containing two-dimensional subalgebras is isomorphic to the Lie algebra of rotations* (8.4).

Proof. Suppose L^3 does not have a two-dimensional subalgebra and $x \in L^3$. Then it is necessary that the rank of L^3 is equal to one and its characteristic equation reduces to the quadratic

$$\lambda^2 - \tau_1(x)\lambda + \tau_2(x) = 0,$$

which does not have real roots. Let $\lambda = \lambda_1 \pm i\lambda_2$ be its complex roots ($\lambda_2 \neq 0$). Then the solution y of the equation $\operatorname{ad}(x)\langle y\rangle = (\lambda_1 + i\lambda_2)y$ will also be complex, for example, $y = y_1 \pm iy_2$, where $y_1, y_2 \in L^3$. This equation, after separation of the real and imaginary parts, gives the two relations

$$\operatorname{ad}(x)\langle y_1\rangle = \lambda_1 y_1 - \lambda_2 y_2, \qquad \operatorname{ad}(x)\langle y_2\rangle = \lambda_2 y_1 + \lambda_1 y_2.$$

From these, it is easy to find

$$\operatorname{ad}(x)\langle[y_1, y_2]\rangle = 2\lambda_1[y_1, y_2],$$

whence the requirement of the absence of a two-dimensional subalgebra necessarily forces $\lambda_1 = 0$ and requires the existence of a $\mu \neq 0$ such that

$$[y_1, y_2] = \mu x.$$

Consequently, the vectors y_1, y_2, x are linearly independent and generate a basis in L^3 with the table of the commutators

$$[y_1, y_2] = \mu x, \qquad [y_2, x] = \lambda_2 y_1, \qquad [x, y_1] = \lambda_2 y_2. \tag{8.5}$$

As a result of the transformations $y_1' = ay_1$, $y_2' = ay_2$, $x' = bx$, with real a, b, the constants μ and λ_2 are transformed into

$$\mu' = (a^2/b)\mu, \qquad \lambda_2' = b\lambda_2.$$

Now it is seen that by choosing suitable values of a and b it is possible to have $\lambda_2' = 1$ and $\mu' = \pm 1$. But if $\lambda_2 = 1$ and $\mu = -1$, then from (8.5) the equation

$$[y_2, x + y_1] = x + y_1$$

follows, which shows that the pair $(y_2, x + y_1)$ generates a two-dimensional subalgebra. Thus only one possibility remains, namely, $\lambda_2 = \mu = 1$, and then (8.5) coincides with (8.4), except for a change in notation. ■

9. Example

The Lie algebra of operators L^4, discussed in the example in Section 5.9, is considered. To simplify the calculations, the basis (5.9.9) is exchanged for the basis

$$u_1 = \zeta_1, \qquad u_2 = 2\zeta_4 - \zeta_2, \qquad u_3 = \zeta_3, \qquad u_4 = \zeta_4.$$

A calculation of the commutators of the basis vectors by formula (7.1.2) gives the table of commutators:

	u_1	u_2	u_3	u_4
u_1	0	$3u_1$	u_2	0
u_3	$-3u_1$	0	$3u_3$	0
u_3	$-u_2$	$-3u_3$	0	0
u_4	0	0	0	0

(9.1)

The general vector $x \in L^4$ is written in the form $x = x^\alpha u_\alpha$. In what follows, the subalgebra N, with the basis $\{x_1, \ldots, x_k\}$, is denoted by $N(x_1, \ldots, x_k)$.

Table (9.1) gives the following information about the structure of the Lie algebra L^4. That the center is $Z = N(u_4)$ and that $L^4 = Z \oplus N(u_1, u_2, u_3)$ are seen at once. As will be seen, the subalgebra $N(u_1, u_2, u_3)$ is semisimple, from which, because of the Maltsev–Harish-Chandra theorem in Section 14.7, it follows that $N(u_4) = R$ is a radical and $L^3 = N(u_1, u_2, u_3)$ is a Levi's factor.

The (inner differentiation) mapping ad(v) is automatically calculated by equation (1.2) for the general vector $v = v^1u_1 + v^2u_2 + v^3u_3 \in L^3$ and it turns out to be

$$\begin{aligned} \mathrm{ad}(v)\langle x\rangle = (3v^2x^1 - 3v^1x^2)u_1 + (v^3x^1 - v^1x^3)u_2 \\ + (3v^3x^2 - 3v^2x^3)u_3. \end{aligned} \tag{9.2}$$

The representation of the mapping ad(v) in matrix form follows:

$$\mathrm{ad}(v) = \begin{bmatrix} 3v^2 & -3v^1 & 0 \\ v^3 & 0 & -v^1 \\ 0 & 3v^3 & -3v^2 \end{bmatrix}. \tag{9.3}$$

The Killing's form (5.2) is found by multiplication of two matrices (9.3),

$$K\langle x, y\rangle = 6(3x^2y^2 - x^1y^3 - x^3y^1),$$

which is obviously not degenerate. Thus the semisimplicity of the Lie algebra L^3 follows from the Cartan's criterion in Section 14.6.

It is convenient to find the general automorphism of the group Int L^3 by the method discussed in Section 14.4, that is, by constructing one-parameter groups of automorphisms $A_\alpha(t)$ for every basis vector u_α ($\alpha = 1, 2, 3$). For example, to find the group $A_1(t)$ corresponding to the vector u_1, it is necessary to solve equation (4.1) with $u = u_1$. In (9.2), where it is necessary to assume $v^1 = 1$, $v^2 = v^3 = 0$, the equation becomes

$$\partial_t x'^1 u_1 + \partial_t x'^2 u_2 + \partial_t x'^3 u_3 = -3x'^2 u_1 - x'^3 u_2.$$

This is equipollent to the system

$$\partial_t x'^1 = -3x'^2, \qquad \partial_t x'^2 = -x'^3, \qquad \partial_t x'^3 = 0,$$

which needs to be solved with the initial conditions $x'^\alpha = x^\alpha$ ($\alpha = 1, 2, 3$). This solution is easily found and gives the matrix of the automorphism

$$A_1(t) = \begin{bmatrix} 1 & -3t & \frac{3}{2}t^2 \\ 0 & 1 & -t \\ 0 & 0 & 1 \end{bmatrix}.$$

The analogous calculations for the vectors u_2 and u_3 give the automorphisms

$$A_2(t) = \begin{bmatrix} e^{3t} & 0 & 0 \\ 0 & 1 & 0 \\ 0 & 0 & e^{-3t} \end{bmatrix}, \qquad A_3(t) = \begin{bmatrix} 1 & 0 & 0 \\ t & 1 & 0 \\ \frac{3}{2}t^2 & 3t & 1 \end{bmatrix}.$$

Then one should operate according to the rule (4.6). Here it is convenient to assume $t = 2a$ in A_1, $t = \frac{1}{3}\ln b$ in A_2, and $t = 2c$ in A_3 and construct the product $A = A_1(2a) \circ A_3(2c) \circ A_2(\frac{1}{3}\ln b)$. As a result, the general automorphism $A \in \mathrm{Int}\ L^3$ is obtained and it depends on three parameters, a, b, c:

$$A = \begin{bmatrix} p^2b & -6pa & 6a^2b^{-1} \\ 2pbc & 1 - 12ac & -2ab^{-1} \\ 6bc^2 & 6c & b^{-1} \end{bmatrix}, \tag{9.4}$$

where $p = 1 - 6ac$, $b > 0$, and the identity isomorphism occurs when $a = c = 0$, $b = 1$.

Now, it is possible to find the components of the vector $x' = A\langle x \rangle$ in the basis $\{u_\alpha\}$:

$$
\begin{aligned}
x'^1 &= p^2bx^1 - 6pax^2 + 6a^2b^{-1}x^3, \\
x'^2 &= 2pbcx^1 + (1 - 12ac)x^2 - 2ab^{-1}x^3, \\
x'^3 &= 6bc^2x^1 + 6cx^2 + b^{-1}x^3.
\end{aligned}
\tag{9.5}
$$

The decomposition of the space L^3, on the classes of similar vectors, can be found here using the invariance property of the Killing's form. From this it is found that the function

$$(1/6)K\langle x, x \rangle = 3(x^2)^2 - 2x^1x^3$$

is an invariant of the group Int L^3. This fact is easily verified by a direct calculation based on (9.5). Thus, all the vectors $u = x^\alpha u_\alpha$ (one-dimensional subalgebra) are separated into three nonintersecting classes

$$(\alpha)\ K\langle x, x)\rangle > 0, \qquad (\beta)\ K\langle x, x \rangle < 0, \qquad (\gamma)\ K\langle x, x \rangle = 0.$$

A representative of the class (α) is the vector u_2. It is easily shown, by an application of transformation (9.5), that any vector $u = x^\sigma u_\sigma$ belonging to class (α) can be transformed into u_2. Analogously, any vector from the class (β) is similar to the vector $u_1 + u_3$. It is not possible for both x^1 and x^3 to be zero in the class (γ). Thus, in (9.5), by a suitable choice of the parameter a or c it is always possible to force $x'^2 = 0$. But then, in the class (γ), there results $x^1x^3 = 0$, so that $x^1 = 0$, $x^3 \neq 0$, or $x^1 \neq 0$, $x^3 = 0$. This gives the vectors u_1 and u_3. But the vector u_3 is similar to the vector u_1. Applying the automorphism (9.5) to u_1 gives a vector with components

$$x'^1 = p^2b, \qquad x'^2 = 2pbc, \qquad x'^3 = 6bc^2,$$

which, for $p = 0$, differs from the vector u_3 by a nonessential multiplier. Thus the vector u_1 can be taken as the unique representative of the class (γ). Finally, the optimal system Θ_1 of the Lie algebra L^3 is

$$N(u_1), \qquad N(u_2), \qquad N(u_1 + u_3). \tag{9.6}$$

Next we construct the system Θ_2. One of the basis vectors of any two-dimensional subalgebra can be selected, say, any vector from (9.6). Let this subalgebra be $N(u_1, u)$. Then the second vector can be taken in the form $u = x^2u_2 + x^3u_3$. Since $[u_1, x^2u_2 + x^3u_3] = 3x^2u_1 + x^3u_2$, equation (8.1) becomes

$$3x^2u_1 + x^3u_2 = \lambda x^2u_2 + \lambda x^3u_3 + \mu u_1,$$

from which $3x^2 = \mu$, $x^3 = \lambda x^2$, $0 = \lambda x^3$ and the unique solution (correct up to a multiplier) is $u = u_2$. Therefore, the subalgebra is $N(u_1, u_2)$.

Further, let the unknown subalgebra be $N(u_2, u)$, where $u = x^1u_1 + x^3u_3$. Here equation (8.1) is

$$-3x^1u_1 + 3x^3u_3 = \lambda x^1u_1 + \lambda x^3u_3 + \mu u_2,$$

from which $\mu = 0$, $(\lambda + 3)x^1 = 0$, $(\lambda - 3)x^3 = 0$. If $x^3 = 0$, then again, the subalgebra $N(u_1, u_2)$ is obtained. When $x^3 \neq 0$, $\lambda = 3$ and $x^1 = 0$. This gives the subalgebra $N(u_2, u_3)$.

An analogous analysis shows that no subalgebras of the form

$$N(u_1 + u_2, u)$$

exist. Consequently, the optimal system for the Lie subalgebra L^3 consists of two subalgebras,

$$N(u_1, u_2), \qquad N(u_2, u_3). \tag{9.7}$$

Finally, the optimal system Θ_3 contains only one subalgebra, the Lie algebra $L^3 = N(u_1, u_2, u_3)$ itself. With this, the problem of constructing the optimal systems is solved for L^3.

Addition of the operator qu_4 to any of those calculated and verification that these are subalgebras finally leads to the following table for the optimal systems Θ_k for the Lie algebra L^4:

k	1	2	3	4	
Θ_k	$u_1 + qu_4$	u_1, u_4	u_1, u_2, u_3	u_1, u_2, u_3, u_4	(9.8)
	$u_2 + qu_4$	u_2, u_4	u_1, u_2, u_4		
	$u_1 + u_3 + qu_4$	$u_1 + u_3, u_4$	u_2, u_3, u_4		
	u_4	$u_1, u_2 + qu_4$			
		$u_2 + qu_4, u_3$			

Here q is an arbitrary real parameter. Nonsimilar subalgebras correspond to different values of q.

§15. CORRESPONDENCE BETWEEN LIE GROUPS AND LIE ALGEBRAS

1. Lie Algebra of Lie Group

According to Section 12.12 there is a one-to-one correspondence between the classes $Q(B, \phi)$ of isomorphic Banach groups of Lie and the classes $Q(B, C)$ of isomorphic Banach algebras of Lie. In this section it will be shown that this correspondence extends to subordinated objects, for example, subgroups and subalgebras, ideals, etc. It is convenient to introduce the following definition.

The Banach algebra of Lie L is called the *Lie algebra of the Banach Group* (B, ϕ) if the vector space L coincides with B and if the commutator $[u, v]$ in L is given in terms of the structural operator C of the group (B, ϕ) by

$$[u, v] = C\langle u, v\rangle. \tag{1.1}$$

To simplify the analysis, only Lie groups and Lie algebras of finite dimension will be considered. For these cases, because of the definition in Section 12.16 of the structural constants for the group G^r and property (13.2.2) of the structural constants for the Lie algebra L^r, the following deduction is a consequence of the previous definition. The Lie algebra L^r is a Lie algebra of the group G^r if and only if L^r and G^r are defined on the same r-dimensional Banach space $B = R^r$, and if in B there exists a basis in which the structural constants of the Lie algebra L^r are equal to the corresponding structural constants of the group G^r.

The correspondence established by the definition between Lie algebras and Lie groups is retained under isomorphisms. Indeed, an isomorphism of the group G^r onto the group G'^r, the structure of which is defined on the space B', is, according to the definition in Section 12.3 and the criterion in Section 12.16, an invertible mapping $\gamma: B \to B'$ that preserves the structural constants of the group G^r. Thus the structural constants of the Lie algebra L'^r of the group G'^r are equal to the corresponding structural constants of the Lie algebra L^r. Then, from the lemma in Section 13.3, the Lie algebra L'^r is isomorphic to L^r. This last isomorphism, according to the formulas (12.10.2) and (13.3.1), is $\partial\gamma(0)$.

In what follows, the vectors $e \in L^r$ of the Lie algebra L^r of the group G^r are considered as directing vectors of one-parameter subgroups of G^r (see Section 12.5). This means that if $\{g(t)\}$ is a one-parameter group in G^r, then $\partial_t g(0) = e$. The description of the correspondence between Lie groups and Lie algebras is connected with the important property of the group commutator $q: B^2 \to B$ operating by $q(a, b) = a \cdot b \cdot a^{-1} \cdot b^{-1}$. Select the one-parameter subgroups $\{g_1(t)\}, \{g_2(t)\} \subset G^r$. For any $t \in R$, $t \geq 0$, the group commutator

$$q(t) = g_1(\sqrt{t}) \cdot g_2(\sqrt{t}) \cdot g_1^{-1}(\sqrt{t}) \cdot g_2^{-1}(\sqrt{t})$$

is discussed in the following lemma.

Lemma. (*About the commutator*) *If* $\partial_t g_i(0) = e_i$ $(i = 1, 2)$, *then*

$$\partial_t q(0) = [e_1, e_2]. \tag{1.2}$$

Proof. From the theorem in Section 12.5, for every one-parameter subgroup $\{g(t)\}$ with directing vector e the representation is

$$g(t) = te + \tfrac{1}{2}t^2\partial A(0)\langle e, e\rangle + t^3F(t),$$

where the function $F: R \to B$ is bounded in the null neighborhood. The calculations, analogous to those of the proof of the lemma in Section 7.10, lead to the expression

$$\partial_t q(0) = \partial A(0)\langle e_1, e_2\rangle - \partial A(0)\langle e_2, e_1\rangle,$$

which, according to definition (12.9.1), is equal to the value of the structural operator $C\langle e_1, e_2\rangle$. Thus equality (1.2) follows from (1.1). ■

2. Subgroups and Subalgebras

Let the Banach group of Lie $G^r = (B, \phi)$ be given and the dimensionality of the space B be equal to r, and let $0 < s \leq r$. Consider s-dimensional manifolds $H \subset B$, given parametrically by $H = \sigma(R^s)$, where σ is the invertible mapping $\sigma: R^s \to R$, of the class $C_\infty(R^s)$, satisfying the conditions $\sigma(0) = 0$ and $\operatorname{rank}(\partial\sigma(0)) = s$. Let O be a null neighborhood of the space R^s and $O' = \sigma(O)$ be its image in H under the mapping σ.

The pair (O', ϕ) is called the *subalgebra kernel of the group* (B, ϕ) if $\phi(a, b) \in H$ for any $a, b \in O'$. In the local theory the subgroup kernels, obtained when the neighborhood O' and the space R^s are varying, do not differ and generate an equivalence class, defined by the properties of the manifold H, in a null neighborhood of the space B as small as desired.

The nonempty class of indistinguishable subgroup kernels (O', ϕ) is called an *s-parametric Lie subgroup of the Lie group* (B, ϕ).

This Lie subgroup of the group (B, ϕ), given as a manifold $H \subset B$, will be denoted by (H, ϕ) or H^s, or merely H.

The space R^s and the mapping σ play an auxiliary role in this construction. In particular, R^s can always be considered to coincide with the space $B' \subset B$, where $B' = \partial\sigma(0)(R^s)$ is tangent to the manifold H at zero, and the mapping σ is assumed to be normalized, that is, $\partial\sigma(0) = I_{B'}$.

On the manifold $H = \sigma(B')$ it is always possible to introduce a vector structure by defining addition # and multiplication $*$ of the vectors $a \in H$ with the numbers $\lambda \in R$ by

$$a \# b = \sigma(\sigma^{-1}(a) + \sigma^{-1}(b)), \qquad \lambda * a = \sigma(\lambda\sigma^{-1}(a)),$$

where the usual addition and multiplication of numbers are made on the subspace $B' \subset B$. It is easy to verify, with this vector structure and norm $\|a\| = |\sigma^{-1}(a)|$, that the manifold H is a Banach space.

Comparison of the definition of the subgroup kernel with the definition in Section 12.1 shows that the pair (O', ϕ_H), where ϕ_H is a reduction of the multiplication ϕ to the manifold H, is a group kernel on the Banach space H. *Thus, any Lie subgroup of the group* (B, ϕ) *is itself a Lie group.*

Since only the multiplication law is essential (the vector structure of the space B is not required) in the definition (12.5.1) of a one-parameter subgroup of the Lie group (B, ϕ), any one-parameter subgroup of the Lie subgroup $H \subset (B, \phi)$ is also a one-parameter subgroup (B, ϕ). Thus either every one-parameter subgroup $\{g(t)\}$ of the group (B, ϕ) is locally or entirely included in H or its intersection with H is $\{0\}$.

Lemma. *If* (B, ϕ) *is a canonical group, then any s-parameter Lie subgroup* H^s *is a "plane" manifold* $H^s \subset B$; *that is, it coincides with an s-dimensional subspace of the space B in the null neighborhood O. In addition,* H^s *is a canonical Lie group.*

Proof. Any one-parameter subgroup $\{g(t)\} \subset H^s$ is a straight line in B and thus is local for any $a \in H^s$, $\lambda \in R$ and $\lambda a \in H^s$. Consequently, if $|t| < 1$, then for any $a, b \in O'$ the quantity $t^{-1}\phi(ta, tb) \in H^s$, and since the subspace H^s is closed, the $\lim_{t\to 0} t^{-1}\phi(ta, tb) \in H^s$. That calculation gives

$$\lim_{t\to 0} t^{-1}\phi(ta, tb) = \partial\phi(0, 0)\langle(a, b)\rangle = a + b,$$

so that $a + b \in H^s$. This shows that H^s is locally a vector space and thus coincides with the s-dimensional subspace B' of the space B. As the vector structures of the manifold H^s and the subspace B' also coincide (using σ, the mapping $H^s \to B$ can be made), all one-parameter subgroups of the group (H^s, ϕ) are straight lines in the Lie group H^s; that is, H^s is a canonical group. ■

Theorem. *Let* L^r *be a Lie algebra of the canonical group* $G^r = (B, \phi)$. *If* H^s *is a Lie subgroup of the group* G^r, *then the Lie algebra* L^s *of the group* H^s *is a subalgebra of the Lie algebra* L^s. *Conversely, if the subalgebra* $L^s \subset L^r$ *is given, then there is a subgroup* $H^s \subset G^r$ *for which* L^s *is its Lie algebra.*

Proof. Let H^s be a Lie subgroup of the group G^r. Because of the lemma, $B' = H^s$ is a vector subspace in B. If $e_1, e_2 \in B'$, then the one-parameter subgroups $\{te_1\}$ and $\{te_2\}$ belong to B'. As H^s is a group, so also is $q(t) = (te_1)\cdot(te_2)\cdot(te_1)^{-1}\cdot(te_2)^{-1} \in B'$ for any t. From the lemma in Section 15.1, it follows from this that $\partial_t q(\sqrt{t})|_{t=0} = [e_1, e_2] \in B'$. Thus $B' = L^s$ is a Lie algebra of the group H^s and the operation of commutation in L^s coincides with the operation of commutation in $B = L^r$. According to the definition in Section 13.4, this means that L^s is a subalgebra in L^r.

On the contrary, let the subalgebra $L^s \subset L^r$ be given. According to Section 12.15, the multiplication ϕ can be given by the SCH-series (12.15.5), which expresses $\phi(a, b)$ by means of values of the "long" commutators with elements a, b. Therefore, if $a, b \in B' = L^s$, then from the definition of the subalgebra in Section 13.4 it also follows that $\phi(a, b) \in B'$. Consequently, (B', ϕ) is a Lie subgroup of the group (B, ϕ). In addition, according to the construction, the quantity C on B' is the structural operator of the subgroup (B', ϕ), whence L^s is its Lie algebra. ■

3. Normal Divisors and Ideals

In Section 12.3 the idea of an automorphism of the group (B, ϕ) has already been defined as a transformation $P: B \to B$, of the class $C_\infty(O)$ on a null neighborhood $O \subset B$, which for any $a, b \in O$ satisfies the relation

$$P(a, b) = P(a) \cdot P(b). \tag{3.1}$$

In the group $\mathrm{Aut}(B, \phi)$ of all automorphisms of the group $G^r = (B, \phi)$, special automorphisms, generated by elements of the group G^r itself, are distinguished.

The transformation $P_a: B \to B$ of the group G^r, defined by

$$P_a(b) = a^{-1} \cdot b \cdot a, \tag{3.2}$$

is called the *inner automorphism of the group* G^r.

It is obvious that for P_a, the relation (3.1) is satisfied, so any inner automorphism is an automorphism of the group G^r. In what follows it will be shown that the union of all inner automorphisms P_a of the group G^r, taken when a assumes values over the whole group G^r, generates a local Lie group. This group is called the *group of inner automorphisms of the group* $G^r = (B, \phi)$ and will be denoted by $\mathrm{Int}(B, \phi)$ or $\mathrm{Int}\, G^r$.

If (B, ϕ) is a canonical group, then, from the theorem in Section 12.7, each of its automorphisms is a linear transformation of the space B. In this case the inner automorphisms can be manifestly expressed by means of the auxiliary mappings $A(a)$ and $V(a)$ of the group (B, ϕ).

Lemma. *If* (B, ϕ) *is a canonical group, then*

$$a^{-1} \cdot b \cdot a = V(a) \circ A(-a)\langle b \rangle. \tag{3.3}$$

Proof. Since P_a is a linear mapping, $P_a\langle b\rangle = \partial P_a(0)\langle b\rangle$. To calculate the derivative it should be noted that in a canonical group $a^{-1} = -a$, whence

$$\partial P_a(0) = \partial_b \phi(\phi(-a, b), a)|_{b=0} = \partial_1 \phi(-a, a) \circ \partial_2 \phi(-a, 0).$$

Now, $\partial_2\phi(-a, 0) = A(-a)$ and differentiation of the identity $\phi(-a, a) = 0$ give $\partial_1\phi(-a, a) = \partial_2\phi(-a, a) = V(a)$ (see Section 12.4). Thus $\partial P_a(0) = V(a) \circ A(-a)$. ■

The subgroup $H^s \subset G^r$ is called a *normal divisor of the group* G^r if it is invariant relative to any inner automorphism $P_a \in \operatorname{Int} G^r$. This means that for every $a \in G^r$ and any $b \in H^s$ the quantity $P_a(b) \in H^s$.

Theorem. *Let H^s be a Lie subgroup of the canonical group $G^r = (B, \phi)$, L^r be a Lie algebra of the group G^r, and L^s be a Lie algebra of the Lie subgroup H^s. Then H^s is a normal divisor of the group G^r if and only if L^s is an ideal in L^r.*

Proof. A repetition of the first part of the proof of the theorem in Section 15.2, with the vectors $e_1 \in B'$, $e_2 \in B$, shows that if $B' = H^s$ is a normal divisor of the group G^r, then $[e_1, e_2] \in B'$. According to the definition in Section 13.4 this means that $B' = L^s$ is an ideal in L^r. Conversely, if L^s is an ideal in L^r and $a \in L^r$, $b \in L^s$, then any "long" commutator constituted from the vectors a, b belongs to the subspace $L^s = B'$. Since the multiplication ϕ can be represented in a form of the SCH-series (12.15.5) also $\phi(a, b) - a \in B'$. Thus the linear mapping $A(a) = \partial_2\phi(a, 0)$ maps B' into B'. But then this is also the case for $V(a)$, $A(-a)$, and $V(a) \circ A(-a)$. According to the lemma this means that the subspace B', which is the subgroup $H^s \subset G^r$ according to the theorem in Section 15.2, is a normal divisor of the group G^r. ■

4. Centers

The element $b \in G^r$ is called a *central element* if for any $a \in G^r$ the equality $a \cdot b = b \cdot a$ is satisfied. Rewritten in the form $b = a^{-1} \cdot b \cdot a$, this shows that the given definition is equipollent to the following: The element $b \in G^r$ is central if it is invariant under any inner automorphism P_a.

The union of all central elements of the group G^r generates a subgroup, called the *center of the group*.

The center is a Lie subgroup of the group G^r. Indeed, if $G^r = (B, \phi)$ is a canonical group, then any of its inner automorphisms is a linear mapping. Therefore, together with b, all elements tb remain invariant, that is, the entire one-parameter subgroup passing through the central element b remains invariant. From a line in the proof of the lemma in Section 15.2, it then follows that if a and b belong to the center, then $a + b$ also belongs to it. Thus, the center locally, in some null neighborhood $O \subset B$, coincides with the vector space $B' \subset B$. Consequently, the center of a canonical group is its Lie subgroup. Finally, it is easy to verify that under a group isomorphism a center transforms into a center.

Obviously the center is invariant relative to any inner automorphisms. Therefore it is a normal divisor. The following theorem determines the correspondence between the centers of the group G^r and its Lie algebra L^r.

Theorem. *If $H^s = B'$ is a center of the canonical group $G^r = (B, \phi)$, then the subspace $B' \subset B$ is a center of the Lie algebra L^r of the group G^r.*

Conversely, if Z is a center of the Lie algebra L^r, then $H^s = Z$ is a center of the group G^r.

Proof. A repetition of the first part of the proof of the theorem in Section 15.3, applied to the center $H^s = B'$ of the group G^r, shows that $[e_1, e_2] = 0$ for any $e_1 \in B'$, $e_2 \in B$. According to the definition in Section 13.4 this means that $B' = L^s$ is a center of the Lie algebra L^r. Conversely, if Z is a center in L^r and $a \in L^r$, $b \in Z$, then all "long" commutators consisting of the vectors a, b are equal to zero and the SCH-series (12.15.5) assumes the form $\phi(a, b) = a + b$. From this, $a \cdot b = b \cdot a$ and b is a central element of the group G^r. Consequently, the elements $b \in Z$ are the only central elements in the group G^r; that is, $H^s = Z$ is its center. ■

5. Factor Groups and Factor Algebras

Let H^s be a subgroup of the group G^r. For the elements, $a, b \in G^r$, the relation $\backsim$ is defined by the following rule: $b \backsim a$ if $ba^{-1} \in H^s$. It is obvious that the relation $\backsim$ is a set-theoretical criterion of equivalence. Using the relation $\backsim$, the group G^r is separated into nonintersecting classes of equivalent elements, relative to the subgroup H^s. Here the classes will be denoted by $\bar{a}, \bar{b}, \ldots$.

If H^s is a normal divisor of the group $G^r = (B, \phi)$, then for the classes $\bar{a}, \bar{b}, \ldots$ the multiplication operation

$$\psi(\bar{a}, \bar{b}) = \overline{\phi(a, b)} \tag{5.1}$$

is defined, where $a \in \bar{a}, b \in \bar{b}$. With this, the value $\psi(\bar{a}, \bar{b})$ does not depend on the choice of the elements a, b—representatives of the classes $\bar{a}, \bar{b}$. It is easy to verify that the multiplication ψ satisfies the axioms of group multiplication, thereby justifying the following definition.

The union of equivalent classes of elements $\{\bar{a}\}$ of the group G^r, relative to its normal divisor H^s, is called *the factor group of the group G^r with respect to H^s*. A factor group is denoted by G^r/H^s.

It will be shown that the factor group G^r/H^s is a Lie group.

Let $G^r = (B, \phi)$ be a canonical group. Then its normal divisor $H^s = B'$ is a subspace in B which can be decomposed into the direct sum $B = B' \oplus B''$. Therefore, every vector $a \in B$ can be uniquely represented in the form

$a = a' + a''$, where $a' \in B'$, $a'' \in B''$. With this, as is shown in Section 15.3, the subspace $B' = L^s$ is an ideal in the Lie algebra $B = L^r$ of the group (B, ϕ). This means that for any $a, b \in B$ the relation

$$[a, b']'' = 0 \tag{5.2}$$

is satisfied.

Lemma. *Let $a \in \bar{a}$, $b \in \bar{b}$. Then $\bar{b} = \bar{a}$ if and only if $b'' = a''$.*

Proof. The relation $a^{-1} = -a$, which holds in a canonical group, is used. First, it is established that the assertion that a subspace $H^s = B'$ is a normal divisor of the group (B, ϕ) is equipollent to the relation

$$\phi''(a, b) = \phi''(a'', b'') \tag{5.3}$$

for any $a, b \in B$. Indeed, if B' is a normal divisor, then $[a, b]'' = [a'', b'']''$ because of (5.2), and the analogous equality is correct for any "long" commutator included in the SCH-series (12.15.5). Thus equation (5.3) holds. Conversely, if (5.3) is satisfied, then for any $a \in B$, $b' \in B'$ the equalities

$$\phi''(\phi(a, b'), -a) = \phi''(\phi''(a, b'), -a'') = \phi''(a'', -a'') = 0$$

result, from which the inclusion $a \cdot b' \cdot a^{-1} \in B'$ follows. That is why B' is a normal divisor of the group (B, ϕ).

If $\bar{b} = \bar{a}$, then $b \cdot a^{-1} \in B'$, or $\phi''(b, -a) = 0$. Because of (5.3) this means that $\phi''(b'', -a'') = 0$, so that $b'' = a''$. Conversely, if $b'' = a''$, then from (5.3) it follows that $\phi''(b, -a) = 0$, and this implies the inclusion $\phi(b, -a) \in B'$, or $b, a^{-1} \in H^s$. Thus $b \backsim a$ and $\bar{b} = \bar{a}$. ■

From equation (5.3) it follows that the elements $a'' \in B''$ and that together with the multiplication law ϕ'' they form a local Lie group. From the lemma, the classes $\bar{a}$ can be considered as elements of this group instead of a'', and the multiplication law (5.1) can be rewritten in the form $\psi(\bar{a}, \bar{b}) = \phi''(a'', b'')$. Now it is shown that the factor group G^r/H^s of the canonical group G^r is a local Lie group. For the arbitrary (not canonical) group G'^r this fact is correct, since under group isomorphisms a normal divisor transforms into a normal divisor and a class of equivalent elements into a class of equivalent elements. Thus, if γ is an isomorphism and $G'^r = \gamma(G^r)$, $H'^s = \gamma(H^s)$, then also $G'^r/H'^s = \gamma(G^r/H^s)$, and if G^r/H^s is a Lie group, then so is G'^r/H'^s.

Theorem. *Let J^s be an ideal in the Lie algebra L^r of the group G^r and a Lie algebra of the normal divisor H^s of the group G^r. Then the factor algebra L^r/J^s is a Lie algebra of the factor group G^r/H^s.*

Proof. Because of the correspondence between Lie groups and Lie algebras under isomorphisms, it is sufficient to give a proof for the canonical group $G^r = (B, \phi)$. In this case the term $B' = J^s$ in the decomposition $B = B' \oplus B''$ is the Lie algebra of the normal divisor $H^s = B'$. From the proof of the lemma it follows that the operation of commutation in the Lie algebra of the factor group G^r/H^s is given by $[\bar{a}, \bar{b}] = [a'', b'']''$. But the classes $\bar{a}$ are the elements of the factor algebra L^r/J^s, and a simple verification of the definition in Section 13.5 shows that their commutator coincides with $[a'', b'']''$. ■

6. Homomorphisms

Let the Lie groups G, G_1 and the homomorphism $\gamma: G \to G_1$ be given. The union G_0 of all elements $b \in G$, for which $\gamma(b) = 0$, is called a *kernel of the homomorphism* γ.

From the definition in Section 12.3 for the group homomorphism, it follows that if $b \in G_0$, then for any $a \in G$

$$\gamma(a^{-1} \cdot b \cdot a) = \gamma(a^{-1}) \cdot \gamma(b) \cdot \gamma(a) = (\gamma(a))^{-1} \cdot \gamma(a) = 0,$$

that is, $a^{-1} \cdot b \cdot a \in G_0$. According to the definition in Section 15.3 this means that G_0 is a normal divisor of the group G and is also a Lie group. Besides, the image $\gamma(G)$ in the group G_1 is a subgroup whose relation to the group G is given by the following lemma.

Lemma. *The group $\gamma(G)$ is isomorphic to the factor group G/G_0.*

Proof. Let $\backsim$ be a relation of equivalence in the group G relative to its normal divisor G_0. If $b \backsim a$, that is, $b \cdot a^{-1} \in G_0$, then $\gamma(b \cdot a^{-1}) = \gamma(b) \cdot (\gamma(a))^{-1} = 0$, from which $\gamma(b) = \gamma(a)$. Conversely, if $\gamma(b) = \gamma(a)$, then $b \backsim a$, so the elements $\gamma(a) \in \gamma(G)$ are in one-to-one correspondence with the elements $\bar{a}$ of the factor group G/G_0. Thus, the mapping $\gamma_1: G/G_0 \to \gamma(G)$, defined for $a \in \bar{a}$ by $\gamma_1(\bar{a}) = \gamma(a)$, is an isomorphism. ■

The elements of the group G_1, which lie outside its subgroup $\gamma(G)$, do not influence the properties of the considered homomorphism γ. Because of this, one can limit consideration of homomorphisms. It is sufficient to study the behavior of the Lie group under homomorphisms for the canonical groups.

Theorem. *Let L^r be a Lie algebra of the canonical group $G^r = (B, \phi)$ and $\gamma: G^r \to G_1^{r_1}$ be a homomorphism of the group G^r on the canonical group $G_1^{r_1} = (B_1, \phi_1)$, the Lie algebra of which is $L_1^{r_1}$. Then γ is a homomorphism of the Lie algebra L^r on the Lie algebra $L_1^{r_1}$, and its kernel is a Lie algebra of the kernel $G_0 \subset G^r$ of the group homomorphism.*

Proof. From the theorem in Section 12.7, the homomorphism γ can be considered as a linear mapping $B \to B_1$. From the lemma in Section 12.10, which holds not only for isomorphisms but for homomorphisms also, it follows that $\gamma = \partial\gamma(0)$ is a homomorphism of L^r into $L_1^{r_1}$. Actually, the equation $\gamma\langle a\rangle = 0$ defines in B the same subspace B_0, regardless of whether γ is considered a group homomorphism or a homomorphism of the Lie algebra. Thus, the ideal $J_0 \subset L^r$, being a Lie algebra of the kernel G_0 of the group homomorphism γ, coincides with B_0, and it is the kernel of the mapping γ, as a homomorphism of the Lie algebra. ■

7. Groups of Inner Automorphisms

The concept of inner automorphism was independently defined for Lie algebras in Section 14.4 and for Lie groups in Section 15.3. Here the correspondence between the groups of inner automorphisms of Lie groups and Lie algebras is established.

Lemma 1. *The inner automorphisms of the group G^r, with the kernel Z, generate the Lie group* Int G^r, *isomorphic to the factor group G^r/Z.*

Proof. From definition (3.2) of the inner automorphism P_a it follows that $P_a \circ P_c = P_{a\cdot c}$. Thus the mapping $\gamma\colon G^r \to \operatorname{Int} G^r$, operating by $\gamma(a) = P_a$, is a homomorphism of the group $G^r = (B, \phi)$ in the group of transformations of the space B. Its kernel is determined by the equation $\gamma(a) = I_B$, that is, by the equation $a^{-1}\cdot b\cdot a = b$ for any $b \in G^r$. This equality is equipollent to the inclusion $b \in Z$. The rest of the proof follows from the lemma in Section 15.6 and the conclusions in Section 15.5. ■

Lemma 2. *Let $\gamma\colon a \to P_a$ be a homomorphism of the canonical group G^r in the group of linear transformations of the finite-dimensional space X and let $E_a = \partial_t P_{ta}|_{t=0}$ be directing vectors of one-parameter subgroups of the group $\gamma(G^r)$. Then, the operation of commutation is the Lie algebra $\{E_a\}$ of the group $\gamma(G^r)$ is given by*

$$[E_a, E_b] = E_a \circ E_b - E_b \circ E_a. \tag{7.1}$$

Proof. From the theorem in Section 15.1, the commutator $[E_a, E_b]$ is equal to the derivative of the group commutator

$$q(t) = P_{\sqrt{t}a} \circ P_{\sqrt{t}b} \circ P_{\sqrt{t}a}^{-1} \circ P_{\sqrt{t}b}^{-1}$$

with respect to t, at $t = 0$. From the definition of E_a, the representation

$$P_{ta} = I_X + tE_a + t^2 F_a + t^3\Phi(t, a)$$

follows, where the norm $|\Phi(t, a)|$ is bounded as $t \to 0$. To calculate $q(t)$ consider the equality $P_a^{-1} = P_{-a}$, which holds because the group G^r is canonical and F_a is even (i.e., $F_{-a} = F_a$), from which the desired expression

$$q(t) = t(E_a \circ E_b - E_b \circ E_a) + t^{3/2}\Phi_1(t, a, b)$$

is obtained, where the norm $|\Phi_1(t, a, b)|$ is bounded as $t \to 0$.

Theorem. *The Lie algebra of the group* Int G^r *is isomorphic to the adjoined algebra* $\mathscr{D}$ *of the Lie algebra* L^r *of the group* G^r.

Proof. Let $G^r = (B, \phi)$ be a canonical group. From the lemma in Section 15.3, its inner automorphisms are linear transformations of the space B. They are expressed through the auxiliary mappings $A(a)$, $V(a)$ of this group by the formula

$$P_a = V(a) \circ A(-a).$$

According to lemma 1, the mapping $\gamma: a \to P_a$ is a homomorphism of the group G^r on the group Int G^r, that is, the conditions of lemma 2 are satisfied. Thus the operation of commutation in the Lie algebra of the group Int G^r can be calculated by equation (7.1), for which it is sufficient to find the directing vectors E_a of the one-parameter subgroups $\{P_{ta}\}$. In these calculations it is necessary to take into consideration two relations: The first is $\partial V(0) + \partial A(0) = 0$, which is true for any group (B, ϕ), and the second means anti-symmetry of the bilinear mapping $\partial V(0): B \times B \to B$,

$$\partial V(0)\langle h, k\rangle + \partial V(0)\langle k, h\rangle = 0,$$

which holds because of canonicality of the group (B, ϕ). The last relation is equipollent to the equality $\partial V(0)\langle a, a\rangle = 0$, which is easily obtained by differentiating with respect to t, at $t = 0$, the canonical condition $V(ta)\langle a\rangle = a$, established in Section 12.5. Taking into account these relations, the calculation in the vector form for any vector $h \in B$ gives

$$\begin{aligned} E_a\langle h\rangle &= \partial_t P_{ta}\langle h\rangle|_{t=0} \\ &= \partial_t(V(ta)\langle A(-ta)\langle h\rangle\rangle)|_{t=0} \\ &= \partial V(0)\langle a, h\rangle - \partial A(0)\langle a, h\rangle = 2\partial V(0)\langle a, h\rangle \\ &= \partial V(0)\langle a, h\rangle - \partial V(0)\langle h, a\rangle \\ &= C\langle h, a\rangle = [h, a] = \mathrm{ad}(a)\langle h\rangle, \end{aligned}$$

where the last equality follows from definition (14.1.2). From this $E_a = \mathrm{ad}(a)$, that is, the elements of the Lie algebra of the group Int G^r are adjoined mappings of the Lie algebra L^r of the group G^r. In addition, from lemma 2

$$[E_a, E_b] = \mathrm{ad}(a) \circ \mathrm{ad}(b) - \mathrm{ad}(b) \circ \mathrm{ad}(a),$$

which, according to Section 14.2, is exactly the commutator of the adjoined algebra $\mathscr{D}$.

So in the case of the canonical group G^r the Lie algebra of the group Int G^r coincides with the adjoined algebra $\mathscr{D}$ of the Lie algebra of the group G^r. The transition to the arbitrary group G'^r is by means of the isomorphism $G^r \to G'^r$, which generates the isomorphisms Int $G^r \to$ Int G'^r and $L^r \to L'^r$, and thus also isomorphisms of the Lie algebras of the groups Int G^r and Int G'^r and of the adjoined algebras $\mathscr{D}$ and $\mathscr{D}'$. Therefore, the Lie algebra of the group Int G'^r is isomorphic to the adjoined algebra $\mathscr{D}'$ (but, generally speaking, does not coincide with it). ■

Consequence 1. *In the case of the canonical group* $G^r = (B, \phi)$, *the group* Int G^r *coincides with the group* Int L^r.

Indeed, they both are groups of linear transformations of the space B with identical, as was shown, Lie algebras coinciding with $\mathscr{D}$.

It is interesting to note that this fact can be established by direct calculation from the mapping $V(a) \circ A(-a)$ using equations (12.14.2) and (12.14.4) and comparing the result with (14.4.2).

Consequence 2. *The groups of inner automorphisms* Int G^r *and* Int L^r *of the group* G^r *and its Lie algebra* L^r *are isomorphic.*

This follows from the theorem and Consequence 2 of the theorem in Section 12.12.

8. Optimal Subgroup Systems

The subgroups H' and H of the group G^r are called *similar* if there is an inner automorphism $P_a \in$ Int G^r under which $H' = P_a(H)$.

This similarity relation of subgroups is obviously a set-theoretical criterion of equivalence, under which all possible subgroups of the group G^r are separated into nonintersecting classes of similar subgroups. Then, the problem of determining the *optimal system of subgroups* of the given group G^r, that is, the set of representatives of the classes of similar groups, is a problem analogous to that problem considered in Section 14.7 for Lie algebras. The optimal system of s-parameter subgroups of the group G^r $(0 < s \leq r)$ will be denoted, as in the case of Lie algebras, by Θ_s.

From the previous results it follows that the optimal systems of subgroups of the group G^r and subalgebras of its Lie algebra L^r are in one-to-one correspondence. Consequently, the problem of determining the optimal system of subgroups is reduced to the corresponding problem for subalgebras.

In applications, one usually constructs the optimal system of subalgebras, from which the optimal system of subgroups is reconstructed. This method seems more algorithmical because the group Int L^r is always a group of linear transformations of the main space, differing from the group Int G^r, to which it is equal, generally speaking, only in the case of the canonical group G^r.

§16. LIE GROUPS OF TRANSFORMATIONS

1. Definition

The concepts dealing with Lie groups of transformations are very analogous to those which were developed in Chapter I, with the change of the additive group R to a group (B, ϕ). The elements of the Lie theory of groups of transformations, depending on a finite number of parameters, will be expounded here, in connection with which it is assumed that the group (B, ϕ) is the r-parameter group G^r.

Let the Banach space X and the group (B, ϕ) be given. The representation of the kernel (O, ϕ) of the group (B, ϕ) in the group $\tau(X)$ of transformations of the space X is considered. Such a representation is the mapping $\pi: O \to \tau(X)$, satisfying the condition of a group homomorphism: for any $a, b \in O$

$$\pi(b) \circ \pi(a) = \pi(\phi(a, b)).$$

The representation π is called *exact* if there exists a null neighborhood $O \subset B$ such that for $a \in O$ the equality $\pi(a) = I_X$ is satisfied only if $a = 0$. It is clear that the exact representation is a group isomorphism of the group (B, ϕ) into the group $\tau(X)$.

If (O_1, ϕ) and (O_2, ϕ) are two group kernels of the same group (B, ϕ), then in the local theory the representations $\pi_i: O_i \to \tau(X)$ $(i = 1, 2)$, coinciding on $O_1 \cap O_2$, are considered indistinguishable. Thereby every representation is locally sufficiently characterized by the group (B, ϕ). This permits one to talk about the representations of the groups (B, ϕ) and not about its group kernels.

The representation π gives the operation of the group (B, ϕ) on the space X, where every element $a \in B$ operates on $x \in X$ through the transformation $\pi(a)$ by $x \to \pi(a)(x)$. Thereby the mapping $f: X \times B \to X$, defined by $f(x, a) = \pi(a)(x)$, appears. This mapping f is said to *generate* the representation π.

In the local theory (see Section 1.8), one considers local representations of the space X which are defined on some open set $U \subset X$. In these terms, the following basic definition is stated.

The *local Lie group of local transformations of the space* X is the image $\pi(O)$ of an exact representation $\pi: O \to \tau(X)$ of the null neighborhood $O \subset B$ of the group (B, ϕ) for which the generating mapping $f: U \times O \to X$ is a mapping of the class $C_\infty(U \times O)$.

This group will be denoted by (B, ϕ, π, X) or (G^r, π, X), or just $G^r(f)$. The mapping f will also be called the *generating mapping* of the group $G^r(f)$.

From the given definition, the generating mapping possesses the following properties, which are analogous to those in Section 1.9:

1° $f(x, 0) = x$ for any $x \in U$.
2° $f(f(x, a), b) = f(x, \phi(a, b))$ for any $a, b \in O$, $x \in U$.
3° If $a \in O$ and $f(x, a) = x$ for all $x \in U$, then $a = 0$.
4° $f \in C_\infty(U \times O)$.

These properties can be taken as axioms defining the group $G^r(f)$. Then the former definition can be restated as follows.

Let the group $G^r = (B, \phi)$ be given. The r-parameter family of transformations of the space X, corresponding to the vectors $a \in B$, and operating by $f_a(x) = f(x, a)$, is called the *group* $G^r(f)$ if there is a null neighborhood $O \subset B$ and an open space $U \subset X$ such that the mapping $f: U \times O \to X$ satisfies axioms 1°–4°.

The conditions in this definition are not minimal. Actually, the requirement that the group (B, ϕ) be given in advance is unnecessary and can be replaced by the weaker requirement that the mapping $\phi: O \times O \to B$ from 2° (whose existence is assumed) belong to the class $C_\infty(O \times O)$ together with a little stronger, as compared to 3°, requirement about "essential parameters": If $f(x, a) = f(x, b)$ for all $x \in U$, then $a = b$. Then it is possible to show that ϕ is a multiplication law in some group (B, ϕ), that is, it satisfies the axioms 1°–3° in Section 12.1. In particular, this version of the definition of a continuous group of transformations was used in the classical works of S. Lie and his followers.

The group (B, ϕ) is called a *parameteric group* of the group of transformations (B, ϕ, π, X). According to the given definition, the group (B, ϕ, π, X) is isomorphic to its parametric group. Thus, the groups of transformations are isomorphic if and only if their parametric groups are isomorphic.

Note. Let γ be an isomorphism of the group (B', ϕ') into the group (B, ϕ). Then the representation $\pi: B \to \tau(X)$ transforms into the representation $\pi' = \pi \circ \gamma: B' \to \tau(X)$, which gives the generating mapping $f \to f'$ defined by

$$f'(x, a') = f(x, \gamma(a')), \tag{1.1}$$

where $a' \in B'$. Therefore the group (B', ϕ', π', X) as a set of transformations of the space X coincides with the group (B, ϕ, π, X). Only the parameterization of the group is changed.

Example 1. The one-parameter groups of transformations, considered in Chapter I, are a particular case of the groups $G^r(f)$, and they are obtained if $G^r = (B, \phi)$ is a canonical group with $r = 1$.

Example 2. For any group (B, ϕ), the multiplication ϕ gives a representation π of this group in the group $\tau(B)$ by means of $\pi(a)(x) = \phi(x, a)$. Indeed, from axiom 2° in Section 12.1,

$$\begin{aligned}\pi(b) \circ \pi(a)(x) &= \pi(b)(\phi(x, a)) = \phi(\phi(x, a), b)\\ &= \phi(x, \phi(a, b)) = \pi(\phi(a, b))(x),\end{aligned}$$

from which $\pi(b) \circ \pi(a) = \pi(\phi(a, b))$. The generating mapping of the resulting group of transformations of the space B is the multiplication law ϕ. This group $G_1^r(\phi)$ is called the first parametric group of the group (B, ϕ). By analogy, the second parametric group $G_2^r(\phi)$ of the group (B, ϕ), obtained by means of the operation $\pi(a)(x) = \phi(a, x)$, is determined.

2. Tangential Mapping

The partial derivative with respect to a of the generating mapping f is put into correspondence with the group $G^r(f)$ in the null neighborhood of the space B by

$$\xi(x) = \partial_a f(x, 0). \tag{2.1}$$

According to the definition of the derivative, for every $x \in U$ the value $\xi(x)$ is a linear mapping of the space B into the space X. This mapping $\xi: U \to \mathscr{L}(B; X)$ is called the *tangential mapping of the group* $G^r(f)$.

For every $a \in B$, the value $\xi(x)\langle a \rangle$ is a vector of the space X, and when x assumes values over the set U, then $\xi(x)\langle a \rangle$ gives a vector field on U. Thus the tangential mapping ξ gives a *family of vector fields*, which is, obviously, a vector space. For these reasons, which will be clarified in what follows, it is convenient to denote this vector space by $L^r(\xi)$. Thereby the space $L^r(\xi)$ of the vector fields $\xi_a: X \to X$, corresponding to elements $a \in B$ and operating by

$$\xi_a(x) = \xi(x)\langle a \rangle, \tag{2.2}$$

is associated with every group $G^r(f)$.

The detailed study of the correspondence between the groups $G^r(f)$ and the spaces $L^r(\xi)$ is a fundamental part of the theory of Lie groups of transformations.

Let $A(a)$, $V(a)$ be auxiliary mappings of the group (B, ϕ). Differentiating the equation of axiom 2° in Section 16.1 with respect to b at the point $b = 0$ and using definition (2.1) give the relation

$$\xi(f(x, a)) = (\partial_a f(x, a)) \circ A(a),$$

equipollent to

$$\partial_a f(x, a) = \xi(f(x, a)) \circ V(a).$$

Together with 1° in Section 16.1 this relation shows that the value

$$x' = f(x, a) \tag{2.3}$$

of the generating mapping f, as a function of the vector $a \in B$, satisfies the differential equation, with its initial condition,

$$\partial_a x' = \xi(x') \circ V(a), \qquad x'(0) = x. \tag{2.4}$$

Equation (2.4) is called the *Lie equation of the group* $G^r(f)$. This equation is analogous to the Lie equation (1.14.2) for the groups G^1 and it plays a key role in the study of the correspondence between the groups $G^r(f)$ and the spaces $L^r(\xi)$.

Note. Under a change of parameterization by the isomorphism $\gamma: B \to B'$ the group $G^r(f)$ becomes the group $G^r(f')$, where the transformation $f \to f'$ is given by formula (1.1). With this, the tangential mapping ξ transforms into ξ' by

$$\xi'(x) = \xi(x) \circ \partial\gamma(0). \tag{2.5}$$

3. Representation Precision Criterion

The property of precision of the representation π used in the formation of the group $G^r(f)$, expressed by axiom 3° in Section 16.1, is stated by means of the tangential mapping ξ.

Lemma. *In the system of axioms* 1°–4° *in Section* 16.1, *axiom* 3° *is equipollent to the following axiom*:

3°_1 *If* $\xi(x)\langle a\rangle = 0$ *for all* $x \in U$, *then* $a = 0$.

Proof. From the note in Section 16.1 and equation (2.5) it is possible to assume that the group (B, ϕ) is canonical. From the Lie equation and the criterion of canonicality in Section 12.5, the equality

$$\partial_t f(x, ta) = \xi(f(x, ta))\langle a\rangle \tag{3.1}$$

is found along the straight line $\{ta\}$. Thus if for some $a \in B$ it is true that $\xi(x)\langle a\rangle = 0$ for all $x \in U$, then it is also true that $\partial_t f(x, ta) = 0$. Consequently, $f(x, ta) = x$, and if axiom 3° in Section 16.1 is satisfied, then $ta = 0$, that is, $a = 0$.

Another differentiation of equation (3.1) gives

$$\partial_t^2 f(x, ta) = \partial\xi(x')\langle \xi(x')\langle a\rangle, a\rangle,$$

where $x' = f(x, ta)$. The derivative $\partial\xi$ is continuous on U. Therefore, for any point $x_0 \in U$ there exists a neighborhood U_0 of the point x_0 and a neighborhood O_ρ (with the radius ρ) of zero in the space B such that for all $(x, a) \in U_0 \times O_\rho$ the inequality

$$\rho\,|f(x, a) - x - \xi(x)\langle a\rangle| \leq |a|\cdot|\xi(x)\langle a\rangle|$$

holds. Thus if for some $a \in O_\rho$ and all $x \in U_0$ the equality $f(x, a) = x$ is correct, then the former inequality takes the form

$$0 \leq (|a| - \rho)|\xi(x)\langle a\rangle|,$$

and since $|a| < \rho$, $\xi(x)\langle a\rangle = 0$. Consequently, if axiom 3_1° is satisfied, then $a = 0$. ■

Consequence. *The dimensionality of the space of vector fields $L^r(\xi)$ of the group $G^r(f)$ is equal to r.*

Proof. Since $\xi(x)$ linearly maps the r-dimensional space B into X, the dimensionality of $L^r(\xi)$ is not more than r. In addition, if $\{e_\alpha\}$ is a basis in B, then the vector fields $\{\xi_{e_\alpha}\}$ are linearly independent. Indeed, the equality $\lambda^2\xi_{e_\alpha} = 0$, equipollent to the equality $\xi(x)\langle\lambda^\alpha e_\alpha\rangle = 0$ for all x, because of the lemma gives $\lambda^\alpha e_\alpha = 0$, from which it follows that $\lambda^\alpha = 0$ ($\alpha = 1, \ldots, r$). ■

4. First Lie Theorem

This theorem states the possibility of reconstructing the generating mapping of the group $G^r(f)$ from its tangential mapping by solving equation (2.4).

Theorem. *The tangential mapping ξ, generating an r-dimensional space of vector fields $L^r(\xi)$ on the space X, corresponds to the group $G^r(f)$; with this ξ, the generating mapping f of the group satisfies the Lie equation* (2.4).

Conversely, if an r-dimensional space of vector fields $L^r(\xi)$ is given on the open set $U \subset X$ of the class $C_\infty(U)$, and the local Lie group (B, ϕ) with auxiliary mapping $V(a)$ is given, and if the Lie equation (2.4) *for given ξ and V has a unique solution for any $x \in U$, then the solution* (2.3) *is the generating mapping of the group $G^r(f)$, for which (B, ϕ) is its parametric group and ξ is its tangential mapping.*

Proof. The first part of the theorem has already been established in Section 16.2. The statement about the dimensionality of the space of vector fields follows from axiom 3° in Section 16.1 because of the consequence of the lemma in Section 16.3.

Conversely, let the mapping $f: U \times B \to X$ define the solution of the Lie equation (2.4) by relation (2.3). Then axiom 1° in Section 16.1 is satisfied because of the initial condition (2.4). Axiom 3° in Section 16.1 holds because of the assumption about the dimensionality of the space $L^r(\xi)$ and the lemma in Section 16.3. Axiom 4° in Section 16.1 is satisfied by virtue of a well-known theorem about differentiability of the solutions of the differential equations with respect to the parameters on which the initial conditions depend. It is left to verify the truth of axiom 2° in Section 16.1. For this, the mappings $b \to y(b)$ and $b \to z(b)$, operating by $y = f(x', b)$, where $x' = f(x, a)$ and $z = f(x, \phi(a, b))$, are introduced. Differentiating these mappings and using (2.4) give the identities

$$\begin{aligned} \partial_b y &= \partial_b f(x', b) = \xi(y) \circ V(b), \qquad y(0) = x', \\ \partial_b z &= \partial_\phi f(x, \phi) \circ \partial_b \phi \overset{*}{=} \xi(z) \circ V(\phi) \circ A(\phi) \circ V(b) \\ &= \xi(z) \circ V(b), \qquad z(0) = x', \end{aligned}$$

where the equality $\overset{*}{=}$ means using the Lie equation (12.4.3) of the parametric group (B, ϕ). These identities show that both mappings y and z are solutions of equation (2.4) with the same initial condition. From uniqueness of the solution it follows that $y = z$, that is, $f(x', b) = f(x, \phi(a, b))$, and this is axiom 2° in Section 16.1. So with the group (B, ϕ) the solution of the Lie equation (2.4) gives the group $G^r(f)$. Directly from equations (2.4), when $a = 0$, it follows that the given ξ is the tangential mapping. ∎

5. Second Lie Theorem

Here the connection between the group $G^r(f)$ and its parametric group (B, ϕ), in terms of their Lie algebra, is determined. This connection is a consequence of the property of complete integrability of the Lie equation.

Theorem. *The space of vector fields $L^r(\xi)$ of the group $G^r(f)$ is a Lie algebra and is isomorphic to the Lie algebra L^r of the group (B, ϕ).*

Proof. The Lie equation (2.4) of the group $G^r(f)$ is completely integrable on the set $U \times O$; that is, it has a solution satisfying the initial condition $x'(b) = x$ for any point $(x, b) \in U \times O$. Indeed, that solution is $x'(a) = f(x, \phi(b^{-1}, a))$. Thus, the mapping $\xi \circ V: V \times O \to \mathscr{L}(B, X)$, on the right-hand side of equation (2.4), must satisfy the conditions of complete integrability in Section 12.8. To obtain these conditions, the vectors $h, k \in B$ are taken and the derivative of both sides of the equation $\partial_a x'\langle h\rangle = \xi(x')\langle V(a)\langle h\rangle\rangle$ is calculated with respect to a [using (2.4)] at the vector k. This gives

$$\begin{aligned} \partial_a^2\langle k, h\rangle &= \xi(x')\langle \partial V(a)\langle k, h\rangle\rangle \\ &\quad + \partial\xi(x')\langle \xi(x')\langle V(a)\langle k\rangle\rangle, V(a)\langle h\rangle\rangle. \end{aligned}$$

Thus the unknown condition, using the symmetry of the second derivative

$$\partial_a^2 x'\langle k, h\rangle = \partial_a^2 x'\langle h, k\rangle$$

and the notation $k' = V(a)\langle k\rangle$, $h' = V(a)\langle h\rangle$, becomes

$$\begin{aligned}&\xi(x)\langle \partial V(a)\langle h, k\rangle - \partial V(a)\langle k, h\rangle\rangle \\ &\quad = \partial\xi(x)\langle \xi(x)\langle k'\rangle, h'\rangle - \partial\xi(x)\langle \xi(x)\langle h'\rangle, k'\rangle.\end{aligned}$$

According to the Maurer–Cartan equation (12.8.2) and definition (15.1.1), of the commutator in the Lie algebra L^r of the group $G^r(\phi)$,

$$\partial V(a)\langle h, k\rangle - \partial V(a)\langle k, h\rangle = C\langle V(a)\langle k\rangle, V(a)\langle h\rangle\rangle = [k', h'].$$

On the other hand, according to the definition (7.1.1) of the commutator of vector fields,

$$\partial\xi(x)\langle \xi(x)\langle k'\rangle, h'\rangle - \partial\xi(x)\langle \xi(x)\langle h'\rangle, k'\rangle = [\xi(x)\langle k'\rangle, \xi(x)\langle h'\rangle].$$

Thus the condition of complete integrability of the mapping $\xi \circ V$ is equipollent to satisfaction of the equality

$$\xi(x)\langle [k, h]\rangle = [\xi(x)\langle k\rangle, \xi(x)\langle h\rangle] \tag{5.1}$$

for any $x \in U$ and $k, h \in B$. Consequently, the space of vector fields $L^r(\xi)$ of the group $G^r(f)$, obtained using the tangential mapping ξ given by equation (2.2), is a Lie algebra. Also, the mapping $\gamma: B \to L^r(\xi)$, operating by $\gamma(h) = \xi(x)\langle h\rangle$, is a one-to-one mapping onto because of the consequence of the lemma in Section 16.3, and it is therefore an isomorphism because of (5.1). ■

6. Third Lie Theorem

According to the second Lie theorem an r-dimensional Lie algebra $L^r(\xi)$ of tangential vector fields, generated by its tangential mapping ξ, is in one-to-one correspondence with the group $G^r(f)$ of transformations of the Banach space X. The following theorem establishes the truth of the converse theorem, in connection with which it is usually called the "converse third Lie theorem for groups of transformations." Here the assignment of the r-dimensional space of vector fields $L^r(\xi)$ on the open set $U \subset X$ is understood as an assignment of the r-dimensional Banach space B and the mapping $\xi: U \to \mathscr{L}(B; X)$ satisfies axiom 3_1° in Section 16.3. With this, the elements of the Lie algebra $L^r(\xi)$ are the vector fields ξ_a, obtained using the vectors $a \in B$ and operating by formula (2.2).

Theorem. *For every r-dimensional Lie algebra $L^r(\xi)$ of vector fields of class $C_\infty(U)$ on the open set $U \subset X$, there corresponds a group $G^r(f)$ of local trans-*

formations of the space X for which $L^r(\xi)$ is its Lie algebra of tangential vector fields.

Proof. By means of the operation of commutation in $L^r(\xi)$, every pair of vector fields (ξ_a, ξ_b) is put into correspondence with the vector fields $\xi_c = [\xi_a, \xi_b]$. The vector c is uniquely defined by the pair of vectors (a, b), for the equality $\xi_c = 0$ means that $\xi(x)\langle c\rangle = 0$ for all $x \in U$, from which $c = 0$ because of axiom 3_1° in Section 16.3. This determines an algebraic structure, with the operation $[a, b] = c$, on the space B. It is easy to verify that this operation satisfies axioms 1°–3° in Section 13.1. Therefore, with the introduction of the commutator $[a, b]$, the space B becomes the Lie algebra L^r. Besides, from the definition of this commutator, the vector fields of the Lie algebra $L^r(\xi)$ satisfy the relation (5.1).

According to the theorem in Section 12.12, there is a group (B, ϕ) whose Lie algebra is L^r. Let $V(a)$ be an auxiliary mapping of this group. Then together with the given mapping ξ, it is possible to form equation (2.4). Because equation (5.1) holds, this equation is completely integrable on the set $U \times B$. By the theorem in Section 16.4, this solution generates the group $G^r(f)$, for which the given ξ is its tangent mapping. Consequently, $L^r(\xi)$ is a Lie algebra of the group $G^r(f)$. ■

7. Canonical Group of the Second Kind

If the parametric group (B, ϕ) of the transformation group $G^r(f)$ is canonical (see Section 12.5), then the group $G^r(f)$ is called *canonical of the first kind.*

For applications it is important to have an effective and simple algorithm to construct the group $G^r(f)$ from its Lie algebra $L^r(\xi)$. From this point of view the algorithm described in the proof of the theorem in Section 16.6 leading to a canonical group of the first kind is inconvenient and is never actually used. In practice, another algorithm connected with determining one-parameter groups and "multiplications" of the resulting mappings, every one of which is taken with its independent parameter, is used. The description of this algorithm and its foundation will be given.

Let the Lie algebra $L^r(\xi)$ of vector fields on the open space $U \subset X$ be given, and let B be an r-dimensional space with elements from $L^r(\xi)$ described as in Section 16.6. Let the basis $\{e_\alpha\} \subset B$, with corresponding basis $\{\xi_\alpha\} \subset L^r(\xi)$, where $\xi_\alpha(x) = \xi(x) < \langle e_\alpha\rangle\}$, be chosen. For every vector field ξ_α one finds the one-parameter group $G^1(\xi_\alpha)$ by integration of the Lie equation (1.14.2), which here has the form

$$\partial_t x' = \xi_\alpha(x'), \qquad x'(0) = x. \tag{7.1}$$

In the case of finite dimensional X, this problem is reduced to integration of ordinary differential equations. In practice these can be solved for vector fields of a simple analytical nature. Consequently, one-parameter groups $G^1(f_\alpha)$ of transformations of the space X, the elements of which are the transformations $f_{\alpha,t}$ with the parameter $t \in R$, are obtained. Finally, for the arbitrary vector $b = b^\alpha e_\alpha \in B$ the "product"

$$f'_b = f_{1,b^1} \circ f_{2,b^2} \circ \cdots \circ f_{r,b^r} \tag{7.2}$$

is put into correspondence with it. Then the set $G^r(f')$ of the "products" (7.2), corresponding to all possible vectors b from a null neighborhood $O \subset B$, is available. This set is generated by the mapping $f' \colon U \times O \to X$, operating by $f'(x, b) = f'_b(x)$.

Theorem. *The set $G^r(f')$ is a group of transformations whose Lie algebra is $L^r(\xi)$.*

Proof. According to the theorem in Section 16.6, there is a group $G^r(f)$ for the given Lie algebra $L^r(\xi)$. Every one-parameter subgroup $G^r(f_\alpha)$, found by the described algorithm, belongs to $G^r(f)$. Thus every mapping f'_b (7.2) also belongs to $G^r(f)$, that is, there exists some mapping f_a. Therefore, only one mapping f_a can correspond to the given f'_b because the equality $f_a = f_c$, equipollent to the equality $f_{a \cdot c^{-1}} = I_X$, is possible only when $a \cdot c^{-1} = 0$, that is, when $a = c$ (axiom 3° in Section 16.1). Thereby the mapping $\gamma \colon B \to B$ defined by $a = \gamma(b)$ appears and the identity

$$f'(x, b) = f(x, \gamma(b)) \tag{7.3}$$

holds. As f' is obviously a mapping of the class $C_\infty(U \times O)$ and so is f, γ is also a mapping of the class $C_\infty(O)$.

When $b = 0$, equation (7.2) gives $f'(x, 0) = x$, from which it follows by (7.3) and axiom 3° in Section 16.1 that $\gamma(0) = 0$. Such a mapping γ is said to be *locally invertible*. Indeed, the calculation of the derivative with respect to b at the point $b = 0$, with the vector $h = h^\alpha e_\alpha \in B$, of the "product" (7.2), using the equality $f_{\alpha, b^\alpha}|_{b=0} = I_X$ and equations (7.1), results in

$$\partial_b f'(x, 0)\langle h \rangle = h^\alpha \xi_\alpha(x) = \xi(x)\langle h \rangle. \tag{7.4}$$

On the other hand, the analogous differentiation of the right-hand side of the identity (7.3), with the mapping f, gives

$$\partial_b f(x, \gamma(b))(0)\langle h \rangle = \xi(x) \circ \partial\gamma(0)\langle h \rangle.$$

A comparison of this equation with (7.4) and an application of the lemma in Section 16.3 leads to $\partial\gamma(0)\langle h \rangle = h$, from which $\partial\gamma(0) = I_B$. Thus γ is invertible.

From this mapping γ, the mapping $\phi': O \times O \to B$ is defined via

$$\phi'(a, b) = \gamma^{-1}(\phi(\gamma(a), \gamma(b))). \tag{7.5}$$

It is easy to verify that ϕ' satisfies all the axioms 1°–3° in Section 12.1 of the multiplication law. Thereby the group (B, ϕ') is determined, and γ is an isomorphism of this group onto the group (B, ϕ) (see Section 12.3).

From the foregoing result the mapping f' satisfies the axioms 1°, 3°, and 4° in Section 16.1. In addition, from definitions (7.3) and (7.5), the relations

$$\begin{aligned} f'(f'(x, a), b) &= f(f(x, \gamma(a)), \gamma(b)) = f(x, \phi(\gamma(a), \gamma(b))) \\ &= f(x, \gamma(\phi'(a, b))) = f'(x, \phi'(a, b)) \end{aligned}$$

follow. From here, with ϕ', it is clear that axiom 2° in Section 16.1 is satisfied for f'. Thus $G^r(f')$ is a group of transformations of the space X, and (7.4) shows that $L^r(\xi)$ is its Lie algebra. ■

The group $G^r(f')$ generated by the mapping f' accordining to equation (7.2) will be called *canonical of the second kind.*

Consequence. *A composition of r transformations belonging to one-parameter subgroups $G^1(f_\alpha)$ $(\alpha = 1, \ldots, r)$ of the group $G^r(f)$, corresponding to r linearly independent vector fields of its Lie algebra $L^r(\xi)$, and considered for independently varying parameters of the groups $G^r(f_\alpha)$, gives all transformations of the group $G^r(f)$.*

Note 1. An important peculiarity of the algorithm for forming a canonical group $G^r(f)$ of the second kind is that there is no way that has general applicability for finding beforehand the parametric group (B, ϕ). The parametric group, as it is seen from the proof of the theorem, is automatically formed and can be found, using formula (7.2), by calculation of the composition $f'_b \circ f'_a$. In particular, this peculiarity makes the algorithm of practical convenience.

Note 2. In the proof of the theorem, it was noted that from equation (7.5) the mapping γ is an isomorphism of the parametric group (B, ϕ') onto (B, ϕ). Thus if the group of transformations (B, ϕ, π, X) is given in advance, then, in its reconstruction as a canonical group of the second kind, the parametric group is reconstructed correctly up to an isomorphism γ satisfying the condition $\partial\gamma(0) = I_B$.

8. Examples

(α) The first general example is that discussed in Section 14.4 for the construction of the group of inner automorphisms Int L^r of a finite-dimensional Lie algebra L^r. Now it is clear that the rule formulated there to obtain any automorphism $A \in \mathrm{Int}\, L^r$ by equation (14.4.6), analogous to (7.2), gives the group $\mathrm{Int}\, L^r$ as canonical of the second kind. This rule is justified by the theorem in Section 16.7.

(β) Another general example gives the first parametric group of the group (B, ϕ) which was considered in Example 2 in Section 16.1. According to definition (2.1), the tangential mapping coincides with the auxiliary mapping A. Thus if the mapping $A: B \to \mathscr{L}(B)$ is known, then the group (B, ϕ) can be reconstructed as canonical of the second kind, up to an isomorphism. For this, it is necessary to choose a basis $\{e_\alpha\} \in B$, having the matrix $(A_\alpha^\beta(a))$ of the mappings $A(a)$ for every $\alpha = 1, \ldots, r$, to solve the system of equations for the vector $x' = x'^\beta e_\beta$:

$$\partial_t x'^\beta = A_\alpha^\beta(x'), \qquad x'^\beta(0) = x^\beta \qquad (\beta = 1, \ldots, r).$$

Its solution gives the transformations $x'(t) = f_{\alpha, t}$, with the aid of which the general transformation of the first parametric group, corresponding to the vector $b = b^\alpha e_\alpha$, is obtained from (7.2):

$$\psi(x, b) = f_{1, b^1} \circ f_{2, b^2} \circ \cdots \circ f_{r, b^r}.$$

On the other hand, transformations of the first parametric group have the form $x \to \phi(x, b)$. Thus, the identity (7.3) here gives the form $\psi(x, b) = \phi(x, \gamma(b))$, from which $\gamma(b) = \psi(0, b)$. Consequently, the multiplication law ϕ' in the canonical group of the second kind (B, ϕ'), isomorphic to the group (B, ϕ), obtained from (7.5) is

$$\phi'(a, b) = \gamma^{-1}(\psi(\gamma(a), b)), \qquad \gamma(b) = \psi(0, b).$$

(γ) As a simple concrete example we can consider the group G^3 of linear-fractional transformations of the straight line $R(x)$, whose Lie algebra $L^3(\xi)$ has the basis

$$\xi_1(x) = 1, \qquad \xi_2(x) = x, \qquad \xi_3(x) = x^2.$$

Here solution of the Lie equations (7.1) gives the following transformations $x \to f_{\alpha, t}(x)$:

$$f_{1,t}(x) = x + t, \qquad f_{2,t}(x) = x \exp t, \qquad f_{3,t}(x) = \frac{x}{1 - tx}.$$

The general transformation of the canonical group $G^3(f)$ of the second kind, obtained from (7.2) with the vector $b = (b^1, b^2, b^3)$, is

$$f_b(x) = \frac{b^1 + (\exp(b^2) - b^1 b^3)x}{1 - b^3 x}.$$

9. Similarity of Transformation Groups

A special kind of transformation groups isomorphism, found with the aid of transformations of the space X, is called a *similarity of groups.* This concept comprises different representations π of the same group (B, ϕ) in the group $\tau(X)$. Analogous to the concept for one-parameter groups of transformations presented in Section 1.11, it is given by the following definition.

The group $G^r(f') = (B, \phi, \pi', X)$ is called *similar to the group* $G^r(f) = (B, \phi, \pi, X)$ if there exists a transformation p of the space X of class $C_\infty(X)$ such that the generating mappings f' and f of these groups, for any $a \in O \subset B$, are related by

$$f'_a = p \circ f \circ p^{-1}. \tag{9.1}$$

The following theorem gives an infinitesimal criterion of group similarity, expressing this property through the corresponding property of their tangential mappings.

Theorem. *The group $G^r(f')$ is similar to the group $G^r(f)$ if and only if there exists a mapping p of the space X under which their tangential mappings are connected by*

$$\xi'(x') = \partial p(x) \circ \xi(x), \qquad x' = p(x). \tag{9.2}$$

Proof. Equation (9.2) is found as a result of differentiating relation (9.1) with respect to a, at the point $a = 0$, and then using definition (2.1). Conversely, let the relation (9.2) be satisfied. Then, for the mapping $\bar{f}$, operating by

$$\bar{f}(x', a) = p(f(x, a)), \qquad x' = p(x),$$

from the Lie equation (2.4) the identities

$$\partial_a \bar{f}(x', a) = \partial p(f) \circ \xi(f) \circ V(a) = \xi'(\bar{f}(x', a)) \circ V(a),$$
$$\bar{f}(x', 0) = p(x)$$

are found. On the other hand, the generating mapping f' also satisfies the Lie equation (2.4):

$$\partial_a f'(x', a) = \xi'(f'(x', a)) \circ V(a), \qquad f'(x', 0) = x' = p(x).$$

Consequently, both mappings $\bar{f}$ and f' satisfy the same differential Lie equation with equal initial conditions. From uniqueness of the solution it follows that the mappings coincide, that is, $f'(x', a) = p(f(x, a))$ for $x' = p(x)$. This is the relation of similarity (9.1). ■

Lie algebras of the vector fields $L^r(\xi')$ and $L^r(\xi)$ whose generating mappings ξ' and ξ are connected by the relation (9.2) are said to be *similar*. Thus the established theorem can be restated by saying that the groups $G^r(f')$ and $G^r(f)$ are similar if and only if their Lie algebras $L^r(\xi')$ and $L^r(\xi)$ are similar.

Note. If the Lie algebra $L^r(\xi)$ of vector fields is given on the space X and if with the aid of some mapping $p: X \to X$ the mapping ξ' is determined by equation (9.2), then the space of vector fields $L^r(\xi')$ generated by the mapping ξ' is a Lie algebra isomorphic to $L^r(\xi)$. This fact can be easily verified by direct calculation of the commutator $[\xi'_a, \xi'_b]$, where $\xi'_a = \partial p\langle \xi_a \rangle$, etc.

10. Transitivity

The property of transitivity is one of the important characteristics of group of transformations as representations. It is stated in terms of orbits, obtained by operation of the group $G^r(f) = (B, \phi, \pi, X)$, generated by the mapping $f: U \times O \to X$. An *orbit of the point* $x \in U$ is the set of points $f(x, a)$ obtained when the vector a assumes values in the null neighborhood $O \subset B$. The orbit of the point x will be denoted by $f_x(O)$.

The group $G^r(f)$ is called *transitive at the point* $x \in U$ (*on the set* U) if the orbit of the point x (any $x \in U$) is an open set in X. A group which is not transitive is called *intransitive.*

This definition can be stated in another way by saying that the group $G^r(f)$ is called transitive if the point $x \in U$ can be transformed to any point of its neighborhood by a suitable group transformation. It is also clear that the group $G^r(f)$ is transitive on U if and only if any two sufficiently close points of the set U can be transfered into each other by some transformation of this group.

If the group $G^r(f)$ is transitive at the point x, then, in general, the orbit $f_x(O)$ covers the neighborhood of the point x repeatedly; that is, some points $x' \in f_x(O)$ can be obtained from x by different transformations of the group. To separate those cases when this situation cannot occur, the idea of a "simply transitive group" is introduced.

The group $G^r(f)$ is said to be *simply transitive at the point* $x \in U$ (*on the set* U) if for the given x (for every $x \in U$) there exists a null neighborhood $O \subset B$ such that the mapping $O \to f_x(O)$ is one-to-one (continuously differentiable of the class C_∞).

It is clear that the group $G^r(f)$ can be transitive only if the dimensionality of the space X is not more than the number r, that is, if $\dim X \leq r$. Exact criteria of transitivity are stated using the tangential mapping ξ.

With this goal, the concept of the *general rank* of a linear mapping is introduced. Let B and X be finite-dimensional spaces. The mapping $\xi: X \to \mathscr{L}(B, X)$ is considered. For every $x \in X$, the quantity $\xi(x)$ is a linear mapping $B \to X$ and, thus, has a definite rank, denoted by rank $\xi(x)$. This number is equal to the rank of the matrix of the linear mapping $\xi(x)$, expressed in any bases of the spaces B and X (the rank does not depend on these bases), and it can be calculated by

$$\operatorname{rank} \xi(x) = \dim \xi(x)(B),$$

where $\xi(x)(B)$ is a vector subspace in X, which is the image of the space B under the mapping $\xi(x)$.

The *general rank of the mapping* $\xi: X \to \mathscr{L}(B; X)$ *on the open set* $U \subset X$ is defined to be the number

$$r_* = r_*(\xi) = \max_{x \in U} \operatorname{rank} \xi(x). \tag{10.1}$$

Obviously, if the mapping ξ of the class $C_\infty(U)$ is given, then the set of points $x \in U$ for which

$$\operatorname{rank} \xi(x) = r_*(\xi) \tag{10.2}$$

is open in the space X. Conversely, the set of points for which $\operatorname{rank} \xi(x) < r_*(\xi)$ is closed. Thus in the local theory it is always possible to limit consideration to such sets U, at all points of which the equality (10.2) is satisfied. In this case it is said that the *mapping* ξ *has the rank of* U (*equal to its general rank*). In these terms, the following criterion of group transitivity is stated.

Theorem. *Let* $\dim X = n$ *and let the tangential mapping* ξ *of the group* $G^r(f)$ *have the rank of the open set* $U \subset X$. *Then the group* $G^r(f)$ *is transitive on* U *if and only if the general rank of the mapping* ξ *on* U *is equal to* n, *that is,*

$$r_*(\xi) = n. \tag{10.3}$$

Proof. For the generating mapping f in the null neighborhood $O \subset B$ the representation

$$f(x, a) = x + \xi(x)\langle a \rangle + F(x, a), \qquad |F(x, a)| < K|a|^2, \tag{10.4}$$

holds, with some constant $K > 0$. Thus the transitivity of the group $G^r(f)$ at the point $x \in U$ is equipollent to solvability of the equation

$$\xi(x)\langle a \rangle = x' - x - F(x, a)$$

relative to a for points x' close to x. But, if the norm $|x' - x|$ is sufficiently small, then because of the estimate (10.4), the solvability of this equation is equipollent to the solvability of the linear equation

$$\xi(x)\langle a \rangle = x' - x.$$

In turn, this equation is solvable relative to a for any $x' \in X$ if and only if $\operatorname{rank} \xi(x) = \dim X = n$. ■

Consequence 1. *The group $G^r(f)$ is transitive if and only if*

$$r_*(\xi) < n. \tag{10.5}$$

As was mentioned in the proof of the theorem, a linear equation has a unique solution if and only if rank $\xi(x) = n = r$, so the following statement holds.

Consequence 2. *The group $G^r(f)$ is simply transitive if and only if*

$$r_*(\xi) = n = r. \tag{10.6}$$

The first (and the second) parametric group of the local Lie groups (B, ϕ), considered in Example 2 in Section 16.1 and (β) in Section 16.8 are examples of simply transitive groups. Consequently, $n = r$. As the tantential mapping of the first parametric group (B, ϕ) is the auxiliary mapping A of the group (B, ϕ) and $A(0) = I_B$, there exists a null neighborhood $O \subset B$ such that rank $A(a) = r$ for all $a \in O$, that is, $r_* = n$.

11. Similarity of Simply Transitive Groups

The following theorem generalizes the property noted in Sections 3.8 and 3.9 of similarity of one-parameter transformation groups to the case of r-parameter groups. In addition, it shows that the classification results of simply transitive groups relative to isomorphism and relative to similarity coincide.

Theorem. *Any two isomorphic simply transitive groups of transformations of the space X are similar.*

Proof. Let ξ' and ξ be tangential mappings of simply transitive groups $G^r(f')$ and $G^r(f)$ of transformations of an r-dimensional space X. To determine similarity of these groups it is sufficient, because of the theorem in Section 16.9, to prove the existence of a transformation p of the space X such that relation (9.2) is satisfied. For simply transitive groups, the linear mapping $\xi(x)\colon B \to X$ is invertible, so the relation for p can be expressed as

$$\partial p(x) = \xi'(p(x)) \circ (\xi(x))^{-1}. \tag{11.1}$$

Consequently, to establish the existence of the transformation p it is necessary to construct this mapping from its derivative given as a function of x and p. The solution of this problem certainly exists if the right-hand side of (11.1) satisfies the condition of complete integrability. As this condition is equipollent to symmetry of the second derivative $\partial^2 p(x)$, calculated from (11.1), it is convenient to derive it by differentiation of the relation

$$\xi'_a(p) = \partial p\langle \xi_a(x)\rangle,$$

equipollent to (11.1). Calculation of the derivative with respect to x, at the vector $k \in X$, gives

$$\partial \xi'_a(p)\langle \partial p\langle k\rangle\rangle = \partial^2 p\langle k, \xi_a(x)\rangle + \partial p\langle \partial \xi_a(x)\langle k\rangle\rangle.$$

Assuming that $k = \xi_b(x)$, a permutation of the vectors a and b and subtraction of the results term by term, taking into consideration equation (11.1), shows that the symmetry condition for $\partial^2 p(x)$ is equipollent to the condition

$$\partial \xi'_a(p)\langle \xi'_b(p)\rangle - \partial \xi'_b(p)\langle \xi'_a(p)\rangle = \partial p\langle \partial \xi_a(x)\langle \xi_b(x)\rangle\rangle - \partial \xi_b(x)\langle \xi_a(x)\rangle\rangle.$$

According to the definition of the vector field commutator (7.1.1), this equality can be rewritten in the form

$$[\xi'_b(p), \xi'_a(p)] = \partial p\langle [\xi_b(x), \xi_a(x)]\rangle.$$

This is an identity for the mapping $\xi \to \xi'$ given by (11.1), which is an isomorphism. Thus the mapping (11.1) is completely integrable and the existence of the solution of equation (11.1) follows from the Frobenius theorem. ■

Consequence 1. *The set of transformations p, transferring the given simply transitive group $G^r(f)$ into the given simply transitive group $G^r(f')$, depends upon r arbitrary constants.*

Indeed, such constants are components of the vector $p_0 \in X$, with the initial condition $p(x_0) = p_0$, to which it is always possible to subordinate the solution of equation (11.1).

Consequence 2. *Any simply transitive group is similar to its first parametric group (B, ϕ).*

Indeed, isomorphism of these groups is obvious and, in addition, as was noted in Section 16.10, the first parametric group (B, ϕ) is simply transitive.

12. Lie Groups Admitted by Differential Equations

The results introduced in Section 5.5 concerning the full group GE, admitted by the system of differential equations E, will be enriched now.

In Section 7.13, it has been established that the set LE of operators (vector fields), admitted by the system E, is a Lie algebra, as is the space of solutions of the determining equations DE. Thus if LE is the finite-dimensional Lie algebra $L^r(\zeta)$, then the system E admits the Lie group of transformations $G^r(h)$ corresponding to this Lie algebra. Indeed, according to the definition, the system E admits any one-parameter group, corresponding to an arbitrary operator from LE, and the group $G^r(h)$ is the union of elements (transformations) belonging to its one-parameter subgroups. So, *in the case of finite-dimensional Lie algebras LE, the full group of the system E is a Lie group.*

If the Lie algebra LE is infinite dimensional, then one ought to adhere to the definition of a full group, which was given in Section 5.5. But in this case the local Lie group of transformations $G^r(h)$, which is admitted by the system E, corresponds to a finite-dimensional subalgebra $L^r(\zeta) \subset LE$. Besides, in applications it often happens that in the infinite-dimensional Lie algebra LE one naturally distinguishes, in some sense, a "trivial" infinite-dimensional ideal J, with respect to which the factor algebra LE/J is infinite dimensional. Usually, a certain Lie group, admitted by the system E, corresponds to this factor algebra.

In the case of infinite-dimensional Lie algebras LE it is said that the group GE is an "infinite" (in the sense of dimensionality) Lie group. General transformations of "infinite" Lie groups do not depend on a finite number of parameters, as was the case for the groups $G^r(h)$, but, generally speaking, depend on arbitrary functions.

However, "infinite" Lie groups are often met and used in applications (see, for example, Section 10.4). But the complete theory of such groups is not yet available.

REFERENCES

The classical Lie theory has been expounded in different forms by many authors.† The basic ideas of the Banach groups of Lie theory were expressed in Ref. [1]. A systematic exposition, based upon the apparatus of differential equations, was given for the first time in Ref. [2], in connection with the study of the possibilities of extending the theory to infinite Lie groups. Some concrete infinite Lie groups are considered in Refs. [3] and [4].

The problem of topological structure of infinite Lie groups is not yet solved, although a few works [5–9] have been devoted to this subject. A special place is occupied by Delsarte's groups [10], whose relation to Lie groups is not yet clear.

[1] Birkhoff G. Analitical groups. — Trans. Amer. Math. Soc. 43, № 1 (1938), 61—101.
[2] Овсянников Л. В. Аналитические группы. Новосибирск, НГУ, 1972.
[3] Овсянников Л. В. О бесконечных группах отображений, задаваемых дифференциальными уравнениями. — ДАН СССР 148, № 1 (1963), 36—39.

† Elaboration of its applications to differential equations is available in Refs. [1–4] cited in Chapter I.

[4] Хабиров С. В. Бесконечномерные непрерывные группы преобразований трехмерного пространства, задаваемые системами дифференциальных уравнений первого порядка. — В сб. «Динамика сплошной среды», вып. 12, Новосибирск (1972), 131—146.
[5] Lie S. Untesuchungen über unendliche kontinuierlice Gruppen. — Ber. Sachs. Ges. **21** (1895), 43—150.
[6] Medolaghi P. Sulla theoria dei gruppi infiniti continui. — Ann. Mat. **25** (1897), 179—217.
[7] Cartan E. Sur la structure des groupes infinis de transformation. — Oeuvres, Partie 2, **2**. Paris (1963), 571—714.
[8] Kuranishi M. On the local theory of continuous infinite pseudogroups, 1—2. — Nagoya Math. **15** (1959), 225—260; **19** (1961), 55—91.
[9] Singer I., Sternberg S. The infinite groups of Lie and Cartan. — J. analise math. **15** (1965), 1—114.
[10] Delsarte J. Les groupes de transformations linéares dans l'éspase de Hilbert. — Mém. Acad. Sci., fasc. **57**, Paris (1932).

CHAPTER

V

Invariant Solutions

The study of the operation of the admitted group on the set of solutions of a differential equation begins, in the present chapter, by consideration of "fixed points" of that operation, which are invariant solutions. These are solutions which transform into themselves under transformations from the admitted group. For that, the general theory of invariants and invariant manifolds, whose preliminary foundations were laid in Chapter I is considered here. The theory is related to the fact that the determination of the invariants for many parameters groups requires the analysis of not one vector field but a family of vector fields. Here new objects appear, for example, special invariant manifolds, which must be taken into account by the theory. One of the important results of the theory of invariants is the theorem about the existence of universal invariants. Like the corresponding theorem from Chapter I this result is proved for Lie groups of transformations of an arbitrary Banach space.

In the theory of invariant manifolds, in addition to the criterion of invariance, ideas and properties of induced Lie groups and Lie algebras are essential. The most important theoretical result is contained in a theorem concerning representation of nonsingular invariant manifolds by equations connecting only invariant groups. The concept of rank of a nonsingular invariant manifold, which is necessary to resolve classification issues, naturally appears here. The geometrical interpretation of the solution $y = u(x)$ as a (local) manifold U in the space of "variables" (x, y) is useful in the study of solutions of the differential equation E. It is convenient in theoretical problems to write the solution, as the fact that U is a manifold permits us to do, by a simple formula of inclusion $U_k \subset E$ for the properly continuous manifold U_k.

For any group H admitted by the given differential equation E, we introduce the fundamental idea of an invariant H-solution as a manifold U, invariant relative to the group H. Applying the former results to the problem of finding the invariant H-solutions leads to a theorem of existence of the factor system E/H, possessing two main properties: It possesses all invariant H-solutions, and the number of independent variables is less than in the initial system E. To construct the factor system, there is a clear and operable algorithm.

Special attention is paid to the case, important for applications, of the extension groups and following from such groups the concept of automodal solutions. It is proposed that the latter be understood in a wide sense, namely, as a solution of the Abel H group. A sufficient foundation is described for this proposal.

If the full group GE of the given differential equation E has more than one parameter, then there appears an abundance of possibilities for constructing the invariant H-solutions from different subgroups $H \subset GE$. It is natural that this fact evokes a necessity to classify such solutions. For the main classification criteria there are the value of the rank of a solution and the property of group similarity. The result of this classification is the optimal system of invariant solutions.

§17. INVARIANTS OF TRANSFORMATION GROUPS

1. Definition and Criterion of Invariant

The terminology and notation introduced in Section 3 will be used here. Let the group $G^r(f) = (B, \phi, \pi, X)$ be an image under the representation $\pi: (B, \phi) \to \tau(X)$ of the local Lie group (B, ϕ) in the group of all transformations of the space X and a Banach space P, on which the group (B, ϕ) operates trivially.

The nonconstant mapping $F: X \to P$ is called an *invariant of the group* (B, ϕ, π, X) if it is a fixed element of the continuous representation π_{PX}. In other words, the mapping $F \neq \text{const}$ is an invariant of the group $G^r(f)$ for any $(x, a) \in X \times B$, that is, the equality

$$F(f(x, a)) = F(x) \tag{1.1}$$

holds.

The definition (1.1) seems to coincide with (3.3.1). The difference is that in (1.1) the "parameter" a is not a real number but an r-dimensional vector $a \in B$. With this, definition (3.3.1) is a special case of definition (1.1).

Using the Lie equation (16.2.4) for any differentiable mapping $F: X \to P$, the analog of equation (3.3.2) is

$$\partial_a F(x') = \partial F(x') \circ \xi(x') \circ V(a), \qquad x' = f(x, a), \tag{1.2}$$

where $V(a)$ is an auxiliary mapping of the group (B, ϕ). The following theorem, being an analog of the theorem in Section 3.4, gives an infinitesimal criterion for the invariance of the mapping F.

Theorem. *The mapping $F: X \to P$, $F \neq$ const, of the class $C_1(X)$ is an invariant of the group $G^r(f)$ with the tangential mapping ξ if and only if for every $x \in X$ the equation*

$$\partial F(x) \circ \xi(x) = 0 \tag{1.3}$$

is satisfied.

Proof. As $V(0) = I_B$, then because of (1.2) it remains to repeat the proof of the theorem in Section 3.4. ■

The definition given in Section 3.5 of the universal invariant is transferred without change to the case of r-parameter groups of transformations. The main problem of the theory of invariants is now to prove the existence of a universal invariant of the given group $G^r(f)$ and to provide an algorithm for its determination. Here, as well as in Section 3.5, this problem is solved for an arbitrary (generally speaking, infinite-dimensional) Banach space X.

2. Complete Families of Vector Fields

Consider now an s-dimensional Banach space B, an arbitrary Banach space X, an open set $U \subset X$, and the mapping $\xi: U \to \mathscr{L}(B; X)$ of the class $C_\infty(U)$.

The image $\xi(x)(B) \subset X$ is called *a family of vector fields on the space X*. Every concrete vector field ξ_a of this family is determined by an assignment of the vector $a \in B$ and operates by $\xi_a(x) = \xi(x)\langle a \rangle$. It is clear that the family of vector fields is a finite-dimensional vector space, namely, a subspace of the space of all vector fields on the set U. As in Section 16.10, one determines the quantity rank $\xi(x)$ and the general rank $r_* = r_*(\xi)$ on U, of the mapping ξ, by the formula (16.10.1). In what follows, it is supposed that the set U is chosen in such a way that the mapping ξ has a rank on U equal to its general rank.

The family of vector fields $\xi(x)(B)$ is said to be *complete* if it satisfies the two conditions: (α) $r_* = s$ and (β) for any $a, b \in B$ there is a mapping $\omega_{ab}: U \to B$ such that

$$[\xi_a, \xi_b](x) = \xi(x)\langle\omega_{ab}(x)\rangle. \tag{2.1}$$

Lemma 1. *The following two transformations $\xi \to \bar{\xi}$ maintain the property of completeness of the vector field family: a composition with any mapping $\omega: U \to GL(B)$,*

$$\bar{\xi}(x) = \xi(x) \circ \omega(x), \tag{2.2}$$

and the change under the operation of the invertible transformation $p: U \to U'$, for which $\partial p(x)$ is a homeomorphism of the space X for any $x \in U$,

$$\bar{\xi}(x') = \partial p(x) \circ \xi(x), \qquad x' = p(x). \tag{2,3}$$

Proof. From the invertibility of the mapping $\omega(x)$ and $\partial p(x)$ the satisfaction of condition (α) is obvious in both cases. Condition (β) is verified by a direct calculation of the commutators. If $a' = \omega(x)\langle a\rangle$, $b' = \omega(x)\langle b\rangle$, then by (2.1)

$$\begin{aligned}[\bar{\xi}_a, \bar{\xi}_b] &= \xi(x)\langle\omega_{a'b'}(x) + \partial\omega\langle\xi_{b'}, a\rangle - \partial\omega\langle\xi_{a'}, b\rangle\rangle \\ &= \bar{\xi}(x)\langle\bar{\omega}_{ab}(x)\rangle,\end{aligned}$$

that is, (2.1) holds for $\bar{\xi}$, given by (2.2). Then, using the equality $\partial_x\bar{\xi}(x') = \partial\bar{\xi}(x') \circ \partial p(x)$ from (2.3), we obtain

$$\partial\bar{\xi}_a(x')\langle\bar{\xi}_b(x')\rangle = \partial^2 p(x)\langle\xi_b(x), \xi_a(x)\rangle + \partial p(x)\langle\xi_a(x)\langle\xi_b(x)\rangle\rangle,$$

from which, because of (2.1),

$$[\bar{\xi}_a(x'), \bar{\xi}_b(x')] = \partial p(x)\langle\xi(x)\langle\omega_{ab}(x)\rangle\rangle = \bar{\xi}(x')\langle\bar{\omega}_{ab}(x')\rangle,$$

that is, (2.1) holds for the $\bar{\xi}$ given by (2.3). ■

The mapping $F: U \to P$ is said to be an *invariant of the family of vector fields* $\xi(x)(B)$ if it satisfies the equation [analog of (1.3)]

$$\partial F(x) \circ \xi(x) = 0 \tag{2.4}$$

for any $x \in U$. Here P is an arbitrary Banach space. In a manner analogous to that for groups of transformations, the idea of a universal invariant of the family of vector fields is defined. It is an invariant $J: X \to Y$ such that for any invariant $F: X \to P$ the mapping $\Phi: Y \to P$ exists with the equality $F = \Phi \circ J$. Under transformations (2.2) and (2.3) invariants go into invariants according to the following lemma.

Lemma. *Let $F: X \to P$ be an invariant of the family of vector fields $\xi(x)(B)$. If the transformation $\xi \to \bar{\xi}$ is made by equation (2.2), then F is also an invariant* *of the family* $\bar{\xi}(x)(B)$. *If the transformation is made by formula* (2.3), *then* $\bar{F} = F \circ p^{-1}$ *is an invariant of the family* $\bar{\xi}(x')$. *In both cases the universal invariant transforms into the universal invariant.*

Proof. For the transformation (2.2) the statement follows from the identity

$$\partial F(x) \circ \bar{\xi}(x) = \partial F(x) \circ \xi(x) \circ \omega(x).$$

For transformation (2.3) the result follows from the identity

$$\partial F(x) \circ \xi(x) = \partial \bar{F}(x') \circ \bar{\xi}(x'), \qquad x' = p(x).$$

The last statement is obvious. ■

Theorem. *For any complete family of vector fields* $\xi(x)(B)$ *satisfying the condition*

$$r_*(\xi) < \dim X, \tag{2.5}$$

there is a universal invariant.

Proof. The induction is with respect to s. For $s = 1$ the statement was established in the theorem in Section 3.5. Let $s > 1$. A nonzero vector $q \in B$ is selected and the vector field $\xi(x)\langle q\rangle = \xi_q(x)$ is considered. From the similarity theorem in Section 3.8 there exists a transformation of the form (2.3), transforming $\xi_q(x)$ into a constant vector field, where the family $\xi(x)(B)$ is transformed into a family still satisfying the conditions of the theorem. Consequently, from the Lemma it follows without loss of generality that it is possible to consider, from the very beginning, the field $\xi_q(x) = e$ to be constant. That is assumed in what follows.

Let $T = \{te\}$ be a one-dimensional subspace in X, generated by the vector e, and let Y be a direct addition, so that $X = T \oplus Y$. Then for any vector field $\xi_a(x) = \xi(x)\langle a\rangle$, where $a \in B$, the representation is

$$\xi_a(x) = c_a(x)e + \eta_a(x) \tag{2.6}$$

with the mapping η_a: $X \to Y$. According to this, for $a = q$

$$c_q(x) = 1, \qquad \eta_q(x) = 0.$$

Further, let B' be a direct addition in B of a one-dimensional subspace $\{\lambda q\} \subset B$, so that for any $a \in B$ the decomposition $a = \lambda_a q + a'$, where $a' \in B'$, holds. As c_a and η_a in (2.6) linearly depend on a, $c_a = \lambda_a + c_{a'}$ and $\eta_a = \eta_{a'}$. From this, without loss of generality, it can be assumed that $c_{a'} = 0$.

Indeed, in the other case it is possible to make a transformation of the form (2.2) with the mapping ω: $X \to GL(B)$, given by $\omega(x)\langle a\rangle =$

$(\lambda_a - c_{a'}(x))q + a'$, as a result of which the family of vector fields $\bar{\xi}(x) = \xi(x) \circ \omega(x)$ is obtained with the required property. From these, the equalities follow:

$$\begin{aligned}\bar{\xi}_a(x) &= \xi(x)\langle(\lambda_a - c_{a'}(x))q + a'\rangle \\ &= (\lambda_a - c_{a'}(x))e + c_{a'}(x)e + \eta_{a'}(x) = \lambda_a e + \eta_{a'}(x).\end{aligned}$$

Thus it is sufficient to prove the theorem in the case in which $c_{a'}(x) = 0$ for any $a' \in B'$, and then the representation (2.6) has the form

$$\xi_a(x) = \lambda_a e + \eta_{a'}(x). \tag{2.7}$$

According to the decomposition $X = T \oplus Y$, it is possible to assume $x = (t, y)$, where $t \in R$, $y \in Y$. Then for every $t \in R$ the family of vector fields $\eta(t, y)(B')$, operating by $\eta(t, y)\langle a'\rangle = \eta_{a'}(x)$, is defined on the space Y by the formula (2.7). It is shown that this family is complete.

Indeed, since $\xi_q(x) = e$, $\eta_{a'}(x) = 0$ causes $\xi_a(x) = \lambda_a \xi_q(x)$, from which $a = \lambda_a q$ because of the completeness of the family $\xi(x)(B)$, that is, $a' = 0$. From this, property (α) defined in Section 17.2 is established. Further, using (2.7), the commutator $[e, \xi_a]$ becomes

$$[e, \xi_a](x) = \partial_t \eta_{a'}(t, y).$$

From the completeness of $\xi(x)(B)$ it follows that this commutator is $\xi\langle\omega_a\rangle = \lambda_{\omega_a} e + \eta_{\omega'_{a'}}$, where $\omega_a = \lambda_{\omega_a} q + \omega'_{a'}$. But the vector $\partial_t \eta_{a'}$ is contained in Y, and thus $\lambda_{\omega_a} = 0$. Therefore the equation $\partial_t \eta_{a'} = \eta\langle\omega'_{a'}\rangle$ results, whence the mapping $\omega' : X \to \mathscr{L}(B')$ for which

$$\partial_t \eta = \eta \circ \omega' \tag{2.8}$$

appears. Calculation of the commutator $[\xi_a, \xi_b]$, using the representation (2.7) (the lower index of the commutator indicates the variable with respect to which the differentiation is made), gives

$$[\xi_a, \xi_b]_x = \lambda_a \partial_t \eta_{b'} - \lambda_b \partial_t \eta_{a'} + [\eta_{a'}, \eta_{b'}]_y.$$

On the other hand, from the completeness of the family $\xi(x)(B)$,

$$[\xi_a, \xi_b]_x = \xi\langle\omega_{ab}\rangle = \lambda_{\omega_{ab}} e + \eta_{\omega'_{ab}},$$

where $\omega_{ab} = \lambda_{\omega_{ab}} q + \omega'_{ab}$. Comparison of the last two expressions shows that $\lambda_{\omega_{ab}} = 0$ and in addition, because of (2.8),

$$[\eta_{a'}, \eta_{b'}]_y = \eta\langle\omega'_{ab} - \lambda_a \omega'_{b'} + \lambda_b \omega'_{a'}\rangle = \eta\langle\bar{\omega}_{a'b'}\rangle,$$

which is equation (2.1) of property (β), defined in Section 17.2.

Now $\dim B' = s - 1$, so according to the assumption of induction the family $\eta(B')$ has a universal invariant, which, generally speaking, does not depend on t. Finally, it is established that it is possible to choose a universal invariant that does not depend on t. With this goal in mind, a transformation of the type (2.2), namely, $\bar{\eta} = \eta \circ \theta$, for some mapping $\theta: X \to GL(B')$, is considered. From (2.8) the relation

$$\partial_t \bar{\eta} = \eta \circ (\alpha_t \theta + \omega' \circ \theta)$$

is established. For θ it is possible to take a solution of the ordinary differential equation

$$\partial_t \theta + \omega' \circ \theta = 0, \qquad \theta(0) = I_{B'},$$

and then $\partial_t \bar{\eta} = 0$, that is, the transformed mapping $\bar{\eta}(B')$ already does not depend on t. From the Lemma, the universal invariant $J: Y \to Z$ is also the universal invariant of the family $\eta(B')$, and since $\partial_t J = 0$, it is also for the family $\xi(x)(B)$. ■

It is pertinent to note that the method of proof of this theorem gives an inductive process for constructing the universal invariant in s steps, in each of which the dimensionality of the space B is decreased by one unit. If the conclusions made in Section 3.6 are taken into consideration, then a useful consequence will be obtained.

Consequence. *If* $\dim X = n$ *and inequality* (2.5) *is satisfied, that is,* $r_* < n$, *then the universal invariant* J *of the family* $\xi(x)(B)$ *can be considered as the mapping* $J: X \to R^{n-r_*}$.

3. Universal Invariant of Group

Let the group $G^r(f) = (B, \phi, \pi, X)$ of the open set $U \subset X$ of transformations of the finite-dimensional space X be given, where $\dim B = r$, $\dim X = n$. From the theorem in Section 17.1 any invariant of this group is an invariant of the family of vector fields $\xi(x)(B)$, which is built with the aid of the group's $G^r(f)$ tangential mapping ξ. Let the general rank of the mapping ξ on U be $r_* = r_*(\xi)$. With these conditions the following result holds.

Theorem. *The group* $G^r(f)$ *has invariants if and only if* $r_* < n$. *If this inequality is satisfied, then there is a universal invariant* J *of the group* $G^r(f)$, *which is a mapping* $J: X \to R^{n-r_*}$.

Proof. From the definition (16.10.1) of the number r_* it follows that $r_* \leq n$ always. The equality $r_* = n$, because of the theorem in Section 16.10, is equipollent to transitivity of the group $G^r(f)$. In that case the invariant F must satisfy $F(x') = F(x)$, where $x' = f(x, a)$ is any point from some neighborhood of the point x, that is, $F = \text{const}$ on this neighborhood. This contradicts the definition of an invariant. Consequently, invariants exist only in intransitive groups.

Let $r_* < n$. Then for every $x \in U$ the image $\xi(x)(B)$ is a subspace in X, which means that $r_* \leq r$. In this image one can choose r_* linearly independent vector fields $\xi_i(x)$, corresponding to linearly independent vectors $e_i \in B$ $(i = 1, \ldots, r_*)$, and $\xi_i(x) = \xi(x)\langle e_i \rangle$. Let B_0 be a subspace in B (proper if $r_* < r$) generated by the set $\{e_i\}$. It is clear that the family of vector fields $\xi(x)(B_0)$ is complete.

Indeed, the condition of completeness (α), defined in Section 17.2, is satisfied since according to the construction $\dim B_0 = r_*$. Also, for any $a \in B$ and every $x \in U$, the representation

$$\xi_a(x) = \lambda_a^i(x)\xi_i(x) \tag{3.1}$$

holds, in which the coefficients, generally speaking, depend on the point x. As the space $\xi(x)(B)$ is a Lie algebra, for any $a, b \in B_0$,

$$\begin{aligned}[\xi_a(x), \xi_b(x)] &= \xi_{[a,b]}(x) = \lambda_{[a,b]}^i(x)\xi_i(x)\\ &= \xi(x)\langle\lambda_{[a,b]}^i(x)e_i\rangle,\end{aligned}$$

that is, the condition of completeness (β) expressed by (2.1) is satisfied.

Any invariant (universal invariant) of the group $G^r(f)$ is an invariant (universal invariant) of the family of vector fields $\xi(x)(B_0)$. This follows from the criterion of the theorem in Section 17.1 and the representation (3.1). Thus, if $r_* < n$, the existence of a universal invariant of the group $G^r(f)$ follows from the theorem in Section 17.2 and its consequence. ■

From this theorem, as was the case in Section 3.6, the following consequence can be drawn.

Consequence. *For any group $G^r(f)$ satisfying the conditions of the theorem there exists a complete set of $n - r_*$ scalar functionally independent invariants*

$$J^k: X \to R \qquad (k = 1, \ldots, n - r_*) \tag{3.2}$$

that possess the property that any invariant of this group can be given as a function of invariants (3.2.).

In practice the complete set of functionally independent invariants is usually constructed using a procedure analogous to the one that was realized in the proof of the theorem.

According to the previous results, any group $G^r(f) = (B, \phi, \pi, X)$ of transformations of the space X with dimensionality $\dim X = n$ appears to have three integral characteristics (r, n, r_*), where r is the dimensionality of the group, n is dimensionality of the space of the representations, and r_* is the general rank of the tangential mappings of the group. Among these numbers the relations

$$r_* \leq 4, \qquad r_* \leq n \tag{3.3}$$

are required. Thus the group for which $r < n$ always has invariants. But the group $G^r(f)$ can have invariants even in the case in which $r \geq n$, provided $r_* < n$.

In general, if $r_* < r$, then the tangential vector fields ξ_a, obtained for all possible $a \in B$, are called *linearly connected*. This means that there exists a decomposition into the direct sum $B = B_* \oplus B'$ such that $\dim B_* = r_*$, the space of vector fields $\xi(x)(B_*)$ has the general rank equal to r_*, and for any vector $a' \in B'$ the representation

$$\xi_{a'}(x) = \xi(x)\langle\omega_{a'}(x)\rangle \tag{3.4}$$

holds for some mapping $\omega_{a'}: X \to B_*$. With this, the number $\dim B' = r - r_*$ is called the *number of independent linear connections of the vector fields* $\xi_a \in \xi(x)(B)$. Thus if $r \geq n$, the group has invariants only if its tangential vector fields ξ_a are linearly connected and the number of independent linear connections is sufficiently large.

4. Invariants of Inner Automorphisms

In constructing the optimal systems of subalgebras (see Section 14.7), it is useful to know the invariants of the inner automorphisms group Int L^r of the given Lie algebra L^r.

According to the definitions in Sections 14.3 and 14.4, the tangential mapping e of the group Int L^r gives, for any $u \in L^r$, the vector field $e_u = ad(u)$, operating by $e_u(x) = ad(u)\langle x\rangle = [x, u]$ [see (14.3.1)]. Now here $B = L^r$, so the maximal dimensionality of the vector space $\mathrm{ad}(L^r)$ is equal to the maximal rank of the linear mapping $ad(u)$ when the vector u takes values in L^r. Consequently,

$$r_* = \max_{u \in L^r} \operatorname{rank} \mathrm{ad}(u). \tag{4.1}$$

But the rank of the linear mapping $\mathrm{ad}(u)\colon L^r \to L^r$ is equal to the number of nonzero roots of the characteristic polynomial of this mapping. This polynomial is Killing's polynomial of the Lie algebra L^r (see Section 14.8). Thus because of the definition given in Section 14.8 for the rank l of the Lie algebra L^r, equation (4.1) gives $r_* = r - l$. As $l > 0$ and $n = r$, the group of automorphisms Int L^r is always intransitive and the complete set of its functionally independent invariants consists of $n - r_* = l$ scalar invariants. So the universal invariant J of the group Int L^r is the mapping $J: L^r \to R^l$, where l is the rank of the Lie algebra L^r.

It is easy to establish that *all nonzero coefficients* $\tau_i(x)$ *of Killing's polynomial* $\det(\lambda I_r - ad(x))$ [see (14.8.2)] *are invariants of the group* Int L^r. Indeed, let $\lambda = \lambda(x)$ be a root of the characteristic polynomial and let y be the corresponding eigenvector of the liner mapping ad(x), that is,

$$[y, x] = \lambda(x)y.$$

Under the operation of the inner automorphism $A \in \text{Int } L^r$, this relation transforms into

$$[A\langle y \rangle, A\langle x \rangle] = \lambda(x)A\langle y \rangle,$$

showing that $A\langle y \rangle$ is an eigenvector of the mapping ad($A\langle x \rangle$) corresponding to the same eigenvalue $\lambda(x)$. This means that every eigenvalue $\lambda(x)$ is equal to an eigenvalue $\lambda(A\langle x \rangle)$. As the converse statement is also correct, the union of eigenvalues of the mapping ad(x) coincides with the union of all eigenvalues of the mapping ad($A\langle x \rangle$). Consequently, all (nonzero) coefficients of Killing's polynomial, as symmetric functions of its roots, are invariants of the group Int L^r.

Now let Z be the center of the Lie algebra L^r and let $L^r = Z \oplus L^s$, where L^s is a subalgebra of L^r. Then if the basis in L^r is chosen so that it contains the basis of the center Z in the decomposition of the arbitrary vector $x \in L^r$ by the basis, the nonzero coefficients of the basis vectors are also invariants of the group Int L^r.

5. Examples

(α) An illustration for the point in Section 17.4 is the example in Section 14.9 in which $L^4 = L^3 \oplus Z$, where the center Z is generated by the vector u_4 and the basis in L^3 is $\{u_1, u_2, u_3\}$. From formula (14.9.3) for ad(v) Killing's polynomial of the Lie algebra L^3 is

$$\det(\lambda I_3 - \text{ad}(x)) = \begin{vmatrix} \lambda - 3x^2 & 3x^1 & 0 \\ -x^3 & \lambda & x^1 \\ 0 & -3x^3 & \lambda + 3x^3 \end{vmatrix} = \lambda^3 - 3(3(x^2)^2 - 2x^1x^3)\lambda.$$

The only nonzero coefficient is $\tau_2(x) = \frac{1}{2}K\langle x, x \rangle$, where K is the Killing's form found in Section 14.9. Here $l = 1$ and one of the scalar invariants is $J_1: R^3 \to R$ of the group Int L^3, operating by $J_1(x) = K\langle x, x \rangle$. Also, if $x = x^\alpha u_\alpha \in L^4$, then x^4 will also be an invariant. Thus the universal invariant $J: R^4 \to R^2$ of the group Int L^4 can be taken in the form

$$J(x) = (J_1(x^1, x^2, x^3), x^4).$$

(β) Let $G^3(f)$ be the group of rotations of the space $R^3(x, y, z)$. The basis of its Lie algebra of operators $L^3(\xi)$ is

$$\xi_1 \cdot \partial = z\partial_y - y\partial_z, \qquad \xi_2 \cdot \partial = -z\partial_x + x\partial_z, \qquad \xi_3 \cdot \partial = y\partial_x - x\partial_y. \quad (5.1)$$

Here $r = n = 3$, but the operators (5.1) are linearly connected by $x\xi_1 + y\xi_2 + z\xi_3 = 0$, and this is their only independent connection. Thus $r_* = 2$, because of which there must be one scalar invariant. For its calculation one of the operators (5.1), for example $\xi_3 \cdot \partial$, can be dropped.

First, the universal invariant of the operator $\xi_1 \cdot \partial$ is found. Obviously, it can be taken in the form

$$J = (x, \sqrt{y^2 + z^2}).$$

With this, the vector $\xi_1(0, 1, 0) = (0, 0, -1)$ is nonzero, and it can be assumed that $T = R(z)$. Then the decomposition $X = T \oplus Y$ can be chosen in the form $R^3(x, y, z) = R(z) \oplus R^2(x, y)$. Now the transformation of the kind (2.3) is made with the mapping $p: (x, y, z) \to (\bar{x}, \bar{y}, \bar{z})$, given by the formulas

$$\bar{x} = x, \bar{y} = \sqrt{y^2 + z^2}, \qquad \bar{z} = z.$$

As a result of this, the first two operators of the basis (2.3) transform into

$$\bar{\xi}_1 \cdot \partial = -y\partial_z, \qquad \bar{\xi}_2 \cdot \partial = -\bar{z}\partial_{\bar{x}} + \frac{\bar{x}\bar{z}}{\bar{y}}\partial_{\bar{y}} + \bar{x}\partial_{\bar{z}}.$$

In turn, by means of a suitable transformation of the kind (2.3), the operator $\bar{\xi}_2 \cdot \partial$ can be transformed into the operator

$$\bar{\bar{\xi}}_2 \cdot \partial = -\bar{y}\partial_{\bar{x}} + \bar{x}\partial_{\bar{y}},$$

with the obvious invariant $J_1(\bar{x}, \bar{y}) = \bar{x}^2 + \bar{y}^2$. Finally, the unknown universal invariant of the rotation group is given by

$$J_1 \circ J(x, y, z) = x^2 + y^2 + z^2.$$

6. Invariants of Extensions

Some natural transitions from the group (B, ϕ, π, X) to the group (B, ϕ, π', X'), based on extension of the space X up to the space X', with corresponding extension of the representation π up to the representation π', will here be understood as an *extension of the group* (B, ϕ, π, X).

The first way of making such an extension is a transition to *multiple groups*—double, triple, and so on. To construct a double group, the space X is extended up to $X^2 = X \times X$ and it is assumed that the same representation π is used for both factors. Thereby the double representation $\pi^2: (B, \phi) \to X^2$ is determined, as is the double group (B, ϕ, π, X^2). If the initial group is generated by the mapping $f: X \times B \to X$ and its trangential mapping is ξ, then the double group is generated by the mapping $f^2: X^2 \times B \to X^2$, operating by

$$f^2(x_1, x_2, a) = (f(x_1, a), f(x_2, a)), \tag{6.1}$$

where $x_1 \in X$ and $x_2 \in X$ are independnetly changing vectors. The tangential mapping ξ^2 of the double group is the mapping $X^2 \to X^2$, operating by

$$\xi^2(x_1, x_2) = (\xi(x_1), \xi(x_2)).$$

To write the operators of a double group, it is necessary to write $\partial = (\partial_1, \partial_2)$, where ∂_1 and ∂_2 are operators of differentiation with respect to the vectors x_1 and x_2, respectively. In this notation the infinitesimal operator of a double group, corresponding to the vector $a \in B$ and to the vector field $\xi_a^2 = (\xi_a, \xi_a)$, has the form

$$\xi_a^2(x_1, x_2) \cdot \partial = \xi_a(x_1) \cdot \partial_1 + \xi_a(x_2) \cdot \partial_2. \tag{6.2}$$

Generally, k-multiple groups (B, ϕ, π^k, X^k) of transformations of the space X^k are defined analogously.

The space B and the group (B, ϕ) are not changed during the transition to a k-multiple group, but the dimensionality $\dim X^k = k \dim X$ (in the case of finite-dimensional X) is k times the original dimension. Thus, for a sufficiently large k, the extended k-multiple group is necessarily intransitive and therefore has invariants. The invariants of a k-multiple group are called *k-point invariants* of the initial group.

For example, the group $G^1(f)$ of transformations of the one-dimensional space $X = R(x)$ generated by the mapping $x' = x + a$ is transitive. The double group $G^1(f^2)$ whose generating mapping is

$$x_1' = x_1 + a, \qquad x_2' = x_2 + a$$

gives a two-point invariant $J = x_1 - x_2$ of the group $G^1(f)$.

The second way of extension, which is the basis for the group analysis of differential equations, is a transition to a *continuation group.* It relates to the groups $G^r(h) = (B, \phi, \pi, Z)$ of transformations of the product $Z = X \times Y$. The continuation of a group is realized by continuation of every transformation h_a belonging to it up to the transformation $\underset{k}{h}_a$, according to the rule described in Section 4.5 and analogous to the continuation of one-parameter groups done in Section 4.7. In this case the space $Z = X \times Y$ is extended up to the continuous space $Z_k = X \times Y \times Y_1 \times \cdots \times Y_k$ and, accordingly, the representation $\pi: (B, \phi) \to \tau(Z)$ up to the representation $\pi_k: (B, \phi) \to \tau(Z_k)$. With this, the continuous transformations form the group $G^r(\underset{k}{h}) = \underset{k}{G^r}(h) = (B, \phi, \pi_k, Z_k)$, isomorphic to the initial one. This is guaranteed by the lemma in Section 4.6.

Simultaneously with the group, its tangential mapping ζ is continued by the rules described in Section 4.8. Then the continuous tangential mapping $\underset{k}{\zeta}$ is a tangential mapping of the continuation group $\underset{k}{G^r}(h)$, and it generates the continuation Lie algebra $\underset{k}{L^r}(\zeta) = L^r(\underset{k}{\zeta})$ of vector fields (or operators) on the continuous space Z_k. The lemma in Section 7.12 guarantees that the Lie algebra $\underset{k}{L^r}(\zeta)$ be isomorphic to the initial Lie algebra $L^r(\zeta)$.

Under continuation of the group $G^r(h)$, the dimensionality of the space Z_k increases without bound, together with k, and thus the continuation group $\underset{k}{G^r}(h)$, for sufficiently large k, is necessarily intransitive and therefore has invariants. Invariants of a continuation group are called *differential invariants* of the initial group. They have already been discussed in Section 4.11. All criteria and rules described in the present paragraph hold for differential invariants. In particular, the idea of a universal differential invariant of the kth order, as a universal invariant of the continuation group $\underset{k}{G^r}(h)$ makes sense.

§18. INVARIANT MANIFOLDS

1. Definition

The concept of a manifold $\Psi \subset X$ invariant relative to the one-parameter group G^1 of transformations of the space X was already considered in Section 3.10. This concept is trivially generalized for the case of an arbitrary group (B, ϕ, π, X) because of the fact that every such group is a union of one-parameter subgroups belonging to it (consequence of the theorem in Section 12.6). But in the case of r-parameter groups new problems appear, which require study. Thus it is suitable to state independently a new definition of the concept of an invariant manifold of the group $G^r(f) = (B, \phi, \pi, X)$.

The manifold $\Psi \subset X$ is called an *invariant manifold* of the group $G^r(f)$ [or *manifold invariant relative to the group* $G^r(f)$] if for every point $x \in \Psi$ there exists a neighborhood of zero $O \subset B$ such that for any $a \in O$ the point $x' = f(x, a)$ also belongs to Ψ.

It is clear that Ψ is invariant relative to $G^r(f)$ if and only if Ψ is an invariant manifold of any of its one-parameter subgroups. Also, if Ψ is an invariant manifold of the group $G^r(f)$, then it is also an invariant manifold of any subgroup (see Section 15.2) of this group.

Manifolds Ψ given implicitly (see Section 3.11) in the finite-dimensional space X for which $\dim X = n$ will be considered below. Thus the open space

$U \subset X$ and the mapping $\psi: U \to R^s$ $(0 < s < n)$ of class $C_\infty(U)$ are given, where $\psi(x_0) = 0$ for some point $x_0 \in U$ and the set $\psi(U_0)$ is open in R^s for any neighborhood $U_0 \subset U$ of the point x_0. Under these conditions, the manifold Ψ is said to be a *point locus* of $x \in U$ for which

$$\psi(x) = 0. \tag{1.1}$$

2. Invariance Criterion

In what follows it is assumed that the regular manifold Ψ is given by equation (1.1) (see Section 3.11). This means that for every $x \in U$ the derivative $\partial\psi(x)$, as a linear mapping $X \to R^s$, has rank on U equal to s. Then this rank is equal to the general rank r_* of the vector field $\partial\psi$ on U (see Section 17.2) and the regular condition of the manifold (1.1) is written as

$$r_*(\partial\psi) = s. \tag{2.1}$$

Under these conditions the dimensionality of the manifold Ψ is $\dim \Psi = n - s$. The infinitesimal criterion of invariance of the manifold is formulated using the tangential mapping ξ of the group $G^r(f)$.

Theorem. *The regular manifold* Ψ, *given by equation* (1.1), *is invariant relative to the group* $G^r(f)$ *if and only if*

$$\partial\psi(x) \circ \xi(x)|_\Psi = 0. \tag{2.2}$$

Proof. Let ξ_a be a vector field on U corresponding to the vector $a \in B$, that is, $\xi_a(x) = \xi(x)\langle a \rangle$. It is obvious that equation (2.2) is equipollent to equation $\partial\psi(x)\langle\xi_a(x)\rangle|_\Psi = 0$ or, what is the same, to the equality $(\xi_a \cdot \partial)\psi|_\Psi = 0$. The latter coincides with (3.12.1) and, because of the theorem in Section 3.12, gives a criterion of invariance of the manifold Ψ relative to a one-parameter group $G^1(\xi_a)$. Thus both statements follow from the fact that the group $G^r(f)$ is the union of all groups $G^1(\xi_a)$, obtained when the vector a assumes values in the space B. ■

There is a simple method for forming invariant manifolds of the given transitive group $G^r(f)$. Take any invariant $F: X \to R^s$ and generate the equation $F(x) = 0$. The manifold in X, given by this equation, is invariant relative to the group $G^r(f)$. This fact is obvious from the definition of an invariant in Section 17.1 [or from criterion (17.1.3)]. In what follows, it will be shown that this way to generate invariant manifolds, seemingly very limited at first sight, indeed is general.

Note. During transition from the group $G^r(f)$ to its similar group $G^r(f')$, under the transformation p of the space X (see Section 16.9), any invariant manifold Ψ of the group $G^r(f)$ transforms into an invariant manifold Ψ' of the group $G^r(f')$. From this, if Ψ is given by the equation (1.1), the equation of the transformed manifold Ψ' is

$$\psi \circ p^{-1}(x) = 0. \tag{2.3}$$

3. Induced Group

According to the definition in Section 17.1 of an invariant manifold Ψ of the group $G^r(f) = (B, \phi, \pi, X)$ for every point $\bar{x} \in \Psi$ and for any $a \in O$, the image $\bar{x}' = f_a(\bar{x}) \in \Psi$ is determined. To find this image, there is no necessity to consider the transformation f_a on the whole space X; it is sufficient to use only its reduction $f_a | \Psi$ on the manifold Ψ. It is clear that this reduction is a local transformation of the manifold Ψ.

The transformations $f_a | \Psi \colon \Psi \to \Psi$ are called *induced transformations of the invariant manifold* Ψ *of the group* $G^r(f)$. The set of all induced transformations is given by the reduction $f | \Psi$ generating the group $G^r(f)$ of the mapping f. This reduction $f | \Psi$ will be called an induced mapping $\Psi \times O \to \Psi$; it can be considered as a mapping of the null neighborhood $O \subset B$ into the group $\tau(\Psi)$ of all transformations of the manifold Ψ, operating by $(f | \Psi)_a = f_a | \Psi$.

It is easy to verify that the induced mapping $f | \Psi$ generates the representation π_Ψ of the group (B, ϕ) in the group $\tau(\Psi)$, operating by $\pi_\Psi(a) = f_a | \Psi$. Indeed, according to the given definitions of $\bar{x} \in \Psi$, this will be $\pi_\Psi(a)(\bar{x}) = f(\bar{x}, a)$, and the equation

$$\pi_\Psi(b) \circ \pi_\Psi(a)(\bar{x}) = f(f(\bar{x}, a), b) = f(\bar{x}, \phi(a, b)) = \pi_\Psi(\phi(a, b))$$

results. The resulting representation $\pi_\Psi \colon O \to \tau(\Psi)$ is a homomorphism of the group (B, ϕ) into te group $\tau(\Psi)$. That is why the image $\pi_\Psi(O)$ is a group of transformations of the manifold Ψ.

The group $\pi_\Psi(O)$ is called an *induced group on the invariant manifold* Ψ *of the group* $G^r(f)$ and will also be denoted by $G(f | \Psi)$.

Actually, the induced group $G(f | \Psi)$ is a local Lie group of transformations of the manifold Ψ (which is considered as a local Banach space; see Section 15.2). To establish this fact, consider the kernel of the representation π_Ψ, which is obviously generated only by those $a \in O$ for which the transformation f_a leaves all points $\bar{x} \in \Psi$ fixed.

Lemma. *The union of the transformations f_a, leaving all points of the invariant manifold Ψ fixed, is a normal divisor of the group $G^r(f)$.*

Proof. Let $f_b \in G^r(f)$ and $f_a(\bar{x}) = \bar{x}$ for all $\bar{x} \in \Psi$. If $\bar{x}' = f_b(\bar{x})$, then $\bar{x}' \in \Psi$ because of invariance of Ψ, and $f_b^{-1} \circ f_a \circ f_b(\bar{x}) = f_b^{-1} \circ f_a(\bar{x}') = f_b^{-1}(\bar{x}') = \bar{x}$. ■

Theorem. *The group $G(f\,|\,\Psi)$, induced by the group $G^r(f)$ on its invariant manifold Ψ, is a Lie group of transformations of the manifold Ψ.*

Proof. The normal divisor of the group $G^r(f)$ is put into correspondence with the normal divisor H^p of its parametric group $G^r = (B, \phi)$. Let $G^r/H^p = (\bar{B}, \bar{\phi})$ be the corresponding factor group (see Section 15.5), with multiplication law of the class $(\bar{a}, \bar{b}) \to \bar{\phi}(\bar{a}, \bar{b})$. It is easy to verify that the mapping $\bar{\pi}\colon (\bar{B}, \bar{\phi}) \to \tau(\Psi)$, operating by $\bar{\pi}(\bar{a}) = f_a|\Psi$ if $a \in \bar{a}$, is a representation of the group $(\bar{B}, \bar{\phi})$. Note that the representation π is exact. For if $\bar{\pi}(\bar{a}) = I_\Psi$, that is, $\bar{\pi}(\bar{a})(\bar{x}) = f(\bar{x}, a) = \bar{x}$ for all $\bar{x} \in \Psi$, then according to the construction, $a \in H^p$ and $\bar{a} = 0$. The remainder follows from the definition in Section 16.1, all of whose conditions are satisfied. ■

The normal divisor H^p and the representation π give the group $(H^p, \phi'$ $\pi, X)$ of transformations of the space X whose elements, and only these elements, leave all points of the invariant manifold Ψ unchanged.

The group (H^p, ϕ', π, X) is called the *stationary group of the manifold* Ψ.

To find the order p of a stationary group, it can be assumed, without loss of generality, that the group (B, ϕ) is canonical. Then the equality $f_{ta}(\bar{x}) = \bar{x}$, for the one-parameter subgroup $G^1(f_{ta}) \subset G^r(f)$, is equipollent to the equality $\xi(\bar{x})\langle a\rangle = 0$. This means that if $\bar{x} \in \Psi$, then the last equality holds for all $a \in H^p$ and only for them. Thus for any $\bar{x} \in \Psi$, $\dim \xi(\bar{x})(B) = r - p$. On the other hand, let $r_*(\xi\,|\,\Psi)$ be the general rank of the tangential mapping ξ of the group $G^r(f)$ on the manifold Ψ. This rank is calculated through the dimensionality of the space of vector fields $\xi(\bar{x})(B)$ by means of $r_*(\xi\,|\,\Psi) = \dim \xi(\bar{x})(B)$. Consequently, the dimensionalities of the stationary subgroup (H^p, ϕ', π, X) of the invariant manifold Ψ of the group $G^r(f)$ and the induced group $(\bar{B}, \bar{\phi}, \pi_\Psi, \Psi)$ on Ψ are given by

$$\begin{aligned} \dim H^p &= p = r - r_*(\xi\,|\,\Psi), \\ \dim \bar{B} &= r - p = r_*(\xi\,|\,\Psi). \end{aligned} \tag{3.1}$$

4. Induced Lie Algebra

Let ξ_a be a vector field on the space X, belonging to the Lie algebra $L^r(\xi)$ of the group $G^r(f)$. Reduction of this vector field on a manifold Ψ is denoted by the symbol $\xi_a|\Psi$. For any two vector fields ξ_a, $\xi_b \in L^r(\xi)$ the commutator of their reduction on Ψ is determined by the formula

$$[\xi_a|\Psi, \xi_b|\Psi] = [\xi_a, \xi_b]|\Psi. \tag{4.1}$$

It is obvious that the set of all reductions $\xi_a|\Psi$, which are obtained when the vector a takes values in the space B, together with the commutation operation

(4.1), generates a Lie algebra. This will be called the *induced Lie algebra* on the manifold Ψ and will be denoted by $L(\xi\,|\,\Psi)$. For the case of an arbitrary manifold Ψ, this concept is not very useful because of the weak connection with the initial group. But the situation is quite different when Ψ is an invariant manifold of the group $G^r(f)$.

Theorem. *If Ψ is an invariant manifold of the group $G^r(f)$, then the induced Lie algebra $L(\xi\,|\,\Psi)$ is a Lie algebra of the induced group $G(f\,|\,\Psi)$ and its dimensionality is equal to $r_*(\xi\,|\,\Psi)$.*

Proof. To realize the Lie algebra $L(\bar{\xi})$ of the group $G(\bar{f})$, where $\bar{f} = f\,|\,\Psi$, the notation and results from the theorem in Section 18.3 are used. It is supposed that the parametric group $G^r = (B, \phi)$ of the group $G^r(f)$ is canonical. Then its normal divisor H^p and the factor group $G^r\,|\,H^p = (\bar{B}, \bar{\phi})$ are also canonical. Thus one-parameter subgroups of the group $G(\bar{f})$ have the form $\{\bar{f}_{t\bar{a}}\}$, where $\bar{a} \in \bar{B}$. But, according to the definition, $\bar{f}_{t\bar{a}} = f(\bar{x}, ta)$ if $a \in \bar{a}$. Consequently, calculation of the tangential vector field $\bar{\xi}_{\bar{a}}$ of the one parameter subgroup $\{\bar{f}_{t\bar{a}}\}$ gives

$$\bar{\xi}_{\bar{a}}(\bar{x}) = \partial_t \bar{f}_{t\bar{a}}(\bar{x})|_{t=0} = \partial_t f(\bar{x}, ta)|_{t=0} = \xi(\bar{x})\langle a\rangle = (\xi_a\,|\,\Psi)(\bar{x}),$$

that is, $\bar{\xi}_{\bar{a}} = \xi_a\,|\,\Psi$. This relation determines a mapping of the Lie algebra $L(\bar{\xi}) \to L(\xi\,|\,\Psi)$ which is one-to-one because if $\xi_a\,|\,\Psi = \xi_b\,|\,\Psi$, then $\xi(\bar{x})\langle a - b\rangle = 0$ for any $\bar{x} \in \Psi$, from which it follows that $a - b \in H^p$ and $\bar{a} = \bar{b}$ because of the lemma in Section 15.5. Consequently, the Lie algebras $L(\bar{\xi})$ and $L(\xi\,|\,\Psi)$ coincide as sets. Also, from the relations

$$[\bar{\xi}_{\bar{a}}, \bar{\xi}_{\bar{b}}] = \bar{\xi}_{[\bar{a}, \bar{b}]} = \bar{\xi}_{\overline{[a, b]}} = \xi_{[a, b]}\,|\,\Psi = [\xi_a, \xi_b]\Psi$$

and definition (4.1), in this Lie algebra the commutation operations coincide. Finally, the statement about the dimensionality of the induced Lie algebra follows from (3.1). ■

5. Least Invariant Manifolds

Let the group $G^r(f)$ be generated by the mapping $f\colon U \times O \to X$ of the product of the open set $U \subset X$ with a null neighborhood $O \subset B$. The *orbit* $f_x(O)$ of a fixed point $x \in U$ is considered. As was already noted in Section 16.10, if this orbit contains a neighborhood of the point x, then the group $G^r(f)$ is transitive at this point. In this case the point x cannot belong to any invariant manifold of the group $G^r(f)$.

Conversely, suppose the orbit $f_x(O)$ does not contain any neighborhoods of the point x. Then this orbit is a manifold in X, given parametrically by the equation $\bar{x} = f(x, a)$, for the vector parameter $a \in O$. In this case the orbit

$f_x(O)$ is an invariant manifold of the group $G^r(f)$. So if $\bar{x} \in f_x(O)$, then $\bar{x} = f(x, a)$ for some $a \in O$, and for any $b \in O$ (if $|b|$ is sufficiently small) it is also true that $f(\bar{x}, b) = f(x, \phi(a, b)) \in f_x(O)$. Also, it is clear that any invariant manifold of the group $G^r(f)$ containing x necessarily also contains the orbit $f_x(O)$. Thus *the orbit $f_x(O)$ is the least invariant manifold of the group $G^r(f)$ containing the point x.* According to the results of Section 16.10, the dimensionality of the orbit $f_x(O)$ is given by

$$\dim f_x(O) = \operatorname{rank} \xi(x). \tag{5.1}$$

Lemma. *The group $G(f \mid f_x(O))$, induced on the orbit $f_x(O)$ of any point $x \in U$, is simply transitive.*

Proof. If $x_1 = f(x, a)$ and $x_2 = f(x, b)$ are any two points from $f_x(O)$, then $x_2 = f(x_1, \phi(a^{-1}, b))$, that is, x_1 is transformed into x_2 by a transformation belonging to the group. Therefore, the group $G(f \mid f_x(O))$ is transitive. Besides, as the dimensionality of the induced group is given by (3.1) and, thereby, is equal to the common rank $r_*(\xi \mid f_x(O))$, the property of simple transitivity of the group $G(f \mid f_x(O))$ follows from Consequence 2 in Section 17.10. ∎

6. Singular Invariant Manifolds

Let $r_* = r_*(\xi)$, defined by (16.10.1), be the general rank of the tangential mapping ξ of the group $G^r(f)$ on the set U. It is not supposed that ξ has a rank on U. [That is, at the different points $x \in U$ the rank $\xi(x)$ can be different.] Using the number r_* all points $x \in U$ are distinguishable as *singular* or *nonsingular* in the following sense.

The point x is called a *singular point* [*of the group* $G^r(f)$] if rank $\xi(x) < r_*$ and a *nonsingular point* [*of the group* $G^r(f)$] if rank $\xi(x) = r_*$.

The manifold $\Psi \subset U$ is called a *singular manifold of the group* $G^r(f)$ [or *of its Lie algebra* $L^r(\xi)$] if all its points are singular and if the tangential mapping ξ has a rank on Ψ. In other words, a singular manifold is characterized by the fact that for any point $\bar{x} \in \Psi$

$$\operatorname{rank} \xi(\bar{x}) = r_*(\xi \mid \Psi) < r_*. \tag{6.1}$$

The manifold $\Psi \subset U$ is called a *nonsingular manifold of the group* $G^r(f)$ [or *its Lie algebra* $L^r(\xi)$] if all its points are nonsingular, that is, if

$$r_*(\xi \mid \Psi) = r_*. \tag{6.2}$$

In particular, these definitions imply that all invariant manifolds of the group $G^r(f)$ are either singular on nonsingular invariant manifolds.

To ascertain whether a manifold is singular, there is a simple algorithm, realizable in any coordinate system of the spaces B and X. Let $\{u_\alpha\}$ be a basis in B ($\alpha = 1, \ldots, r$) and $\{e_i\}$ be a basis in X ($i = 1, \ldots, n$), where the decompositions of the vectors $x, \xi \in X$ have the forms $x = x^i e_i$, $\xi = \xi^i e_i$ and the basis of the Lie algebra $L^r(\xi)$ is generated by the vectors $\xi_a(x) = \xi(x)\langle u_\alpha \rangle$. Then the tangential mapping ξ is completely described by an $r \times n$ matrix

$$M(x) = (\xi_\alpha^i(x)) \qquad (i = 1, \ldots, n; \alpha = 1, \ldots, r), \tag{6.3}$$

where i is the column index and α the row index. From this it follows that

$$\operatorname{rank} \xi(x) = \operatorname{rank} M(x). \tag{6.4}$$

Because of this the concepts "M has a rank on U," "general rank $r_*(M)$," "general rank $r_*(M \mid \Psi)$" make sense for matrices of the type (6.3). The convenience of this representation arises because the rank of a matrix is calculated by means of simple algebraic operations, which are easily accomplished for values of r and n that are not too large.

To determine the common rank of the matrix M, it is necessary to find its minor of highest order that is not identically zero; the order of this minor is equal to $r_*(M)$.

To find the singular manifolds, it is necessary to calculate all minors of the matrix M with order $r_*(M)$ and set them equal to zero. The union S of all solutions (set of components $\{x^i\}$ of the point x) of the resulting system of equations is the set of all singular points of the matrix M, and also of the group $G^r(f)$. Any manifold $\Psi \subset S$ on which M has a rank is a singular manifold of the group $G^r(f)$. In the general case, the manifold Ψ is not required to be invariant. Verification of the manifold's invariance is made by using the criterion in Section 18.2.

7. The Representation Theorem

The algorithm described in the previous section does not permit determination of nonsingular invariant manifolds of the group $G^r(f)$. Their description is given in the following theorem concerning representation of nonsingular invariant manifolds Ψ satisfying the condition

$$\dim \Psi < n. \tag{7.1}$$

Theorem. *Nonsingular invariant manifolds of the group $G^r(f)$ exist if and only if this group is intransitive. If a nonsingular manifold of the group $G^r(f)$ is given regularly by equation (1.1), then there exists an invariant of this group $F: X \to R^s$ such that this manifold is given by the equation $F(x) = 0$.*

Proof. If the group $G^r(f)$ is transitive and Ψ is its nonsingular invariant manifold, then the orbit $f_{\bar{x}}(O)$ of the point $\bar{x} \in \Psi$ belongs to Ψ and, at the same time, it contains a neighborhood of the point $\bar{x}$ in the space X. Thus $\dim \Psi = \dim X = n$, which contradicts condition (7.1). If this group is intransitive, then it has invariants, which means it has nonsingular manifolds, obtained as discussed in Section 18.2.

Let Ψ be a nonsingular invariant manifold of the group $G^r(f)$, regularly given by equation (1.1). The regularity of this assignment means that $\dim \Psi = n - s$, and for any point $\bar{x} \in \Psi$ its orbit $f_{\bar{x}}(O) \subset \Psi$, so the inequality $r_* < n - s$ is correct.

Since the group $G^r(f)$ is intransitive, it has a universal invariant $J: X \to R^{n-r_*}$. Further, if $r_* < n - s$, then there is a universal invariant $\bar{J}$ of the induced group $G(f \mid \Psi)$, which, because of the nonsingularity of Ψ and equality (6.2), is the mapping $\bar{J}: \bar{\psi} \to R^{n-s-r_*}$. On the other hand, the reduction $J \mid \Psi$ is, obviously an invariant of the induced group $G(f \mid \Psi)$ and thus can be given in the form $J \mid \Psi = \Phi \circ \bar{J}$ for some mapping $\Phi: R^{n-s-r_*} \to R^{n-r_*}$. Because of that the image $\bar{J}(\Psi)$ contains an open set $W \subset R^{n-s-r_*}$ whose manifold $\Phi(W) \subset R^{n-r_*}(y)$ of dimensionality $n - r_* - s' \leq n - r_* - s$ is given by the mapping Φ, where $s \leq s'$. The manifold $\Phi(W)$ can also be given implicitly by the equation $\omega(y) = 0$ for some mapping $\omega: R^{n-r_*} \to R^{s'}$, where $r_*(\partial\omega) = s'$. Thereby the invariant F of the group $G^r(f)$ is given as the mapping $X \to R^{s'}$, operating by $F(x) = \omega \circ J(x)$. Next it will be shown that the manifold $\Psi' \subset X$, given by the equation $F(x) = 0$, coincides with Ψ.

Indeed, according to the construction, for any $\bar{x} \in \Psi$ $y = \bar{J}(\bar{x}) \in W$ and $\Phi(y) \in \Phi(W)$, because of which $F(\bar{x}) = \omega(\Phi(y)) = 0$. This means that $\Psi \subset \Psi'$ and, therefore, $\dim \Psi \leq \dim \Psi'$. But $\dim \Psi' = n - s'$ because $r_*(\partial\omega) = s'$. Hence the inequality $n - s \leq n - s'$ is correct, so that $s' < s$. Together with the earlier established inequality, $s \leq s'$, this gives $s' = s$. Consequently, both manifolds Ψ and Ψ' have the same dimensionality, and from the inclusion $\Psi \subset \Psi'$ their coincidence follows. If $r_* = n - s$, then the induced group $G(f \mid \Psi)$ is simply transitive because of (3.1) and Consequence 2 in Section 16.10. Thus Ψ is the least invariant manifold of the group $G^r(f)$ and coincides (locally) with the orbit of some point $x_0 \in \Psi$. But the same property is possessed by the manifold Ψ' given by the equation $J(x) = J(x_0)$, which also has dimensionality $n - s$. Thus Ψ' coincides with Ψ and the statement of the theorem is correct with the mapping $x \to F(x) = J(x) - J(x_0)$. ■

8. Rank of Nonsingular Invariant Manifolds

During the proof of the theorem in Section 18.7 it was established that the dimensionality of any nonsingular invariant manifold of the group $G^r(f)$ is not smaller than the number $r_* = r_*(\xi)$. It is equal to this number only for nonsingular least invariant manifolds (orbits of nonsingular points), which are all given by means of the universal invariant J of the group $G^r(f)$ by the equation $J(x) = J(x_0)$.

The theorem in Section 18.7 provides every nonsingular invariant manifold with an important numerical characteristic—its rank. *The rank of the nonsingular invariant manifold* Ψ *of dimensionality* $\dim \Psi = n - s$, *of the group* $G^r(f)$, is the number

$$\rho = \rho(\Psi) = n - s - r_*. \tag{8.1}$$

The geometric meaning of rank is clarified by consideration of the space $R^{n-r_*}(y)$, in which the universal invariant J of the group $G^r(f)$ operates. This space will be called the *space of invariants*. The points (vectors) of the space of invariants are the values of the universal invariant $y = J(x)$, obtained when the point x takes values in the open space $U \subset X$. From the representation theorem in Section 18.7, the nonsingular manifold Ψ is given by the equation $\omega \circ J(x) = 0$, which can be considered as an equation in the space of the invariants $\omega(y) = 0$. Assume it to be the equation of the manifold Ψ in this space. From the equality $r_*(\partial\omega) = s$, the dimensionality of the latter manifold is equal to $n - r_* - s$, that is, to the number (8.1). Consequently, the rank of a nonsingular invariant manifold is equal to the dimensionality of this manifold in the space of invariants.

In practice, the universal invariant J of the group $G^r(f)$ is usually presented in the form of the complete set $\{J^1, \ldots, J^{n-r_*}\}$ of functionally independent scalar invariants (see Section 17.3). In these terms the representation theorem for a nonsingular invariant manifold of dimensionality $n - s$ asserts that such a manifold can always be defined by the system of equations, in the space of invariants,

$$\omega^\sigma(y^1, \ldots, y^{n-r_*}) = 0 \qquad (\sigma = 1, \ldots, s). \tag{8.2}$$

Equations (8.2) are independent in the sense that $r_*(\partial_\nu \omega^\sigma) = s$. The equations of this manifold in the space X are obtained if it is assumed that $y^\nu = J^\nu(x)$ $(\nu = 1, \ldots, n - r_*)$, whence they become

$$\omega^\sigma(J^1(x), \ldots, J^{n-r_*}(x)) = 0 \qquad (\sigma = 1, \ldots, s).$$

9. Example

Consider the Lie algebra $L^3(\xi)$ of the group $G^3(f)$ of transformations of the space $R^3(x, y, z)$, generated by the operators

$$\xi_1 \cdot \partial = \partial_x, \qquad \xi_2 \cdot \partial = x\partial_x + y\partial_y + z\partial_z,$$

$$\xi_3 \cdot \partial = x^2\partial_x + 2xy\partial_y + 2xz\partial_z.$$

The corresponding matrix M is

$$M(x, y, z) = \begin{bmatrix} 1 & 0 & 0 \\ x & y & z \\ x^2 & 2xy & 2xz \end{bmatrix}$$

It is obvious that its general rank $r_*(M) = 2$. It is also seen that this matrix does not have a rank in any neighborhood of the origin of the coordinates. Here $r_* < n = 3$ and $n - r_* = 1$, so the group $G^3(f)$ is intransitive and its universal invariant is the mapping $J\colon R^3(x, y, z) \to R$. This scalar invariant is easily found as a solution of the system of equations

$$\partial_x J = 0, \qquad x\partial_x J + y\partial_y J + z\partial_z J = 0,$$

$$x^2\partial_x J + 2xy\partial_y J + 2xz\partial_z J = 0,$$

which can be taken in the form of $J = y/z$. The set S of singular points of the group $G^3(f)$ is found if all minors of the second order of the matrix M are calculated and set equal to zero. This system is equipollent to

$$y = 0, \qquad z = 0, \tag{9.1}$$

that is, the x axis is given. At all points of this axis $\text{rank}(M\,|\,S) = 1$, because of which $M\,|\,S$ has a rank and (9.1) is an invariant manifold of the group $G^3(f)$. But it cannot be given by any equation of the form $F(y/z) = 0$, with a mapping F of the class C_∞, in any neighborhood of the axis x!

It is easily verified that the plane Ψ given by the equation

$$y - z = 0$$

is a nonsingular (in addition to the points of the x axis) manifold which is invariant relative to the group $G^3(f)$. Using the universal invariant J, one can show that it is also given by the equation $J = 1$. The rank of this invariant manifold is equal to zero.

§19. INVARIANT SOLUTIONS OF EQUATIONS

1. Summary of Initial Concepts

The differential equation E (system of s scalar equations) of type (5.1.1), satisfying the conditions described in Section 5.1, is considered. This equation is given with the aid of the mapping $\varepsilon\colon Z_k \to R^s$ of the continuous space Z_k (see Section 4.2), obtained by continuation of the space $Z = X \times Y$, with $X = R^n(x)$, $Y = R^m(y)$. Here the vector $x \in X$ is called the "independent variable" and the vector $y \in Y$ the "unknown function."

A Lie group $G^r(h) = (B, \phi, \pi, Z)$ of transformations h_a of the space Z, admitted by equation E (see Section 16.12), is supposed to be known and fixed. This group can coincide with the full group GE of the equation E or can only be a subgroup of the full group. Only the Lie algebra $L^r(\zeta)$ of the group $G^r(h)$, whose elements are the vector fields $\zeta_a\colon Z \to Z$ operating by $\zeta_a(z) = (\xi_a(z), \eta_a(z))$ and written in the form of the operators

$$\zeta_a \cdot \partial = \xi_a \cdot \partial_x + \eta_a \cdot \partial_y \qquad (a \in B), \tag{1.1}$$

takes part in most of the calculations.

According to the concepts of the continuation group $\underset{k}{G^r}(h) = G^r(\underset{k}{h})$, as a group of continuous transformations $\underset{k}{h_a}\colon Z_k \to Z_k$, and the continuation Lie algebra $\underset{k}{L^r}(\zeta) = L^r(\underset{k}{\zeta})$ as a Lie algebra of the operators $\underset{k}{\zeta_a} \cdot \partial$, and with the definition in Section 5.2, the equation E, as a manifold $E \subset Z_k$, is an invariant manifold of the group $\underset{k}{G^r}(h)$ [of the Lie algebra $\underset{k}{L^r}(\zeta)$] or a differential invariant manifold (see Section 4.11) of the initial group $G^r(h)$ [Lie algebra $L^r(\zeta)$].

In what follows it is assumed that the manifold E is a nonsingular (*in the sense of Section* 18.6) *invariant manifold of the continuation group* $\underset{k}{G^r}(h)$.

Every solution u of the equation E is a mapping $X \to Y$, operating by $y = u(x)$. The set of all solutions of class $C_\infty(V)$ on the open set $V \subset X$ will be denoted by the symbol SE. The manifold $U \subset Z$, given by the equation

$$y = u(x), \tag{1.2}$$

corresponds to the solution u. With the aid of the mapping $\underset{0}{u}\colon X \to Z$, operating by $\underset{0}{u}(x) = (x, u(x))$, the manifold U can also be given parametrically, namely, as $U = \underset{0}{u}(V)$. The continuation of the solution is done with equation (4.4.1), because of which the manifold U is also continued up to the manifold $U_k \subset Z_k$, determined by (5.6.2). As in Section 7.6, the property that the mapping $u: V \to Y$ is to be a solution of the equation E is equipollent to the inclusion $U_k \subset E$.

The fundamental value of the statement "The equation E admits the group $G^r(h)$" is that the group $G^r(h)$ operates on the set of solutions SE (see Section 5.6). This means that because of the operation of any transformation $h_a \in G^r(h)$, every solution $u \in SE$ transforms into a solution of the same equation E. The transformation of the solutions $u \to u_a$ is done by means of equation (5.7.3), which can be rewritten in the form

$$u_a = (g_a \circ \underset{0}{u}) \circ (f_a \circ \underset{0}{u})^{-1} \tag{1.3}$$

if the transformation $h_a = (f_a, g_a)$ operates by $h_a(z) = (f_a(z), g_a(z))$.

2. Definition

If the differential equation E admits the group $G^r(h)$, then it also admits any of its subgroups H. In accordance with this, the concept of an invariant solution is defined in such a way that it also covers all subgroups $H \subset G^r(h)$.

The solution $u \in SE$ is called an *invariant H-solution of the equation E if* the manifold U, corresponding to it, is an invariant manifold of the group H.

It is clear that the manifold solution U has the dimensionality $\dim U = n = \dim X$. On the other hand, it is obvious that $r_*(\partial(y - u(x))) = m$. Consequently, as $\dim Z = n + m$, the manifold U is given regularly (see Section 18.2) by the equation (1.2). Let ζ be a tangential mapping of the group H, with representation (1.1). Applying the criterion of invariance of a regularly given manifold in Section 18.2 to U, we find the following criterion of invariance of the solution $u \in SE$: For any $x \in V$

$$\eta(x, u(x)) - \partial_u(x)\langle\xi(x, u(x))\rangle = 0. \tag{2.1}$$

The invariant H-solution u is called *singular* or *nonsingular* depending on whether the corresponding manifold U is a singular or nonsingular manifold of the group H.

3. Necessary Conditions

Let $r_* = r_*(\zeta)$ be a general rank of the tangential mapping of the group H. To find whether the given invariant H-solution is singular or nonsingular, it is necessary, according to Section 18.6, to find the number $r_*(\zeta \mid U)$ and to compare it with r_*. But from the invariance condition (2.1),

$$r_*(\zeta \mid U) = r_*(\xi(x, u(x)), \eta(x, u(x))) = r_*(\xi(x, u(x))) = r_*(\xi \mid U).$$

Thus the solution u is singular if $r_*(\xi\,|\,U) < r_*$ and nonsingular if $r_*(\xi\,|\,U) = r_*$.

So *the solution u is nonsingular if and only if*

$$r_*(\xi\,|\,U) = r_*. \tag{3.1}$$

If u is a nonsingular invariant H-solution, then the corresponding invariant manifold U has a rank equal to its dimensionality in the space of invariants. But the universal invariant of the group H is the mapping $J\colon Z \to R^{n+m-r_*}$, and the dimensionality of U is equal to $(n+m) - m = n$. Thus formula (18.8.1) gives the following value of the rank of the nonsingular invariant H-solution:

$$\rho = \rho(H) = n - r_*. \tag{3.2}$$

Besides, because of the theorem of representation in Section 18.7, a nonsingular invariant manifold U, regularly given by equation (1.2), can be regularly given by an equation of the kind

$$\omega \circ J(z) = 0, \tag{3.3}$$

with some mapping $\omega\colon R^{n+m-r_*} \to R^m$ of class C_∞, where $r_*(\partial\omega) = m$.

Lemma. *For the equation E, admitting the group H, to have nonsingular invariant H-solutions, the relations*

$$r_* \leq n, \qquad r_*(\partial_y J) = m \tag{3.4}$$

must be satisfied.

Proof. As equation (3.3) is equipollent to equation (1.2), it is possible to find the "unknown" m-dimensional vector $y = u(x)$ from it. Thus $r_*(\partial_y(\omega \circ J)) = m$. But $\partial_y(\omega \circ J) = \partial\omega(J) \circ \partial_y J$, and as $r_*(\partial_y J) \leq m$, the second relation of (3.4) follows from the well-known theorem about the rank of compositions of linear mappings and from the equality $r_*(\partial\omega) = m$. The first relation was actually proved above; it follows from equation (3.2) because $\rho \geq 0$. ■

It is not very convenient to use (3.4) in applications for it requires a preliminary knowledge of a universal invariant. A more delicate infinitesimal condition for the invariant H-solution to exist, whose verification is much simpler, is of the form

$$r_* \leq n, \qquad r_*(\xi) = r_*. \tag{3.5}$$

This result follows from the next theorem.

Theorem. *For nonsingular invariant H-solutions to exist, it is necessary that relations* (3.5) *be satisfied. Then equality* (3.4) *is a consequence of* (3.5).

Proof. Since $r_*(\xi \mid U) \le r_*(\xi) \le r_*$, the equality (3.5) follows from (3.1) and the necessity of (3.5) is obvious. Let (3.5) be satisfied. Then from $r_*(\zeta) = r_*$ the linear mapping $\zeta(z)\colon B \to Z$, with $B = R^r$, can be contracted to a mapping $\bar{\zeta}(z)\colon B_0 \to Z$, where $B_0 = R^{r_*} \subset B$. The mapping $\bar{\zeta}(z)$ is complete and the universal invariant for $\bar{\zeta}$ coincides with J (see the proof of the theorem in Section 17.3). It is clear that the space B_0 can be chosen such that $r_*(\bar{\xi}) = r_*$, where $\bar{\zeta} = (\bar{\xi}, \bar{\eta})$. Then $\bar{\xi}$ can be given in the form $\bar{\xi} = (\xi', \xi'')$, where $r_*(\xi') = r_*$ and the linear mapping $\xi'(z)\colon B_0 \to \xi'(z)(B_0)$ is invertible. Let the decomposition of the space X into the product $X = X' \times X''$ correspond to this representation, where $X' = \xi'(Z)(B_0)$, and let $x = (x', x'')$. In this notation, the condition that J is an invariant becomes

$$\xi' \cdot \partial_{x'} J + \xi'' \cdot \partial_{x''} J + \bar{\eta} \cdot \partial_y J = 0,$$

and after operating on it with the inverse mapping ξ'^{-1}, it becomes

$$\partial_{x'} J + (\xi'' \circ \xi'^{-1}) \cdot \partial_{x''} J + (\bar{\eta} \circ \xi'^{-1}) \cdot \partial_y J = 0.$$

Consequently,

$$\rho + m = r_*(\partial J) = r_*(\partial_{x'} J, \partial_{x''} J, \partial_y J) = r_*(\partial_{x''} J, \partial_y J).$$

But the general inequality $r_*(\partial_{x''} J, \partial_y J) \le r_*(\partial_{x''} J) + r_*(\partial_y J)$ holds, where, obviously, $r_*(\partial_{x''} J) \le \rho$ and $r_*(\partial_y J) \le m$. Thus the previous equality is possible only if $r_*(\partial_y J) = m$. ■

4. Projections

Let $\bar{Z} = R^{m+\rho}(\bar{z})$ be a space of invariants, where the universal invariant J of the group H operates. From the theorem of representation in Section 18.7, for any nonsingular invariant manifold $\Psi \subset Z$ of the group H, its *projection* $\bar{\Psi} = J(\Psi)$ *in the space of invariants* $\bar{Z}$ is defined. The correspondence between nonsingular invariant manifolds and their projections is one-to-one because of the following lemma.

Lemma. *A nonsingular invariant manifold* Ψ *is uniquely determined by its projection* $\bar{\Psi}$ *in the space of invariants.*

Proof. Let $\Psi' = J^{-1}(\bar{\Psi})$ be a manifold which is a complete inverse image of the manifold $\bar{\Psi}$ in the space Z. Then $\bar{\Psi} = J(\Psi')$, and thus the manifold Ψ' is invariant. Indeed, if $z \in \Psi'$ and $z' = h_a(z)$ under some transformation $h_a \in H$, then $J(z') = J(z)$, that is, $z' \in \Psi'$. Also, $\Psi \subset \Psi'$ and therefore the manifold Ψ is nonsingular. So Ψ' has rank ρ', equal to its dimensionality in the space of invariants (see Section 18.8), that is, equal to the dimensionality of the manifold $\bar{\Psi}$ in $\bar{Z}$. But this dimensionality is equal to the rank ρ of the nonsingular invariant manifold Ψ, so that $\rho' = \rho$. From the formula (18.8.1) it follows that the dimensionalities of the manifolds Ψ' and Ψ are equal. Consequently, $\Psi' \equiv \Psi$. ■

Consequence. *For any given manifold $\bar{\Psi} \subset \bar{Z}$ there is one and only one nonsingular invariant manifold $\Psi \subset Z$, whose projection is $\bar{\Psi}$.*

***Note* 1.** *The rule of inclusion of projections holds: $\bar{\Psi}_1 \subset \bar{\Psi}_2$ if and only if $\Psi_1 \subset \Psi_2$.*

***Note* 2.** If Ψ is given by $\psi(z) = 0$, then, according to the representation theorem in Section 18.7, it is also given by the equation $\bar{\psi} \circ J(z) = 0$ or, if it is assumed that $\bar{z} = J(z)$, by the equation $\bar{\psi}(\bar{z}) = 0$. The equation $\bar{\psi}(\bar{z}) = 0$ will be called *a projection of the equation $\psi(z) = 0$ into the space of invariants.* The projection $\bar{\psi}(\bar{z}) = 0$ gives the manifold $\bar{\Psi} \subset \bar{Z}$, which is a projection of the manifold Ψ. Indeed, if $z \in \Psi$, then $\psi(z) = 0$ is equipollent to the equality $\bar{\psi} \circ J(z) = 0$, that is, $\bar{z} = J(z) \in \bar{\Psi}$. As this is correct for any $z \in \Psi$, so $J(\Psi) = \bar{\Psi}$.

In what follows the space $\bar{Z}$ is considered as the product $\bar{Z} = \bar{X} \times \bar{Y}$, where $\bar{X} = R^{\rho}(\bar{x})$, $\bar{Y} = R^m(\bar{y})$, so that $\bar{z} = (\bar{x}, \bar{y})$. According to this, the universal invariant J takes the form $J = (J_1, J_2)$, where its components, the invariants $J_1: Z \to \bar{X}$ and $J_2: Z \to \bar{Y}$, operate by $\bar{x} = J_1(x, y)$ and $\bar{y} = J_2(x, y)$. It is assumed that these representations are chosen such that

$$r_*(\partial_x J_1) = \rho, \qquad r_*(\partial_y J_2) = m \tag{4.1}$$

hold, which is always possible under conditions (3.4).

In this notation, the nonsingular invariant mapping U of equation (1.2) can be given in the space of invariants by an equation of the form $\bar{y} = \bar{u}(\bar{x})$, with some mapping $\bar{u}: \bar{X} \to \bar{Y}$. This equation gives the projection $\bar{U}$ of the manifold U in $\bar{Z}$. The mapping $\bar{u}$ will be called a *projection of the mapping u in the space of invariants.*

The inverse transition from the projection $\bar{u}$ to u is done by solving the equation

$$J_2(x, y) = \bar{u}(J_1(x, y)), \tag{4.2}$$

relative to y. In the general case the solvability of equation (4.2) is not guaranteed by the previous conditions. The mapping $\bar{u}: \bar{X} \to \bar{Y}$ will be called *restored* if equation (4.2) is solvable, that is, it determines (implicitly) the mapping $u: X \to Y$, operating by (1.2). So every restored mapping $\bar{u}$ is a projection of a mapping u, which provides a nonsingular invariant manifold of the group H with the aid of equation (1.2). Because of the lemma, the correspondence between u and $\bar{u}$ is one-to-one.

5. Continuation of Invariant Manifolds

Let the invariant manifold U of the group $G^r(h)$ be given by the equation $\Phi(z) = 0$, where the mapping $\Phi: Z \to R^m$ satisfies the equation

$$(\zeta \cdot \partial)\Phi = \lambda \cdot \Phi, \tag{5.1}$$

for some coefficient $\lambda: Z \to \mathscr{L}(R^m)$, as well as the condition

$$r_*(\partial_y \Phi) = m. \tag{5.2}$$

For some $k \geq 1$, the continuation manifold $U_k \subset Z_k$, given by the system of equations $\Phi = 0$ and

$$D\Phi = 0, D^2\Phi = 0, \qquad \ldots, \qquad D^k\Phi = 0, \tag{5.3}$$

is generated, where D is an operator of total differentiation (4.3.1). Because of condition (5.2), equations (5.3) can be solved relative to $\underset{1}{y}, \ldots, \underset{k}{y}$, so the system (5.3) is equipollent to a system of the form

$$\underset{1}{y} = \underset{1}{v}(x, y) \qquad \underset{2}{y} = \underset{2}{v}(x, y), \qquad \ldots, \qquad \underset{k}{y} = \underset{k}{v}(x, y). \tag{5.4}$$

Lemma. *The manifold U_k is a differential invariant manifold of the group $G^r(h)$.*

Proof. From condition (5.2), the manifold U_k of equation (5.3) is given regularly. Let $\zeta_a \cdot \partial$ be any operator of the Lie algebra $L^r(\zeta)$ of the group $G^r(h)$. Applying the standard commutator (4.9.5), generated by the operator $\underset{\infty}{\zeta_a} \cdot \partial$, to the mapping Φ, the relation

$$D^s(\underset{\infty}{\zeta_a} \cdot \partial\Phi) - (\underset{\infty}{\zeta_a} \cdot \partial)D^s\Phi = \sum_{l=0}^{s-1} \binom{s}{l} D^{s-l}\xi_a \cdot D^{l+1}\Phi$$

holds for every $s = 1, \ldots, k$. As $\underset{\infty}{\zeta_a} \cdot \partial\Phi = \lambda \cdot \Phi$, the previous relation on the manifold U_k, given by the system of equations (5.3) (and also $\Phi = 0$), becomes

$$(\underset{\infty}{\zeta_a} \cdot \partial)D^s\Phi|_{U_k} = 0 \qquad (s = 1, 2, \ldots, k).$$

Consequently, U_k is invariant because of the criterion in Section 18.2. ■

Let the differential invariant $F: Z_k \to P$, that is, the invariant of the continuation group $\underset{k}{G^r}(h)$, and its reduction be given. From the assignment of the manifold U_k by the system of equations (5.4), this reduction can be considered as the mapping $F | U_k: Z \to P$.

Consequence 1. *The mapping $F | U_k$ is an invariant of the group $G^r(h)$.*

Proof. As U_k is an invariant manifold of the group $\underset{k}{G^r}(h)$, the reduction $F|U_k$ of its invariant F is an invariant of the induced group $\underset{k}{G^r}(h)|U_k$. But from the representation U_k by the system of equations (5.4), all the transformations $\underset{k}{h_a}|U_k$ coincide with the transformations h_a belonging to the group $G^r(h)$, so the induced group $\underset{k}{G^r}(h)|U_k$ coincides with $G^r(h)$. Thus $F|U_k: Z \to P$ is an invariant of the group $G^r(h)$. ■

The established lemma, with H satisfying the conditions of the lemma in Section 19.3, can be applied, in particular, to the nonsingular invariant manifold U with the equation $y = u(x)$. Here the system of equations of the form (5.4), giving the continuation manifold U_k, is

$$\underset{1}{y} = \partial u(x), \qquad \underset{2}{y} = \partial^2 u(x), \qquad \ldots, \qquad \underset{k}{y} = \partial^k u(x). \tag{5.5}$$

It is easy to determine, from conditions (4.1), that the same manifold U_k can be given by a system of equations in the form (5.3) using the invariant Φ operating by $\Phi(x, y) = J_2(x, y) - \bar{u}(J_1(x, y))$, where $\bar{u}$ is a projection of the mapping u and the equality $\Phi(z) = 0$ is equipollent to the equality $y = u(x)$. Thus it follows from the lemma that the manifold (5.5) is a differential invariant manifold of the group H. Moreover, every equation $D^s\Phi = 0$ $(s = 1, \ldots, k)$ is solvable relative to $\underset{s}{\bar{y}} = \partial^s \bar{u}(\bar{x})$. This means that the system (5.5) is equipollent to the system of equations in the continuation space of invariants $\bar{Z}_k$, obtained by continuation of the space $\bar{Y}$ with the aid of $\bar{X}$ (see Section 4.2):

$$\underset{1}{\bar{y}} = \partial \bar{u}(\bar{x}), \qquad \underset{2}{\bar{y}} = \partial^2 \bar{u}(\bar{x}), \qquad \ldots, \qquad \underset{k}{\bar{y}} = \partial^k \bar{u}(\bar{x}), \tag{5.6}$$

where $\underset{s}{\bar{y}}$ is an element of the space $\bar{Y}_s$. But the system (5.6) only relates the invariants of the continuation group $\underset{k}{H}$. Because of Note 2 in Section 19.4, this system is a projection of the system (5.5) and thereby gives the projection $\bar{U}_k$ of the continuation manifold U_k in the space of the continuation group $\underset{k}{H}$ invariants. Considering the continuation rule of mappings in Section 4.4, this result is summarized as another consequence of the lemma.

Consequence 2. *If $y = u(x)$ is a nonsingular manifold of the group H and $\bar{u}$ is a projection of the mapping u in the space of invariants of this group, then the continuation manifold U_k is an invariant manifold of the continuation group $\underset{k}{H}$ and its projection $\bar{U}_k$ is a continuation of the projection $\bar{u}$ in the space of invariants of the group $\underset{k}{H}$.*

6. The Existence Theorem

The main result of invariant solution theory here, namely, a theorem about their conditional existence, is established.

Theorem. *Let the system of differential equations E admit the group H, for which the necessary conditions* (3.5) *are satisfied. Then there is a factor system E/H possessing the following property: The projection $\bar{u}$ of any nonsingular invariant H-solution u is a solution of the factor system E/H. Conversely, every restored solution $\bar{u}$ of the factor system E/H is a projection of some invariant H-solution of the system E.*

Proof. According to the agreement in Section 19.1, the manifold E is a nonsingular invariant manifold of the continuation group $\underset{k}{H}$ (here k is the order of the equation E). From the theorem in Section 18.7 there is a projection $\bar{E}$ of the manifold (and the equation) E in the space of invariants of the group $\underset{k}{H}$. If u is a nonsingular invariant H-solution of the equation E, then the formula of inclusion $U_k \subset E$ [here U_k is the manifold (5.5), supplemented by the equation $y = u(x)$] is satisfied. Moreover, from Consequence 2 in Section 19.5, the manifold U_k is invariant relative to $\underset{k}{H}$ and its projection $\bar{U}_k$ [the manifold (5.6), supplemented with the equation $\bar{y} = \bar{u}(\bar{x})$] is a continuation of the manifold $\bar{U}$, given by the equation $\bar{y} = \bar{u}(\bar{x})$ using the projection $\bar{u}$ of the solution u. Thus from Note 2 in Section 19.4 the inclusion $\bar{U}_k \subset \bar{E}$ is correct, that is, $\bar{u}$ is a solution of the equation E. If $\bar{u}$ is restorable, then this reasoning is invertible. Consequently, the statement of the theorem is correct for the factor system $E/H = \bar{E}$. ■

If the system E is given by the equation $\underset{k}{\varepsilon(z)} = 0$, with the mapping $\varepsilon\colon Z_k \to R^s$, then the factor system E/H is given by the equation $\underset{k}{\bar{\varepsilon}(\bar{z})} = 0$, with the mapping $\bar{\varepsilon}\colon \bar{Z}_k \to R^s$, which is a projection of the equation E in the space of invariants of the continuation group $\underset{k}{H}$.

The theorem just established is called *the theorem of conditional existence of the invariant H-solutions* because the existence of even one solution of the factor system E/H is not guaranteed by it. An example in which E/H does not have solutions will be discussed later.

7. Factor System

According to the theorem in Section 19.6, the transformation of the system E into the factor system E/H, considered as a mapping $(E, H) \to E/H$, is defined for any group H, admitted by the system E, which satisfies the necessary conditions (3.4). Summarizing the former results leads to the following algorithm for determination of this transformation.

Algorithm. (α_1) Verification of the necessary conditions (3.5). (α_2) Construction of the universal invariant of the group H. (α_3) Representation of the space of invariants as the product $\bar{Z} = \bar{X} \times \bar{Y}$ and of the universal invariant in the form $J = (J_1, J_2)$ so that conditions (4.1) are satisfied. (α_4) Introduction of new variables by $\bar{x} = J_1(x, y)$, $\bar{y} = J_2(x, y)$. (α_5) Writing the general form (4.2) of the projection $\bar{y} = \bar{u}(\bar{x})$ of the invariant H-solution in terms of the undetermined mapping $\bar{u}: \bar{X} \to \bar{Y}$. ($\alpha_6$) Continuation of the mapping u through calculation of the derivatives $\underset{1}{y}, \ldots, \underset{k}{y}$ from equations obtained by differentiation of equation (4.2), and expressing these in terms of x, y and the derivatives $\underset{1}{\bar{y}}, \ldots, \underset{k}{\bar{y}}$. ($\alpha_7$) Substitution of the resulting expressions for the derivatives into the system E. (α_8) Finally, conditioning the resulting system into the projection $\bar{E} = E/H$.

From the results in Sections 19.4–19.6 it follows that if the initial system has the form $E(n, m, k, s)$, then the factor system E/H is a system of the form $E(\rho, m, k, s)$, where $0 \leq \rho < n$ because of (3.2). Consequently, the *factor system possesses this very important property*: *Its "unknown functions" $\bar{y} \in \bar{Y}$ depend on a smaller number of "independent variables" $\bar{x} \in \bar{X}$ than the original system has.* The number of these is equal to the rank $\rho < n$ of the unknown invariant H-solutions. This number ρ is given by (3.2) and is determined only by the group H.

Using this algorithm it is possible, generally speaking, to find the invariant H-solutions with different ranks—zero, one, two, and so on—at the expense of the choice of the subgroups $H \subset G^r(h)$. It is useful to note that when the rank is zero the factor system E/H is a system of algebraic (not differential) equations, and when the rank is one it is a system of ordinary differential equations.

8. The Separation of Variables Case

The realization of the transition $(E, H) \to E/H$ is considerably simplified if the universal invariant J of the group H has the following property: There exists a representation $J = (J_1, J_2)$ [see stage (a_3) of the algorithm in Section 19.7] in which the component J_1 depends only on x, that is, it is the mapping $X \to \bar{X}$ given by $\bar{x} = J_1(x)$. Then it is said that a *separation of variables* is possible for the invariant J. Simplification is achieved owing to the fact that in this case the expression of the general form for the invariant H-solution [see stage (α_5) of the algorithm in Section 19.7] is an explicit form.

Indeed, if the equation $\bar{y} = J_2(x, y)$ is solved relative to y in the form $y = v(x, \bar{y})$, then the general form of the invariant H-solution becomes

$$y = v(x, \bar{u}(\bar{x})), \qquad \bar{x} = J_1(x). \tag{8.1}$$

It is clear that continuation of this mapping only requires satisfaction of the differentiation operation. This case deserves to be noted since it is often met in practice.

Note. From a theoretical point of view, this property of separation of variables can generally be considered not only in one fixed system E but on a class of systems obtained from E by transformations of the space Z. Indeed, the space of invariants $\bar{Z}$ can always be supplemented up to the space Z' isomorphic to Z and expressed in the form $Z' = \bar{Z} \times T$, where $T = R^{r_*}(t)$. Then it is possible to find a mapping $\bar{t}: Z \to T$ such that the mapping $p: Z \to Z'$, operating by

$$p(z) = (J(z), \bar{t}(z)) = (\bar{z}, \bar{t}), \tag{8.2}$$

is an invertible (locally) transformation of the class $C_\infty(Z)$. Under such a transformation, the system E will transform into its equipollent system E' and the group H into its similar group H'. But the universal invariant of the group H' is, according to the construction, $\bar{z} = (\bar{x}, \bar{y})$, and for it the property of separation of variables is evident.

In spite of this note, to emphasize the property of separation of variables makes sense because the development of transformation (8.2) appears to be very laborious and not justified in practice.

The general form (8.1) of the invariant H-solution permits certain conclusions to be drawn about the structure of the factor system. For example, the following statement is correct: *If the system E is quasi-linear and the group H is such that the variables are separated, then the factor system E/H is also quasi-linear.* This follows from the fact that for every $k = 1, 2, \ldots$ the quantity $\underset{k}{y}$, found by differentiation D^k of equation (8.1), is linearly expressed by $\underset{k}{\bar{y}}$ (heterogeneously) with coefficients depending on $x, \bar{y}, \underset{1}{\bar{y}}, \ldots, \underset{k-1}{\bar{y}}$.

9. Examples

(α) A simple example of a factor system that may not have solutions is given by the equation E for which $X = R^2(t, x)$, $Y = R(y)$,

$$t\partial_t y + x\partial_x y = 1. \tag{9.1}$$

It is easily verified that equation (9.1) admits the one-parameter group H of transformations of the space $Z = R^3$ (t, x, y), with the operator $\zeta \cdot \partial = t\partial_t + x\partial_x$. Here, in the domain $\{(t, x) | t > 0, x > 0\}$, the vector field ζ has the general rank $r_* = 1 < n = 2$. The universal invariant $J: Z \to R^2$ of this group can be taken in the form $J = (t/x, y)$ and for it, evidently, $r_*(\partial_y J) = 1 = m$. Thus the necessary conditions (3.4) are satisfied. Here the variables are separated, and according to the algorithm in Section 19.7 the general form of the invariant H-solution is $y = \bar{u}(t/x)$. But for this expression

$$t\partial_t y + x\partial_x y = t\frac{1}{x}\,\partial\bar{u} - x\frac{t}{x^2}\,\partial\bar{u} \equiv 0,$$

and thus its substitution into (9.1) gives for the factor system E/H the contradictory statement $0 = 1$. Thus equation (9.1) does not actually have invariant H-solutions.

Further examples relate to the sysem E of the boundary layer equations (10.1.2) with $\theta = 0$, admitting the Lie algebra $L^5(\zeta)$ of operators whose basis was given in (10.6.4). The admitted group $G^5(h)$ of transformations of the space $Z = X \times Y$ corresponds to $L^5(\zeta)$, where $X = R^3(t, x, y)$, $Y = R^2(u, v)$. The subgroup $H \subset G^5(h)$, whose Lie algebra of operators has a basis generated by the operators $\zeta_{\alpha_1} \cdot \partial, \ldots, \zeta_{\alpha_k} \cdot \partial$, will be denoted below by $H(\zeta_{\alpha_1}, \ldots, \zeta_{\alpha_k})$.

(β) The subgroup $H(\zeta_1)$, where $\zeta_1 \cdot \partial = \partial_t$, has the universal invariant $J = (x, y, u, v)$ and satisfies the necessary conditions (3.5). The variables are separated, and the general form of the invariant $H(\zeta_1)$-solution is

$$u = \bar{u}(x, y), \qquad v = \bar{v}(x, y).$$

It is evident that here the factor system E/H coincides with the system (7.2), taken when $\theta = 0$. Consequently, the invariant $H(\zeta_1)$-solutions are simply stationary solutions of the system E. Here the rank is $\rho(H) = 2$.

(γ) The subgroup $H(\zeta_2, \zeta_5)$, with the operators

$$\zeta_2 \cdot \partial = 2t\partial_t + y\partial_y - 2u\partial_u - v\partial_v,$$
$$\zeta_5 \cdot \partial = t\partial_x + \partial_u,$$

has the universal invariant $J = (y^2/t, tu - x, yv)$. Here $r_* = 2$, $n = 3$, and $r_*(\partial_{(u,v)}J) = 2 = m$, so that the necessary conditions (3.4) are satisfied. The variables are separated, and if it is assumed that $\bar{x} = y^2/t$, $\bar{u} = tu - x$, $\bar{v} = yv$, then the general form of the invariant $H(\zeta_2, \zeta_5)$-solution is

$$u = \frac{1}{t}(x + \bar{u}(\bar{x})), \qquad v = \frac{1}{y}\,\bar{v}(\bar{x}), \qquad \bar{x} = \frac{y^2}{t}.$$

The rank of this solution is $\rho = 1$. As a result the calculation of derivatives and their substitution into the system E give the factor system E/H of ordinary differential equations

$$2\bar{v}' = \bar{v}/\bar{x} - 1, \qquad 4\bar{x}\bar{u}'' = (2\bar{v} - 2 - \bar{x})\bar{u}',$$

where the corresponding derivatives with respect to $\bar{x}$ are marked by primes. This system is easily integrated by quadratures.

(δ) For the subgroup $H(\zeta_3, \zeta_4, \zeta_5)$, with operators

$$\zeta_3 \cdot \partial = x\partial_x + u\partial_u, \qquad \zeta_4 \cdot \partial = \partial_x, \qquad \zeta_5 \cdot \partial = t\partial_x + \partial_u,$$

it turns out that $r_* = 2 < n = 3$. But here the universal operator is $J = (t, y, v)$ and for it $r_*(\partial_{(u,v)}J) = 1 < m = 2$. Thus the first necessary condition holds but the second does not. Consequently, there are no invariant H-solutions.

10. Extension Groups

One-parameter extension groups were already considered in Sections 1.6 and 1.13, and also 3.7, where their invariants were found. The issue of invariant extension groups is studied here for the general case of an r-parameter group.

The space $Z = X \times Y$, where $X = R^n(x)$, $Y = R^m(y)$, is considered with the goal of applications to differential equations. The bases $\{e_i\} \subset X$ and $\{f_k\} \subset Y$ are fixed; then the basis in Z will be $\{e_i, f_k\}$. To simplify the following equations, it is convenient to give the space Z as a direct sum $Z = X \oplus Y$. Then if $x = x^i e_i \in X$ and $y = y^k f_k \in Y$, the decomposition of the vector $z \in Z$ relative to the basis $\{e_i, f_k\}$ can be written in the form

$$z = x^i e_i + y^k f_k. \tag{10.1}$$

An *extension of the space* Z *in the basis* $\{e_i, f_k\}$ is the transformation $h: Z \to Z$, operating on the vector (10.1) by

$$h(z) = \sum_{i=1}^{n} c^i x^i e_i + \sum_{k=1}^{m} d^k y^k f_k \qquad (c^i > 0, d^k > 0). \tag{10.2}$$

Consequently, an extension of the space Z in the basis $\{e_i, f_k\}$ is a linear transformation whose matrix in this basis is diagonal and all of whose main diagonal elements are positive. It is clear that the set of all extensions (in the given basis) generates Abel's group of transformations of the space Z. This group will be denoted by Diag $Z\{e_i, f_k\}$. It is obvious that it is an $(n + m)$-parameter Lie group.

Any subgroup $H \subset \operatorname{Diag} Z\{e_i, f_k\}$ is called an *extension group of the space Z*. In what follows, extension groups H^r, which are Lie groups with dimensionality r, where $0 < r \leq n + m$, are considered.

In connection with the fact that the identity of the group $\operatorname{Diag} Z\{e_i, f_k\}$ is a transformation with a unit matrix, it is convenient to choose the parametric group (B, ϕ) of the extension group H^r in such a way that its group kernel $B^+\{u_\alpha\}$ in the fixed basis $\{u_\alpha\} \subset B$, where $a = a^\alpha u_\alpha$, satisfies the inequalities $a^\alpha > 0$ $(\alpha = 1, \dots, r)$. Then the multiplication law in $B^+\{u_\alpha\}$ can be given by the formula $\phi(a, b) = \sum_{\alpha=1}^r a^\alpha b^\alpha u_a$, where the group identity is the vector $1_0 = \sum_{\alpha=1}^r u_\alpha$.

With these conditions, the general transformation $h_a \in H^r$ has the form

$$\left.\begin{aligned} x'^i &= x^i \prod_{\alpha=1}^{r} (a^\alpha)^{\lambda^i_\alpha} \\ y'^k &= y^k \prod_{\alpha=1}^{r} (a^\alpha)^{\mu^k_\alpha} \end{aligned}\right\} \qquad (i = 1, \dots, n;\ k = 1, \dots, m). \tag{10.3}$$

Because of the joint parametrization, the tangential mapping of the group H^r should be calculated by the formula $\zeta = \partial_a h_a|_{a=1_0}$. If it is assumed that $\zeta\langle u_\alpha\rangle = \zeta_\alpha$, then the basis of the Lie algebra $L^r(\zeta)$ of operators will be

$$\zeta_\alpha \cdot \partial = \sum_{i=1}^{n} \lambda^i_\alpha x^i \partial_{x^i} + \sum_{k=1}^{m} \mu^k_\alpha y^k \partial_{y^k} \qquad (\alpha = 1, \dots, r). \tag{10.4}$$

The corresponding matrix $M(z)$ formed from the components of the operators (10.4) [see (18.6.3)] can be presented in the form

$$M(z) = M_1 \circ \operatorname{diag}\{x^1, \dots, x^n, y^1, \dots, y^m\},$$

where

$$M_1 = \begin{bmatrix} \lambda^1_1 & \cdots & \lambda^n_1, \mu^1_1 & \cdots & \mu^m_1 \\ \vdots & \cdots & \vdots & \cdots & \vdots \\ \lambda^1_n & \cdots & \lambda^n_r, \mu^1_r & \cdots & \mu^m_r \end{bmatrix} \tag{10.5}$$

is a constant numerical matrix. It is evident that $r_*(M) = \operatorname{rank} M_1$. In addition, it is easy to verify that any linear relation between the operators (10.4) (see Section 17.3) causes their linear dependence. Thus the operators (10.4), being linearly independent by definition, are not linearly related. Consequently, *for the extension group H^r, $r_* = r$ always.*

Because of the equality $r_* = r$, for the group H^r to be intransitive it is necessary and sufficient that $r < n + m$. In this case determination of the invariants for the group H^r leads to a problem of linear algebra.

Theorem. *For the extension group H^r of the space $Z = R^{n+m}(x, y)$, under the condition $r < n + m$, there exists a universal invariant $J: Z \to R^{n+m-r}$ whose independent components have the monomial form*

$$J^\tau(z) = \prod_{i=1}^{n} (x^i)^{\theta_i^\tau} \cdot \prod_{k=1}^{m} (y^k)^{\sigma_k^\tau} \qquad (\tau = 1, \ldots, n + m - r). \tag{10.6}$$

Proof. From the criterion of invariance (theorem in Section 17.1) the monomial (10.6) is an invariant of the Lie algebra $L^r(\zeta)$, with the operator basis (10.4), if and only if

$$\lambda_\alpha^i \theta_i^\tau + \mu_\alpha^k \sigma_k^\tau = 0 \qquad (\alpha = 1, \ldots, r). \tag{10.7}$$

Equations (10.7) generate a system of r linear algebraic equations in $n + m$ unknowns θ_i^τ, σ_k^τ whose matrix is M_1 (10.5). As rank $M_1 = r < n + m$, this system has $n + m - r$ linearly independent solutions. Let these solutions be

$$(\theta_1^\tau, \ldots, \theta_n^\tau, \sigma_1^\tau, \ldots, \tau_m^\tau) \qquad (\tau = 1, \ldots, n + m - r). \tag{10.8}$$

It is asserted that the monomials (10.6), generated by (10.8), are independent. Evidently, it is sufficient to prove their exponential independence, that is, that the equality

$$\prod_{\tau=1}^{n+m-r} (J^\tau(z))^{c_\tau} = 1,$$

with some real c_τ, is possible only if all $c_\tau = 0$. But the previous equation, because of (10.6) and the independence of the variables x^i, y^i, is equipollent to the system

$$\theta_i^\tau c_\tau = 0, \qquad \sigma_k^\tau c_\tau = 0 \qquad (i = 1, \ldots, n; k = 1, \ldots, m).$$

This means that the solutions (10.8) are linearly independent, in contradiction to the construction of these solutions. ■

Consequence. *The extension group H^r, whose elements are given by equalities* (10.3), *satisfies the conditions of existence for the invariant H^r-solution* (3.5) *if and only if*

$$\text{rank}(\lambda_\alpha^i) = r \leq n. \tag{10.9}$$

Proof. Both statements follow from the theorem in Section 19.3 because $r_* = r$, and for the basis operators (10.4) the equality $r_*(\xi) = r_*(\lambda_\alpha^i x^i) = \text{rank}(\lambda_\alpha^i)$ holds. ■

Note. If the extension group H^r satisfies conditions (10.9), then its universal invariant with components (10.6) can always be chosen in such a way that its variables are separated. Indeed, in this case the system $\lambda_\alpha^i \theta_i = 0$ $(\alpha = 1, \ldots, r)$ has exactly $n - r$ linearly independent solutions, which can be taken as a portion of the whole set (10.8) of $n + m - r$ solutions.

11. Dimension Theory

The concept of an extension group is closely connected with the *dimension theory of physical quantities.* Every physical quantity Φ is characterized by giving *a unit of its measurement E and a numerical value* $|\Phi|$, so that $\Phi = |\Phi| E$. If E_α $(\alpha = 1, \ldots, r)$ are some independent units of measurement and if there is an expression of E using them, having the form of a monomial $E_1^{\lambda_1} \cdots E_r^{\lambda_r}$, then this expression is called the *dimensionality of the physical quantity* Φ *in the units* $\{E_\alpha\}$ *and is denoted by* $[\Phi]$; so

$$E = [\Phi] = E_1^{\lambda_1} \cdots E_r^{\lambda_r}.$$

Thereby the representation

$$\Phi = |\Phi| [\Phi] = |\Phi| E_1^{\lambda_1} \cdots E_r^{\lambda_r} \tag{11.1}$$

holds for the physical quantity Φ. From (11.1) it follows that multiplying and raising to a (real) power of physical quantities are reduced to the same operations with their numerical values and their dimensionalities. In addition, if the physical quantity Φ is a function of the physical quantity x, then the dimensionality of a derivative is calculated by the rule $[\partial_x \Phi] = [\Phi]/[x]$.

The quantity Φ with null dimensionality $(\lambda_1 = \cdots = \lambda_2 = 0)$ is said to be *dimensionless.* The quantities $\Phi_1, \ldots, \Phi_N$ are said to be *dependent* (from the point of view of their dimensionalities) if there are numbers $\chi_1, \ldots, \chi_N$, not all equal to zero, with which a "combination" of these quantities

$$\Phi_1^{\chi_1} \cdots \Phi_N^{\chi_N}$$

is dimensionless. In the opposite case the quantities $\Phi_1, \ldots, \Phi_N$ are said to be *independent.*

The main problem of dimension theory is to determine how many independent physical quantities are in the given set $\Phi_1, \ldots, \Phi_N$, to find them, and then to express the other quantities in terms of these independent ones.

Example. In the classical CGS, E_1 = cm (centimeter), E_2 = g (gram mass) and E_3 = sec (second) are taken as independent units of measurements. Then the dimensionality of velocity v, acceleration G, density ρ, and pressure p are

$$[v] = \text{cm sec}^{-1} = E_1 \cdot E_3^{-1}, \qquad [G] = \text{cm sec}^{-2} = E_1 \cdot E_3^{-2},$$
$$[\rho] = \text{g cm}^{-3} = E_1^{-3} \cdot E_2, \qquad [p] = \text{g cm}^{-1}\,\text{sec}^{-2} = E_1^{-1} \cdot E_2 \cdot E_3^{-2}.$$

Here v, g, ρ are independent, and the quantity $p/\rho v^2$ is dimensionless.

Change of a measure scale through transition from the units E_α to new ones E'_α (not summed with respect to α),

$$E_\alpha = a^\alpha E'_\alpha \qquad (\alpha = 1, \ldots, r), \tag{11.2}$$

causes a change in the numerical value of the quantity Φ in accordance with the relation from (11.1):

$$|\Phi|[\Phi] = |\Phi|(a^1)^{\lambda_1} \cdots (a^r)^{\lambda_r} E_1'^{\lambda_1} \cdots E_r'^{\lambda_r} = |\Phi|'[\Phi]',$$

where $[\Phi]' = E_1'^{\lambda_1} \cdots E_r'^{\lambda_r}$ is the dimensionality of Φ in the new units and

$$|\Phi|' = |\Phi| \prod_{\alpha=1}^{r} (a^\alpha)^{\lambda_\alpha}. \tag{11.3}$$

A comparison of the last equation with (10.3) shows that the change in numerical value of the quantity Φ, during the transition to new units of measure, is a group transformation belonging to the dilatation group H^r (of one-dimensional space). With this the group H^r and the dimensionality $[\Phi]$ are uniquely determined by the same set of numbers $\lambda_1, \ldots, \lambda_r$. If there is a finite set $\{\Phi_\nu\}$, consisting of N physical quantities, then for every quantity Φ_ν the set $\lambda_1^\nu, \ldots, \lambda_r^\nu$, determining its dimensionality by (11.1), is put into correspondence with it. Thus, by analogy, the dilatation group H^r of the N-dimensional space $R^N(|\Phi_1|, \ldots, |\Phi_N|)$ is obtained. Finally, the connection between dimensionalities and dilatation groups is given by the following statement.

Conclusion. *The group H^r of dilatations of the space R^N, defined only by the dimensions of the quantities Φ_ν in the system $\{E_\alpha\}$, has a one-to-one correspondence with every finite set $\{\Phi_\nu\}$ of N physical quantities, which can be measured in the system $\{E_\alpha\}$, consisting of r independent measurement units. The transformations belonging to the group H^r give the rule of change, in the form* (11.3), *of the numerical values $|\Phi_\nu|$ as a result of the transition from the system of units $\{E_\alpha\}$ to the system $\{E'_\alpha\}$ by means of equation* (11.2).

An important consequence of the established connection is that the quantity Φ is dimensionless if and only if its numerical value $|\Phi|$ is an invariant of the corresponding group H^r. Thus any result of dimension theory can be obtained as a result of dilatation group theory.

For example, the above-mentioned problem of dimension theory is easily solved by construction of a universal invariant of the group H^r.

In addition, the following theorem, known in dimension theory as the *Pi-theorem*, is obtained at once from the theorems in Sections 19.10 and 18.7.

Any dimensionless function of physical quantities is a function of dimensionless "combinations" of these quantities; any relation among physical quantities is equipollent to some relation among their dimensionless "combinations."

It is also useful to note the influence of dimension theory on group analysis, frequently in applications of the latter. The point is that the differential equations of mathematical physics are always relations among physical quantities. Because of the previous results, they always admit the dilatation group H^r determined by the dimensionalities of all variables (independent, dependent, and parameters) included in the considered equations. If differential equations are given abstractly, then the admitted dilatation group can be found, introducing independent units by the definition and determination of the dimensionalities of the remaining quantities. This is usually done using the equations themselves, based on the requirement that the dimensionality of all the different terms must be the same in these equations. This method is called the *method of dimension comparison.*

For example, $[u] = E_1$ and $[v] = E_2$ can be taken as independent units of measurement for the equations of the boundary layer (10.1.1). If it is assumed that the dimensionalities of the remaining variables are

$$[t] = E_1^{\alpha_1}E_2^{\alpha_2}, \qquad [x] = E_1^{\beta_1}E_2^{\beta_2}, \qquad [y] = E_1^{\gamma_1}E_2^{\gamma_2}, \qquad [p] = E_1^{\delta_1}E_2^{\delta_2},$$

then a comparison of the dimensionalities will lead to a system of linear equations for the powers, with a unique solution. As a result, the following table of powers is obtained, thereby determining the dimensionalities of all variables included in system (10.1.1):

	t	x	y	u	v	p
E_1	0	1	0	1	0	2
E_2	-2	-2	-1	0	1	0

According to the table, the operators of the Lie algebra of the dilatation group, admitted by this system, are written at once:

$$x\partial_x + u\partial_u + 2p\partial_p, \qquad -2t\partial_t - 2x\partial_x - y\partial_y + v\partial_v.$$

These operators coincide with the dilatation operators found in Section 10.4 by solving the determining equations.

12. Similar Solutions

Let the system of differential equations E be given in the space $Z = X \times Y$, and let the basis $\{e_i, f_k\} \subset Z$ (see Section 19.10) be fixed.

The invariant H^r-solution of the system E is called *similar* (*in a narrow sense*) if H^r is a dilatation group of the space Z in the given basis, that is, if $H^r \subset \operatorname{Diag} Z\{e_i, f_k\}$.

The criterion for conditional existence of a similar H^r-solution for the dilatation group H^r admitted by the system E is given by the consequence in Section 19.10. With this, if the relations (10.9) are satisfied, then the universal invariant J of the group H^r can be chosen in such a way that the variables are separated. Moreover, in this case the components of J in the form (10.6) can be taken as

$$\begin{aligned} J_1^\beta(x) &= \prod_{i=1}^{n} (x^i)^{\theta_i^\beta} \\ J_2^k(x, y) &= y^k \prod_{i=1}^{n} (x^i)^{\theta_i^{\rho+k}} \end{aligned} \qquad (\beta = 1, \ldots, \rho; k = 1, \ldots, m), \qquad (12.1)$$

where $\rho = n - r$. Thus the equations, of general form, for the similar (in a narrow sense) H^r-solutions are

$$\begin{aligned} y^k &= \prod_{i=1}^{n} (x^i)^{-\theta_i^{\rho+k}} \cdot \bar{u}^k(\bar{x}^1, \ldots, \bar{x}^\rho), \\ \bar{x}^\beta &= J_1^\beta(x) \end{aligned} \qquad (\beta = 1, \ldots, \rho; k = 1, \ldots, m), \qquad (12.2)$$

The given concept of a similar (in a narrow sense) solution is not very satisfactory because it is not invariant relative to transformations of the space Z. Under transformations of the space Z the dilatation group H^r is transformed into its similar group, which is not, generally, speaking, a dilatation group. For example, the transformation given in the basis $\{e_i, f_k\} \subset Z$ by

$$x'^i = \ln x^i, \qquad y'^k = \ln y^k \qquad (i = 1, \ldots, n; k = 1, \ldots, m), \qquad (12.3)$$

transforms dilatation operators (10.4) into the operators of translation

$$\zeta'_\alpha \cdot \partial = \lambda_\alpha^i \partial_{x'^i} + \mu_\alpha^k \partial_{y'^k} \qquad (\alpha = 1, \ldots, r) \qquad (12.4)$$

and, consequently, the dilatation group H^r into the group of translations H'^r.

Based on the definition of the similar solution, the idea arises that is is suitable to select the structural property such that *any dilatation group is Abel's*. Then invariance relative to transformations of the space Z is guaranteed by the fact that similar groups are isomorphic. So the following definition is made.

The invariant H^r-solution of the system E is called *similar* (*in a broad sense*) if H^r is Abel's group.

The following theorem, generalizing the theorem in Section 3.8 for the case of an r-dimensional space of vector fields, shows that this expansion of the definition of the similar solution retains the possibility of giving it in the form (12.2), after some transformation of the space Z.

Theorem. *Abel's group H^r, whose tangential mapping satisfies the condition $r_*(\zeta) = r$, is similar to the dilatation group.*

Proof. Because of the similarity between a dilatation group and a translation group, noted above, and of the correspondence between the groups and their Lie algebras of operators, it is sufficient to prove that under the conditions of the theorem the Abel's Lie algebra $L^r(\zeta)$ is similar to the Lie algebra of translations.

Let $\bar{Z}$ be the space of invariants of the group H^r and $J: Z \to \bar{Z}$ be its universal invariant. According to the note in Section 19.8, the space $\bar{Z}$ can be supplemented up to the space Z', isomorphic to Z, where $Z' = \bar{Z} \times T$, and the invariant J up to the invertible mapping $p: Z' \to Z'$ given by (8.2). By this mapping the Lie algebra $L^r(\zeta)$ transforms into the Lie algebra $L^r(\zeta')$, isomorphic to it, where $\zeta' = (\bar{\zeta}, \bar{\tau})$. But here $\bar{\zeta} = \partial J \circ \zeta = 0$ because of the criterion (17.1.3) of the invariance of J. Thus the operators of the Lie algebra $L^r(\zeta')$ have the form

$$\zeta'_\alpha \cdot \partial' = \bar{\tau}_a \cdot \partial_{\bar{t}} \qquad (a \in B), \tag{12.5}$$

where B is a parameter space of the group H^r. With this the linear mapping $\bar{\tau}(\bar{z}, \bar{t}): B \to T$ is a mapping *onto*, and it is invertible because $r_*(\tau) = r = \dim B$ and $\dim T = r$. It is now shown that for every fixed $\bar{z} \in \bar{Z}$ there exists a transformation $q: T \to T$, for which

$$\partial q(\bar{t}) \circ \bar{\tau}(\bar{z}, \bar{t}) = \sigma, \tag{12.6}$$

where σ, independent of $(\bar{z}, \bar{t})$, is a linear isomorphism of the space B on T. Owing to the property just noted of $\bar{\tau}$, equation (12.6) can be rewritten in the form $\partial q(\bar{t}) = \sigma \circ (\bar{\tau}(\bar{z}, \bar{t}))^{-1}$. Consequently, the determination of q is reduced to the problem of restoration of a mapping from its derivative. According to the Frobenius's theorem (see Section 12.8), this problem is solvable if the second derivative $\partial^2 q(\bar{t})$, calculated from equation (12.6), is symmetric. But if (12.6) is written, for some $a \in B$, in the form $\partial q\langle\bar{\tau}_a\rangle = \sigma\langle a\rangle$ and the derivative is calculated with respect to $\bar{t}$ at the vector $\bar{\tau}_b$, then the equation

$$\partial^2 q\langle\bar{\tau}_b, \bar{\tau}_a\rangle + \partial q\langle\partial\bar{\tau}_a\langle\bar{\tau}_b\rangle\rangle = 0$$

is found. From this equation, because of the commutator definition (7.1.1) and the fact that the Lie algebra $L^r(\zeta')$ is Abel's,

$$\partial^2 q\langle\bar{\tau}_b, \bar{\tau}_a\rangle - \partial^2 q\langle\bar{\tau}_a, \bar{\tau}_b\rangle = \partial q\langle[\bar{\tau}_a, \bar{\tau}_b]\rangle = 0.$$

So equation (12.6) has the solution q, which, owing to the choice of σ, is such that $r_*(\partial q) = r$. Thus q is a transformation of the space T. Under the operation of q the operators (12.5) will transform, according to the theorem in Section 16.9, into the operators

$$\partial q(\bar{\tau})\langle \tau_a(\bar{z}, \bar{t})\rangle \cdot \partial_{t'} = \sigma_a \cdot \partial_{t'}, \qquad t' = q(\bar{t}),$$

where $\sigma_a = \sigma\langle a\rangle$. It is evident that these are operators of translations. ■

Note. If the tangential mapping ζ of Abel's group H^r is such that $r_*(\zeta) < r$, then in H^r there exists a subgroup H^{r_*} satisfying the condition of the theorem. Obviously, the universal invariants of the groups H^r and H^{r_*} coincide. Thus the similar (in a broad sense) H^r-solutions and H^{r_*}-solutions are the same.

Consequence. *For any similar (in broad sense) H^r-solution there exists a subgroup $H' \subset H^r$ and a transformation of the space Z such that in the transformed space this solution is a similar (in a narrow sense) H'solution.*

§20. CLASSIFICATION OF INVARIANT SOLUTIONS

1. Classification by Rank

Let a system of differential equations E, admitting the Lie group G of transformations of the space Z, be given. Using the algorithm in Section 19.7, it is possible to find solutions of the system E, as some invariant H-solutions of any subgroup $H \subset G$, satisfying the necessary conditions (19.3.5).

An important numerical characteristic of an invariant H-solution is its rank ρ, equal to the number of independent variables in the factor system E/H. Thus the main classification of invariant H-solutions is made by their ranks. According to (19.3.2), rank is given by the formula $\rho = n - r_*$, where $r_* = r_*(\zeta) > 0$. Consequently, the rank of an invariant H-solution can have, generally speaking, values $\rho = 0, 1, 2, \ldots, n - 1$.

When ρ is fixed, different H-solutions will be obtained using different groups, which have the same value of r_*. Thereby a problem of enumeration of the subgroups $H \subset G$, with a given r_*, arises. With this, some inconvenience appears connected with the fact that the number r_* is, mainly, a characteristic of the group H as a representation of some abstract Lie group, but not its structural characteristic. Indeed, if the group G is a representation (exact!) of the Lie group (B, ϕ), then these groups are isomorphic and their subgroups are in one-to-one correspondence, while the numbers r_* of the corresponding subgroups, considered as representations, generally speaking, are different.

Thus from the group point of view, it is more convenient to work with the number r, which is the order of the group H, rather than with r_*. Then the difference $r - r_*$ is equal to the number of independent linear relations among the basis operators of the Lie algebra of the group H, because of which a knowledge of the subgroups of a given order r permits a classification of them by the values r_* without any problems.

So the classification problem of invariant H-solutions, according to their ranks, is reduced to an enumeration problem of the order of all subgroups of the group G and a subsequent sorting of those subgroups according to their rank.

2. Orbits of Solutions

Another main classification of invariant solutions is a fundamental fact, established in Section 6.5 and repeated in Section 19.1: The group G admitted by the equation E operates on the set SE of its solutions. Because of this fact, *for every solution $u \in SE$ its orbit, that is, the set of all those solutions that transform into solutions under the operation of all possible transformations of the group G*, (see also Sections 1.13 and 16.10), *is put into correspondence.* Here an orbit of the solution u will be denoted by (u, G). The analogous notation (u, H), for any subgroup $H \subset G$, has a similar meaning.

Invariant H-solutions $u \in SE$ are characterized by the fact that for them the equality $u = (u, H)$ holds.

Let h be some transformation of the space $Z = X \times Y$. In Section 4.5 it is explained how the transformation h operates on any mapping $u: X \to Y$. For short, an image under this operation will be denoted here by hu. Also, if H is a group of transformations of the space Z, then a set of transformations $\{h \circ h_a \circ h^{-1}\}$, which is obtained when h_a takes values in the group H, is a group which is similar (conjugate) to the group H; this group will be denoted by hHh^{-1}.

Lemma. *The equation of an orbit transformation is*

$$h(u, H) = (hu, hHh^{-1}). \tag{2.1}$$

Proof. If $v \in h(u, H)$, then $v = (h \circ h_a)u$ for some $h_a \in H$. But then $v = (h \circ h_a \circ h^{-1})(hu)$, which means that $v \in (hu, hHh^{-1})$. It is clear that this reasoning is reversible. ■

3. Essentially Different Solutions

Two solutions $u_1, u_2 \in SE$ are said to be *essentially different* (relative to the group G) if $u_2 \notin (u_1, G)$ [of course, also, $u_1 \notin (u_2, G)$]. In the opposite case, these solutions are said to be *nonessentially different* (relative to the group G). In other words, two solutions u_1, u_2 are nonessentially different if one is transformable into the other by a transformation belonging to the group G, and they are essentially different if there is no such transformation.

Let G be a Lie group of transformations of the space Z, admitted by the system E, and let the phrase "to find an invariant H-solution" be understood in the sense of constructing the corresponding factor system E/H. In these terms the following problem is put: Find all possible essentially different (relative to G) invariant H-solutions of the system E which have a given rank ρ and can be obtained from the subgroups $H \subset G$.

During the solution of the problem, stated in this way, one should keep in mind that the invariant H-solution u can be, at the same time, some invariant H'-solution of the same or lesser rank. To take this possibility into account, the following auxiliary concepts are useful.

An invariant H-solution u of rank ρ is called *degenerate* if there is a subgroup $H' \subset G$, with $H \subset H'$, and u is an invariant H'-solution of rank $\rho' < \rho$. The subgroup $\overline{H}$ is called a *maximal extension* of the subgroup H (in the group G), if $H \subset \overline{H}$, the universal invariants of the groups H and $\overline{H}$ coincide, and any other subgroup with these properties is contained in $\overline{H}$. It is clear that the maximal extension of the Lie algebra of operators of the subgroup $\overline{H}$, in the Lie algebra of the group G, corresponds to the maximal extension of the subgroup H. The former can easily be found through consideration of linear relations among the operators.

Theorem. *For the given subgroup $H \subset G$ there are nonessentially different nondegenerate solutions among invariant H-solutions if and only if the maximal extension $\overline{H}$ is a proper normal divisor in some subgroup $H' \subset G$.*

Proof. (Direct statement) Let $\overline{H}$ be a normal divisor in H'. If the invariant H-solution is found by the transformation $h \in H'$, which is not included in $\overline{H}$, then a solution hu, which is different from u because of the maximal nature of $\overline{H}$, will be obtained. Also, because of (2.1) and the equality $\overline{H} = h\overline{H}h^{-1}$, the relations

$$hu = h(u, \overline{H}) = (hu, h\overline{H}h^{-1}) = (hu, \overline{H}) = (hu, H)$$

and found, that is, hu is also invariant H-solution nonessentially different from u.

(Converse statement) If two nondegenerate H-solutions u_1 and u_2 are nonessentially different, then $u_2 = hu_1$ for some transformation $h \in G$, where $h \notin \overline{H}$. Then for any $h_a \in \overline{H}$ the equality $(h \circ h_a \circ h^{-1})u_2 = u_2$ holds. It is asserted that $h \circ h_a \circ h^{-1} \in \overline{H}$. For otherwise there would exist a wider group $H_1 \supset \overline{H}$ such that u_2 would be an invariant H_1-solution. Let ρ be the rank of u_2 as an invariant $\overline{H}$-solution and ρ_1 be its rank, too, but as an invariant H_1-solution. Then the inequality $\rho_1 \leq \rho$ holds. But the inequality $\rho_1 < \rho$ means that u_2 is a degenerate $\overline{H}$-solution, and the equality $\rho_1 = \rho$ means that $\overline{H}$ and H_1 have equal universal invariants. The first possibility contradicts the statement about the nondegenerate nature of the H-solution u_2, and the second one contradicts the definition of $\overline{H}$. Therefore, $h \circ h_a \circ h^{-1} \in \overline{H}$. So there are transformations $h \notin \overline{H}$ for which $h\overline{H}h^{-1} = \overline{H}$. The union of all such h generates a subgroup $H' \subset G$, where $\overline{H}$ is a normal divisor in H'. ■

4. The Factor System Property

During the proof of the previous theorem it was determined that some manifolds, not belonging to the group H, transform invariant H-solutions into invariant H-solutions. This fact leads to an idea that such transformations must be admitted by the factor system E/H. Exactly the same formulation for the corresponding result of group analysis is given by using the concept of a normalizer.

The *normalizer of the group* $H \subset G$ is the widest subgroup $H' \subset G$ possessing the property that H is a normal divisor in H'. Obviously, the normalizer H' of the subgroup H is a set of all elements $h \in G$ for which the equality $hHh^{-1} = H$ holds.

Theorem. *If H' is a normalizer of the subgroup H in G, then the factor system E/H admits the factor group H'/H.*

Proof. Let $h \in H'$; then for any $h_a \in H$ the equality $h \circ h_a = h_b \circ h$, for some $h_b \in H$, is correct. Thus if F is any invariant of the group H, then

$$(F \circ h) \circ h_a = (F \circ h_b) \circ h = F \circ h,$$

that is, $F \circ h$ is also an invariant of the group H. Obviously, this property holds for continuation groups and their invariants also. Thus the factor system E/H, connecting only the invariants of the group H, under any transformation $h \in H'$ will transform into a system, also connecting only invariants of the group H. But the last system is equipollent to the system E/H and, consequently, coincides with it. ■

5. Optimal Systems of Solutions

From the equation for orbit transformations (2.1) it follows that $u = (u, H)$ gives $hu = (hu, hHh^{-1})$. Consequently, every invariant H-solution u transforms into an invariant hHh^{-1}-solution under the operation of any transformation $h \in G$. It is clear from this that rank of the solution is not changed. As the group hHh^{-1} is similar to the subgroup H, the following conclusion can be drawn: *Nonessentially different invariant solutions are found from similar subgroups.* Thereby the problem set in Section 20.3 concerning classification of invariant H-solutions is reduced to the problem of classification of subgroups of the group G, up to similarity. This problem was already discussed in Section 15.8, where it was shown that, in its turn, it is reduced to the corresponding problem for Lie algebras.

Consequently, taking into account the note in Section 20.1 about changes in the rank of an invariant H-solution with the order of the group H, the problem posed of classification is finally reduced to the problem of constructing optimal systems Θ_s of similar algebra classes of the Lie algebra of the group G. The algorithm for solving the last problem is considered in Sections 14.7 and 14.8. If the optimal systems Θ_s are found, then only those subgroups H should be chosen for which the necessary conditions (19.3.5) for the existence of invariant H-solutions are satisfied. Finally, for every group H satisfying these conditions the universal invariant must be found to determine the general form of the invariant H-solution and to build the corresponding factor system E/H.

As a result, the optimal system of the invariant H-solutions of the equation E, denoted by $S_G E$, is found. By optimality of the system $S_G E$ is meant that it is the smallest subspace in the set SE possessing the following property: Any invariant H-solution, which can be found from any subgroup $H \subset G$, is contained in the orbit (u, G) of the solution $u \in S_G E$,

6. Example

The system of equations (5.9.1), for the transonic flow of a gas, admits the group G^4, whose basis of operators for the Lie algebra L^4 is found in (5.9.9). The optimal systems of subalgebras for L^4 are found in Section 14.9 and detailed in table (14.9.8).

The invariant H-solutions of rank $\rho = 1$, obtained from one-parameter subgroups H, are interesting because here $n = m = 2$. Table (14.9.8) shows that there are four classes of similar one-parameter subgroups, the first three of which contain an infinite number of subgroups. Below, two specific examples are considered. For these calculations, equations (5.9.1) are written in the following notation: (x, y) are the independent variables, (u, v) the detpendent variables, $\partial_x u = u_x$, etc., and the system E is

$$u_y - v_x = 0, \qquad uu_x + v_y = 0. \tag{6.1}$$

(α) $H(u_4)$. Here $u_4 \cdot \partial = x\partial_x + y\partial_y$ and the universal invariant is $J = (x/y, u, v)$. The variables are separated and the general form of the invariant $H(u_4)$-solution is

$$u = \bar{u}(t), \qquad v = \bar{v}(t), \qquad t = x/y.$$

A substitution into equation (6.1) leads to the factor system $E/H(u_4)$:

$$t\bar{u}' + \bar{v}' = 0, \qquad \bar{u}\bar{u}' - t\bar{v}' = 0,$$

where the derivative ∂_t is denoted by a prime. All nonconstant ($\bar{u}' \neq 0$) invariant $H(u_4)$-solutions are easily found. They depend on one arbitrary constant C and have the form

$$u = -t^2 = -(x/y)^2, \qquad v = (\tfrac{2}{3})t^3 + C = (\tfrac{2}{3})(x/y)^3 + C.$$

(β) $H(u_1)$. Calculation of the universal invariant of the operator

$$u_1 \cdot \partial = (-xv + yu^2)\partial_x - (xu + 2yv)\partial_y + 2uv\partial_u + (-\tfrac{2}{3}u^3 + \tfrac{3}{2}v^2)\partial_v$$

leads to its representation in the form

$$J = (J^1, J^2, J^3),$$

where

$$J^1 = u^{-3/2}(4u^3 + 9v^2), \qquad J^2 = u^{1/4}(2ux + 3vy), \qquad J^3 = u^{-1/4}(3vx - 2u^2y). \tag{6.2}$$

The variables are not separated. But if equations (6.1) are transformed to their equipollents, at the expense of a change of variables roles, that is, by considering (x, y) as unknown functions of the independent variables (u, v), then the system E will transfer into the following (linear) system ($x_v = \partial_v x$, etc.):

$$x_v - y_u = 0, \qquad uy_v + x_u = 0. \tag{6.3}$$

Relative to the system (6.3), the variables in (6.2) are separated and the invariant $H(u_1)$-solution can be looked for in the form

$$J^2 = \bar{x}(t), \qquad J^3 = \bar{y}(t), \qquad t = J^1. \tag{6.4}$$

As the invariants J^2 and J^3 depend linearly on (x, y), relation (6.4) can easily be found relative to x and y. As a result, the general form of the invariant $H(u_1)$-solution is found to be

$$x = 2u^{1/4}\phi(t) + 3vu^{-5/4}\psi(t), \qquad y = 3vu^{-7/4}\phi(t) - 2u^{-1/4}\psi(t), \qquad t = u^{-3/2}(4u^3 + 9v^2), \tag{6.5}$$

where the functions (ϕ, ψ) are related to the functions $(\bar{x}, \bar{y})$ by $\bar{x} = t\phi$, $\bar{y} = t\psi$. For the system E, taken in the form (6.3), the determination of the factor system $E/H(u_1)$ is free of difficulty, whence the factor system is

$$6t\phi' + 7\phi = 0, \qquad 6t\psi' + 5\psi = 0, \tag{6.6}$$

where the derivative ∂_t is denoted by a prime. Equations (6.6) are integrated in an elementary fashion and the following family of invariant $H(u_1)$-solutions of system (6.3) is found:

$$x = 2C_1u^2r^{-7/6} + 3C_2vr^{-5/6}, \qquad y = 3C_1vr^{-7/6} - 2C_2ur^{-5/6}, \qquad r = 4u^3 + 9v^2, \tag{6.7}$$

where C_1 and C_2 are arbitrary constants.

REFERENCES

For actual applications, the search for partial-solution classes, under which the initial differential equations are essentially simplified, has usually been accomplished either on the basis of "considerations of symmetry" or with the aid of dimensional analysis [1–4]. A long time ago this problem drew the attention of investigators who considered its group nature [5]. G. Birkhoff, especially, urgently insisted on the use of group theory in hydrodynamics and in other applied areas [6]. The possibility of constructing classes of partial solutions with the aid of any admitted group was first pointed out in Ref. [4] cited in Chapter II. After appearance of the monograph [5] cited in Chapter II, activity in this field sharply increased, so that a number of interesting works, determining invariant solutions by group methods in a variety of applied problems, are available [7–19]. Toward this goal, the papers cited in Chapter III, [13], [17], [18], [21], and [22] were done. References [20] and [21] are devoted to constructing optimal systems of subgroups by the method of Section 20, which was first published in Ref. [5] cited in Chapter II.

[1] Buckingham E. Model experiments and the forms of empirical equations. — Trans. Amer. Soc. Mech. Eng. **37** (1915), 263—296.

[2] Birkhoff G. Dimensional analysis of partial differential equations. — Electr. Enging. **67** (1948), 1185—1188.

[3] Седов Л. И. Методы подобия и размерности в механике. М., «Наука», 1967.

[4] Коробейников В. П., Мельникова Н. С., Рязанов Е. В. Теория точечного взрыва. М., Физматгиз, 1961.

[5] Morgan A. J. A. The reduction by one of the number of independent variables in some systems of partial differential equations. — Quart. J. Math. Oxford. **3**, № 12 (1952), 250—259.

[6] Биркгоф Г. Гидродинамика. М., ИЛ, 1963.

[7] Катков В. Л. Один класс точных решений уравнений прогноза геопотенциала. — Изв. АН СССР. Физ. атмосф. и океана **1**, № 10 (1965), 1088—1090.

[8] Катков В. Л. Групповая классификация решений уравнения Хопфа. — Ж. прикл. мех. и техн. физ. № 6 (1965), 105—106.

[9] Катков В. Л. Точные решения некоторых задач конвекции. — Прикл. матем. и техн. **32**, № 3 (1968), 482—486.

[10] Сыровой В. А. Инвариантно-групповые решения уравнений пространственного и стационарного пучка заряженных частиц. — Ж. прикл. мех. и техн. физ. № 3 (1963), 26—35.

[11] Сыровой В. А. О некоторых новых решениях, полученных при помощи инвариантных преобразований. — Ж. прикл. мех. и техн. физ. № 3 (1965), 56—62.

[12] Сухарев М. Г. Инвариантные решения уравнений, описывающие движение жидкости и газа в длинных трубопроводах. — ДАН СССР **175**, № 4 (1967), 781—784.

[13] Далькович А. И. Инвариантно-групповые решения гидродинамической модели циркуляции атмосферы для экваториальной области. — Изв. АН СССР. Физ. атмосф. и океана **4**, № 6 (1968), 579—585.

[14] Филиппов Ю. Г. Применение инвариантно-группового метода к решению задачи определения течений неоднородного океана. — Метеорол. и гидрология. № 9 (1968), 53—62.

[15] Хабиров С. В. Одно инвариантное решение уравнений мелкой воды. — В сб. «Динамика сплошной среды», вып. 3 (1969), Новосибирск, 82—90.

[16] Хабиров С. В. О структуре псевдогруппы, допускаемой уравнениями идеальной несжимаемой жидкости. — В сб. «Динамика сплошной среды», вып. 24 (1967), Новосибирск, 105—114.

[17] Bluman G. W., Cole J. D. The general similarity of the heat equation. — J. Math. and Mech. **18**, № 11 (1969), 1025—1042.

[18] Жмудский А. А., Таранов В. Б. О симметрии движения двухкомпонентной плазмы. — ЖТФ **45** (1975), 158—161.

[19] Таранов В. Б. О симметрии одномерных высокочастотных движений бесстолкновительной плазмы. — ЖТФ **46** (1976), 1271—1277.

[20] Ибрагимов Н. Х. Классификация инвариантных решений уравнений двумерного нестационарного движения газа. — Ж. прикл. мех. и техн. физ. № 4 (1966), 19—22.

[21] Лапко Б. В. Построение оптимальных систем подгрупп группы Ли преобразований, допускаемых уравнениями газовой динамики. — В сб. «Динамика сплошной среды», вып. 15 (1973), Новосибирск, 122—129.

CHAPTER

VI

Partial Invariance

In the search for solutions of differential equations further possibilities arise if the property of solution invariance is not required. In that case, by what should that property be replaced to maintain those advantages which characterized invariant solutions? As it turns out, that question is answered in the present chapter by partial invariance of a manifold.

This concept is naturally formulated in terms of orbits. If the group H is given, then the orbit described by the manifold Ψ under the operation of all possible transformations from H is the smallest invariant manifold of the group H containing the given manifold Ψ. The rank ρ of the indicated orbit is called the rank of the manifold Ψ. Also, Ψ is provided with one more numerical characteristic, showing "how much" Ψ is not invariant relative to the group H. This number, denoted by δ and called the *defect of invariance* (or just *defect*) of the manifold Ψ, is defined as the difference between the dimension of the manifold Ψ's orbit and that of Ψ itself. It is essential that there be an explicit infinitesimal formula for the defect. The manifold Ψ is called *partially invariant relative to the group H* if its orbit does not coincide with the whole space and its defect $\delta > 0$.

To find solutions by applying the concept of partial invariance leads to an algorithm similar to that one which is used in the case of invariant solutions. The essential difference is that a passive system P appears, whose integration conditions belong to the system E/H, in addition to the factor system E/H.

Partial invariant solutions yield to classification, and a new classification parameter is their defect. A new interesting problem, called the problem of reduction of partially invariant solutions, arises in connection with classification issues. A reduction takes place, during the transition from the group H to its subgroup, when the rank of the given partially invariant manifold Ψ does not change but its defect decreases. The importance of this problem arises because it is easier to search for solutions with the smallest defect. A theorem of reduction, giving sufficient conditions for reduction of partially invariant solutions of first-order differential equations, is established in Section 22. The conditions imposed in the statement of this theorem appear to be useful in applications.

One of the applications of partial invariance methods is a theory of so-called multiple waves—the theory of degenerate (in a certain sense) solutions of quasi-linear homogeneous systems of first-order differential equations. These solutions can be interpreted as partially invariant relative to a natural and very simply describable group, admitted by those equations. It is generated by transformations of translation and by one dilatation of the space of independent variables.

In Section 23 a complete description of simple waves (waves of multiplicity one) is given, and double waves (waves of multiplicity two) are also considered. A new theory about reduction is established for double waves, which is used for the equations of gas dynamics. The result obtained consists of the fact that the known isentropic property of gas dynamic simple waves also holds, in a way, for double waves.

§21. RANK AND DEFECT OF MANIFOLDS

1. Manifold Orbit

Let $G = G^r(f)$ be a local Lie group of transformations of the n-dimensional space X, generated by the mapping $f\colon V \times O \to X$ of the product between the open set $V \subset X$ and the null neighborhood $O \subset B$ of the group parameter space B. In what follows it is assumed that the set V does not contain singular points of the group G. In that case any manifold $\Psi \subset V$ is nonsingular relative to the group G (see Section 18.6). The following definition is connected with the concept of orbits of the point $x \in V$, introduced in Section 16.10.

A *manifold orbit* $\Psi \subset V$ is a set $f(\Psi, O)$ of points that belong to the orbits $f_x(O)$ of all possible points $x \in \Psi$. In other words, the orbit $f(\Psi, O)$ of the manifold Ψ is a union of the orbits of all the points of this manifold.

There are two possibilities. The orbit $f(\Psi, O)$ either coincides (locally) with the space X (contains some open set of the space X) or is a manifold in X with dimensionality less than $\dim X = n$. In general, if some manifold $M \subset X$ is such that $\dim M < \dim X$, then M is called a *characteristic manifold* of the space X. So the orbit $f(\Psi, O)$ of the manifold Ψ is a characteristic manifold in X if

$$\dim f(\Psi, O) < n. \tag{1.1}$$

Evidently, if the orbit of the manifold Ψ satisfies the inequality (1.1), then it is a nonsingular invariant manifold (as a union of invariant manifold orbits $f_x(O)$ of points $x \in \Psi$) of the group G. From the theorem in Section 18.7 it follows that the group G is intransitive. Thus the existence of manifolds Ψ, whose orbits are characteristic manifolds in X, is possible only for intransitive groups. Certainly the converse statement is not true, that is, for the group G the property of intransitivity alone does not guarantee that for any manifold Ψ its orbit will be a characteristic manifold in X.

For any invariant manifold Φ of the group G, containing the manifold Ψ, the inclusion $f(\Psi, O) \subset \Phi$ is correct. Thus *the orbit* $f(\Psi, O)$ *is the least invariant manifold of the group* G *containing* Ψ. In the particular case when $\Psi = \{x\}$ consists of one point $x \subset V$, this fact has been mentioned in Section 18.5. It is clear that if Ψ is an invariant manifold of the group G itself, then $f(\Psi, O) = \Psi$.

2. Rank and Defect

Let the inequality (1.1) be satisfied for the given manifold Ψ. Its orbit $f(\Psi, O)$, being a nonsingular invariant manifold of the group G, has a certain rank (see Section 18.8). The following definition assigns the same rank to the manifold (generally speaking, not invariant) Ψ.

The *rank of the manifold* Ψ, *relative to the group* G, is the rank of its orbit $f(\Psi, O)$. Considered as a function of the pair (Ψ, G), this rank is denoted by $\rho(\Psi, G)$.

Another important numerical characteristic of the manifold Ψ, the defect of invariance (or merely defect), is defined in the following way.

The *defect of the manifold* Ψ, *relative to the group* G, is the difference between the dimension of its orbit $f(\Psi, O)$ and the dimension of the manifold Ψ. Considered as a function of the pair (Ψ, G), the defect is denoted by $\delta(\Psi, G)$, so

$$\delta(\Psi, G) = \dim f(\Psi, O) - \dim \Psi. \tag{2.1}$$

The term "defect" emphasizes that the number $\delta(\Psi, G)$ gives a quantitative characteristic of how noninvariant is the given manifold Ψ. It is clear that $\delta(\Psi, G) = 0$ is an invariance criterion of the manifold Ψ. The manifold Ψ, for which $\delta(\Psi, G) > 0$, is called *a partially invariant manifold* of the group G with the invariance defect equal to $\delta(\Psi, G)$.

The rank $\rho = \rho(\Psi, G)$ and the defect $\delta = \delta(\Psi, G)$ are not independent. It is convenient to describe the relation connecting them using the space of invariants $\bar{X}$ of the group G whose dimensionality is equal to $n - r_*$ (theorem in Section 17.3), where r_* is the general rank (on the set V) of the tangential mapping ξ of the group G. In what follows the dimensionality of the space of invariants $\bar{X}$ is denoted by t, so that $t = n - r_*$. Then the dimensionality of the orbit $f(\Psi, O)$ can be expressed, through the rank ρ, by means of (18.8.1), which gives $\dim f(\Psi, O) = \rho + n - t$. Also, let the considered manifold Ψ be regularly given by the equation

$$\psi(x) = 0, \tag{2.2}$$

with the mapping $\psi\colon X \to R^s$ (that is, it is given by a system of s scalar equations). Then $\dim \Psi = n - s$. Substitution of these expressions for the dimensionalities into (2.1) gives the relation

$$\delta = \rho + s - t. \tag{2.3}$$

Lemma. *The inequality*

$$\delta(\Psi, G) \leq r_* \tag{2.4}$$

holds.

Proof. According to (18.5.1), the dimensionality of the orbit $f_x(O)$, for any point $x \in \Psi$, is equal to r_*. Thus the location for every point of the orbit $f(\Psi, O)$ can be given by not more than $r_* + \dim \Psi$ parameters. Consequently, $\dim f(\Psi, O) \leq r_* + \dim \Psi$, because of which (2.4) follows from (2.1). ■

The inequalities necessarily satisfied by the invariance defect of any manifold Ψ, whose orbit satisfies (1.1), are obtained from (2.3) and (2.4) as

$$\max\{s - t, 0\} \leq \delta \leq \min\{n - t, s - 1\}. \tag{2.5}$$

For example, if $s = 1$ and $t = 1$, that is, the manifold Ψ is given by one scalar equation (2.2) and the group G has only one independent invariant, then the inequalities (2.5) give $\delta = 0$. This means that either Ψ is an invariant manifold of the group G or its orbit coincides (locally) with the space X.

3. Calculation of Defect

Formula (2.1) or (2.3) is inconvenient for calculation of the defect of a given manifold Ψ because each requires that the dimensionality of the orbit $\dim f(\Psi, O)$, or the rank of this orbit, be known. It will be shown that there is a simpler *infinitesimal formula* for the defect, expressed directly by means of the mapping ψ, regularly given for the manifold Ψ by (2.2), and the tangential mapping $\xi: V \to \mathscr{L}(B; X)$ of the group G.

Theorem. *The equality*

$$\delta(\Psi, G) = r_*((\partial\psi \circ \xi) | \Psi) \tag{3.1}$$

holds.

Proof. Using the invariance property of each part of (3.1) relative to transformations of the space X, it is possible to make the manifold Ψ "flat," that is, to represent the space X in the form $X = X_1 \times X_2$, where $\Psi = \{O\} \times X_2$ and $\dim X_2 = \dim \Psi = n - s$. According to this representation, let all the vectors be given in the form of the component pairs of X_1 and X_2, that is, $x = (x_1, x_2)$, $f = (f_1, f_2)$, and $\xi = (\xi_1, \xi_2)$, so that the mapping f generating the group G operates by $x_1' = f_1(x_1, x_2, a)$, $x_2' = f_2(x_1, x_2, a)$. Let $y = (0, x_2, a) \in \Psi \times B$. Then the orbit $f(\Psi, O)$ can be presented as a manifold given by the equation

$$x' = f(y) = (f_1(y), f_2(y)), \; y = \Psi \times B.$$

Since f is a mapping of the class C_∞, the dimensionality of the manifold $f(\Psi, O)$ is equal to the general rank of the derivative $\partial_y f$, which can be calculated with $a = 0$ and an arbitrary $x_2 \in X_2$. Let $y_0 = (0, x_2, 0)$. Then because of the tangential mapping definition, this derivative is given in component form by

$$\partial_y f(y_0) = \begin{bmatrix} 0 & \xi_1(y_0) \\ I_{R^{n-s}} & \xi_2(y_0) \end{bmatrix}.$$

From this is follows that

$$\operatorname{rank}(\partial_y f(y_0)) = (n - s) + \operatorname{rank}(\xi_1(y_0)),$$

or in terms of dimensionalities,

$$\dim f(\Psi, O) = \dim \Psi + r_*(\xi_1(y_0)). \tag{3.2}$$

According to the definition of the defect (2.1), the relation

$$\delta(\Psi, G) = r_*(\xi_1(y_0)) \tag{3.3}$$

follows from (3.2).

It is left to note that in the chosen representation the equation for Ψ is $x_1 = 0$, that is, $\psi = x_1$, and thus $\partial\psi \circ \xi = \xi_1$, from which (3.3) coincides with (3.1). ■

Consequence. *The manifold* Ψ, *given by equation* (2.2), *has the given invariance defect* δ *relative to the group* $G(\xi)$ *if and only if* $r_*((\partial\psi \circ \xi)|\Psi) = \delta$.

For use in applications, this form of the theorem is most convenient.

It is useful to note that the invariance criterion of a manifold (18.2.2) is a particular case of (3.1). Indeed, from (3.1) the equality $\delta(\Psi, G) = 0$ is equipollent to the equality $(\partial\psi \circ \xi)|\Psi = 0$.

4. Transition to Subgroup

Invariant manifolds of the group G are also invariant manifolds of any subgroup $G' \subset G$. Thus any partially invariant manifold Ψ of the group G is, at the same time, a partially invariant manifold of its subgroup G'. For purposes of the group analysis of a family of solutions of differential equations, it is important to ascertain what happens to the rank and defect of the given manifold Ψ in a transition from the group G to its subgroup G'. Let ρ and δ be the rank and defect of the pair (Ψ, G) and ρ' and δ' the rank and defect of the pair (Ψ, G').

Lemma. *In the transition to a subgroup, the rank of the manifold* Ψ *does not decrease and its defect does not increase, that is,*

$$\rho' \geq \rho, \qquad \delta' \leq \delta. \tag{4.1}$$

Proof. Let $t = \dim \bar{X}$, where $\bar{X}$ is the space of invariants of the group G, and $t' = \dim \bar{X}'$, where $\bar{X}'$ is the space of invariants of the subgroup $G' \subset G$. Then $t' \geq t$, and it can be assumed that $\bar{X} \subset \bar{X}'$. If Φ is the least invariant manifold of the group G containing Ψ, then the rank $\rho = \rho(\Psi, G)$, according to the definition in Section 21.2, is equal to the rank $\rho(\Phi, G)$. The latter is equal to the dimensionality of the invariant manifold Φ's projection $\bar{\Phi}$ in the space of invariants $\bar{X}$, that is, $\rho = \dim \bar{\Phi}$. Let μ be the number of independent equations which give Φ (or $\bar{\Phi}$); then $\rho = t - \mu$. Analogously, the least invariant manifold of the subgroup G' containing Ψ is Φ' and $\rho' = \dim \bar{\Phi}'$, where $\bar{\Phi}'$ is the projection of Φ' in the space of invariants $\bar{X}'$. If μ' is the number of equations which give Φ' (or $\bar{\Phi}'$), then $\rho' = t' - \mu'$. As $\Phi' \subset \Phi$, which implies $\bar{\Phi}' \subset \bar{\Phi}$, the system of equations giving $\bar{\Phi}$ is included as a subsystem in the system of equations giving $\bar{\Phi}'$. With this an increment of the equation number $\mu' - \mu$ can occur only at the expense of an increment $t' - t$ in the dimensionality of the space of invariants. Otherwise, the manifold

Φ would not be the least invariant manifold which contains Ψ. Thus $\mu' - \mu \leq t' - t$, from which it follows that

$$\rho' - \rho = t' - t - (\mu' - \mu) \geq 0,$$

that is, the first inequality of (4.1) is correct. The second inequality of (4.1) is obvious because of the inclusion $\Phi' \subset \Phi$. ■

5. Reduction

The case of a transition from the group G to the subgroup G' in which the rank of the manifold Ψ is unchanged but its defect decreases is especially emphasized. In this case it is said that a *reduction of the pair* (Ψ, G) relative to the subgroup $G' \subset G$ takes place, or that the pair (Ψ, G) is reduced to the pair (Ψ, G') with less defect. The reduction of (Ψ, G) relative to the subgroup $G' \subset G$ is described by the relations

$$\rho(\Psi, G') = \rho(\Psi, G), \qquad \delta(\Psi, G') < \delta(\Psi, G). \tag{5.1}$$

Example. The group G^r of the space of transformations of $X = R^4(x, y, z, v)$, having the Lie algebra basis

$$\xi_1 \cdot \partial = x\partial_y, \qquad \xi_2 \cdot \partial = \partial_z,$$

is considered. The two-dimensional manifold $\Psi \subset X$, given by the equations

$$y = \lambda xz, \qquad v = 0 \qquad (\lambda = \text{const}),$$

is not invariant relative to the group G^2. Here $\psi = (y - \lambda xz, v)$ and $(\xi_1 \cdot \partial)\psi = (x, 0)$, $(\xi_2 \cdot \partial)\psi = (-\lambda x, 0)$, so that $r_*((\partial\psi \circ \xi)|\Psi) = 1$. Thus $\delta = 1$ and as $s = 2$, $t = 2$, by (2.3) the rank is $\rho = 1$. For the subgroup $G^1 \subset G^2$, with the operator $(\lambda\xi_1 + \xi_2) \cdot \partial = \lambda x\partial_y + \partial_z$, the equality $(\lambda x\partial_y + \partial_z)\psi = (0, 0)$ is obtained; that is, ψ is an invariant. Thus the manifold Ψ is invariant relative to G^1 and the defect $\delta(\Psi, G^1) = 0$. But the space of invariants of the group G^1 is three dimensional and its universal invariant is $J = (x, y - \lambda xz, v)$, because of which $\rho(\Psi, G^1) = 1$. Consequently, the pair (Ψ, G^2) with the defect $\delta = 1$ has been reduced to the pair (Ψ, G^1) with the defect $\delta = 0$. This example is interesting because it contains a subgroup, relative to which the reduction takes place, which depends on a parameter λ from the equation of the manifold Ψ.

6. Essential Parameters

Let the Banach spaces $X = R^n(x)$, $Y = R^m(y)$, $B = R^r(a)$ and the mapping $\phi: X \times B \to Y$ of the class $C_\infty(X \times B)$ be given. The components of the vector $a \in B$ are called *parameters* on which the mapping ϕ, considered as the mapping $\phi_a: X \to Y$, depends.

The parameters a are called *nonessential* if there exist mappings $\gamma: B \to B$ and $\bar{\phi}: X \times \gamma(B) \to Y$ of the class C_∞ which satisfy the relations [for all $(x, a) \in X \times B$]

$$\phi(x, a) = \bar{\phi}(x, \gamma(a)), \qquad r_*(\partial_a \gamma) = q < r. \tag{6.1}$$

If there are no mappings γ with the indicated properties, then the parameters a are said to be *essential*.

If the parameters a are nonessential, then among all possible mappings γ satisfying relations (6.1) there will exist one for which the number $q = r_*(\partial_a \gamma)$ is the smallest. In this case the dimensionality of the image $\gamma(B)$ is equal to q and it is said that *ϕ depends on q essential parameters.*

The problem of determining the number of the essential parameters, on which the given mapping ϕ depends, appears in connection with these definitions. This is an analysis problem which has a known algorithmical solution. An important auxiliary statement is introduced beforehand.

Lemma. *If the number of essential parameters of the mapping ϕ is equal to $q < r$, then there exists a complete $(r - q)$-dimensional family ω of vector fields on the space B such that*

$$\partial_a \phi(x, a) \circ \omega(a) = 0. \tag{6.2}$$

Conversely, if there exists a complete $(r - q)$-dimensional family ω of vector fields with property (6.2) *for which q is minimal, then the number of essential parameters of the mapping ϕ is equal to q.*

Proof. If some mapping $\gamma: B \to B$ is such that $r_*(\partial_a \gamma) = q < r$, then there exists a space B_0, with $\dim B_0 = r - q$, and a mapping $\omega: B \to \mathcal{L}(B_0; B)$ of the class $C_\infty(B)$ such that

$$\partial_a \gamma(a) \circ \omega(a) = 0, \qquad r_*(\omega) = r - q. \tag{6.3}$$

Differentiation of equation (6.1) and generation of its composition with $\omega(a)$ lead to (6.2). The completeness of the family of vector fields $\omega(a)(B_0)$ (see Section 17.2) follows from the equality $r_*(\partial_a \gamma) = q$. Conversely, if with the mapping $\omega: B \to \mathcal{L}(B_0; B)$ a complete family of vector fields $\omega(a)(B_0)$ is determined on B, so that (6.2) is satisfied, then for every fixed $x \in X$ the partial mapping $\phi_x: B \to Y$ is an invariant of this family. Let $\gamma: B \to \bar{B}$ be a universal invariant of the vector fields ω. Its existence is guaranteed by the theorem in Section 17.2. Then for every $x \in X$ there exists a mapping $\bar{\phi}_x: \bar{B} \to Y$ such that $\phi_x = \bar{\phi}_x \circ \gamma$, that is, (6.1) is satisfied. Since q is minimal, the number of essential parameters on which ϕ depends is equal to q. ■

The previously mentioned algorithm for determining the number of essential parameters is as follows. The mapping $\phi_a: X \to Y$ is continued, as is described in Section 4.4, up to the mapping $\underset{k}{\phi_a}: X \to Z'_k$, and the general ranks $r_k = r_*(\partial_a \underset{k}{\phi_a})$ for $k = 0, 1, 2, \ldots$ are calculated (with $\underset{0}{\phi_a} = \phi_a$). This generates a nondecreasing sequence of ranks

$$r_0 \leq r_1 \leq \cdots \leq r_k \leq \cdots \leq r, \tag{6.4}$$

which must stabilize, beginning with some r_k, so $r_k = r_{k+1} = \cdots$. That unknown rank is found by applying the following results.

Theorem. *The number of essential parameters of the mapping ϕ is equal to the smallest of the numbers r_k of the sequence* (6.4) *for which $r_k = r_{k+1}$.*

Proof. Let the number of essential parameters of the mapping ϕ be equal to q. From (6.2) it follows that $r_k \leq q$ $(k = 0, 1, 2, \ldots)$. To establish the inverse inequality, it is necessary to note the existence of the mapping $\omega: X \times B \to \mathscr{L}(B_0, B)$, where $\dim B_0 = r - r_{k+1}$, for which

$$\partial_a \underset{k+1}{\phi}(x, a) \circ \omega(x, a) = 0 \tag{6.5}$$

follows from the equality $r_*(\partial_a \underset{k+1}{\phi}{}_a) = r_{k+1}$. Use of the condition $r_k = r_{k+1}$ shows that, indeed, ω does not depend on x, that is, it is a mapping $\omega: B \to \mathscr{L}(B_0; B)$. Besides, from the equality $r_k = r_*(\partial_a \underset{k}{\phi_a})$ it follows that the family of vector fields $\omega(a)(B_0)$ is complete. Thus if γ is a universal invariant of this family (theorem in Section 17.2), then under the mapping $\gamma: B \to \bar{B}$, where $\dim \bar{B} = r_k$, the equality (6.1) is satisfied, that is $q \leq r_k$. ■

Note. *The number of essential parameters a, on which the generating mapping $f: X \times B \to X$ of the group $G^r(f)$ depends, is equal to* $r = \dim B$. Indeed, if that number is less than r, then there exists a nonzero vector field $\chi: B \to B$ for which $\partial_a f(x, a)\langle \chi(a) \rangle = 0$. But because of the Lie equation $\partial_a f(x, a) = \xi(f) \circ V(a)$ and if $\chi_1(a) = V(a)\langle \chi(a) \rangle$, then $\xi(f)\langle \chi_1(a) \rangle = 0$. According to the exactness criterion, the last equation implies $\chi_1(a) = 0$, whence $\chi(a) = 0$, which contradicts the assumption.

§22. PARTIALLY INVARIANT SOLUTIONS

1. Necessary Conditions

A generalization of the invariant H-solution concept, studied in Sections 19 and 20, is considered here. This generalization is achieved by rejection of the requirement of complete invariance of the solution u and the manifold

$U \subset Z$. The notations and assumptions given in Section 19 are used in what follows.

The solution $u \in SE$ is called a *partially invariant H-solution of equation E* if the manifold U is a partially invariant manifold of the group H admitted by the equation E.

According to the results of Section 21, every solution manifold U can be provided with integer characteristics, namely, with the rank ρ and the defect δ, relative to the group H. Accordingly, the rank $\rho = \rho(U, H)$ and the defect $\delta = \delta(U, H)$ are called the *rank and defect (of invariance) of the corresponding partial invariant H-solution u.*

Because of the specificity of the solution manifold U, which is situated in $(n + m)$-dimensional space $Z = X \times Y$ and is given by the equation in the form $y = u(x)$, it always has a fixed dimensionality $n = \dim X$, and the rank and defect of any pair (U, H) must satisfy some necessary conditions. In the main, these conditions follow from relations (21.2.3) and (21.2.5), to which the following condition is added. As the rank ρ is equal to the number of independent variables in the equations which are derived from E, and the only interesting equations are those for which the number of independent variables of the auxiliary equations is less than n, it is worthwhile to superpose on the rank one more condition: $\rho \leq n - 1$. In the aforementioned relations one should change the number n to $n + m$ and assume $s = m$. Besides, it is convenient to introduce into consideration the quantity μ, equal to the number of equations which give the orbit $h(U, O)$ of the manifold U. Finally, t, as in Section 21.2, denotes the dimensionality of the space of invariants $\bar{Z}$ of the group H, and r_* denotes the general rank of its tangential mapping ζ. The numbers introduced satisfy the relations

$$t = n + m - r_*, \qquad \rho = \delta + n - r_*, \qquad \mu = m - \delta. \tag{1.1}$$

Additional limitations are imposed on the numbers of (1.1) by the fact that if $\Phi(\bar{z}) = 0$ is an equation (system of μ scalar equations) for the projection of the orbit $h(U, O)$ in the space $\bar{Z}$, then its consequence cannot be an equation relating only the independent variables x. This means that satisfaction of the condition $r_*(\partial_y \Phi) = \mu$ or

$$r_*(\partial_z \Phi \circ \partial_y J) = \mu$$

is necessary, where J is the universal invariant of the group H. Since $r_*(\partial_z \Phi) = \mu$, then according to the theorem about the rank of compositions of linear mappings, the inequality $\mu \leq \min\{\mu, r_*(\partial_y J)\}$, equipollent to the inequality $\mu \leq r_*(\partial_y J)$ or $m - r_*(\partial_y J) \leq \delta$, follows. Adjunction of this inequality to (21.2.5) leads to the following conclusion.

For partially invariant H-solutions of rank $\rho < n$ to exist it is necessary that the defect of these solutions satisfy the inequalities

$$\max\{r_* - n, 0\} \leq \delta \leq \min\{r_* - 1, m - 1\} \tag{1.2}$$

and that the universal invariant J of the group H be such that

$$m - \delta \leq r_*(\partial_y J). \tag{1.3}$$

2. Description of Algorithm

Let the group H admitted by the equation (system) E, for which the number r_* is determined and its universal invariant J is found, be given. Any value of δ satisfying the inequalities (1.2) is chosen according to r_*, and the numbers ρ and μ are determined from (1.1). Then inequality (1.3) is satisfied. If this is accomplished, then the following algorithm can be put into operation.

The space of invariants $\bar{Z} = R^t(\bar{z})$ is given in the form of the product $\bar{Z}_1 = \bar{X} \times \bar{Y}$, where $\bar{X} = R^\rho(\bar{x})$, $\bar{Y} = R^\mu(\bar{y})$, and the universal invariant is separated into two parts $J = (J_1, J_2)$, where $J_1\colon Z \to \bar{X}$, $J_2\colon Z \to \bar{Y}$, with the goal of satisfying the equation $r_*(\partial_y J_2) = \mu$. With the aid of this separation, the general form of the equations for the unknown partially invariant H-solutions orbits is formed as an equation of its projections in the space of invariants, namely, $\bar{y} = \bar{u}(\bar{x})$, or in expanded form

$$J_2(x, y) = \bar{u}(J_1(x, y)) \tag{2.1}$$

with the undetermined mapping $\bar{u}$. This equation looks like (19.4.3), but if the defect $\delta > 0$, then it is impossible, in general, to find expressions for y from (2.1) because there are only $\mu = m - \delta < m$ scalar equations. That is why the present procedure differs from the one which was used in the case of invariant solutions (see Section 19.7).

The space Y is considered as a set of pairs $y = (y_1, y_2)$, i.e., it is presented in the form of the product $Y = Y_1 \times Y_2$, where $\dim Y_1 = \mu$, with the goal that $r_*(\partial_{y_1} J_2) = \mu$. The variables y_1 are called *main variables* and y_2 are said to be *parametric*. Expressions for the main variables and their derivatives up to the required order are determined from equation (2.1), and the resulting expressions are substituted into the system of differential equations E. As a result, some overdetermined system E_2 of equations, which relate the parametric variables y_2 (using the known mapping $\bar{u}$ from (2.1)), is obtained.

The system E_2, generally speaking, will be "active," i.e., it can generate independent equations as conditions for the equations of E_2 to be consistent. From the general theory of differential equations it is known that the process of reduction of an "active" system to a "passive" one, where the conditions of consistency are identically satisfied in terms of the equations of the system itself, can theoretically always be realized. To realize this process in practice, in general form, is impossible. In the case considered it is possible only to confirm that the resulting "passive" system will consist of some unique passive system P, relating the parametric variables y_2, independent variables x, and mapping $\bar{u}$, and of some factor system E/H which guarantees the passive property of the system P. The system E/H does not explicitly contain the parametric variables but relates only the functions $\bar{u}$ and the variables $\bar{x} = J_1(x, y)$ from representation (2.1). So *the indicated algorithm gives an equipollent representation of the given system E on the family of solutions, satisfying* (2.1), *in the form of a union of two systems: the system E/H in the space of invariants of the group H and the passive system P*. Then from every solution of the system E/H it is possible to find a solution of the passive system P, i.e., the mapping $u_2: X \to Y_2$, giving the values of the parametric variables $y_2 = u_2(x)$. Following this, the mapping $u_1: X \to Y_1$, giving the values of the main variables $y_1 = u_1(x)$, is obtained from (2.1). Combining these, the solution $u = (u_1, u_2)$ of system E is obtained. This solution, according to its construction, is a partially invariant H-solution of rank ρ and defect δ.

It is clear that the problem of finding partially invariant H-solutions is simpler than the problem of finding arbitrary solutions of the system E. This is true because only the part y_2 of the unknown y depends, generally speaking, on all n independent variables x, and the mapping $\bar{u}$, determining the main variables y_1, is a solution of the factor system E/H, containing only $\rho < n$ independent variables $\bar{x}$.

3. Types of Partially Invariant Solutions

The algorithm for finding partially invariant solutions contains an arbitrariness related to the choice of the defect δ, which satisfies inequalities (1.2). A natural question is, How important is this arbitrariness? To answer, it is expedient to give a list of all possible types of solutions which are independent of the group H and the equations E. Such lists are called *tables of types* of partially invariant solutions. Note that because of relations (1.2) there are four independent numbers among the seven numbers n, m, r_*, δ, ρ, t, μ characterizing the solution. If it is assumed that the numbers m, n are fixed, then only two numbers will be left "free." It is convenient to choose these as r_* and δ.

For example, if $n = 2$ and $m = 3$, then these numbers will be subject to the inequalities

$$1 \leq r_* \leq 4, \qquad \max\{r_* - 2, 0\} \leq \delta \leq \min\{r_* - 1, 2\}.$$

Given below is the table of types of partially invariant H-solutions (Table 3.1), where the last column, giving the type of orbit $H(U, O)$ (equation (2.1)), is schematically indicated by showing the set of scalar invariants generating J_2 (to the left of the double vertical bars), and the set of scalar invariants J_1 (to the right of the double vertical bars). If the subspace $\bar{X}$ is null dimensional, i.e., $\bar{u} = C = \text{const.}$, then to the right of the double vertical bars a C is found in Table 3.1.

TABLE 3.1

Number	r_*	δ	ρ	t	μ	$h(U, O)$
1	1	0	1	4	3	$J^1, J^2, J^3 \parallel J^4$
2	2	0	0	3	3	$J^1, J^2, J^3 \parallel C$
3	2	1	1	3	2	$J^1, J^2 \parallel J^3$
4	3	1	0	2	2	$J^1, J^2 \parallel C$
5	3	2	1	2	1	$J^1 \parallel J^2$
6	4	2	0	1	1	$J^1 \parallel C$

In the general case, the number of types which are obtained for given n and m, associated with the different pairs (r_*, δ), depends only on n, m. A simple calculation shows that the *maximum possible number of types of partially invariant solutions is equal to mn.* Of course, for a specific equation $E(n, m, k, s)$, the actual number of types of partially invariant solutions admitted by the equation, which can be sought using the subgroup H of the group G, can be less than mn. This can occur if the dimensionality of the group G is not sufficiently large or because of the necessary condition (1.3), which was not considered during the construction of the table.

Generally, the classification of the partially invariant solutions of the given differential equation E is done on the basis of the same principles which were used to classify the invariant solutions in Section 20.5. However, here it is necessary to consider an additional characteristic—the defect. Obviously, the defect is invariant under a similarity transformation of the group. Thus the criterion for similarity of subgroups, applied to partially invariant solutions, does not give anything new, and thus optimal systems of subgroups are used to classify partially invariant solutions. But here a new possibility appears. It is connected with the phenomenon in Section 21.5 of reduction of partially invariant manifolds. The influence of this phenomenon on the classification of partially invariant solutions will be discussed in Section 22.5.

4. Example

Equations (20.6.1) of the transonic flow of a gas are considered and partially invariant solutions, relative to the group $H^2 = H(\zeta_3, \zeta_4)$ with the bais of operators of the Lie algebra (see (5.9.9))

$$\zeta_3 \cdot \partial = \partial_v, \qquad \zeta_4 \cdot \partial = x\partial_x + y\partial_y,$$

are sought. Here $n = m = 2$, $r_* = 2$ and the inequalities (2.1) assume the form $0 \le \delta \le 1$ and present two possibilities: $\delta = 0$ and $\delta = 1$. But the universal invariant of this group is $J = (x/y, u)$ and $r_*(\partial_{u,v} J) = 1$, so the necessary condition (1.3) gives $\delta \ge 1$. Thus only $\delta = 1$ remains, and from equations (1.1) $\rho = 1$ and $\mu = 1$ are found. Consequently, only the partially invariant H^2-solutions with rank $\rho = 1$ and defect $\delta = 1$ are possible, relative to the considered group. The general form of the orbit of such solutions is

$$u = \phi(\lambda), \qquad \lambda = x/y. \tag{4.1}$$

Here the variable u is a main one and v is parametric. The calculation of derivatives, with subsequent substitution into equations (20.6.1), gives

$$\partial_x v = -\frac{1}{y}\lambda\phi', \qquad \partial_y v = -\frac{1}{y}\phi\phi', \tag{4.2}$$

where the derivative ∂_λ is denoted by a prime. The system (4.2) is overdefined relative to the unknown function v. The compatibility condition of equations (4.2),

$$\partial_y\left(\frac{1}{y}\lambda\phi'\right) = \partial_x\left(\frac{1}{y}\phi\phi'\right),$$

after simplification becomes

$$((\lambda^2 + \phi)\phi')' = 0. \tag{4.3}$$

Equation (4.3) is a unique compatibility (passiveness) condition of system (4.2). So here (4.3) generates the factor system E/H and (4.2) generates the passive system P.

Because this example will be used in what follows, it is useful to construct the solutions of the foregoing equations. The first integral of the equation (4.3) is

$$(\lambda^2 + \phi)\phi' = c \qquad (c = \text{const}). \tag{4.4}$$

Regardless of further integration of (4.4) the function v can be expressed in terms of ϕ by means of

$$v = -c \ln y - \int \lambda\phi'(\lambda)\, d\lambda. \tag{4.5}$$

5. Reduction Problem

The concept of partially invariant solutions permits an organization of the search process for partial solutions of equation E, admitting the group G, with the aid of different subgroups $H \subset G$. It is expedient in this search to begin with the lowest possible rank and move to successively higher ranks of the studied H-solutions. This holds since, generally speaking, it is easier to find solutions with lower rank because the rank is equal to the number of independent variables in the factor system E/H. Growth of rank is mainly connected with decrease of the order of the group H, i.e., with a transition from some group H to a subgroup H'. It is possible to require that, during the transition from the group H to the subgroup $H' \subset H$, the rank does not increase. But because of the lemma in Section 21.4 the rank does not decrease with that operation. Consequently, the stated requirement means that the rank must be unchanged, i.e., $\rho' = \rho$. On the other hand, if during the transition to the group $H' \subset H$ the defect δ is also unchanged, i.e., $\delta' = \delta$, then from (1.1) it follows that $t' = t$. In other words, the group H and its subgroup H' have the same universal invariant. In this case the transition to a subgroup does not give new solutions. But if $\delta' < \delta$ then, according to Section 21.5, a reduction to partially invariant H'-solutions will take place for some partially invariant H-solutions. Therefore, the question of reduction of partially invariant solutions naturally appears and it is discussed in the following statement.

Reduction Problem. *Let a partially invariant H-solution U, having rank ρ and defect δ, be given. It is required to determine whether there is a subgroup $H' \subset H$ for which the pair (U, H) can be reduced to the pair (U, H'), i.e., U is also a partially invariant H'-solution and the relations $\rho' = \rho$, $\delta' < \delta$ hold.*

The importance of this problem is demonstrated by the fact that in the aforementioned search process for invariant solutions the determined reductions permit us to avoid a great deal of useless work, since in the process when $H' \subset H$, the partially invariant H'-solutions are found before the H-solutions.

Unfortunately, any general and practically convenient reduction criteria for partially invariant solutions have not yet been found. There are only partial theorems which give sufficient reduction conditions. One of those theorems, of a general character, is given in what follows.

6. The Rank Lemma

One property of the continuation of the tangential mapping ζ, for the group $G^r(\zeta)$ of transformations of the space $Z = X \times Y$, will be required.

The first continuation $\underset{1}{\zeta}$, determined in Section 4.8, is considered in the following lemma.

Lemma. *If the group $G^r(\zeta)$ is intransitive and*

$$r_*(\underset{1}{\zeta}) = r_*(\zeta), \tag{6.1}$$

then the general rank of the mapping ζ is equal to the order of the group, i.e., $r_*(\zeta) = r$.

Proof. If $r_*(\zeta) = r_0 < r$, then in the parameter space B of the group $G^r(\zeta)$ there exists a subspace B_0 such that $\dim B_0 = r_0$ and the family of vector fields $\zeta(z)(B_0)$ is complete (see the proof of the theorem in Section 17.3). Therefore any vector field $\zeta_a(z)$, for $a \in B$, can be expressed in the basis $\{\zeta_\alpha(z)\}$ ($\alpha = 1, \ldots, r_0$) by the formula analogous to (17.3.1):

$$\zeta_a(z) = \omega_a^\alpha(z)\zeta_a(z).$$

Because of (6.1), the same expression (with the same coefficients ω_a^α), also holds for the construction of vector fields (see Section 4.8)

$$\underset{1}{\zeta}_a(\underset{1}{z}) = \omega_a^\alpha(z)\underset{1}{\zeta}_\alpha(\underset{1}{z}),$$

for otherwise the rank $r_*(\underset{1}{\zeta})$ would be larger than the rank $r_*(\zeta)$. It is shown that for an intransitive group the last relation is possible only if all coefficients ω_a^α are the constant mappings $Z \to R$. But if this is the case, then the family of vector spaces $\zeta(z)(B)$ coincides with $\zeta(z)(B_0)$, which contradicts the inequality $r_0 < r$.

So it is left to establish the following statement. If a complete family of vector fields with the basis $\{\zeta_\alpha\}$ ($\alpha = 1, \ldots, r$) is given on the space $Z = X \times Y$, where $r < \dim Z = n + m$, then for some vector field ζ the equality

$$\underset{1}{\zeta}(\underset{1}{z}) = \omega^\alpha(z)\underset{1}{\zeta}_a(\underset{1}{z}) \tag{6.2}$$

is possible only for all ω^α constant.

In the components of $\underset{1}{\zeta} = (\xi, \eta, \underset{1}{\eta})$, equation (6.2) is equipollent to the system

$$\xi = \omega^\alpha \xi_\alpha, \qquad \eta = \omega^\alpha \eta_\alpha, \qquad \underset{1}{\eta} = \omega^\alpha \underset{1}{\eta}_a, \tag{6.3}$$

where $\underset{1}{\eta}$ is expressed through ξ, η by the formula of continuation (4.8.4) and has the form $\underset{1}{\eta} = D\eta - \underset{1}{y} \cdot D\xi$. Using equations (6.3) gives

$$\underset{1}{\eta} = \omega^\alpha(D\eta_\alpha - \underset{1}{y} \cdot D\xi_\alpha) + (\eta_\alpha - \underset{1}{y}\langle\xi_\alpha\rangle)D\omega^\alpha = \omega^\alpha \underset{1}{\eta}_\alpha + (\eta_\alpha - \underset{1}{y}\langle\xi_\alpha\rangle)D\omega^\alpha.$$

Once again from (6.3) and the equation $D\omega^\alpha = \partial_x \omega^\alpha - \underset{1}{y} \cdot \partial_y \omega^\alpha$ a final reduction gives

$$(\eta_\alpha - \underset{1}{y}\langle \xi_\alpha \rangle)(\partial_x \omega^\alpha + \underset{1}{\dot{y}} \cdot \partial_y \omega^\alpha) = 0. \tag{6.4}$$

In (6.4) the mappings ξ_α, η_α, and ω^α do not depend on $\underset{1}{y}$, so this identity can be decomposed relative to $\underset{1}{y}$. It is convenient to do further analysis in vector form by introducing the vectors $x = \{x^i\}$, $y = \{y^k\}$, $\underset{1}{y} = \{y_i^k\}$, $\xi_\alpha = \{\xi_\alpha^i\}$, $\eta_\alpha = \{\eta_\alpha^k\}$ and to write $\partial_{x^i}\omega^\alpha = p_i^\alpha$, $\partial_{y^k}\omega^\alpha = q_k^\alpha$. In this notation the decomposition of identity (6.4) relative to the variables y_i^k gives three systems of equations:

(a) $\eta_\alpha^k p_i^\alpha = 0$, (b) $\delta_i^j \eta_\alpha^k q_l^\alpha = \delta_l^k \xi_\alpha^j p_i^\alpha$, (c) $\xi_\alpha^j q_l^\alpha = 0$,

where $i, j = 1, \ldots, n$; $k, l = 1, \ldots, m$. It is left to prove that $p_i^\alpha = q_l^\alpha = 0$ follow from these equations.

In the matrix $M(\xi_\alpha^i, \eta_\alpha^k)$ let the index α enumerate the columns, so it has r columns and $n + m$ rows. Because of the completeness of the family of vector fields, with the basis $\{\zeta_\alpha\}$, the general rank of the matrix M is equal to $r < n + m$. Thus in M there will be found a row after whose deletion there remains a matrix M' still having the rank r. Let this row be designated by the number i_0 (for the number k_0 the explanation is analogous). In equations (a) and (b) assume that $i = i_0$ and $j \neq i_0$. Then for the set $\{p_{i_0}^\alpha\}$ there is the system of linear homogeneous equations

$$\eta_\alpha^k p_{i_0}^\alpha = 0, \qquad \xi_\alpha^j p_{i_0}^\alpha = 0 \qquad (j \neq i_0),$$

whose matrix of coefficients is M'. As the rank of $M' = r$, then it follows that $p_{i_0}^\alpha = 0$ ($\alpha = 1, \ldots, r$). If it is assumed that $i = j = i_0$ in equations (b), then with equations (c) the system of linear homogeneous equations for q_l^a,

$$\eta_\alpha^k q_l^\alpha = 0, \qquad \xi_\alpha^j q_l^\alpha = 0,$$

is obtained, with the matrix M of coefficients. From these $q_l^\alpha = 0$ ($\alpha = 1, \ldots, r$; $l = 1, \ldots, m$). Finally, if it is assumed that $k = l$ in (b), then together with (a) the system

$$\eta_\alpha^k p_i^\alpha = 0, \qquad \xi_\alpha^j p_i^\alpha = 0$$

is obtained, from which it follows that $p_i^\alpha = 0$ ($\alpha = 1, \ldots, r$; $i = 1, \ldots, n$). ∎

The requirement that the group $G^r(\zeta)$ be intransitive is essential in this lemma. For transitive groups the lemma statement about the rank may not be correct. For example, for the group G^3 of transformations of the space $R^2(x, y)$ (x, y scalars) with the Lie algebra basis

$$\zeta_1 \cdot \partial = \partial_x, \qquad \zeta_2 \cdot \partial = \partial_y, \qquad \zeta_3 \cdot \partial = x\partial_x + y\partial_y,$$

it is easy to show that $\underset{1}{\zeta} = \zeta$ and that condition (6.1) is satisfied. But here $r_*(\zeta) = 2$, and the group order is $r = 3$, i.e., the conclusion of the lemma is not correct. This occurs because the group $G^3(\zeta)$ is transitive.

7. Reduction Theorem

According to Section 22.2, on the class of partially invariant H-solutions, the system E is given in a form equipollent to the union of the factor system E/H and the passive system P. Of course, the system P also admits the group H. The reduction theorem for partially invariant H-solutions will be established here under the following additional condition.

On a partially invariant H-solution the passive system P contains a subsystem of the first order such that its equations, together with the equations of E/H, can be solved relative to the first derivatives of all unknown functions (i.e., relative to the vector $\underset{1}{y} \in Y_1$) and can be given in the form

$$\underset{1}{y} = F(x, y). \tag{7.1}$$

Theorem. *For any partially invariant H-solution of rank ρ, for which equations of the form* (7.1) *are obtained from the system E, there exists a subgroup $H' \subset H$ such that this solution is an invariant H'-solution of rank $\rho' = \rho$.*

Proof. Since (7.1) is an invariant manifold of the continuation group $\underset{1}{H}$, according to the representation theorem in Section 18.7 it can be given in the equipollent form $\Phi(x, y, \underset{1}{y}) = 0$, where Φ is an invariant of the group $\underset{1}{H}$. Thus the universal invariant $\underset{1}{J}: Z_1 \to \bar{Z}_1$ of the group $\underset{1}{H}$ must be such that $r_*(\partial_{\underset{1}{y}} \underset{1}{J}) = \dim Y_1 = nm$. Consequently, the dimensionality of the space of invariants of the continuation group $\underset{1}{H}$ is equal to $t + nm$, where t is the dimensionality of the space of invariants of the group H. These dimensionalities are related to the general ranks of the tangential mappings by (1.1). Therefore

$$r_*(\underset{1}{\zeta}) = n + m + nm - (t + nm) = n + m - t = r_*(\zeta);$$

i.e., condition (6.1) of the lemma about rank holds. Also, the group H, according to the condition, is intransitive. (Only for such groups can partially invariant solutions be determined.) Thus, the order of the group H is necessarily equal to $r = n + m - t$.

Let U be the considered partially invariant H-solution with the equation $y = u(x)$. Under operation of the transformation $h_a \in H$ the manifold U transforms into the manifold U_a with the equation

$$y = \phi(x, a), \tag{7.2}$$

where $\phi(x, 0) = u(x)$. Then the number $\delta = r_*(\partial_a \phi)$ is equal to the defect of the manifold U relative to the group H. Indeed, if (7.2) is written in the form of m scalar equations, then δ can be determined from them relative to the same number of vector components $a = (a^1, \ldots, a^r)$, for example, relative to $a' = (a^1, \ldots, a^\delta)$. If the resulting expressions for a' are substituted into the $m - \delta$ equations on the left, then they generate an equipollent subsystem, which will not contain the parameter a (for otherwise $r_*(\partial_a \phi) > \delta$). Consequently, this is the equation of the least invariant manifold of the group H containing U, i.e., the equation of the orbit $h(U, O)$. The number of these equations is $\mu = m - \delta$, and thus the orbit dimensionality is $n + \delta$. According to definition (21.2.1) the defect of the manifold U is $n + \delta - n = \delta$.

Let the manifold U_a be continued on the space Z_1 and equations (7.2) be added to

$$\underset{1}{y} = \partial_x \phi(x, a).$$

When these new equations are substituted into the earlier expressions for the parameters a', the parameters a will not be contained in the result at all. From this it follows that the resulting equations must be equipollent (on the solution U) to the equations of system (7.1), which is invariant relative to H and does not contain the parameter a. Thus the ranks $r_*(\partial_a \phi)$ and $r_*(\partial_a \phi, \partial_x \partial_a \phi)$ are equal. This means that here the sequence (21.6.4) has the form $r_0 = r_1$. Because of the theorem in Section 21.6, about the number of essential parameters, it follows that this number is exactly equal to the defect δ. According to the lemma in Section 21.6 there exists a mapping $\gamma: B \to B_0$, where $\dim B_0 = \delta$, such that the $\gamma(a)$ will be essential parameters and the noninvariant part of system (7.2) can be rewritten in a form such that only the parameters $\gamma(a)$ are included in it. Moreover, these equations can be solved relative to $\gamma(a)$ and presented in the form

$$\psi(x, y) = \gamma(a), \tag{7.3}$$

with a mapping $\psi: Z \to B_0$, which necessarily must be an invariant of the group H. Let H' be the union of all such transformations $h_a \in H$ for which

$$\gamma(a) = \gamma(0). \tag{7.4}$$

Evidently H' is a group, namely, a subgroup $H' \subset H$. When $a = 0$ the initial solution $U_0 \equiv U$ is found, so all the transformations $h_a \in H'$ leave the manifold U invariant. Consequently, U is an invariant H'-solution. Further, the dimensionality of the manifold (7.4) in the space B of the group H is equal to $r - \delta$, and thus the order of the subgroup H' is equal to $r - \delta$. In addition, the subgroup H' is also intransitive and (6.1) is satisfied for its tangential mapping. According to the lemma about the rank, its order is connected with the number t' of its invariants by the relation $r - \delta = n + m - t'$. Thus the rank of U as an invariant H'-solution is equal to

$$\rho' = t' - m = n + \delta - r = n + \delta - r_* = \rho,$$

where, according to (1.1), the number ρ is the rank of the solution U as a partially invariant H-solution. ∎

In applications this theorem works well for the case of first-order systems of differential equations E and is usually used, a priori, to determine a possible form of nonreduced partially invariant solutions. Thus it appears that such solutions are rather rare and, at same time, interesting from the viewpoint of physical quantities described by the system E.

8. Examples

Equations (20.6.1) of the transonic flow of a gas are considered in the following two illustrative examples.

The first example relates to the solution constructed in Section 22.4. Here all the first-order derivatives of u, v with respect to x, y can be found from (4.1) and (4.2). Thus, from the theorem in Section 22.6, every solution of (4.1), (4.5) must be invariant relative to some one-parameter subgroup H^1 of the group H^2. This subgroup can be found by searching for its operator in the form

$$\zeta' \cdot \partial = (\zeta_4 - c_1\zeta_3) \cdot \partial = x\partial_x + y\partial_y - c_1\partial_v \qquad (c_1 = \text{const}). \qquad (8.1)$$

The universal invariant of the group H^1 is $J = (x/y, u, v + c_1 \ln y)$, and the invariant H^1-solution can be found in the form

$$u = \phi(\lambda), \qquad v = -c_1 \ln y + \psi(\lambda), \qquad \lambda = x/y.$$

The corresponding factor system appears to be

$$-\lambda\phi' = \psi', \qquad \lambda\psi' + c_1 = \phi\phi',$$

where the derivative ∂_λ is denoted by a prime. Elimination of ψ leads to the equation

$$(\lambda^2 + \phi)\phi' = c_1,$$

which coincides with (4.4) when $c = c_1$. Also, $\psi(\lambda) = -\int \lambda\phi' \, d\lambda$, and exactly the same solution (4.5) is obtained. So every partially invariant H^2-solution (4.1), (4.5) is an invariant $H^1(\zeta')$-solution, where ζ' is the operator (8.1) with $c_1 = c$.

The problem of determining all possible nonreduced partially invariant solutions of the system (20.6.1), which rank $\rho = 1$ and defect $\delta = 1$, is solved in the second example. The equation of the smallest invariant manifold containing such a solution can be taken in the form

$$v = f(x, y, u). \tag{8.2}$$

Substituting this into system (20.6.1) leads to the equations

$$f_u u_x - u_y = -f_x, \qquad u u_x + f_u u_y = -f_y, \tag{8.3}$$

where $f_u = \partial_u f$ and so on. If these equations can be solved relative to u_x, u_y, then condition (7.1) will be satisfied. Thus

$$f_u^2 + u = 0 \tag{8.4}$$

must hold for nonreduced solutions of the form (8.2). From this a conclusion can be drawn: For any nonreduced partially invariant solution of the type considered, $u < 0$. In the theory of transonic flow of a gas this corresponds to supersonic flows.

If relation (8.4) is satisfied, then the form of relation (8.2) is easily determined from equations (8.3) as

$$v = \pm\tfrac{2}{3}(-u)^{3/2} + c \qquad (c = \text{const}). \tag{8.5}$$

With this function f equations (8.3) are reduced to one equation

$$\pm(-u)^{1/2}u_x + u_y = 0,$$

whose general solution can be found by a well-known method (of characteristics) and is given implicitly by

$$x \mp (-u)^{1/2}y = F(u), \tag{8.6}$$

with the arbitrary function F. Solutions of the form (8.5), (8.6) are called *simple waves* in the theory of transonic flow of gases. Thereby the following remarkable fact is established.

Any nonreduced partially invariant solution of system (20.6.1), *with rank* $\rho = 1$ *and defect* $\delta = 1$, *is a simple wave.*

§23. MULTIPLE WAVES

1. Quasilinear Systems

The general form of the first-order quasi-linear system, in the space $Z = X \times Y$, is

$$e\langle \underset{1}{y} \rangle = f, \tag{1.1}$$

with given mappings $e: Z \to \mathscr{L}(Y_1; R^s)$ and $f: Z \to R^s$, where Y_1 is the first continuation of the space Y with the aid of X (see Section 4.2). It is also supposed that $r_*(e) = s$. To write equation (1.1) in coordinate form, in the spaces $X = R^n(x)$ and $Y = R^m(y)$, systems of coordinates are chosen and the components of the vectors $x = \{x^i\}$, $y = \{y^k\}$, and $\underset{1}{y} = \{y_i^k\}$ are introduced. The mapping e is also given by the components set $\{e_{k\sigma}^i\}$, where the mapping $e_{k\sigma}^i: Z \to R$ operates by $z \to e_{k\sigma}^i(z)$. In this notation equation (1.1) assumes the form of a system of scalar equations

$$e_{k\sigma}^i(x, y)y_i^k = f_\sigma(x, y) \qquad (\sigma = 1, \ldots, s). \tag{1.2}$$

The concept of characteristics, which are defined below, plays an important role in the qualitative theory of systems (1.2). Let X^* be a space conjugate to X. To every vector $\lambda = (\lambda_1, \ldots, \lambda_n) \in X^*$ it is possible to associate a matrix $(\lambda_i e_{k\sigma}^i(z))$ whose rows are labeled with the index σ and whose columns are given by the index k. The vector λ is called a *normal characteristic vector* of the system (1.2), at the point $z \in Z$, if the rank $(\lambda_i e_{k\sigma}^i(z)) < s$. Let the manifold U, with the equation $y = u(x)$, be a solution of system (1.2). The smooth hypersurface $\Gamma \subset X$ is called a *characteristic hypersurface* (or merely *characteristic*) *of the solution* U if the vector of a normal to Γ at every point $x \in \Gamma$ is a normal characteristic vector of the system (1.2) at the point $z = (x, u(x))$.

The system (1.2) is said to be *underdefined*, *defined*, or *overdefined* if $s < m$, $s = m$, or $s > m$, respectively.

System (1.2) is said to be *autonomous* if the mappings e and f do not depend on x and *homogeneous* if $f = 0$.

In what follows autonomous homogeneous quasi-linear systems having the form

$$e_{k\sigma}^i(y)y_i^k = 0 \qquad (\sigma = 1, \ldots, s) \tag{1.3}$$

will be considered.

Note. It is natural to study system (1.3) up to equivalent similarities (see Section 6.4), which here are any *invertible linear mappings in the spaces* X, Y, *and* R^s. Obviously, such transformations maintain the differential structure of equation (1.3), changing only the functions $e^i_{k\sigma}$. This equivalence group will be denoted by $G(e)$.

2. Solutions of the Traveling Waves Types

Let $X = R^2(t, x)$, where the variable t is interpreted as time and the variable x as distance. The mapping $u: X \to Y$, given by the function $v: R \to Y$ and operating by $u(t, x) = v(x - ct)$, where $c =$ const., is called a *traveling wave*. This form shows that the graph of $y = u(t, x)$ is found at any time $t > 0$ from the graph $y = u(0, x) = v(x)$ by a shift of the latter, as a whole, by the distance ct in the positive direction (if $c > 0$) of the x axis. This shift depends upon the time t, and its velocity is equal to $c = \partial_t x$, so it seems that the wave $y = v(x)$ "travels" with velocity c.

The idea of a traveling wave is easily generalized for the case of an n-dimensional space $X = R^n(x)$. The mapping $u(x) = v(\lambda)$, where $\lambda = \lambda_i x^i$, given by the vector $(\lambda_1, \ldots, \lambda_n) \in X^*$ and the function $v: R \to Y$, is called by this term. Moreover, for every $p < n$ it is possible to consider the mapping $v: R^p \to Y$, which gives p-multiple traveling waves by the equation $u(x) = v(\lambda^1, \ldots, \lambda^p)$, where the linear forms $\lambda^\pi = \lambda^\pi_i x^i$ are generated from the linearly independent vectors $(\lambda^\pi_1, \ldots, \lambda^\pi_n) \in X^*$ $(\pi = 1, \ldots, p)$. With this, every component wave, corresponding to some λ^π, "travels" in its subspace X_π of dimensionality $n - p + 1$, defined by giving the constant values of the other $(p - 1)$ forms. This terminology is used to describe solutions of the system (1.3).

The solution $U\{y = u(x)\}$ of system (1.3) is called a *p-multiple traveling wave* if the mapping $u: X \to Y$ is the above described p-multiple traveling wave.

The following property of traveling waves is essential. If the solution U, with the equation $y = u(x)$, is a p-multiple traveling wave and $p < m$, then the functions u^k $(k = 1, \ldots, m)$ are functionally dependent and the number of those functional connections is not less than $m - p$. In the general case it can be assumed that the number of such connections is equal to $m - p$. An equipollent expression of this property is given in terms of the Jacobi matrix $(\partial_{x^i} u^i(x))$, whose rank must be equal to p. So for a p-multiple traveling wave U the relations

$$r_*(\underset{1}{y} \,|\, U) = p < \min\{n, m\} \tag{2.1}$$

hold.

It is clear that relations (2.1) are necessary, but certainly not sufficient, for the solution U to be a p-multiple wave. Thus a generalization of the traveling wave concept can be introduced by the following definition.

The solution U of system (1.3) is called a *p-multiple wave* if it satisfies (2.1). Then a 1-multiple wave is called a simple wave and a 2-multiple wave a double wave and so on.

In other words, the solution $y = u(x)$ is a p-multiple wave if there exist mappings $\lambda: X \to R^p$ and $v: R^p \to Y$ such that $u = v \circ \lambda$, or in component form

$$u^k(x) = v^k(\lambda^1(x), \ldots, \lambda^p(x)) \qquad (k = 1, \ldots, m). \tag{2.2}$$

The variables $\lambda^1, \ldots, \lambda^p$ are said to be *parameters of the wave.*

So the generalization achieved by the given definition consists of the fact that the parameters of a p-multiple wave, in contrast to a p-multiple traveling wave, must not be linear forms on the space X.

The aim of this paragraph is to study multiple waves as partially invariant solutions of system (1.3).

3. Admitted Group

It is evident that any system of the form (1.3) admits the group G^{n+1} with the Lie algebra L^{n+1}, whose operator basis is

$$\xi_0 \cdot \partial = x^i \partial_i, \qquad \xi_i \cdot \partial = \partial_i \qquad (i = 1, \ldots, n), \tag{3.1}$$

where $\partial_i = \partial_{x^i}$. Here $\xi_0 \cdot \partial$ is a dilatation operator and $\xi_i \cdot \partial$ are operators of translation in the directions of the coordinate axes of the space X. It is possible to show in the general case that G^{n+1} is the full group (see Section 5.5) of the system (1.3), i.e., this system does not admit any other transformations (belonging to one-parameter groups) of the space Z.

The algebraic structure of the group G^{n+1} is studied in an elementary fashion. It is clear that the subspace $L^n \subset L^{n+1}$, with the basis $\{\xi_i\}$, is an Abel subalgebra, to which there corresponds the subgroup $G^n \subset G^{n+1}$ of all translations of the space X. Moreover, as $[\xi_i, \xi_0] = \xi_i$, L^n is an Abel ideal and at the same time a derived algebra. From this it follows that G^n is an Abel normal divisor in G^{n+1} and that the group G^{n+1} is solvable (see Section 13.7). In addition, the equality $[\xi_i, \xi_0] = \xi_i$ also shows that the Lie algebra L^{n+1} (and the group G^{n+1}) has a zero (unit) center.

Like a representation in the group of transformations of the space Z, the group G^{n+1} is also very simply described. Actually, it is a group of transformations of the space X, and for its tangential mapping ξ, given by (3.1), the equality $r_*(\xi) = n$ holds. Thus the group G^{n+1} is transitive in X and its universal invariant in Z is $y = (y^1, \ldots, y^m)$. Being a canonical group of the second kind (see Section 16.7), it is generated by the mapping $x \to x' = ax + b$, or in the components

$$x'^i = ax^i + b^i \qquad (i = 1, \ldots, n), \tag{3.2}$$

where the group parameters a and b^i are real numbers.

4. Subgroups and Their Invariants

To classify the Lie subalgebra L^{n+1} (and therefore the group G^{n+1}) let any vector field $\xi_p \in L^{n+1}$ be given in its components $p = (p^0, p^1, \ldots, p^n)$ in the basis (3.1), so that

$$\xi_p = p^0\xi_0 + p^i\xi_i. \tag{4.1}$$

If $\xi_q = q^0\xi_0 + q^i\xi_i$ is another vector field from L^{n+1}, then the commutator $[\xi_q, \xi_p]$ is given by

$$[\xi_q, \xi_p] = (q^ip^0 - q^0p^i)\xi_i. \tag{4.2}$$

The matrix expression of the adjoined mapping (see Section 14.1) of the Lie algebra $L^{(n-1)}$, in the basis (3.1), is written at once from (4.2) as

$$\operatorname{ad}(\xi_p) = \begin{bmatrix} 0 & 0 & \cdots & 0 \\ -p^1 & p^0 & \cdots & 0 \\ \vdots & \vdots & \ddots & \vdots \\ -p^n & 0 & \cdots & p^0 \end{bmatrix}.$$

By this expression the group of inner automorphisms Int L^{n+1} is easily constructed using the algorithm given in Section 14.4. The operation of a general automorphism $A \in \operatorname{Int} L^{n+1}$, on the vector (4.1), by the rule $A\langle\xi_p\rangle = p'^0\xi_0 + p'^i\xi_i$ is determined by the equations

$$p'^0 = p^0, \qquad p'^1 = ap^i + b^ip^0 \qquad (i = 1, \ldots, n), \tag{4.3}$$

where a and b^i are the group parameters of the group Int L^{n+1}. It should be noted that since the group G^{n+1} has a unit center, it must be isomorphic to the group Int L^{n+1} (see Section 15.7). This fact is well demonstrated by the comparison of (3.2) with (4.3).

All the subalgebras $N \subset L^{n+1}$ are separated into two types, depending on whether they contain the vector field with the nonzero component p^0 or not. Any subalgebra $N \subset L^n$ is *of type* (α) if all operators of the subalgebra are the operators of translation ($p^0 = 0$). The remaining subalgebras are said to be *of type* (β).

Now it is useful to note that any invertible linear transformation V of the space X induces a similar transformation of the algebra L^{n+1} *in the algebra of operators of the same kind* (in the new coordinate system). Indeed, if $x' = V\langle x \rangle$, or in the components $x'^i = V^i_j x^j$, then

$$\xi_i \cdot \partial = V^j_i \xi'_j \cdot \partial', \qquad \xi_0 \cdot \partial = \xi'_0 \, 0 \, \partial', \tag{4.4}$$

where $\partial' = \partial_{x'}$, and if $\partial_{x'^i} = \partial'_i$, then $\xi'_j \cdot \partial' = \partial'_j$, $\xi'_0 \cdot \partial' = x'^i \partial'_i$. From this, because of invertibility of the matrix (V^j_i), the basis (3.1) is found again in the components $x' = (x'^1, \ldots, x'^n)$. But the transformation $x' = V\langle x \rangle$ is an element of system (1.3)'s equivalence transformation group $G(e)$ (see note in Section 23.1). Thus it is sufficient to carry out classification of subalgebras of the Lie algebra L^{n+1} up to a similar transformation V.

It is useful to note *the decomposition* $L^{n+1} = N^1(\xi_0) \oplus L^n$, *every component of which is invariant relative to any similar transformation* V. This follows directly from equations (4.4).

Let N^{n-p} be a subalgebra of type (α), and let its basis of vector fields be

$$\bar{\xi}_\mu = a^i_\mu \xi_i \qquad (\mu = 1, \ldots, n - p).$$

Then rank $(a^i_\mu) = n - p$ and, evidently, it is possible to find a similar transformation V such that under it $\xi'_\mu = \bar{\xi}_\mu$. Consequently, any type (α) subalgebra is characterized by the number p, where $1 \leq p \leq n - 1$, and it is similar to the standard subalgebra N^{n-p}_α of this type with the basis

$$\xi_\mu \cdot \partial = \partial_\mu \qquad (\mu = 1, \ldots, n - p). \tag{4.5}$$

Let N^{n-p} be a type (β) subalgebra. Then in it there exists a vector $\xi_{q'}$ of the form (4.1), such that $q^0 \neq 0$. With a suitable similar transformation (4.3) it is possible to determine a transformed vector $\xi_{q'}$ with $q'^i = 0$ ($i = 1, \ldots, n$). Thus it is possible to assume that the subalgebra N^{n-p} contains the vector ξ_0. Then it can be expanded in the direct sum $N^{n-p} = N^1(\xi_0) \oplus N^{n-p-1}$, where $N^{n-p-1} \subset L^n$. Taking into account the previous note concerning the subspaces $N^1(\xi_0)$ and the L^n invariance leads to the observation that the operation of the exterior automorphism on N^{n-p} is reduced to its operation on the component $N^{n-p-1} \subset L^n$. But N^{n-p-1} is a type (α) subalgebra and it is possible to use the previous results. Consequently, any type (β) subalgebra is characterized by the number p, where $1 \leq p \leq n - 1$, and is similar to a standard subalgebra N^{n-p}_β of the same type with the basis

$$\xi_0 \cdot \partial = x^i \partial_i, \qquad \xi_\mu \cdot \partial = \partial_\mu \qquad (\mu = 1, \ldots, n - p - 1). \tag{4.6}$$

Every subgroup of the group G^{n+1} corresponds to a standard subalgebra, namely, the standard subgroup H^{n-p}_α corresponds to the subalgebra L^{n-p}_α and the standard subgroup H^{n-p}_β corresponds to the subalgebra L^{n-p}_β.

The invariants of these standard subgroups (in the space X) are easily found. The universal invariant of the subgroup H_α^{n-p} is

$$J(x) = (x^{n-p+1}, x^{n-p+2}, \dots, x^n) \tag{4.7}$$

and for the subgroup H_α^{n-p} can be taken as

$$J(x) = (x^{n-p}/x^n, x^{n-p+1}/x^n, \dots, x^{n-1}/x^n). \tag{4.8}$$

In both cases the dimensionality of the space of invariants is equal to p.

5. Simple Waves

From the definition in Section 23.2 every p-multiple wave is a partially invariant G^{n+1}-solution of rank $\rho = p$. Also, for the group G^{n+1} the equality $r_* = n$ holds, whence it follows from (22.1.1) that $\delta = \rho$, i.e., the defect of this solution is also equal to the number p. The space of invariants of the group G^{n+1} is Y. Thus the least invariant manifold Φ of this group containing a p-multiple wave is a p-dimensional manifold in the space Y. This manifold can be given by a system with $(n - p)$ equations which relate the variables y, or parametrically by expressing all the y in terms of the p parameters of the wave (see (2.2)). The representation of Φ in parametric form is

$$y^k = v^k(\lambda^1, \dots, \lambda^p) \qquad (k = 1, \dots, m). \tag{5.1}$$

To find the multiple waves, it is necessary to substitute representation (5.1) into system (1.3) and investigate the compatibility of the resulting equations relative to the unknown parameters of the wave λ. In the general case such an analysis is difficult. Only simple and double waves will be considered in detail.

A simple wave is a partially invariant G^{n+1}-solution with rank $\rho = 1$ and defect $\delta \leq 1$. Its parametric representation is

$$y^k = v^k(\lambda) \qquad (k = 1, \dots, m). \tag{5.2}$$

Upon substituting this into system (1.3), a system of equations relating the unknown functions $v^k(\lambda)$ and the unknown parameter λ are obtained:

$$e_{k\sigma}^i v^{k\prime} \lambda_i = 0 \qquad (\sigma = 1, \dots, s), \tag{5.3}$$

where the prime means the derivative ∂_λ and $\lambda_i = \partial_{x^i} \lambda$.

The solution structure of system (5.3), in large measure, depends on the number $r_*(e_{k\sigma}^i v^{k\prime}) = r \leq \min\{n, s\}$, which is defined not only by the initial system (4.3) but also by the mapping $v: R \to R^m$. For example, if $r = n$, then the unique solution will be $\lambda_i = 0$ $(i = 1, \ldots, n)$, i.e., $\lambda = \text{const}$. In what follows, this trivial case is excluded, and the condition

$$r_*(e^i_{k\sigma} v^{k\prime}) = r < n \tag{5.4}$$

is assumed to hold.

The hypersurfaces $\Lambda \subset X$, on which $\lambda = \text{const.}$, are called *level surfaces of the simple wave*. A description of all simple waves is given by the following theorem.

Theorem. *Level surfaces of a simple wave, which satisfy condition* (5.4), *are generated by r-dimensional planes, and the parameter is defined implicitly by the equation*

$$\lambda = f(a(\lambda, x)), \tag{5.5}$$

where f is an arbitrary mapping $R^{n-r} \to R$, $a = (a^1, \ldots, a^{n-2})$, $a^\tau(\lambda, x) = a^\tau_i(\lambda) x^i$ $(\tau = 1, \ldots, n - r)$, and the vectors $a^\tau = (a^\tau_1, \ldots, a^\tau_n)$ form a complete set of linearly independent solutions of the system of linear equations

$$e^i_{k\sigma} v^{k\prime} a^\tau_i = 0 \qquad (\sigma = 1, \ldots, s). \tag{5.6}$$

Proof. Because of condition (5.4) the system (5.3) contains only $r \le s$ linearly independent equations, and thus, without loss of generality, it is possible to assume that $r = s$. Let Λ_0 be a level surface corresponding to the value $\lambda = \lambda_0$ and let the point $x_0 \in \Lambda_0$. The auxiliary space $T = R^r(t)$, where $t = (t^1, \ldots, t^r)$, is introduced. It is now confirmed that for any $t \in T$ the point x_t, with the components

$$x^i_t = x^i_0 + t^\sigma e^i_{k\sigma} v^{k\prime}(\lambda(x_t)) \qquad (i = 1, \ldots, n),.$$

also belongs to Λ_0. Indeed, for every $\sigma = 1, \ldots, r$ the quantity $\lambda = \lambda(x_t)$ is a unique solution of the Cauchy problem

$$\partial_{t^\sigma} \lambda = \lambda_i e^i_{k\sigma} v^{k\prime}(\lambda), \qquad \lambda(x_0) = \lambda_0,$$

which, because of equation (5.3), has a constant solution $\lambda = \lambda_0$. As $t \in T$ varies the point x_t, because of condition (5.4), generates an r-dimensional plane in X. Consequently, the level surface Λ_0 consists of these r-dimensional planes.

Owing to condition (5.4) the parameters t^σ are independent of Λ_0 and can be explicitly given by $x \in \Lambda_0$ and the solution of the equations $e^i_{k\sigma} v^{k\prime} a^{n-r+\theta}_i = \delta^\theta_\sigma$ $(\sigma, \theta = 1, \ldots, r)$, namely, by

$$t^\theta = a^{n-r+\theta}_i (x^i - x^i_0).$$

Let the vectors a^τ $(\tau = 1, \ldots, n - r)$ form a complete set for the solutions of system (5.6). Then all the constructed vectors a^j $(j = 1, \ldots, n)$ are linearly independent, and in the space X it is possible to choose the variables $x'^j = a^j_i(x^i - x^i_0)$ as the coordinates. In these coordinates, according to the construction, points of the type $(0, \ldots, 0, x'^{n-r+1}, \ldots, x'^n)$ belong to Λ_0. Thus the value λ is determined only by values of the coordinates $x'^\tau = a^\tau_i x^i$ $(\tau = 1, \ldots, n - r)$, i.e., the representation of the form (5.5) is necessarily correct. The fact that the function λ, determined implicitly by equation (5.5), is a solution of system (5.3) is easily verified by direct calculation of the derivatives and substitution into the system. ■

The case when $r = n - 1$ is very often met in applications. Then there is only one vector $(a_1, \ldots, a_n)$—the solution of the system (5.6) and the representation of the solution (5.5) can be obtained via

$$a_i(\lambda)x^i = F(\lambda), \tag{5.7}$$

where F is an arbitrary function $R \to R$.

Note 1. *In the case of a defined or overdefined system* (1.3) (*i.e., if* $m \leq s$) *the level surfaces of a simple wave are the characteristics of this system.* Indeed, because of equations (5.3), the rank of the matrix $(e^i_{k\sigma}\lambda_i)$ is necessarily less than the number m.

Note 2. Condition (5.4) of the existence of nontrivial simple waves may not be satisfied for an arbitrary mapping v. Then this condition becomes an additional requirement on v and leads to some equations which this mapping must satisfy. Such equations are called *equations of simple waves.* Equation (22.8.4) is such an equation in the second example in Section 22.8. If the functions v are not completely determined by the equations of simple waves, then the remaining arbitrariness can be used to subordinate the simple waves to some additional conditions.

6. Wave Equation Example

The wave equation for the function $\phi\colon R^3(x, y, t) \to R$ (the derivatives are denoted by subscripts as $\partial_x \phi = \phi_x$, etc.),

$$\phi_{xx} + \phi_{yy} = \phi_{tt}, \tag{6.1}$$

is equipollent to the system of equations, for the functions $u = \phi_x$, $v = \phi_y$, $w = \phi_t$,

$$v_x - u_y = 0, \qquad w_x - u_t = 0, \qquad w_y - v_t = 0, \qquad u_x + v_y - w_t = 0. \tag{6.2}$$

Here $n = m = 3$ and $s = 4$. It is convenient to look for simple waves choosing the functions w as a parameter. Then for the functions $w \to u(w)$ and $w \to v(w)$ the following system of the form (5.3) is found:

$$v'w_x - u'w_y = 0, \qquad w_x - u'w_t = 0,$$

$$w_y - v'w_t = 0, \qquad u'w_x + v'w_y - w_t = 0,$$

where the derivative ∂_w is denoted by a prime. It is easy to verify, with arbitrary functions u and v, that the rank (5.4) of this system is $r = 3$. And $r = 2$ if and only if

$$u'^2 + v'^2 = 1. \tag{6.3}$$

The vector $a = (a_1, a_2, a_3)$, being the solution of the corresponding system (5.6), here can be taken as $a_1 = u'$, $a_2 = v'$ and $a_3 = 1$. Thus the parameter of the simple wave will be determined by the equation of the form (5.7)

$$xu'(w) + yv'(w) + t = f'(w), \tag{6.4}$$

where $f' = \partial_w f$ and f is an arbitrary function. The potential ϕ (the solution of equation (5.7)) is found by integration of the total differential $d\phi = u\,dx + v\,dy + w\,dt$ and is

$$\phi(x, y, t) = xu(w) + yv(w) + tw - f(w). \tag{6.5}$$

The family of simple waves obtained from (6.3) and (6.4)—partially invariant solutions of system (6.2)—depends on two arbitrary functions f and u, and the function v is determined from (6.3). The present arbitrariness can be used to satisfy some additional conditions (initial, boundary, and so on).

7. Double Waves

A double wave is a partially invariant solution with rank $\rho = 2$ and defect $\delta \leq 2$. For $n = 1$, double waves cannot exist at all, and if $n = 2$, then any solution is a double wave. Thus it makes sense to search for double waves only if $n > 2$.

The parametric representation of a double wave has the form

$$y^k = v^k(\lambda, \mu) \qquad (k = 1, \ldots, m). \tag{7.1}$$

As the result of substituting this into system (1.3), the system of equations relating the functions v of the parameters λ, μ and the unknown parameters λ, μ,

$$e^i_{k\sigma} v^k_\lambda \lambda_i + e^i_{k\sigma} v^k_\mu \mu_i = 0 \qquad (\sigma = 1, \ldots, s), \tag{7.2}$$

is found, where the subscripts λ and μ denote the derivatives ∂_λ and ∂_μ.

Generally speaking, system (7.2) is active. The compatibility conditions for equations (7.2) give new equations, relative to the unknowns λ, μ. For example, if equations (7.2) are solved with respect to λ_i and λ_j $(i \neq j)$ in the form $\lambda_i = p$, $\lambda_j = q$, then the new equation

$$\partial_j p = \partial_i q \tag{7.3}$$

must be adjoined to system (7.2). It can happen that conditions of the form (7.3) generate new first-order equations, because of which the total number of such equations, obtained from the (5.3), can be more than s. To analyze the total passive system, in its general form, is impossible. Nevertheless, it turns out that its first-order subsystem can give useful information. Let this system consist of N independent equations of the form

$$a_\nu^i(\lambda, \mu)\lambda_i + b_\nu^i(\lambda, \mu)\mu_i = 0 \qquad (\nu = 1, \ldots, N). \tag{7.4}$$

The following theorem about the reduction of double waves is established.

Theorem. *If in the system* (7.4) *the number of independent equations is* $N = 2n - 1$, *then the double wave* (7.1) *is an invariant solution relative to a subgroup* $H^{n-2} \subset G^{n+1}$.

Proof. Since the total number of unknowns λ_i, μ_i is equal to $2n$, within the conditions of the theorem the system (7.4) can be reduced to a form in which all these unknowns are expressed in terms of only one, say λ_1. Let this reduced system for (7.4) be $(p_1 = 1)$

$$\lambda_i = p_i(\lambda, \mu)\lambda_1, \qquad \mu_i = q_i(\lambda, \mu)\lambda_1 \qquad (i = 1, \ldots, n). \tag{7.5}$$

Initially system (7.5) is converted to the simplest form by means of an invertible change of variables $(\lambda, \mu) \to (\bar{\lambda}, \bar{\mu})$, where

$$\bar{\lambda} = \phi(\lambda, \mu), \qquad \bar{\mu} = \psi(\lambda, \mu). \tag{7.6}$$

As the result of this transformation the coefficients p, q become

$$\bar{p}_i = \frac{\phi_\lambda p_i + \phi_\mu q_i}{\phi_\lambda + \phi_\mu q_1}, \qquad \bar{q}_i = \frac{\psi_\lambda p_i + \psi_\mu q_i}{\phi_\lambda + \phi_\mu q_1} \qquad (i = 1, \ldots, n). \tag{7.7}$$

This change of parameters can be used to make $\bar{q}_1 = 0$. To do that it is sufficient to take the solution of the equation $\psi_\lambda + \psi_\mu q_1 = 0$ as ψ. The equality $q_1 = 0$ is preserved under any transformation (7.6) if and only if $\psi_\lambda = 0$. Using such a transformation it is possible to have $\bar{q}_2 = 1$ by choosing the solution of the equation $\phi_\lambda = \psi_\mu q_2$ as ϕ. The equalities $q_1 = 0$ and $q_2 = 1$ are both preserved in the change of variables (7.6) if and only if $\psi_\lambda = 0$ and $\phi_\lambda = \psi_\mu$, i.e., when

$$\bar{\lambda} = \lambda\psi'(\mu) + \omega(\mu), \qquad \bar{\mu} = \psi(\mu). \tag{7.8}$$

With these the equations for the coefficients of the transformations (7.7) assume the form

$$\bar{p}_i = p_i + \frac{1}{\psi'}(\lambda\psi'' + \omega')q_i, \qquad \bar{q}_i = q_i. \tag{7.9}$$

Further simplifications are connected with considerations of the compatibility conditions for equations (7.5). First of all, since $\mu_1 = 0$ and $\mu_2 = \lambda_1$, $\lambda_{11} = 0$. Taking these into account yields the equations

$$\lambda_{i1} = p_{i\lambda}\lambda_1^2, \qquad q_{i\lambda} = 0 \qquad (i = 2, \dots, n). \tag{7.10}$$

Calculation of the derivatives λ_{ij} and μ_{ij} $(i \neq j)$ and use of equations (7.5) and (7.10) lead to the conclusion that functions $a(\lambda, \mu)$ and $b(\lambda, \mu)$ exist which satisfy the equations

$$p_{i\mu} = aq_i, \qquad p_{i\lambda} = q_{i\mu} + bq_i \qquad (i = 2, \dots, n). \tag{7.11}$$

In particular, when $i = 2$, these equations have the form $p_{2\mu} = a$, $p_{2\lambda} = b$, from which the relation $a_\lambda = b_\mu$ follows. With this relation and (7.10) the compatibility condition for equations (7.11) reduces to $q_{i\mu\mu} + bq_{i\mu} = 0$, leading to two possibilities: (α) the function b depends only on μ, and (β) $q_{i\mu} = 0$ $(i = 2, \dots, n)$, and then all $q_i = C_i$ are constants.

***Case* (α).** If $b = b(\mu)$, then by choosing a suitable transformation of the forms (7.8), (7.9), it is possible to show that $b = 0$. Then $a_\lambda = 0$ and the function a depends only on μ. It is also possible to have $a = 0$ by using one of the transformations (7.8), (7.9). As a result the system of equations (7.10), (7.11) is simplified to

$$p_{i\mu} = 0, \qquad p_{i\lambda} = q_{i\mu}, \qquad q_{i\lambda} = 0 \qquad (i = 2, \dots, n).$$

The solution of this system (for every i) is

$$p_i = \lambda A_i + B_i, \qquad q_i = \mu A_i + C_i,$$

where A_i, B_i, C_i are arbitrary constants. So in case (α) system (7.5) is reduced to the following simple form:

$$\lambda_i = (\lambda A_i + B_i)\lambda_1, \qquad \mu_i = (\mu A_i + C_i)\mu_2, \qquad \mu_2 = \lambda_1 \qquad (i = 2, \dots, n), \tag{7.12}$$

where $A_2 = 0$ and $C_2 = 1$.

***Case* (β).** If $q_{i\mu} = 0$, then $q_i = C_i = \text{const}$ $(i = 1, 2, \dots, n)$. Also, the existence of a function $h = h(\lambda, \mu)$ such that $a = h_\mu$ and $b = h_\lambda$ follows from the equation $a_\lambda = b_\mu$. Thus system (7.11) becomes

$$p_{i\mu} = h_\mu C_i, \qquad p_{i\lambda} = h_\lambda C_i,$$

which integrates to the form $p_i = hC_i + D_i$. So in case (β) system (7.5) is reduced to the simplest form:

$$\lambda_i = (h(\lambda, \mu)C_i + D_i)\lambda_1, \qquad \mu_i = C_i\lambda_1 \qquad (i = 1, \dots, n), \tag{7.13}$$

where $C_1 = 0$, $D_1 = 1$, and $C_2 = 1$.

The second stage consists of a transformation of the determined system by means of a suitable linear transformation of the space X. In case (α), remembering that $B_1 = 1$ and $C_1 = 0$, the transformation $x \to \bar{x}$ (summed with respect to $i = 1, \dots, n$) is used:

$$\bar{x}^1 = B_i x^i, \qquad \bar{x}^2 = C_i x^i, \qquad \bar{x}^j = x^j \qquad (j = 3, \dots, n).$$

As a result of this transformation the system becomes

$$\lambda_2 = 0, \qquad \mu_1 = 0, \qquad \mu_2 = \lambda_1,$$
$$\lambda_j = A_j\lambda\lambda_1, \qquad \mu_j = A_j\mu\mu_2 \qquad (j = 3, \dots, n). \tag{7.14}$$

Then two subcases are distinguished: (α_1) all $A_j = 0$ ($j = 3, \dots, n$), and (α_2) not all A_j are equal to zero.

In subcase (α_1) the general solution of system (7.14) is

$$\lambda = Kx^1 + L, \qquad \mu = Kx^2 + M,$$

with arbitrary constants K, L, M, and $K \neq 0$. By additional transformations of translation and dilatation (or changes of the form (7.6)), this solution is reduced to the standard one: $\lambda = x^1$, $\mu = x^2$. Consequently, λ and μ are invariants (up to component enumeration) of the subgroup $H_\alpha^{n-2} \subset G^{n+1}$ with the operator basis

$$\xi_j \cdot \partial = \partial_j \qquad (j = 3, \dots, n). \tag{7.15}$$

In subcase (α_2), without loss of generality, it is possible to assume that $A_3 \neq 0$. One more linear transformation of the space X is used, namely,

$$\bar{x}^1 = x^1, \qquad \bar{x}^2 = x^2, \qquad \bar{x}^3 = A_i x^i, \qquad \bar{x}^j = x^j \qquad (j = 4, \dots, n).$$

As a result system (7.14) becomes

$$\lambda_2 = 0, \qquad \lambda_3 = \lambda\lambda_1, \qquad \mu_1 = 0, \qquad \mu_3 = \mu\mu_2,$$
$$\mu_2 = \lambda_1, \qquad \lambda_j = \mu_j = 0 \qquad (j = 4, \dots, n).$$

Solution of this linear subsystem has the form

$$\lambda = \lambda(x^1, x^3), \qquad \mu = \mu(x^2, x^3),$$

where, because of $\mu_2 = \lambda_1$, the equality $\mu_2(x^2, x^3) = \lambda_1(x^1, x^3) = \gamma(x^3)$ must hold. Thus the general form of the solution of the linear subsystem is

$$\lambda = \gamma(x^3)x^1 + \alpha(x^3), \qquad \mu = \gamma(x^3)x^2 + \beta(x^3). \tag{7.16}$$

Substitution into the linear equations $\lambda_3 = \lambda\lambda_1$ and $\mu_3 = \mu\mu_2$ leads to the system (the prime notates a derivative with respect to x^3)

$$\gamma' = \gamma^2, \qquad \alpha' = \gamma\alpha, \qquad \beta' = \gamma\beta,$$

whose general solution is

$$\gamma = \frac{1}{C - x^3}, \qquad \alpha = \frac{A}{C - x^3}, \qquad \beta = \frac{B}{C - x^3},$$

with arbitrary constants A, B, C. Upon substitution into (7.16) and by the use of obvious additional transformations of translation this solution is reduced to the standard one: $\lambda = x^1/x^3$, $\mu = x^2/x^3$. These are invariants of the standard subgroup $H_\beta^{n-2} \subset G^{n+1}$ with the basis of operators

$$\xi_0 \cdot \partial = x^i\partial_i, \qquad \xi_j \cdot \partial = \partial_j \qquad (j = 4, \ldots, n). \tag{7.17}$$

In case (β), where system (7.5) was reduced to (7.13), the linear transformation of the space X is

$$\bar{x}^1 = D_i x^i, \qquad \bar{x}^2 = C_i x^i, \qquad \bar{x}^j = x^j \qquad (j = 3, \ldots, n).$$

As a result (7.13) transforms into the system

$$h_2 = h\lambda_1, \qquad \mu_1 = 0, \qquad \mu_2 = \lambda_1, \qquad \lambda_j = \mu_j = 0 \qquad (j = 3, \ldots, n),$$

from which it follows at once that $\lambda = \lambda(x^1, x^2)$, $\mu = \mu(x^1, x^2)$. But by means of transformations of the form (7.6), this solution is reduced to $\lambda = x^1$, $\mu = x^2$. So λ, μ are invariants of the standard subgroup $H_\alpha^{n-2} \subset G^{n+1}$ with the basis (7.15). ∎

The practical significance of this theorem is that the construction problem of invariant double waves is much simpler than that problem for double waves with defect $\delta > 0$. Thus non-reduced waves are especially interesting. In solving the existence problem, during the process of forming compatibility for the parameters of a double wave, it was necessary to add a *restriction on the reduction*, namely, that $N = 2n - 1$ independent equations are obtained. Analytically, this restriction leads to the requirement that the functions (7.1) must satisfy some additional equations, which must be added to the factor system E/G^{n+1}. This explains the peculiarity of the nonreduced double waves.

8. Example

Double waves for the system (6.2), which is equipollent to the wave equation (6.1), are considered. Here $2n - 1 = 5$. According to the theorem in Section 23.7, every time five equations of the form (7.4) are found from the compatibility conditions for a double-wave parameter, this wave will be an invariant solution of the second rank relative to the group G^4. Therefore, nonreduced double waves exist for the system (6.1). In particular, so-called functionally invariant solutions of a wave equation, to be analyzed, belong to this class.

The solution ϕ of equation (6.1) is said to be *functionally invariant* if for any function $F: R \to R$ the composition $F \circ \Phi$ is also a solution of this equation.

It is easy to establish that the solution ϕ is functionally invariant if and only if $\phi_x^2 + \phi_y^2 = \phi_t^2$, i.e., in the notation of Section 3.6 the relation $u^2 + v^2 = w^2$ is satisfied. *When applied to system* (6.1) *this means that a functionally invariant solution is a double wave* (*of a special kind*).

An analytic description of these double waves is simplest when w and θ are taken as parameters, where

$$u = w \cos \theta, \qquad v = w \sin \theta. \tag{8.1}$$

As the result of substitution of expressions (8.1) into system (6.1) the latter, after using some identities, is reduced to

$$\begin{aligned} w_x &= w_t \cos \theta - w\theta_y, \qquad & w_y &= w_t \sin \theta + w\theta_x, \\ \theta_x &= \theta_t \cos \theta, & \theta_y &= \theta_t \sin \theta. \end{aligned} \tag{8.2}$$

The compatibility conditions of equations (8.2) are identically satisfied. Thus here it is impossible to have five equations of the form (7.4) in the general case.

To find the general solution of system (8.2), the notation

$$\sigma = x \sin \theta - y \cos \theta, \qquad \tau = x \cos \theta + y \sin \theta + t \tag{8.3}$$

is used. With this the general solution is given by

$$\theta = f(\tau), \qquad w = \frac{g'(\tau)}{1 + \sigma f'(\tau)}, \tag{8.4}$$

where f and g are arbitrary functions and the derivative ∂_τ is denoted by the prime. The corresponding functionally invariant solution of the wave equation (6.1) is $\phi = g(\tau)$.

Note. The first-order system, equipollent to the wave equation in three-dimensional space ($n = 4$)

$$\phi_{xx} + \phi_{yy} + \phi_{zz} = \phi_{tt}, \tag{8.5}$$

found for the functions $u = \phi_x$, $v = \phi_y$, $w = \phi_z$, $p = \phi_t$ consists of seven equations. Thus in the system corresponding to (7.4) there will be $N = 7 = 2n - 1$ equations. So according to the theorem in Section 23.6 *any double wave of general character* (i.e. for which the functions v (7.1) are arbitrary) *reduces to an invariant solution.*

9. Gas Dynamics Equation

The system of equations for the flow of an adiabatic gas (Section 11.1) in two-dimensional space is considered, so that $X = R^3(x, y, t)$ and the velocity vector has the two components (u, v). If instead of the density ρ (see Section 11.1) the specific volume $\tau = 1/\rho$ is introduced, then the system is given in the form (partial derivatives with respect to the corresponding arguments are denoted by subscripts)

$$\begin{aligned} u_t + uu_x + vu_y + \tau p_x &= 0, \\ v_t + uv_x + vv_y + \tau p_y &= 0, \\ \tau_t + u\tau_x + v\tau_y - \tau(u_x + v_y) &= 0, \\ S_t + uS_x + vS_y &= 0, \end{aligned} \tag{9.1}$$

where the function $\tau = \tau(p, S)$ is assumed to be given such that $\tau_p \neq 0$.

The solution of system (9.1) is said to be *isentropic* if, in it, the entropy $S = \text{const}$ and *isobaric* if $p = \text{const}$.

It is easy to establish a well-known property of system (9.1): *Any non-isobaric simple wave is an isentropic solution.* The question arises, In what measure does this property transfer to double waves? The following theorem gives an answer, where the term "wave of general character" means that the function $\tau(p, S)$ is not subject to any special limitations (arbitrary).

Theorem. *A nonisobaric double wave of general character, having the invariance defect $\delta = 2$, is isentropic.*

Proof. Suppose the double wave is not isentropic. Then the variables p and S can be taken as parameters for the double wave of general character, i.e., it can be assumed that

$$u = u(p, S), \qquad v = v(p, S), \qquad \tau = \tau(p, S), \tag{9.2}$$

where the function $\tau(p, S)$ is prescribed beforehand. In this representation the statement of the theorem is equipollent to the fact that the double wave of general character certainly must be reducible.

To establish this fact, the following system of the form (7.2), arising from substituting (9.2) into (9.1),

$$\begin{aligned} u_p p_t + (\tau + uu_p)p_x + vu_p p_y &= 0, \\ v_p p_t + uv_p p_x + (\tau + vv_p)p_y &= 0, \\ \tau_p p_t + (u\tau_p - \tau u_p)p_x + (v\tau_p - \tau v_p)p_y &= \tau u_S S_x + \tau v_S S_y, \\ S_t + uS_x + vS_y &= 0, \end{aligned} \tag{9.3}$$

is considered. If it is assumed that the double wave cannot be reduced then, since $n = 3$ and $2n - 1 = 5$, the prohibition on reduction, which is dictated by the theorem in Section 23.7, comes into effect: Any compatibility condition of equation (9.3), having the form of a linear relation between the first-order derivatives of p and S, must be an algebraic consequence of equations (9.3) (*prohibition of a fifth equation*).

For example, the first two equations of (9.3) can be written as

$$p_x = Ap_t, \qquad p_y = Bp_t.$$

Their compatibility condition,

$$\partial_x(Bp_t) = \partial_y(Ap_t),$$

after simplification using equations (9.3) is found to be

$$(\tau u_{pS} - u_p \tau_S - u_p^2 u_S - u_p v_p v_S)S_x$$
$$-(\tau v_{pS} - v_p \tau_S - u_p v_p u_S - v_p^2 v_S)S_y = 0.$$

Prohibition of the fifth equation requires the coefficients of S_x and S_y to be zero, i.e., satisfaction of the conditions

$$\tau u_{pS} = u_p(\tau_S + u_p u_S + v_p v_S), \qquad \tau v_{pS} = v_p(\tau_S + u_p u_S + v_p v_S). \quad (9.4)$$

Two more compatibility conditions for (9.3), which contain the first and second derivatives of the function S, are found for the function p when S is given. From them, using the fourth equation of (9.3), all second derivatives of S can be excluded. As a result, only one nontrivial equation relating S_x and S_y remains. The prohibition of a fifth equation gives two more relations between the functions (9.2), one of which is not identically satisfied because of (9.4). But then the functions u, v, τ must satisfy a system of three differential equations and the function τ cannot be arbitrary. This means that the considered double wave is not a wave of general character. Thus for waves of general character the prohibition of a fifth equation cannot be satisfied; consequently it can be reduced. ■

One should note that there are double waves of general character with defect $\delta = 2$. Also, for some special functions $\tau(p, S)$ there are nonisobaric and nonisentropic double waves, for which $\delta = 2$. Double waves with defect $\delta < 2$ can be found as partially invariant solutions of rank $\rho = 2$ relative to one-parameter or two-parameter subgroups of the group G^4 (with the operator basis (23.3.1) when $n = 3$).

REFERENCES

The concept of a partially invariant solution was introduced for the first time in Ref. [4] in Chapter II and also was discussed in Ref. [1]. It was generalized to arbitrary manifolds in Ref. [2], where the term "defect of invariance" was suggested. Also there, the problem of reduction of partially invariant manifolds, which was also in Ref. [3], was proposed. The search for partially invariant solutions for concrete equations was realized in Refs. [4–9]. A systematic study of the traveling waves concept was begun in Ref. [10]. Functionally invariant solutions of wave equation were found in Ref. [11]. A complete proof of the theorem in Section 23.7, about the reduction of double waves, was given here for the first time.

[1] Овсянников Л. В. Ovsyannikov L. V. Partly invariant solutions of the equations addmitting a group. — Proc. 11-th Int. Congr. Appl. Mech., Munich (1964), 868—870.
[2] Овсянников Л. В. Частичная инвариантность. — ДАН СССР **186**, № 1 (1969), 22—25.
[3] Бытев В. О. К задаче о редукции. — В сб. «Динамика сплошной среды», вып. 5 (1970), Новосибирск, 146—148.
[4] Ибрагимов Н. Х. Обобщенные движения в римановых пространствах. — ДАН СССР **178**, № 1 (1968), 27—30.
[5] Ибрагимов Н. Х. Группы обобщенных движений. — ДАН СССР **187**, № 1 (1969), 25—28.
[6] Меньщиков В. М. Решения уравнений двумерной газовой динамики типа простых волн. — Ж. прикл. мех. и техн. физ. № 3 (1969), 129—134.
[7] Меньщиков В. М. К теории частично инвариантных решений дифференциальных уравнений. — В сб. «Динамика сплошной среды», вып. 11 (1972), Новосибирск, 82—93.
[8] Пухначев В. В. Неустановившиеся движения вязкой жидкости со свободной границей, описываемые частично инвариантными решениями уравнений Навье—Стокса. — В сб. «Динамика сплошной среды», вып. 10 (1972), Новосибирск, 125—137.
[9] Верещагина Л. И. Об одном решении уравнений пространственного нестационарного пограничного слоя. — Вестн. Ленингр. ун-та, вып. 2, № 7 (1974), 102—106.
[10] Яненко Н. Н. Бегущие волны системы квазилинейных уравнений. — ДАН СССР, **109**, № 1 (1956), 44—47.
[11] Смирнов В. И., Соболев С. Л. Новый метод решения плоской задачи упругих колебаний. — Труды сейсмологического института, вып. 20 (1932), 1—37.

CHAPTER

VII

Differential Invariants

The present chapter is devoted to the theory and applications of differential invariants of Lie transformation groups, which can be infinite. The concept of a differential invariant was already introduced in Section 17.6; an invariant of a continuous group is called by this term. Those differential equations, which admit the given group, are expressed by means of the differential invariants. Using them, such equations are constructed and classified, and a special transformation of a differential equation, the so-called group stratification, is also discussed.

The central theorem of this chapter is the theorem in Section 24.4 about the finiteness of bases for differential invariants of arbitrary Lie groups, and also its generalization in Section 24.8 for infinite groups. In the proposed variant Tresse's theorem is completely established in infinitesimal language by means of invariant differential operators.

An equation E is called *automorphic* relative to the given group G if all its solutions are situated on the orbit of one of them. In other words, from one given solution U it is possible to find all solutions by applying to U the transformations of the group G. Classical problems of the construction and classification of automorphic systems provide excellent examples of applications of the general theory, and these play an important role in problems of invariant transformations for differential equations.

The discussion about a group *stratification* of a given differential equation E, using the group G admitted by it, goes back to S. Lie.

This problem was completely solved at the beginning of this century and then was forgotten. Its solution is discussed here because of the interest that has developed in the past few years concerning the study of the differential equations of mathematical physics and mechanics by methods of group analysis. Also, this provides a good reason for the reader's attention to be focused on an underdeveloped field of study of differential invariant solutions.

§24. GENERAL THEORY

1. Preliminary Considerations

The results discussed in Chapters I, V and VI are connected with the concept of differential invariants of the group G^r of transformations of the space $Z = X \times Y$. According to Section 17.6 the invariants of the continuation groups $\underset{\chi}{G^r}$ for $\chi = 1, 2, \ldots$ are denoted by this term. The importance of this concept is emphasized by the fact that the differential equations, which admit the group G, are expressible by means of differential invariants (the theorem of representation in Section 18.7). This fact generates some new problems. They include, for example, the construction of differential equations which admit a group G^r given beforehand, and the problem must be solved on the basis of preliminary knowledge of the differential invariants of this group.

The present section is devoted to the study of the fuctional algebraic structure of the union of all differential structures of the given group G^r. The main problem is to determine the possibility of distinguishing some minimal quantity of differential invariants, through which it would be possible to express the rest using certain operations.

It is essential that differential invariants of a sufficiently high order always exist for any group G^r (see Section 17.6). With that every differential invariant of order χ, resulting from a χth continuation $\underset{\chi}{G^r}$ of the group G^r, is also an invariant of any continuation $\underset{\chi+\sigma}{G^r}$, when $\sigma > 0$. Thus for every $\chi \geq 0$ a universal differential invariant of order $\chi + 1$ is obtained by extension of a χth-order universal differential invariant. But this results in an extension of the representation space Z_χ up to $Z_{\chi+1}$.

Let $\zeta = (\xi, \eta)$ be a tangential mapping of the group $G^r(\zeta)$. In what follows, the universal invariant of the group $G^r(\zeta)$ is denoted by J and the universal invariant of the continuation group $\underset{\chi}{G^r}(\zeta) = G^r(\underset{\chi}{\zeta})$ is denoted by $\underset{\chi}{J}$ (with that, $\underset{0}{J} = J$). To find the dimensionality of the space $\overline{Z}_\chi$ of differential invariants whose order is not more than χ, the general ranks $r_\chi = r_*(\underset{\chi}{\zeta})$, which form a nondecreasing bounded sequence

$$r_0 \leq r_1 \leq \cdots \leq r_\chi \leq r_{\chi+1} \leq \cdots \leq r, \tag{1.1}$$

are introduced. Also, the dimensionality $v_\chi = \dim Z_\chi$ of the continuous space is found by a simple calculation to be

$$v_\chi = n + m\binom{n+\chi}{n} \qquad (\chi = 0, 1, 2, \ldots). \tag{1.2}$$

Consequently (see Section 17.3), the dimensionality of the space of invariants is equal to

$$\dim \bar{Z}_\chi = v_\chi - \mathrm{r}_\chi \qquad (\chi = 0, 1, 2, \ldots). \tag{1.3}$$

Because of the boundedness, the sequence (1.1) is certainly convergent, i.e., $r_\chi = r_{\chi+1} = r_{\chi+2} = \cdots$ for some $\chi \geq 0$. From general considerations based upon the fact that the number of essential parameters in the mapping $(z, a) \to h(z, a)$, generating the group $G^r(h)$, is equal to its order r (see Section 21.6), it is clear that the value of r_χ must be equal to r. A direct proof of this fact will follow.

Lemma. *If* $r_\chi = r_{\chi+1} = r_{\chi+2}$, *then* $r_\chi = r$.

Proof. The continuation vector fields $\underset{\chi}{\zeta}$ have the following structure (see Section 4.8):

$$\underset{\chi}{\zeta_\alpha} = (\xi_\alpha, \eta_\alpha, \underset{1}{\eta_\alpha}, \ldots, \underset{\chi}{\eta_\alpha}) \qquad (\alpha = 1, \ldots, \mathrm{r}), \tag{1.4}$$

where $\underset{k}{\eta_\alpha}$ is obtained from the continuation formulas

$$\underset{\sigma+1}{\eta_\alpha} = D\underset{\sigma}{\eta_\alpha} - \underset{\sigma+1}{y} \cdot D\xi_\alpha \qquad (\sigma = 0, 1, 2, \ldots), \tag{1.5}$$

where D is an operator of total differentiation.

Suppose $r_\chi = r_{\chi+1} < r$. The last inequality means that the vector fields $\underset{\chi+1}{\zeta_\alpha}$ are linearly related, and the number of independent linear relations is equal to $r - r_\chi$. Without loss of generality it is possible to assume that the linear relations have the form (index v assumes the values $1, \ldots, \mu = r_\chi$, and the index τ the values $1, \ldots, r - r_\chi$)

$$\underset{\chi}{\xi_{\mu+\tau}} = \phi_\tau^v \underset{\chi}{\zeta_v} \qquad (\tau = 1, \ldots, r - \mu),$$

with some mappings $\phi_\tau^v : Z_\chi \to R$. Because of (1.4) this means that the equalities

$$\xi_{\mu+\tau} = \phi_\tau^v \xi_v, \qquad \eta_{\mu+\tau} = \phi_\tau^v \eta_v, \qquad \underset{\sigma}{\eta_{\mu+\tau}} = \phi_\tau^v \underset{\sigma}{\eta_v} \qquad (\sigma = 1, \ldots, \chi) \tag{1.6}$$

hold. With the same coefficients ϕ_τ^v, the equations

$$\underset{\chi+1}{\eta_{\mu+\tau}} = \phi_\tau^v \underset{\chi+1}{\eta_v} \qquad (\tau = 1, \ldots, r - \mu)$$

must be satisfied, for otherwise $r_\chi < r_{\chi+1}$. By (1.5) and (1.6) these equalities reduce to

$$D\phi^\nu_\tau \cdot \underset{\sigma}{\eta_\nu} = \underset{\sigma+1}{y} \cdot D\phi^\nu_\tau \cdot \xi_\nu \qquad (\sigma = 0, 1, \ldots, \chi;\ \tau = 1, \ldots, r-\mu). \quad (1.7)$$

But for the mapping $\phi^\nu_\tau : Z_\chi \to R$,

$$D\phi^\nu_\tau = \underset{\chi}{D}\phi^\nu_\tau = \underset{\chi-1}{D}\,\phi^\nu_\tau + \underset{\chi+1}{y} \cdot \underset{w}{\partial}\phi^\nu_\tau,$$

where $w = \underset{\chi}{y}$. Now it can be seen that the vector $\underset{\chi+1}{y}$ is included only in equations (1.7),where it is explicitly displayed. Thus equations (1.7) can be easily decomposed relative to $\underset{\chi+1}{y}$, thereby generating the following system:

$$\begin{array}{lll} p^\nu_\tau \cdot \xi_\nu = 0 & q^\nu_\tau \cdot \xi_\nu = A_\tau \partial^\xi_D & \\ p^\nu_\tau \cdot \underset{\sigma}{\eta_\nu} = 0 & q^\nu_\tau \cdot \underset{\sigma}{\eta_\nu} = \underset{\sigma+1}{y} \cdot A_\tau & (\sigma = 0, 1, \ldots, \chi - 1), \quad (1.8) \\ p^\nu_\tau \cdot \underset{\chi}{\eta_\nu} = A_\tau \delta^\nu_w, & q^\nu_\tau \cdot \underset{\chi}{\eta_\nu} = 0 & \end{array}$$

where $p^\nu_\tau = \partial_w \phi^\nu_\tau$ and $q^\nu_\tau = \underset{\chi-1}{D}\,\phi^\nu_\tau$ with some mappings $A_\tau : Z_\chi \to R$. The symbols δ^ξ_D and δ^v_w denote the identity mappings of the spaces R^n and Y_χ, respectively.

So the system of relations (1.8) is a consequence of the fact that $r_\chi = r_{\chi+1} < r$. If also $r_{\chi+1} = r_{\chi+2}$, then an analogous system of relations, obtained by changing χ to $\chi + 1$ and the mappings A_τ to some mappings B_τ, results. But then from (1.8) it follows that $A_\tau = 0$ ($\tau = 1, \ldots, r - \mu$), whence the system (1.8) is homogeneous with respect to p^ν_τ and q^ν_τ. Since $r_*(\underset{\chi}{\zeta}) = r_\chi = \mu$, the unique solution of this system is the zero solution, i.e., $p^\nu_\tau = q^\nu_\tau = 0$ or

$$\partial_w \phi^\nu_\tau = 0, \qquad \underset{\chi-1}{D}\,\phi^\nu_\tau = 0 \qquad (w = \underset{\chi}{y}).$$

From these results it follows that

$$\partial_x \phi^\nu_\tau = 0, \qquad \partial_y \phi^\nu_\tau = 0, \qquad \underset{\sigma}{\partial_y}\phi^\nu_\tau = 0 \qquad (\sigma = 1, \ldots, \chi);$$

i.e., all quantities ϕ^ν_τ are constants. But this existence of linear relations, with constant coefficients, contradicts the linear independence of the vector fields $\underset{\infty}{\zeta_\alpha}$. ■

2. Invariant Differentiation

Vector fields δ, over the continuous space Z_∞, generated with the aid of linear differential operators $\delta \cdot \partial$ are considered. Operators of this kind, distinguished by the following definition, are put into correspondence with every group $G^r(\zeta)$.

The operator $\delta \cdot \partial$ is said to be an *operator of invariant differentiation* [*of the group* $G^r(\zeta)$] if for any differential invariant F of the group $G^r(\zeta)$ the mapping $(\delta \cdot \partial)F$ is also a differential invariant of this group.

The union of all scalar invariants of the group $G^r(\zeta)$ (including the constants $c \in R$) generates an algebraic field. Thus the indicated union will be called a *field of invariants of the group* $G^r(\zeta)$. The following lemma is established.

Lemma. *The set of all operators of invariant differentiation of the group* $G^r(\zeta)$ *is a Lie algebra over the field of invariants of this group.*

Proof. Evidently the operators of invariant differentiation generate a vector space over the field of invariants of the group $G^r(\zeta)$. Also, if two such operators $\delta_1 \cdot \partial$ and $\delta_2 \cdot \partial$ are given, then for any differential invariant F of the group $G^r(\zeta)$ these operators give the two differential invariants $(\delta_1 \cdot \partial)F$ and $(\delta_2 \cdot \partial)F$. Thus

$$(\delta_1 \cdot \partial) \circ (\delta_2 \cdot \partial)F - (\delta_2 \cdot \partial) \circ (\delta_1 \cdot \partial)F = ([\delta_1, \delta_2] \cdot \partial)F$$

is also a differential invariant of the group $G^r(\zeta)$. ∎

Note. *Any operator* $\delta \cdot \partial$ *which commutes with every operator* $\underset{\infty}{\zeta} \cdot \partial$ *of the Lie algebra of the group* $G^r(\underset{\infty}{\zeta})$ (*i.e.*, $[\underset{\infty}{\zeta}, \delta] = 0$) *is an operator of invariant differentiation.*

Indeed, if

$$(\underset{\infty}{\zeta} \cdot \partial) \circ (\delta \cdot \partial) = (\delta \cdot \partial) \circ (\underset{\infty}{\zeta} \cdot \partial)$$

then $(\underset{\infty}{\zeta} \cdot \partial)((\delta \cdot \partial)F) = 0$ for any differential invariant F. Consequently, $(\delta \cdot \partial)F$ is an invariant of the group $G^r(\underset{\infty}{\zeta})$ because of criterion (17.1.3).

Theorem. *For any group* $G^r(\zeta)$ *of transformations of the space* $Z = X \times Y$ *there are* $n = \dim X$ *operators of invariant differentiation which are linearly independent over the field of invariants of this group.*

Proof. Let the number $k \geq 0$ be chosen such that $r_k = r$. The operator of total differentiation D (see Section 4.3) is considered, and it is asserted that there exist nonzero mappings $\lambda: Z_k \to R^n$ such that the operator $\lambda \cdot D$ commutes with the operator $\underset{k}{\zeta} \cdot \partial$. Evaluating $[\lambda \cdot D, \underset{k}{\zeta} \cdot \partial]$ by means of the standard commutator (4.9.3) gives

$$(\lambda \cdot D) \circ (\underset{k}{\zeta} \cdot \partial) - (\underset{k}{\zeta} \cdot \partial) \circ (\lambda \cdot D) = \lambda \cdot (D \circ (\underset{k}{\zeta} \cdot \partial) - (\underset{k}{\zeta} \cdot \partial) \circ D) - ((\underset{k}{\zeta} \cdot \partial)\lambda) \cdot D$$
$$= ((\lambda \cdot D)\xi - (\underset{k}{\zeta} \cdot \partial)\lambda) \cdot D.$$

Because of this relation the mapping λ possesses the required property if and only if

$$(\underset{k}{\zeta_\alpha} \cdot \partial)\lambda = (\lambda \cdot D)\xi_\alpha \qquad (\alpha = 1, \ldots, r) \tag{2.1}$$

for any basis operator $\zeta_\alpha \cdot \partial$ of the Lie algebra $L^r(\zeta)$. It remains to prove the existence of a solution of this system of equations for the unknown λ. Here it is convenient to search for λ in an implicit form, specifying it by the equation

$$\phi(\underset{k}{z}, \lambda) = 0, \tag{2.2}$$

where the mapping $\phi: Z_k \times R^n \to R^n$ satisfies the condition $r_*(\partial_\lambda \phi) = n$. Application of the operator $\underset{k}{\zeta_\alpha} \cdot \partial$ to the identity $\phi(\underset{k}{z}, \lambda(\underset{k}{z})) = 0$ and use of the previous condition establish that the system of equations (2.1) is equipollent to the system

$$(\underset{k}{\zeta_\alpha} \cdot \partial + (\lambda \cdot D)\xi_\alpha \cdot \partial_\lambda)\phi = 0 \qquad (\alpha = 1, \ldots, r). \tag{2.3}$$

In other words the mapping ϕ must be an invariant of the family $\{\bar{\zeta}_\alpha\}$ of vector fields $\bar{\zeta}_\alpha = (\underset{k}{\zeta_\alpha}, (\lambda \cdot D)\xi_\alpha)$ over the space $Z_k \times R^n$.

For any vector field $\zeta = (\xi, \eta)$ it is convenient to write $\xi = \zeta_X$. Then by direct calculation the equation

$$[\bar{\zeta}_\alpha, \bar{\zeta}_\beta] = ([\underset{k}{\zeta_\alpha}, \underset{k}{\zeta_\beta}], (\lambda \cdot D)[\zeta_\alpha, \zeta_\beta]_X),$$

showing that the family $\{\bar{\zeta}_\alpha\}$ is complete, is easily verified.

Because $r_k = r$, it also follows that $r_*(\{\bar{\zeta}_\alpha\}) = r$. Thus the universal invariant $\bar{J}$ of the family of vector fields $\{\bar{\zeta}_\alpha\}$ (whose existence is guaranteed by the theorem in Section 17.2) must be a mapping in the space $\bar{Z}_k \times R^n$ and can be represented in the form $\bar{J} = (\underset{k}{J}, J')$, where $J': Z_k \times R^n \to R^n$ and $r_*(\partial_\lambda J') = n$. Using J', equation (2.2) can be given in the form

$$J'(\underset{k}{z}, \lambda) = c, \tag{2.4}$$

with an arbitrary vector $c \in R^n$. According to the construction every vector λ, determined from (2.4) for some c, is a solution of system (2.3) and gives the operator $\lambda \cdot D$, which commutes with any operator $\underset{k}{\zeta} \cdot \partial$, i.e., according to the previous note it is an operator of invariant differentiation. Because $r_*(\partial_\lambda J') = n$ it is always possible to choose n such vectors c, for which the operators $\lambda \cdot D$ obtained from (2.4) will be linearly independent over the field of invariants. ■

It is useful to note that if the vectors $\lambda_1, \ldots, \lambda_n$ are independent solutions of system (2.1) and $\Phi^1, \ldots, \Phi^n$ are any differential invariants $Z_k \to R$, then the vector

$$\lambda = \Phi^i \lambda_i \tag{2.5}$$

is also a solution of the system. Moreover, *because of the linearity of systems* (2.1), *equation* (2.5) *is a representation of its general solution.*

3. Examples

In the following examples $X = R^2(x^1, x^2)$, $Y = R(y)$. The notation to be used is $\partial_1 = \partial_{x^1}$, $\partial_2 = \partial_{x^2}$ and also $y_1 = \partial_1 y$, $y_2 = \partial_2 y$, $y_{11} = \partial_1^2 y$, and so on. The operator of total differentiation has two components $D = (D_1, D_2)$, where

$$D_i = \partial_i + y_i \partial_y + y_{i1}\partial_{y_1} + y_{i2}\partial_{y_2} + \cdots \qquad (i = 1, 2).$$

The operators of invariant differentiation are to be found in the form

$$\delta \cdot D = \lambda D_1 + \mu D_2 .$$

(α) The group G^4 whose basis of the Lie algebra is the operators

$$\zeta_1 \cdot \partial = \partial_1, \qquad \zeta_2 \cdot \partial = \partial_2,$$
$$\zeta_3 \cdot \partial = x^1\partial_1 + x^2\partial_2, \qquad \zeta_4 \cdot \partial = x^2\partial_1 - x^1\partial_2 .$$

Here $r_0 = 2$ and $v_0 = n + m = 3$, whence the group has one scalar invariant, namely, $J = y$. After the first continuation, the group $\underset{1}{G^4}$, for which $r_1 = 4$, is obtained. Since $v_1 = 5$, $\underset{1}{G^4}$ has the same invariant $\underset{1}{J} = y$ and there are no others (independent of y). The corresponding system of operators (2.3) is (here $\bar\partial = (\partial, \partial_\lambda, \partial_\mu)$)

$$\bar\zeta_1 \cdot \bar\partial = \partial_1, \qquad \bar\zeta_2 \cdot \bar\partial = \partial_2,$$
$$\bar\zeta_3 \cdot \bar\partial = \zeta_3 \cdot \partial - y_1\partial_{y_1} - y_2\partial_{y_2} + \lambda\partial_\lambda + \mu\partial_\mu,$$
$$\bar\zeta_4 \cdot \bar\partial = \zeta_4 \cdot \partial - y_2\partial_{y_1} + y_1\partial_{y_2} + \mu\partial_\lambda - \lambda\partial_\mu .$$

The universal invariant of this operator system can be taken in the form $\bar{J} = (y, p^2, \phi)$, where

$$p^2 = (\lambda^2 + \mu^2)(y_1^2 + y_2^2), \qquad \phi = \arctan\frac{\mu}{\lambda} + \arctan\frac{y_1}{y_2}.$$

From this λ and μ can easily be found as

$$\lambda = pq^{-1}\cos\left(\phi - \arctan\frac{y_1}{y_2}\right), \qquad \mu = pq^{-1}\sin\left(\phi - \arctan\frac{y_1}{y_2}\right),$$

where

$$q = \sqrt{y_1^2 + y_2^2}.$$

Here the set of operators of invariant differentiation is found by taking p and ϕ as any functions of y. In particular, the choice $(p, \phi) = (1, 0)$ or $(1, \pi/2)$ gives two independent operators of invariant differentiation

$$\delta_1 \cdot \partial = q^{-2}(y_1 D_1 + y_2 D_2), \qquad \delta_2 \cdot \partial = q^{-2}(y_2 D_1 - y_1 D_2). \tag{3.1}$$

(β) The group G^4 whose basis of the Lie algebra is the operators

$$\zeta_1 \cdot \partial = \partial_1, \qquad \zeta_2 \cdot \partial = \partial_2, \qquad \zeta_3 \cdot \partial = \partial_y,$$
$$\zeta_4 \cdot \partial = x^1\partial_1 + x^2\partial_2 + y\partial_y.$$

Here $r_0 = 3$ and after the first continuation $r_1 = 3$ is also found. The universal invariant of the group $\underset{1}{G^4}$ is $\underset{1}{J} = (y_1, y_2)$. The corresponding system (2.3), constructed for the first continuation, gives only one scalar invariant λ/μ. After the second continuation it is found that $r_2 = 4$, $v_2 = 8$, and the universal invariant of the group $\underset{2}{G^4}$ is easily found to be $\underset{2}{J} = (y_1, y_2, y_{12}/y_{11}, y_{22}/y_{11})$. Here the corresponding system of operators (2.3) is

$$\bar{\zeta}_1 \cdot \bar{\partial} = \partial_1, \qquad \bar{\zeta}_2 \cdot \bar{\partial} = \partial_2, \qquad \bar{\zeta}_3 \cdot \bar{\partial} = \partial_y,$$
$$\bar{\zeta}_4 \cdot \bar{\partial} = \zeta_4 \cdot \partial - y_{11}\partial_{y_{11}} - y_{12}\partial_{y_{12}} - y_{22}\partial_{y_{22}} + \lambda\partial_\lambda + \mu\partial_\mu.$$

The component J' [see (2.4)] of the universal invariant for this system of operators can be taken in the form $J' = (\lambda y_{11}, \mu y_{11})$. If it is assumed that $\lambda y_{11} = p$, $\mu y_{11} = q$, then the choice $(p, q) = (1, 0)$ or $(0, 1)$ gives two independent operators of invariant differentiation

$$\delta_1 \cdot \partial = y_{11}^{-1} D_1, \qquad \delta_2 \cdot \partial = y_{11}^{-1} D_2. \tag{3.2}$$

4. Basis of Invariants

Let $\underset{\infty}{J}: Z_\infty \to \bar{Z}_\infty$ be a universal invariant of the infinitely continued group $\underset{\infty}{G^r}$. The space $\bar{Z}_\infty$ can be given in the product form $\bar{Z}_\infty = \bar{Z} \times \bar{Y}_1 \times$

$\overline{Y}_2 \times \cdots$, where $\overline{Y}_k$ is a space of differential invariants of order $k > 0$ (the space of invariants of zero order is $\overline{Z}$). According to this representation the universal $\underset{\infty}{J}$ can be given by its components from every multiplier as

$$\underset{\infty}{J} = (J, J'_1, J'_2, \ldots). \tag{4.1}$$

Thereby J'_k *is a universal differential invariant of order k in the following sense: For any differential invariant* $F: Z_k \to P$ *of order k there exists a mapping* $\Phi: \overline{Z}_k = \overline{Z}_{k-1} \times \overline{Y}_k \to P$ *such that with the partial mapping* $\Phi_M: \overline{Y}_k \to P$ *(where* $M = \underset{k-1}{J}$*) the representation* $F = \Phi_M \circ J'_k$ *holds.*

Because of the lemma in Section 24.1 there must exist a k such that $\overline{Y}_k = Y_k$. Thus for a sufficiently large k, the universal differential invariant J'_k can be considered as a mapping $Z_k \to Y_k$, and if it is assumed that $w = \underset{k}{y}$, then $r_*(\partial_w J'_k) = \dim Y_k = v_k - v_{k-1}$ (see (1.2)). Here "sufficiently large k" means any $k > \chi$, where χ is the least number for which $r_\chi = r$ in the sequence (1.1).

Let λ denote the set $\{\lambda_i\}$ of linearly independent vector fields $\lambda_i: Z_\chi \to R^n$ $(i = 1, \ldots, n)$ over the field of invariants, for which $\lambda_i \cdot D$ are the operators of invariant differentiation (existence of such a set is guaranteed by the theorem in Section 24.2), and let $\lambda \cdot D = \{\lambda_i \cdot D\}$.

Lemma. *If* $k > \chi$ *and* J'_k *is a universal differential invariant of order k, then* $(\lambda \cdot D)J'_k$ *is a universal differential invariant of order* $k + 1$.

Proof. From the theorem in Section 24.2 it follows that $(\lambda \cdot D)J'_k$ is an invariant. Also, since λ only depends on a vector of the space Z_χ and D operates from Z_k into Z_{k+1}, this invariant has order $k + 1$. To establish its universality it is sufficient to prove that

$$r_*(\partial_v((\lambda \cdot D)J'_k)) = \dim Y_{k+1} \qquad (v = \underset{k+1}{y}), \tag{4.2}$$

which means that the image of the mapping $(\lambda \cdot D)J'_k: Z_{k+1} \to Y_{k+1}$ contains an open set $U \subset Y_{k+1}$. Indeed, on U it is possible to express the vector $\underset{k+1}{y}$ through $(\lambda \cdot D)J'_k$, and from that, also any invariant of order $k + 1$ (using invariants of order $\leq k$).

To establish equation (4.2) it is useful to note that from the viewpoint of J'_k it is always possible to change the operator $\lambda \cdot D$ to

$$\lambda \cdot \underset{k}{D} = \lambda \cdot \underset{k-1}{D} + \lambda \cdot \underset{k+1}{y} \cdot \partial_w \qquad (w = \underset{k}{y}).$$

Thus

$$(\lambda \cdot D)J'_k = (\lambda \cdot \underset{k-1}{D})J'_k + \lambda \cdot \underset{k+1}{y} \cdot \partial_w J'_k,$$

and if we write $v = \underset{k+1}{y}$, then for any vector $a \in Y_{k+1}$ it is found that

$$\partial_v((\lambda \cdot D)J'_k)\langle a\rangle = \lambda \cdot a \cdot \partial_w J'_k.$$

Consideration of the equation $\lambda \cdot a \cdot \partial_w J'_k = b$, as an equation for the vector a, shows that it has a unique solution for any $b \in Y_{k+1}$ because of the invertibility of the linear mappings $\lambda(z)\colon R^n \to R^n$ and $\partial_w J'_k(\underset{\chi}{z})\colon Y_k \to Y_k$. This solution is $a = \lambda^{-1} \cdot b \cdot (\partial_w J'_k)^{-1}$. Thus the linear mapping

$$\partial_v((\lambda \cdot D)J'_k)(z_k)\colon Y_{k+1} \to Y_{k+1}$$

is also invertible, from which equation (4.2) follows. ■

In Section 3.6 a *functional basis* of invariants (of null order) has already been discussed. Now this concept can be extended to the union of all differential invariants of the given group. Clearly, to obtain a finite basis here it is impossible to restrict attention only to a functional operation, i.e., to a transformation of the invariant $F\colon Z_k \to P$, using the mapping $\Phi\colon P \to Q$, to the invariant $\Phi \circ F$, because then the order of the invariant is preserved. But by using the operation of invariant differentiation the order of invariants can be increased, and, therefore, it appears to be sufficient to determine the existence of such bases.

Theorem. *For any group $G^r(\zeta)$ there is a finite* basis of differential invariants, *i.e., there exists a finite set of scalar differential invariants such that any differential invariant of this group is obtained from the invariants of the basis with a finite number of functional operations and invariant differentiations.*

Proof. As before let χ be the least number for which $r_\chi = r$ in the sequence (1.1). Then the universal differential $\underset{\chi+1}{J}$ consists of a finite set of scalar invariants of order $\leq \chi + 1$, and these invariants are functionally independent. It is established using $\underset{\chi+1}{J}$ that it is possible to construct a functional basis of invariants (i.e., the universal invariants J'_k) of any order $k > \chi$. This follows from the previous lemma and is proved by induction on k: As the result of applying the operators of invariant differentiation $\lambda \cdot D$ to the universal invariant $J'_{\chi+1}$ the universal invariant $J'_{\chi+2}$ is obtained, and so on. But then for any $k \geq 0$ the universal invariant of the kth continuation $\underset{k}{J}$ is formed. The remaining differential operators of order not more than k are expressed through functional operations by means of the invariant $\underset{k}{J}$ because of its universality. ■

Consequence. *The basis of differential invariants of the group $G^r(\zeta)$ is contained in its universal differential invariant $\underset{\chi+1}{J}$, where χ is the least number for* *which* $r_\chi = r$ *in the sequence* (1.1). *In particular, if* $r_0 = r$, *then the basis is contained in* $\underset{1}{J}$.

Naturally the question arises, Will the basis of differential invariants not already be contained in J_χ? The following examples show that the answer is sometimes positive and sometimes negative. In the general case this situation has apparently not been studied.

5. Examples of Bases

Example (α) in Section 24.3 had $\chi = 1$. If the operators of invariant differentiation (3.1) are applied to the universal invariant $\underset{1}{J} = y$, then $(\delta_1 \cdot \partial)y = 1$ and $(\delta_2 \cdot \partial)y = 0$ are found, i.e., new invariants will not be found. Thus the basis only gives the universal $\underset{2}{J}$. But here $\dim \bar{Z}_2 = 4$, so $\underset{2}{J}$ has four scalar components, which can be taken as

$$y, \qquad q^{-2}y_{11}, \qquad q^{-2}y_{22}, \qquad y_{12}/(y_{11} + y_{22}) - \arctan(y_1/y_2). \tag{5.1}$$

Example (β) in Section 24.3 had $\chi = 2$. Here the universal invariant has the components

$$J^1 = y_1, \qquad J^2 = y_2, \qquad J^3 = y_{12}/y_{11}, \qquad J^4 = y_{22}/y_{11}. \tag{5.2}$$

As a result of applying the operators (3.2) to J^1 and J^2 the invariants J^3 and J^4 are obtained. Using the same operators on J^3 and J^4, we obtain

$$\begin{aligned}
(\delta_1 \cdot \partial)J^3 &= y_{11}^{-3}(y_{11}y_{112} - y_{12}y_{111}),\\
(\delta_1 \cdot \partial)J^4 &= y_{11}^{-3}(y_{11}y_{122} - y_{22}y_{111}),\\
(\delta_2 \cdot \partial)J^3 &= y_{11}^{-3}(y_{11}y_{122} - y_{12}y_{112}),\\
(\delta_2 \cdot \partial)J^4 &= y_{11}^{-3}(y_{11}y_{222} - y_{22}y_{112}),
\end{aligned}$$

and it is easy to verify that all the components y_{ijk} of the vector y_3 can be determined from here, i.e., the resulting invariants of the third order are independent. Consequently, in this example, the basis of differential invariants coincides with $\underset{1}{J} = (y_1, y_2)$.

6. Determining Equations

The basis of differential invariants can be used to construct the determining equations of the Lie algebra for the given group G^r. They were already discussed in Section 5.4 in connection with finding the group admitted by the given system of differential equations. The abstract system of determining equations is called an *AS*-system. Below, the problem of determining an *AS*-system for a given group (or a Lie algebra of operators) is considered.

Let the group G^r and any of its differential invariants F be given. With any vector field $\zeta: Z \to Z$ the equation (system of equations)

$$(\underset{\infty}{\zeta} \cdot \partial)F = 0 \tag{6.1}$$

is generated. The criterion of the theorem in Section 17.1 states that if a vector field from the Lie algebra L^r of operators of the group G^r is taken as ζ, then equation (6.1) is an identity. Thus, if equation (6.1) is considered as an equation relative to ζ (with the given F), then any operator from L^r is its solution. The question arises, Does equation (6.1) have "extraneous" solutions, i.e., are there vector fields which satisfy equation (6.1) but are not contained in L^r? As it turns out, if the given invariant F contains a basis of the invariants of the group G^r, then the answer is negative, i.e., only vector fields from the Lie algebra L^r can satisfy equation (6.1).

Lemma. *If the invariant F of the group $\underset{\infty}{G^r}$ contains a basis of the differential invariants, then any solution of the system* (6.1) *is contained in the Lie algebra L^r of operators of the group G^r.*

Proof. It is useful to note that if F contains a basis of the differential invariants of the group G^r, then regardless of the solution ζ of system (6.1), for any differential invariant Φ of the group G^r the identity $(\underset{\infty}{\zeta} \cdot \partial)\Phi = 0$ holds.

From Sections 7.12, 7.13 it follows that the solution set of system (6.1) is the Lie algebra L', where, because of previous considerations, $L^r \subset L'$. Let $r'_0 \leq r'_1 \leq \cdots$ be a sequence of ranks (1.1) for the family of vector fields which generate L'. If $L^r \neq L'$, then there exists a k for which $r_k < r'_k$. This inequality means that the universal invariant $\underset{k}{J}$ is essentially larger than $\underset{k}{J'}$, i.e., there is a kth-order differential invariant F_k of the Lie algebra L^r which is not a differential invariant of the Lie algebra L^1. But then equality (6.1) with $F = F_k$ is not possible even for one operator $\zeta \cdot \partial \in L'$, for otherwise F_k would be an invariant of $\underset{k}{L'}$ because of the criterion in Section 17.1. But this fact contradicts the note made at the beginning of the proof. ■

Consequence. *A basis of the differential invariants of the group G^r uniquely defines the group.*

Proof. If in (6.1) the basis of differential invariants is taken as F, then, because of the preceding lemma, the set of solutions of equations (6.1) is exactly the Lie algebra L^r of operators of the group G^r, and it is uniquely determined by it (theorem in Section 16.6). ■

For example, for the group G^4 of Example (β) in Section 24.3, the basis of the invariants, as was determined in Section 24.5, is $\underset{1}{J} = (y_1, y_2)$, and the system (6.1) consists of two equations:

$$(\underset{1}{\zeta} \cdot \partial)y_1 = 0, \qquad (\underset{1}{\zeta} \cdot \partial)y_2 = 0.$$

Since

$$\underset{1}{\zeta} \cdot \partial = \xi^1\partial_1 + \xi^2\partial_2 + \eta\partial_y + \eta^1\partial_{y_1} + \eta^2\partial_{y_2},$$

the system really consists of the equations $\eta^1 = 0$, $\eta^2 = 0$. By calculating the additional components η^1, η^2 (first continuation) by formulas (4.8.4) and taking into account that the components of the operator $\zeta \cdot \partial$ do not depend on the vector $\underset{1}{y} = (y_1, y_2)$ (decomposition with respect to the variables y_1, y_2), the following *AS*-system, determining the group G^4,

$$\xi^1_2 = \xi^1_y = \xi^2_1 = \xi^2_y = \eta_1 = \eta_2 = 0, \qquad \xi^1_1 = \xi^2_2 = \eta_y, \tag{6.2}$$

is obtained. Here the derivatives $\partial_1, \partial_2, \partial_y$ are denoted by subscripts.

But it is useful to note that the construction of the determining equations of the given Lie algebra L^r, by means of a basis of the differential invariants, is not efficient in practice. There is a simpler method, which consists of *the exclusion of parameters from the representation by means of the basis operators of the Lie algebra L^r general operator*. This method is realized as follows. Let $\{\zeta_\alpha\}$ be a basis of L^r. The representation of an arbitrary vector $a = (a^1, \ldots, a^r) \in R^r$ is taken to be

$$\zeta = a^\alpha \zeta_\alpha, \tag{6.3}$$

and from it, and the results obtained by its differentiation, all the parameters a^α ($\alpha = 1, \ldots, r$) are determined; then the parameters a^α are excluded by setting all the first-order derivatives with respect to z equal to zero. (This guarantees the independence of a from z.) The resulting equations form the system of determining equations for the Lie algebra L^r.

For example, if for the group G^4 from Example (β) in Section 24.3 it is assumed that $\zeta = (\xi^1, \xi^2, \eta)$, then the representation (6.3) will be

$$\xi^1 = a^1 + a^4x^1, \qquad \xi^2 = a^2 + a^4x^2, \qquad \eta = a^3 + a^4y.$$

As the result of applying the described method, the system of determining equation (6.2) is exactly obtained.

7. Infinite Lie Groups

Development of the theory for differential invariants for so-called *infinite Lie groups* (or infinite Lie algebras of operators) has practical interest.

This is because the infinite Lie groups naturally appear as the full groups admitted by differential equations (see Section 16.12 and also the examples in Sections 5.9, 6.8, 7.14, 8.3, 9.5, and 10.4). In what follows, the infinite Lie group will be denoted by the symbol G^∞ and the corresponding Lie algebra of operators by the symbol L^∞.

One general way of describing the Lie algebra L^∞ consists in giving the system of abstract determining equations. According to the definition given in Section 7.15, the system of differential equations with the unknown mapping $\zeta: Z \to Z$ is called an AS-system if it is linear and homogeneous and if, together with its two solutions ζ_1 and ζ_2, the commutator $[\zeta_1, \zeta_2]$ is also a solution of the system.

Generally speaking, the general solution of an AS-system contains some arbitrary functions. To predict the number and structure of these functions is not simple. It can only be confirmed that the number of independent arbitrary functions, which are mappings $\phi: Z \to R$, must be less than $\nu_0 = \dim Z$ (otherwise ζ would be an arbitrary mapping $Z \to Z$ and the AS-system would be empty). So it is expedient to develop a theory of differential invariants for the Lie algebra L^∞ directly from the basis of the AS-systems of the determining equations.

In what follows an essential role is played by the already determined representation of the AS-system's differential structure. First of all, this system consists of a finite number of equations, and without loss of generality it can be considered as *passive*. The passive property means that any conditions of compatibility of the equations of the AS-system are identically satisfied because only equations of the AS-system may be differentiated. The existence of such a differential structure of the AS-system is guaranteed by known results of the general theory of differential equations.

Further, a passive AS-system will be called *reduced* if all its equations are solved relative to some derivatives such that these derivatives, called *fundamental*, are expressed as functions of the independent variables and other derivatives and the following condition is satisfied. None of the fundamental derivatives and derivatives of them are contained in the right-hand sides of the equations in the reduced system. Also, any derivative on the right-hand side of any equation has order not more than the order of the fundamental derivative, which is on the left-hand side of the same equation.

Evidently, the assumption that the AS-system is reduced is not a limitation since the reduction of a passive AS-system can always be accomplished.

Let the AS-system, which is passive and reduced, be given; it will merely be denoted by S. All derivatives (of any order) of the fundamental derivatives, which are on the left-hand side of the equation S, are also called *fundamental derivatives*. All other derivatives are called *parametric*. The following important fact applying to S, which follows from the general theory of integration of passive systems, is established here:

Lemma. *Any fundamental derivative of a passive reduced system S is expressed linearly and homogeneously through parametric derivatives.*

Proof. All the equations of the system S are separable into classes, where the class S^p of index p generates those equations in which the order of the fundamental derivatives is equal to p.

This separation gives the representation

$$S = S^{p_1} \cup S^{p_2} \cup \cdots \cup S^{p_l}, \tag{7.1}$$

where $p_1 < p_2 < \cdots < p_l$. If $p_{k+1} > p_k + 1$, then there is an omission between the classes S^{p_k} and $S^{p_{k+1}}$. These omissions can be filled in consecutive order, beginning with S^{p_1}, using the following operation of increasing the class.

Let the first omission be between S^{p_k} and $S^{p_{k+1}}$, i.e., $p_{i+1} = p_i + 1$ if $1 \le i \le k - 1$ and $p_{k+1} > p_k + 1$. All the equations of class S^{p_k} are differentiated once with respect to all the independent variables and all the fundamental derivatives of order less than $p_k + 1$, which appear on the right-hand sides, are replaced by their expressions from the equations of the orders S^{p_i} $(1 \le i \le k)$. As a result a system of linear equations for the fundamental derivatives of order p_{k+1} arises. None of the relations between the parametric derivatives can be a consequence of this system because of the passiveness of the system S. Thus all the fundamental derivatives of order $p_k + 1$ are expressed linearly in terms of the parametric ones, which gives the class S^{p_k+1}.

If in representation (7.1) all omissions are already filled, then the operation of increasing the class can be applied to the higher class S^{p_l}. This gives the class S^{p_l+1} whose addition to the system S leads to the continuation system S_1. Finally, as a result of an infinite number of repetitions of the continuation operation, the system S_∞, which gives the required expression for all the fundamental derivatives, is obtained. ■

8. Invariants of Infinite Groups

The problem of the existence of differential invariants of the groups G^∞ is solved, as was done for the groups G^r, by comparison between the dimensionality v_k of the continuation space Z_k and the common rank r_k of the continuation mapping $\underset{k}{\zeta}$. To find the rank r_k one should note that, in the case of the group G^∞, it is determined not by the number of linearly independent rows in the matrix representation of the mapping $\underset{k}{\zeta}$ (in a finite-dimensional case the rows were numbered with the index $\alpha = 1, \ldots, r$), as this number here is infinite, but with the number of linearly independent columns, which correspond to all possible different derivatives of the components of the vector ζ. But these derivatives are connected by the system S_∞ of determining equations, because of which all the fundamental derivatives can be eliminated by replacing them with expressions in terms of the parametric derivatives. As the latter are independent, the common rank $r_k = r_*(\underset{k}{\zeta})$ of the group G^∞ is equal to the number of parametric derivatives of order $\leq k$.

In the general case, as k increases, the number of parametric derivatives of order $\leq k$ increases faster than the dimensionality ν_k of the continuation space Z_k. Thus for a sufficiently large k there cannot be differential invariants. But it should be noted that if for some k the difference $\nu_k - r_k > 0$, then necessarily $\nu_{k+1} - r_{k+1} \geq \nu_k - r_k$. This follows from the facts that the difference $\nu_k - r_k$ is equal to the number of scalar invariants of order not more than k and that the universal invariant $\underset{k+1}{J}$ contains the universal invariant J_k. Thus the mentioned situation, when $r_k > \nu_k$ for a sufficiently large k, can be realized only if $r_0 \geq \nu_0$. From these considerations it follows that the criterion for the existence of an infinite number of differential invariants of the group G^∞ is expressed by

$$\lim_{k \to \infty} (\nu_k - r_k) = \infty. \tag{8.1}$$

It should be noted that the fast growth of the number r_k is mainly caused by the fact that the unknown components in AS-systems do not depend on the $n + m = \dim Z$ independent variables. At the same time the growth of the number ν_k is mainly determined by the number $n = \dim X$ of the independent point coordinates of the space X. Thus equation (8.1) can only be guaranteed when the components of the vector ζ depend on variables whose number, not less than m, is fixed beforehand. It turns out, analogous to the theorem in Section 24.4 for the groups G^r, that the basis of differential invariants is finite in all cases for the groups G^∞.

Theorem. *For any group G^∞ there is a finite basis for its differential operators.*

Proof. It is clear that the statement of the theorem is not trivial only if equation (8.1) holds. Then there will exist a $k \geq 1$, for which $\nu_k > r_k$. Consider the vector field $\underset{k}{\bar{\zeta}} = (\underset{k}{\zeta}, (\lambda \cdot D)\xi)$ on the space $Z_k \times R^n(\lambda)$ together with the continuation vector field $\underset{k}{\zeta}$. For these $r_*(\underset{k}{\bar{\zeta}}) = r_*(\underset{k}{\zeta}) = r_k$ since all first-order derivatives of ξ are included in $\underset{k}{\zeta}$. As the field $\underset{k}{\bar{\zeta}}$ is given on the space of dimensionality $\nu_k + n$, it has the universal invariant $\underset{k}{\bar{J}}$ consisting of $\nu_k + n - r_k > n$

independent scalar invariants. Consequently, this invariant can be given in the form $\underset{k}{\bar{J}} = (\underset{k}{J}, J')$, where the invariant $J'\colon Z_k \times R^n \to R^n$ possesses the property that $r_*(\partial_\lambda J') = n$. Thus, as in the case of groups G^r, it is possible to construct n independent invariants, over the field of invariant differential operators, in the form $\lambda \cdot D$ (analogue of the theorem in Section 24.2).

Let J_p be a universal invariant of order p. From the previous result J_{k+1} exists. It is shown that there exists $\chi \geq k + 1$ such that for all $p \geq \chi$ the invariant $(\lambda \cdot D)J_p$ is functionally equivalent to the invariant J_{p+1}, i.e., it is related to J_{p+1} by an invertible mapping. Obviously, if this is so, then a finite basis of the differential invariants exists and is contained in the universal invariant $\underset{\chi}{J}$.

The truth of the previous statement follows from the fact that the invariant differentiation operator $\lambda \cdot D$ is equivalent to the operator of differentiation ∂_x from the viewpoint of the results obtained by its application to the functionally independent scalar components. Thus, if for every $q > 1$ the universal invariant J_{k+q} is given in the form

$$J_{k+q} = ((\lambda \cdot D)J_{k+q-1}, J'_{k+q}) \qquad (q = 2, 3, \ldots), \tag{8.2}$$

then beginning with some $q = q_*$ the term J'_{k+q} will disappear and not be present for $q > q_*$. ■

9. Examples

In the notation introduced in Section 24.3 the operators of the Lie algebra L^∞ over the space $Z = X \times Y$ are given by

$$\zeta \cdot \partial = \xi^1 \partial_1 + \xi^2 \partial_2 + \eta \partial_y. \tag{9.1}$$

(α) The following example was considered by S. Lie. Let $\xi^1 = \alpha(x^1)$, $\xi^2 = 0$, $\eta = \alpha'(x^1)y$, with the arbitrary function $\alpha\colon R \to R$ and $\alpha' = \partial_1 \alpha$, i.e.,

$$\zeta \cdot \partial = \alpha \partial_1 + \alpha' y \partial_y. \tag{9.2}$$

This Lie algebra has the invariant $J = x^2$. To find the differential invariants, it is convenient to operate directly with the operator (9.2) (although it is not difficult to write the corresponding *AS-system*). The first continuation will be

$$\underset{1}{\zeta} \cdot \partial = \alpha \partial_1 + \alpha' y \partial_y + \alpha'' y \partial_{y_1} + \alpha' y_2 \partial_{y_2}.$$

If this is written in the form

$$\underset{1}{\zeta} \cdot \partial = \alpha \partial_1 + \alpha'(y \partial_y + y_2 \partial_{y_2}) + \alpha'' y \partial_{y_1},$$

then it is seen that any invariant is an invariant of the system of operators

$$\partial_1, \qquad y\partial_y + y_2 \partial_{y_2}, \qquad y \partial_{y_1}.$$

From here it follows that there is only one first-order invariant, namely, $J_1 = y_2/y$. To find the differential invariant operators of the form $\lambda D_1 + \mu D_2$ an operator of the form (2.3)

$$\overline{\zeta} \cdot \overline{\partial} = \alpha \partial_1 + \alpha' y \partial_y + \lambda \alpha' \partial_\lambda$$

is considered. Its invariants μ and λ/y give two independent operators of differential invariants:

$$\delta_1 \cdot \partial = y D_1, \qquad \delta_2 \cdot \partial = D_2,$$

with the aid of which, starting from the invariant J_1, it is possible to find all the differential invariants. Consequently, the basis of the latter is generated by the invariants x^2 and y_2/y.

(β) It is easy to verify for operators of the form $\zeta \cdot \partial = \xi^1 \cdot \partial_1 + \xi^2 \partial_2$, where ξ^1 and ξ^2 do not depend on y, that the equation

$$\xi_1^1 + \xi_2^2 = 0$$

is the determining equation. The corresponding group G^∞ is generated by all transformations of the plane $X = R^2(x^1, x^2)$ which *maintain constant area.* For such transformations, the problem of finding the differential invariants of the mappings $y: X \to Y$, under the condition that the group G^∞ operates trivially on the space Y, is solved.

First, let $Y = R(y)$. Then $\eta = 0$ in the operator (9.1) and the corresponding passive reduced system S of the determining equations is

$$\xi_2^2 = -\xi_1^1, \qquad \xi_y^1 = \xi_y^2 = 0, \qquad \eta = 0. \tag{9.3}$$

Here all the derivatives of ξ^1 and also all the derivatives $\partial_1^k \xi^2$ ($k = 0, 1, \ldots$) are parametric. The number of parametric derivatives of order $\leq k$ is equal to

$$p_k = k + 1 + \tfrac{1}{2}(k + 1)(k + 2).$$

At the same time the dimensionality of the space Z_k is (formula (1.2))

$$v_k = 2 + \tfrac{1}{2}(k + 1)(k + 2).$$

By more detailed analysis of the continuation operators it can be determined that the rank of the mapping $\underset{k}{\zeta}$ is

$$r_k = 1 + \tfrac{1}{2}(k + 1)(k + 2).$$

Thus for any $k \geq 0$ there is only one independent scalar invariant $J = y$ which generates the basis.

Now let $Y = R^2(y^1, y^2)$. In this case the determining equation $\eta = 0$ in the system (9.3) is replaced by the equations $\eta^1 = \eta^2 = 0$, and the number of parametric derivatives p_k is the same. But now the dimensionality of the space Z_k is equal to

$$\nu_k = 2 + (k+1)(k+2),$$

and $\nu_k > p_k$ for any $k \geq 0$ and property (8.1) is satisfied. The ranks r_k are also changed and it is found that $r_1 = 5$. As $\nu_1 = 8$ the universal invariant, $\underset{1}{J}$ consists of three independent components, which can be taken in the form

$$J^1 = y^1, \qquad J^2 = y^2, \qquad J^3 = y_1^1 y_2^2 - y_2^1 y_1^2. \tag{9.4}$$

Two independent operators of invariant differentiation are found from the first continuation $\underset{1}{\zeta}$ and they are

$$\delta_1 \cdot \partial = y_2^2 D_1 - y_1^2 D_2, \qquad \delta_2 \cdot \partial = y_2^1 D_1 - y_1^1 D_2. \tag{9.5}$$

Their application to the invariants J^1 and J^2 does not give any new invariants. As a result of operating on J^3 with operators (9.5) there are obtained two independent differential invariants of the second order. Further analysis shows that (9.4) is a basis for the differential invariants.

§25. AUTOMORPHIC SYSTEMS

1. Definition

Let the Lie group G (possibly infinite) of transformations of the finite-dimensional space $Z = X \times Y$ be given. The differential equations E (system), given on Z, are considered.

The system E is called *automorphic relative to the group* G if any solution of this system is obtained from one fixed solution by a transformation belonging to the group G.

In connection with this definition a number of problems arise about the description of the differential structure of an automorphic system, about construction of an automorphic system which possesses an a priori given solution, and so on. Also, an application of the concept of an automorphic system to a problem about some equipollent transformation of the given system of differential equations admitting the group G will be given in Section 25.2.

In what follows, automorphic (relative to G) systems will be denoted by the symbol AG.

It appears that the general theory of AG-systems is closely connected with the concept of partial invariance of a manifold. On this basis that theory will be developed below. In that discussion the term "partially invariant differential manifold" will be used to denote the partially invariant manifolds which are contained in the continuation spaces Z_k.

2. Automorphic System Structure

The Lie group $G^r(h)$, generated by the mapping $h: Z \times B \to Z$, is considered. Let the mapping $u: X \to Y$ be given; it is required to determine the form of the AG^r-system which has u as its solution.

With this goal in mind the orbit $h(U, O)$ of the manifold $U\{y = u(x)\}$ (see Section 21.1) is introduced. Generally speaking this orbit is not a characteristic manifold in the space Z. But if the continued manifolds $\underset{k}{U}$, given by the equations

$$\underset{l}{y} = \partial^l u(x) \qquad (l = 0, 1, \ldots, k), \tag{2.1}$$

and the corresponding continued spaces Z_k are considered, then it is possible to confirm that there will exist a $k \geq 0$ for which the *continued* orbit $\underset{k}{h}(\underset{k}{U}, O)$ will be a characteristic manifold in Z_k. This follows from the fact that the dimensionality of a continuous orbit is not more than the number $n + r$ (where $n = \dim X$) and $\dim Z_k$ increases indefinitely with k.

Let k be the least number for which $\underset{k}{h}(\underset{k}{U}, O)$ is a characteristic manifold in Z_k. Now it must be noted that the continued orbit $\underset{k}{h}(\underset{k}{U}, O)$, according to its construction, is an invariant manifold of the continued group $\underset{k}{G^r}$. In the general case one can assume that this manifold is nonsingular. Thus, according to the representation theorem in Section 18.7, the manifold $\underset{k}{h}(\underset{k}{U}, O)$ has a projection in the space $\bar{Z}_k$ of differential invariants and can be given by the system of equations connecting only the differential invariants of the group G^r of order $\leq k$, i.e., by the system

$$\Phi^\sigma(\underset{k}{J}) = 0 \qquad (\sigma = 1, \ldots, s), \tag{2.2}$$

where $\underset{k}{J}$ is the universal invariant of the group $\underset{k}{G^r}$. System (2.2), according to its construction, is such that the given mapping u is its solution.

System (2.2) *is automorphic.* Indeed, if u' is a solution of this system, then its orbit $\underset{k}{h}(\underset{k}{U'}, O)$ is an invariant manifold in Z_k and the projection of the former must be in the manifold (2.2) (by the definition of the solution), i.e., must be in the projection of the orbit $\underset{k}{h}(\underset{k}{U}, O)$. Because of the rule about the insertion of projections (Note in Section 19.4) it follows from here that the orbit $\underset{k}{h}(\underset{k}{U}', O) \subset \underset{h}{h}(\underset{k}{U}, O)$. In particular, $U'_k \subset \underset{k}{h}(\underset{k}{U}, O)$ and further $U' \subset h(U, O)$. This means that the solution u' is obtained from u by some transformation of the group G^r.

So any AG^r-system has the form of system (2.2) *and satisfies the following condition: The continued orbit $\underset{k}{h}(\underset{k}{U}, O)$ of some solution u, as an invariant manifold in Z_k, coincides with the manifold given by system* (2.2).

3. Construction of Automorphic Systems

The above presentation about the general structure of AG^r-systems shows that situations, analogous to those considered in Section 22.1, arise in continued spaces. That is, the continued manifold $\underset{k}{U}$ is contained in the invariant characteristic manifold of the group $\underset{k}{G^r}$, and thus it is a partially invariant manifold of this group. With that, the orbit $\underset{k}{h}(\underset{k}{U}, O)$ is, evidently, the least invariant manifold of the group $\underset{k}{G^r}$ containing $\underset{k}{U}$. This permits the determination of the defect and rank of the manifold $\underset{k}{U}$.

To calculate the defect δ_k of the manifold $\underset{k}{U}$ it is necessary to use formula (21.3.1), where, according to (2.1), the set $\{\underset{l}{y} = \partial^l u(x)\}$ should be taken as ξ—the continuation mapping $\underset{k}{\zeta}$, obtained from (4.8.5) and (4.8.6). This gives the expression

$$\partial\psi\langle\xi\rangle = (\eta - \xi\cdot\partial u, \underset{1}{\eta} - \xi\cdot\partial^2 u, \dots, \underset{k}{\eta} - \xi\cdot\partial^k u). \tag{3.1}$$

During transition to the manifold $\underset{k}{U}$ this expression is simplified and the final result is given in the following lemma.

Lemma. *For the mapping $\omega\colon X \to Y$, operating by*

$$\omega(x) = (\eta - \underset{1}{y}\cdot\xi)\circ\underset{1}{u}(x) = \eta(x, u(x)) - \partial u(x)\langle\xi(x, u(x))\rangle, \tag{3.2}$$

the defect of the invariant manifold $\underset{k}{U}$ is equal to

$$\delta_k = r_*(\omega, \partial\omega, \dots, \partial^k\omega). \tag{3.3}$$

Proof. According to the lemma in Section 4.4, for any mapping v: $Z_k \to P$ the relation (4.4.2), which can be generalized by induction for any l as

$$(D^l v)\circ\underset{l+k}{u}(x) = \partial^l(v\circ\underset{k}{u})(x) \qquad (l = 1, 2, \dots), \tag{3.4}$$

holds. From (3.4), equation (4.8.5), and definition (3.2), for any $l \leq k$

$$\begin{aligned}(\underset{l}{\eta} - \xi\cdot\partial^{l+1}u)|\underset{k}{U} &= (D^l(\eta - \underset{1}{y}\cdot\xi))|\underset{k}{U} \\ &= (D^l(\eta - \underset{1}{y}\cdot\xi))\circ \underset{l+1}{u} = \partial^l((\eta - \underset{1}{y}\cdot\xi)\circ\underset{1}{u}) = \partial^l\omega.\end{aligned}$$

Thus (3.3) follows from (3.1) and (21.3.1). ■

So far only automorphic systems having the given solution u have been discussed. If no solution is given beforehand, then the classification problem of enumerating all automorphic systems of the given group G^r arises. The solution of this problem is based on the theory of partial invariance and is connected with the following algorithm for constructing AG^r-systems.

The defect δ_k must satisfy conditions (22.1.2) and (22.1.3), in which $r_* = r_k = r_*(\underset{k}{\zeta})$ and $m = \nu_k - n$ should be assumed, where $\nu_k = \dim Z_k$. Here it is necessary to note that the right inequality of (22.1.2) is excessively restrictive and can be replaced with the inequality $\delta \leq \min\{r_*, m\}$. This is connected with the fact that the number $r_* - 1$ was introduced in Section 22 to decrease the number of independent variables on which the considered partially invariant solutions depend, and the number $m - 1$ to fill the entire space Z (locally) with the orbit $h(U, O)$. For automorphic systems both of these requirements are not necessary. Thus for any $k = 0, 1, 2, \ldots$ the necessary conditions for the AG^r-system of order k to exist are

$$\max\{r_k - n, 0, \nu_k - n - \bar{r}_k\} \leq \delta_k \leq \min\{r_k, \nu_k - n\}, \qquad \bar{r}_k > 0, \quad (3.5)$$

where $\bar{r}_k = r_*(\partial_v \underset{k}{J})$ and $v = \underset{k}{\tilde{z}} = (y, \underset{1}{y}, \ldots, \underset{k}{y})$. The rank ρ_k is expressed through the defect δ_k by the formula (22.1.1), so

$$\rho_k = \delta_k + n - r_k \qquad (k = 0, 1, 2, \ldots). \quad (3.6)$$

If for some $k \geq 0$ there is a number δ_k satisfying inequalities (3.5) then the automorphic system whose continued solutions have defect δ_k and rank ρ_k is constructed using the algorithm described in Section 22.2. This system is given by the equations of the projections of the orbits $h(\underset{k}{U}, \underset{k}{O})$ in the space of invariants

$$J''(\underset{k}{z}) = \phi(J'(\underset{k}{z})), \quad (3.7)$$

where $\underset{k}{J} = (J', J'')$ is a decomposition of the universal invariant $\underset{k}{J}$ according to the value (3.6) of the rank ρ_k (i.e., J' consists of exactly ρ_k independent scalar invariants). With that, the automorphism property is guaranteed by the fact that (3.7) are the equations of the least invariant manifold in Z_k containing the kth continuation of any solution of system (3.7).

In contrast to Section 22.2, there is no system E in which the expressions for the fundamental derivatives, found from equations (3.7), should be substituted. But system (3.7) can be *active* and thus the fundamental derivatives must satisfy some "inner" compatibility conditions. These conditions have the form of differential equations in which the mapping ϕ is unknown.

So any kth-order AG^r-system has the form of a system of equations to which the equations for the unknowns ϕ, which express the compatibility conditions of equations (3.7), *are added.*

In what follows, the number ρ_k, equal to the number of independent variables in equations (3.7), will be called the *rank of the automorphic system.*

The algorithm described for the construction of AG-systems can be applied to the case of infinite Lie groups, taking into account the singularities noted in Section 24.7.

4. Classification of Automorphic Systems

According to the previous considerations, the kind of an automorphic system is determined by the order of the continuation k, by the rank ρ_k, and by the universal invariant $\underset{k}{J}$. The most important result of the AG-system's classification is the fact that the number of different kinds of automorphic systems is finite for any group G (possibly infinite). The *regular case*, when the least k for which $\underset{k}{J}$ contains the basis of differential invariants coincides with the least k for which the inequalities (3.5) are satisfied, is especially to be noted.

For the classification it is convenient to introduce into consideration the number

$$n_0 = v_0 - r_0 - \bar{r}_0, \tag{4.1}$$

which, roughly speaking, is equal to the number of those independent zero-order invariants not depending on y. Using n_0 inequalities, (3.5) can be essentially simplified. Indeed, according to the definition of the number $\bar{r}_k$, the dimensionality of the space of invariants is equal to $\bar{r}_k + n_0$, because of which the equality $v_k = r_k + \bar{r}_k + n_0$ holds. From here, in particular, $v_k - \bar{r}_k \geq n$. Thus, the left-hand side of inequality (3.5) is equal to $v_k - n - \bar{r}_k = r_k - n + n_0$, and the right side is equal to r_k, and this inequality has the form $r_k - n + n_0 \leq \delta_k \leq r_k$. From this, according to (3.6), the inequalities

$$n_0 \leq \rho_k \leq n \tag{4.2}$$

are found for rank ρ_k.

Theorem. *For any Lie group G there is a finite number of different kinds of AG-systems. In the regular case this number is equal to $n - n_0 + 1$.*

Proof. The theorem in Section 24.8 concerning finiteness of the basis of differential invariants is used. If the number of arbitrary order independent scalar differential invariants of the group G is finite, then the statement is trivial. Let condition (24.8.1) hold and let χ be the number determined in the proof of the theorem in Section 24.8 and possessing the property that for any $k \geq \chi$ the universal invariant J_{k+1} is obtained from J_k by applying the operators of invariant differentiation. Then it is possible to have $\bar{r}_k + n = v_k - r_k + n - n_0 \geq \chi$.

With these conditions the invariant $J'\underset{k}{(z)}$ can be taken as $J'_\chi = J'\underset{\chi}{(z)}$ found during the process of constructing the automorphic systems of given rank ρ_k and of any orders $k \geq \chi$ given in the form (3.7). Let ψ_k be that part of the mapping ϕ, which in the k-order AG-system gives the expression for the invariant J_k, having the form

$$J_k = \psi_k(J'_\chi). \tag{4.3}$$

Then these equations will also be included in the $(k + 1)$th-order AG-system which has the same rank. Also, the equations

$$J_{k+1} = \psi_{k+1}(J'_\chi)$$

will be in this AG-system. But because of the definition of the number χ, it can be assumed that $J_{k+1} = (\lambda \cdot D)J_k$, from which it follows that the last equations necessarily must be equipollent to those equations obtained by applying the operator $\lambda \cdot D$ to equations (4.3). From (4.3) this means that the relation $\psi_{k+1} = (\lambda \cdot D)\psi_k$ holds, i.e., the equations of order $k + 1$ are merely differential consequences of equations (4.3). The same thing will take place if the order of the AG-system is increased. It remains to note that, because of (4.2), the ranks ρ_k are limited in total and the number of kinds of AG-systems of order $\leq \chi$ is finite.

In the regular case the value χ can be taken equal to the least k for which inequalities (3.5) are satisfied. Then different kinds of AG-systems will be found by choosing different values of the rank ρ. That number is exactly equal to $n - n_0 + 1$ because of (4.2). ■

A more detailed classification of AG-systems must take into account the group structure of the group G and the possible reduction to a smaller defect.

5. Example

The group G^∞ from Example (α) in Section 24.9 with the tangential mapping (24.9.1) is considered. Here $v_0 = 3, r_0 = 2, \bar{r}_0 = 0$ and, according to formula (4.1), $n_0 = 1$ is found. Also $\bar{r}_1 = 1$, so the necessary conditions for the existence of automorphic systems are satisfied, beginning with $k = 1$. The inequality (4.2) becomes $1 \le \rho_k \le 2$, i.e., AG-systems with ranks $\rho_k = 1$ and $\rho_k = 2$ are possible.

The automorphic system of the first rank is generated using the universal invariant $\underset{1}{J} = (x^2, y_2/y)$ and has the form

$$y_2 = \omega(x^2)y, \tag{5.1}$$

with the arbitrary function $\omega: R \to R$. All automorphic systems with the first rank and with the order more than one are obtained merely by applying the operators of invariant differentiation yD_1 and D_2 to (5.1).

To construct the automorphic systems of rank two the second-order invariants, which are easily calculated using invariant differentiation,

$$yD_1(y_2/y) = y_{12} - (y_1y_2/y), \qquad D_2(y_2/y) = y_{22}/y - (y_2/y)^2,$$

and which give the universal invariant

$$\underset{2}{J} = (x^2, y_2/y, y_{12} - y_1y_2/y, y_{22}/y),$$

are required. Consequently, the automorphic system of second rank and second order has the form

$$y_{12} = qy_1 + \phi(p, q), \qquad y_{22} = \psi(p, q)y, \tag{5.2}$$

where the notation $p = x^2$ and $q = y_2/y$ has been introduced. With that, the functions ϕ and ψ must satisfy the compatibility condition obtained from $(y_{12})_2 = (y_{22})_1$. After simplification using system (5.2) this condition becomes

$$\phi_p + (\psi - q^2)\phi_q - \phi\psi_q + q\phi = 0. \tag{5.3}$$

All other automorphic systems of the second rank are found from (5.2) by applying the operators yD_1 and D_2.

§26. GROUP SPLITTING

1. Statement of Problem

Here again the differential equation E (system) and the Lie group G admitted by it (which can be infinite) are taken.

As was noted in Section 20, the operation of the group G on the set SE of solutions of equation E leads to subdivision of this set into classes of equivalent solutions: The solutions u_1 and u_2 are equivalent relative to G if there is a transformation belonging to G under which u_1 transforms into u_2. (Such solutions were said to be not essentially different in Section 20.3.) It is clear that the solutions u_1 and u_2 are equivalent, relative to the group G, if and only if they belong to the same orbit (u, G) of some solution u. So the set of equivalent solution classes is the set of all possible orbits (u, G).

The following problem was posed by S. Lie: For the given equation E and the group G admitted by it find the equations of its solution orbits and the equations whose solutions determine the compatibility of the orbits. In other words, it is desired to find the representation of the general solution of E in the form of a composition consisting of some mapping of the set SE into the orbit set of solutions $u \in SE$, and of a transformation from the group G operating within every orbit.

The solution of this problem is based on results from Section 25, according to which the orbits of solutions, as invariant manifolds in some Z_k, are described using the automorphic system of the group G. Thus, one of the components on which the equation E must be decomposed is an AG-system. Generally speaking, it contains arbitrary functions (of the form ϕ and ψ in (25.5.2)), which should be determined in such a way that all solutions of the AG-system are solutions of equation E. With this goal in mind, equation E is written in the form of its projection $\bar{E}$ into the space of invariants and all differential invariants are replaced by their independent expressions. In this fashion the second component of the decomposition of equation E appears. It is a combination of $\bar{E}$ with the projection of the AG-system into the space of invariants and the compatibility conditions of the AG equations. This component is called the *resolving system* and it will be denoted by the symbol RE.

So the solution of this S. Lie problem must lead to the pair (E, G) representation, expressed by the symbolic formula

$$(E, G) \Rightarrow AG \cup RE. \tag{1.1}$$

A representation of the form (1.1), in which the first part is equipollent to the equation E, will be called a *group splitting of equation E relative to the group G admitted by it.*

2. Existence of Group Splitting

The equipollent requirement of representation (1.1) places a limitation on the AG-system included in it. Indeed, in the space of invariants the orbits of different solutions can have different dimensionalities, which are determined by the corresponding AG-system. To guarantee an inclusion of all solutions of equation E it is necessary to look for the orbit of maximal dimensionality. From the viewpoint of the results in Section 25.4 this means that group splitting of the form (1.1) can be obtained only in the case when an AG-system of rank $\rho_k = n$ is taken. In these considerations and in the following theorem condition (24.8.1) is supposed to be satisfied.

Theorem. *If the differential equation E admits the Lie group G, then there is a uniquely determined group splitting in the form* (1.1) *of this equation into the* automorphic *system AG and the* resolving *system RE.*

Proof. According to the results in Section 25.4 the value $\rho_k = n$ gives one definite type of the AG-system in the form (25.3.7). Let $\bar{E}$ be a projection of equation E into the space of differential invariants $\bar{Z}_\infty$ of the group G, from which the corresponding resolving system RE is constructed, where the mapping ϕ is unknown. From any solution ϕ of the system RE it is possible to find some solution of the AG-system, which, according to the construction, is a solution of equation E. It is left to prove that any solution $u \in SE$ can be obtained in this way.

For that it is sufficient to establish that some mapping ϕ, included in AG-system (25.3.7), corresponds to the solution u. Indeed, with the determined ϕ the given u is a solution of the AG-system and of the equation E, which means that it must satisfy all consequences of these equations, particularly, all the equations RE.

To determine ϕ from the given u the mapping $p = J' \mid \underset{k}{U}: X \to X$ is introduced into the considerations. In the general case, $r_*(\partial p) = n$ and there is an inverse mapping p^{-1}, from which the unknown ϕ is uniquely determined by the equation

$$\phi = (J'' \mid \underset{k}{U}) \circ p^{-1}. \tag{2.1}$$

In those solutions for which $r_*(\partial p) < n$ the scalar components of the invariant J' are functionally independent, i.e., the rank of the AG-system is reduced ($\rho_k < n$). Taking this fact into account, the repetition of the previous consideration shows that such solutions are also produced as solutions of the system $AG \cup RE$. ■

The practical significance of group splitting consists in the fact that solutions of the AG-system are very simply found at the expense of its automorphic property (by operation with a group transformation on any of its solutions), and the resolving system RE turns out to be simpler when compared with the initial equation E. The latter occurs because the system RE has fewer solutions than E does because of removal of those excesses which were introduced by the existence of the admitted group G. In particular, it is expedient to use the method of group splitting in the case of an infinite Lie group G. But it should be noted that the exact meaning of the term "simplification of an equation" is not sufficiently clear from the point of view of its application to the differential structure. This problem needs further investigation.

3. Differential Invariant Solutions

In connection with group splitting the question arises: For what other AG-systems of ranks $\rho_k < n$ is it also possible to construct the union of the form (1.1)? Of course, the equipollence between the equation E and the union $AG \cup RE$ will not take place. Indeed, this union defines only some class of partially invariant solutions of equation E. Such solutions will be called *differential invariant solutions of rank* ρ_k.

These solutions, generally speaking, are not invariant: Any invariant (in the sense of Section 19.2) solution, being continuous, is an invariant manifold in the continuous space (see Section 19.5), and a differential invariant solution in the continuous space is only a partially invariant manifold.

At the present time, the concept of differential invariant solutions is not completely treated. First of all there does not exist a sufficiently wide set of substantial examples of such solutions. Also, the classification problems, in particular reduction to a smaller defect, and so on, need detailed investigations.

4. Example of Splitting

The system of equations for a stationary boundary layer (10.7.2) with the given function $x \to \theta(x)$ is considered. In Section 10.7 it is shown that this system admits the group G^∞ with the operator

$$\zeta \cdot \partial = \tilde{\phi}\partial_y + \tilde{\phi}' u \partial_v, \tag{4.1}$$

where $\tilde{\phi}: R(x) \to R$ is an arbitrary function and $\tilde{\phi}' = \partial_x \tilde{\phi}$. The generating mapping of this group is

$$\begin{aligned} x' &= x, & y' &= y + \tilde{\phi}(x), \\ u' &= u, & v' &= v + \tilde{\phi}'(x)u. \end{aligned} \tag{4.2}$$

The group G^∞ has a universal invariant (of first order) $J = (x, u)$. Since here $n = 2$, the automorphic system of rank two cannot be constructed using J, and therefore it is necessary to consider the first continuation of this group. Calculation by means of the formulas (4.8.4) gives

$$\underset{1}{\zeta} \cdot \partial = \tilde{\phi}\partial_y + \tilde{\phi}' u \partial_v - \tilde{\phi}' u_y \partial_{u_x} + (\tilde{\phi}'' u + \tilde{\phi}'(u_x - v_y))\partial_{v_x} + \tilde{\phi}' u_y \partial_{v_y},$$

where $u_x = \partial_x u$ and so on. Here $r_1 = 3$ and the universal invariant $\underset{1}{J}$ must be an invariant of the system of operators (which are obtained as coefficients of $\tilde{\phi}$, $\tilde{\phi}'$, and $\tilde{\phi}''$)

$$\partial_y, \qquad u\partial_v - u_y\partial_{v_x} + u_y\partial_{v_y} + (u_x - v_y)\partial_{v_x}, \qquad u\partial_{v_x},$$

and it is easily found to be

$$\underset{1}{J} = (x, u, u_x + v_y, u_y, uu_x + vu_y).$$

Thus the AG-system of rank $\rho_1 = 2$ can be written in the form

$$u_x + v_y = \omega, \qquad u_y = \phi, \qquad uu_x + vu_y = \psi, \tag{4.3}$$

with functions ω, ϕ, ψ: $R^2(x, u) \to R$ so far unknown. The compatibility conditions of the system (4.3), with the initial system (10.7.2), are

$$\omega = 0, \qquad \psi + \theta = \phi\phi_u,$$

from which the functions ω and ψ are determined. Because of these equations, which are part of the resolving system, the initial equations are equipollent to the system (4.3). One more equation of the resolving system is given by the compatibility condition of equations (4.3), which can be conveniently found in the following way. It can be assumed that the function v is determined from the third equation of (4.3), namely,

$$v = \phi_u - (\theta + uu_x)/\phi. \tag{4.4}$$

Substitution of this expression into the first equation of (4.3) gives, considering the second equation of (4.3), the unknown resolving equation

$$u\phi_x = \phi^2\phi_{uu} + \theta\phi_u. \tag{4.5}$$

Finally, the group splitting of system (10.7.2), relative to the group G^∞, with the operator (4.1) is a union of the automorphic AG^∞-system

$$u_x + v_y = 0, \qquad u_y = \phi(x, u), \qquad uu_x + vu_y = \phi\phi_u - \theta(x) \tag{4.6}$$

and the resolving system RE, consisting of one equation (4.5).

It is interesting to note that the formal (without using the group property) transition from the system (10.7.2) to the equation (4.5) is known in hydrodynamics by the name "transformation of Crocco." It is possible to show that the solution of the main boundary value problem of the boundary layer theory is actually reduced to the construction of a certain solution of equation (4.5), with which the solution of the automorphic system is reduced to the integration of one first-order ordinary differential equation.

REFERENCES

In the classical theory of differential invariants, whose foundation was laid by S. Lie [1], constructions connected with finite transformations of a group were used. The main result, the theorem about finiteness of the basis, belongs to Tresse [2]. The proof of this theorem using the invariant differentiation operators is given here for the first time. S. Lie's problem about the splitting of a given differential equations system into automorphic and resolving systems was solved in general form by Vessiot [3], who used a parametric representation of all the variables by means of auxiliary invariant parameters. An example of group splitting, according to Vessiot, of a concrete system of equations appearing in gas dynamics, was given in Ref. [4]. The possibilities and advantages of the explicit representation of group splitting were discussed in Ref. [5], from which the example in Section 26.4 was taken. The works [6, 7] were also devoted to these problems.

[1] Lie S. Über Differentialinvarianten. — Math. Ann. **24**, № 1 (1884), 52—89.
[2] Tresse A. Sur les invariants differentiels des groupes continus de transformations. — Acta math. **18** (1894), 1—88.
[3] Vessiot E. Sur l'integration des sistem differentiels qui admittent des groupes continus de transformations. — Acta math. **28** (1904), 307—349.
[4] Кухарчик П. Групповые свойства уравнений коротких волн в газовой динамике. — Bull. Acad. pol. sci. sér. sci. techn. **13**, № 5 (1969), 469—477.
[5] Овсянников Л. В. Групповое расслоение уравнений пограничного слоя. — В сб. «Динамика сплошной среды», вып. 1 (1969), Новосибирск, 24—35.
[6] Субоч Ю. Ф. Об одном групповом свойстве систем дифференциальных уравнений. — В сб. «Динамика сплошной среды», вып. 6 (1971), Новосибирск, 201—207.
[7] Верещагина Л. И. Групповое расслоение уравнений пространственного нестационарного пограничного слоя. — Вестн. Ленингр. ун-та, вып. 3, № 13 (1973), 82—86.

CHAPTER

VIII

Special Problems

There are a number of group analyses for special differential equations whose exposition would undoubtedly be of interest to those who have mastered the previous chapters and would like to move further into the theory and its applications. The choice of the problems presented in the chapter is dictated, on one hand, by inevitable limitations and, on the other, by a desire to show a variety of possibilities presented by the methods of group analysis.

The group analysis of a second-order differential equation with any number n of independent variables (the values $n = 2, 3, 4$ are most often met in applications), contrary to the case $n = 2$ considered in Section 9, meets with considerable difficulties. At the present time there does not exist a complete group classification even for $n = 3$. Most valuable are those general results which can be established relatively simply. The most remarkable fact here is the connection existing between the group admitted by the equation and a group of conformal transformations (or motions) of a Riemannian space associated with the equation. One should pay attention to so-called standard equations, admitting a maximal group, which so far have not been investigated in detail from the viewpoint of boundary value problems.

Tangential transformations are a natural continuation of the point transformations considered in this book. The foundations for that theory were also laid in the works by S. Lie. But the possibilities presented by this generalization are very limited since the generalization turns out to be consistent only in the case of a one-dimensional space of the unknown functions, and only when the transformation is of the first order. In other cases the tangential mappings reduce to continuous point (or first-order tangential) transformations.

Wider possibilities are created by so-called Lie–Bäcklund transformations in which the order of derivatives is not limited beforehand. At the present time the attention of a number of investigators is focused on this field.

One of the important problems is the extension of the group analysis methods to boundary value problems for differential equations. Here certain difficulties appear which are caused by the local character of the theory. These are not always compatible with defining the solution "in general" for the boundary value problem. So far there exists only isolated examples in which the properties of invariant solutions of such boundary value problems can be established locally a priori. Some possibilities include the study of invariant characteristics and strong discontinuities for a special class of differential equations. These are demonstrated in Section 29. Also, the property of "free boundary" invariance is determined for a problem involving the Navier–Stokes equations.

A classical field of application for the group analysis methods is that of conservation laws, whose determination is important for the solution of different problems in physics and mechanics. The fundamental theorem, establishing a connection between invariant properties of variational functionals and conservation laws, belongs to E. Noether. It is essential in this problem that the group, admitted by the functional, is also admitted by Euler's differential equation for its extremal. An important generalization of Noether's theorem was obtained relatively recently. It is based upon a remarkable identity which connects Euler's equation with an invariance condition for a functional and with a conservation law. From that a sufficient condition for the converse of Noether's theorem to hold was also obtained. The detailed general results are given in Section 30. There are numerous interesting examples of these results applied to concrete problems whose exposition, unfortunately, is beyond the scope of this book.

§27. SECOND-ORDER LINEAR EQUATIONS

1. The Equation and Its Transformation

The linear differential equation of the form $E(n, 1, 2, 1)$, with second-order partial derivatives, in the unknown mapping $u: X \to R$

$$F[u] \equiv a^{ij}u_{ij} + b^{i}u_{i} + cu = 0, \tag{1.1}$$

is considered. In this exposition it is assumed that in the independent variables space $X = R^n(x)$ the components of the vector $x = (x^1, \ldots, x^n)$ are introduced, according to which the vectors $\underset{1}{u} \in Y_1$ and $\underset{2}{u} \in Y_2$ are presented in the form $\underset{1}{u} = (u_1, \ldots, u_n)$ and $\underset{2}{u} = (u_{11}, u_{12}, \ldots, u_{nn})$ as elements of the first and the second continuations of the space $Y = R(u)$ using X (see Section 4.2), and $u_i = \partial_i u$, $u_{ij} = \partial_i \partial_j u$, where $\partial_i = \partial_{x^i}$ $(i = 1, \ldots, n)$. The coefficients $a^{ij} = a^{ji}$, b^i, c are given functions $X \to R$ of the class $C_\infty(X)$.

The problem of group classification (see Section 6) relative to the arbitrary elements—the set $\{a^{ij}, b^i, c\}$ of all coefficients of the equation—is considered for equation (1.1). This problem has been considered in Section 8.8 for $n = 1$ and in Section 9 for $n = 2$. In what follows, it will be assumed that $n > 2$, where necessary.

According to Section 6.4 the problem of group classification should be solved, up to equivalence transformations, for a general differential structure of equation (1.1). For the following geometric interpretation it is useful to consider transformations of the linear operator F, defined by equation (1.1), and to distinguish three types of such transformations.

(α) The space X of transformations of the form $x' = p(x)$ [in coordinates $x'^i = p^i(x)$, $i = 1, \ldots, n$], which generates the transformation $u \to u'$ defined by $u = u' \circ p$ and, correspondingly, the transformation $F \to F'$, where $F[u] = F'[u']$. With these equation (1.1) transforms into

$$F'[u'] = a'^{ij} u'_{ij} + b'^i u'_i + c' u', \tag{1.2}$$

where $u'_i = \partial'_i u'$, etc., and $\partial'_i = \partial_{x'^i}$. The new (primed) coefficients are expressed in terms of the old ones by

$$a'^{kl} \circ p = a^{ij} p^k_i p^l_j, \qquad b'^k \circ p = a^{ij} p^k_{ij} + b^i p^k_i, \qquad c' \circ p = c. \tag{1.3}$$

(β) The linear transformations of the unknown function $u \to e^\theta u$ with any function $\theta \colon X \to R$ of class $C_\infty(X)$ and at same time multiplication of the operator F by the multiplier $e^{-\theta}$, i.e., the transformation

$$F'[u] = e^{-\theta} F[e^\theta u], \tag{1.4}$$

which transforms equation (1.1) into the equation

$$F'[u] \equiv a^{ij} u_{ij} + b'^i u_i + c' u = 0, \tag{1.5}$$

with the same a^{ij} but with new

$$b'^i = b^i + 2a^{ij}\theta_j, \qquad c' = c + b^i \theta_i + a^{ij}\theta_{ij} + a^{ij}\theta_i \theta_j. \tag{1.6}$$

(γ) The multiplication of the operator F by a multiplier, for example, $F' = e^{-2\theta}F$ with the function $\theta: X \to R$, i.e., transformation to the equation

$$F'[u] = \bar{a}^{ij}u_{ij} + \bar{b}^i u_i + \bar{c}u = 0, \tag{1.7}$$

where

$$a_{ij} = e^{2\theta}\bar{a}^{ij}, \qquad b^i = e^{2\theta}\bar{b}^i, \qquad c = e^{2\theta}\bar{c}. \tag{1.8}$$

2. Invariance Condition

The operator of the group $G^1(\zeta)$, of transformations of the space $Z = X \times Y$ by equation (1.1), is to be found in the form

$$\zeta \cdot \partial = \xi \cdot \partial_x + \eta\partial_u = \xi^i\partial_i + \eta\partial_u, \tag{2.1}$$

with the vector field $\zeta = (\xi, \eta): Z \to Z$. To write the invariance conditions for equation (1.1), relative to the operator (2.1), the second continuation of this operator,

$$\underset{2}{\zeta} \cdot \partial = \zeta \cdot \partial + \underset{1}{\eta} \cdot \underset{1}{\partial} + \underset{2}{\eta} \cdot \underset{2}{\partial},$$

is required, where $\underset{1}{\eta} = (\eta_1, \ldots, \eta_n)$ and $\underset{2}{\eta} = (\eta_{11}, \eta_{12}, \ldots, \eta_{nn})$ are calculated using the complete differentiation operator

$$D_i = \partial_i + u_i\partial_u + u_{ij}\partial_{u_j} + \cdots \qquad (i = 1, \ldots, n)$$

by the formulas (4.8.4) of continuation, namely,

$$\eta_i = D_i\eta - u_jD_i\xi^j, \qquad \eta_{ik} = D_k\eta_i - u_{ij}D_k\xi^i.$$

Since ζ does not depend on $\underset{1}{u}$ there results

$$\begin{aligned} \eta_i &= \partial_i\eta + u_i\partial_u\eta - u_j\partial_i\xi^j - u_iu_j\partial_u\xi^j, \\ \eta_{ik} &= \partial_k\eta_i + u_k\partial_u\eta_i + u_{kl}\partial_{u_l}\eta_i - u_{ij}(\partial_k\xi^j + u_k\partial_u\xi^j). \end{aligned} \tag{2.2}$$

Evidently the result of the operator $\underset{2}{\zeta} \cdot \partial$ operating on $F[u]$ is a linear nonhomogeneous form in $\underset{2}{u}$. Thus the invariance condition

$$(\underset{2}{\zeta} \cdot \partial)F[u]\,|_{F[u]=0} = 0$$

can be written using an indefinite Lagrange multiplier as

$$(\underset{2}{\zeta} \cdot \partial)F[u] = \lambda F[u]. \tag{2.3}$$

From that it is possible a priori to confirm only that λ does not depend on $\underset{2}{u}$ but that λ should be considered as a function of $(x, u, \underset{1}{u})$. Equation (2.3) is easily decomposed relative to $\underset{2}{u}$ and gives the equations

$$\begin{aligned} &a^{kj}\partial_{u_i}\eta_j - a^{ij}(\partial_j\xi^k + u_j\partial_u\xi^k) + \xi^j\partial_j a^{ik} = \lambda a^{ik} \qquad (i, k = 1, \ldots, n), \\ &a^{ik}(\partial_k\eta_i + u_k\partial_u\eta_i) + b^i\eta_i + c\eta + u_i\xi^k\partial_k b^i + u\xi^k\partial_k c = \lambda b^i u_i + \lambda cu. \end{aligned} \tag{2.4}$$

Also η_i from (2.2) must be substituted here. The following decomposition of equations (2.4) relative to $\underset{1}{u}$ can be accomplished very easily if it is known beforehand that λ does not depend on $\underset{1}{u}$. Indeed, it appears that this property holds if equation (1.1) only satisfies some natural assumptions:

It is assumed that the matrix (a^{ij}) *is nonzero and the case of a* very degenerate *equation* (1.1) *is excluded. This occurs when* (1.1) *is equivalent to the equation*

$$u_{11} + \bar{b}u_1 + \bar{c}u = 0; \tag{2.5}$$

i.e., it is actually equal to an ordinary second-order differential equation, which has already been studied in Section 8.8.

Lemma. *If the matrix* (a^{ij}) *is nonzero and equation* (1.1) *is not very degnerate, then the component* ξ *of the operator* (2.1) *does not depend on u and the multiplier* λ *does not depend on* $\underset{1}{u}$.

Proof. From the first equation of (2.4) the multiplier λ is, at most, a linear function of $\underset{1}{u}$, and without loss of generality it is possible to assume

$$\lambda = \lambda^i u_i + \lambda_0. \tag{2.6}$$

The substitution of this expression into the first equation of (2.4) and the decomposition of the result relative to $\underset{1}{u}$ gives the equations

$$a^{jk}\partial_u\xi^i + a^{ki}\partial_u\xi^j + a^{ij}\partial_u\xi^k = -a^{ki}\lambda^j \qquad (i, j, k = 1, \ldots, n), \tag{2.7}$$

$$a^{ik}\partial_u\eta - a^{kj}\partial_j\xi^i - a^{ij}\partial_j\xi^k + \xi^j\partial_j a^{ik} = \lambda_0 a^{ik} \qquad (i, k = 1, \ldots, n). \tag{2.8}$$

The left-hand side of equation (2.7) is not changed under a permutation of the indices i, j, k. Therefore the identities

$$a^{ki}\lambda^j = a^{kj}\lambda^i \qquad (i, j, k = 1, \ldots, n)$$

follow.

Suppose that all the multipliers λ^i are not equal to zero, for example, $\lambda^1 \neq 0$. Then upon setting $q^i = \lambda^i/\lambda^1$ the previous relation becomes

$$a^{ik} = q^i a^{k1} \qquad (i, k = 1, \ldots, n).$$

Repeating the application of this relation leads to

$$a^{ik} = q^i q^k a^{11} \qquad (i, k = 1, \ldots, n).$$

Therefore $a^{11} \neq 0$ since otherwise the matrix (a^{ik}) would be zero. Thus, the combination of the terms in the second-order derivatives of equation (1.1) leads to the form $q^i q^k u_{ik}$. According to the theorem in Section 3.8 there exists a similarity transformation $y: X \to X$ which transforms the vector field $q = \{q^i\}$ into a constant one, for example, into $(1, 0, \ldots, 0)$. Then in the coordinates $y = (y^1, \ldots, y^n)$ the sum of all the terms having second order derivatives in equation (1.1) will consist only of one term u_{11}.

So without loss of generality in the derivation, it is possible to assume, in the considered case (i.e., when not all of λ^i are equal to zero) that the initial equation (1.1) is such that $a^{11} = 1$ and $a^{ik} = 0$ $(ik > 1)$ Then (2.7) when $k = i = j = 1$ and $k = j = 1, i > 1$ gives

$$3\partial_u \xi^1 = -\lambda^1, \qquad \partial_u \xi^i = -\lambda^i = 0 \qquad (i = 2, 3, \ldots, n). \tag{2.9}$$

Also, from (2.8) when $i = 1, k > 1$ the relations

$$\partial_1 \xi^k = 0 \qquad (k = 2, 3, \ldots, n) \tag{2.10}$$

follow.

Then it is necessary to refer to the second equation (2.4) and substitute expressions (2.2) and (2.6) into it. In the general case the decomposition of this equation relative to $\underset{1}{u}$ leads to

$$\partial_u^2 \xi^j = 0 \qquad (j = 1, \ldots, n), \tag{2.11}$$

$$a^{ik}\partial_u^2 \eta - a^{ij}\partial_u \partial_j \xi^k - a^{jk}\partial_u \partial_j \xi^i - b^i \partial_u \xi^k = \lambda^k b^i \qquad (i, k = 1, \ldots, n), \tag{2.12}$$

$$2a^{ik}\partial_u \partial_k \eta - a^{jk}\partial_j \partial_k \xi^i - b^i \partial_u \eta - b^k \partial_k \xi^i + \xi^k \partial_k b^i$$
$$= \lambda_0 b^i + \lambda^i cu \qquad (i = 1, \ldots, n), \tag{2.13}$$

$$a^{ik}\partial_i \partial_k \eta + b^i \partial_i \eta + c\eta + \xi^k u \partial_k c = \lambda_0 cu. \tag{2.14}$$

In the case considered here, from (2.12) when $k = 1, i > 1$, using (2.10), the equations

$$-b^i \partial_u \xi^1 = \lambda^1 b^i \qquad (i = 2, \ldots, n)$$

follow, or as a result of (2.9)

$$\lambda^1 b^i = 0 \qquad (i = 2, \ldots, n).$$

As the equality $\lambda^1 = 0$ contradicts the assumption that the vector $\{\lambda^i\}$ is nonzero, the equalities $b^i = 0$ $(i = 2, \ldots, n)$ follow. This means that equation (1.1), transformed to the variables y, can be reduced to the form (2.5), i.e., it is strongly degenerate.

Consequently, if (1.1) is not strongly degenerate, then $\lambda^i = 0\,(i = 1, \ldots, n)$, and then $\partial_u \xi^i = 0\ (i = 1, \ldots, n)$ is obtained from (2.7) and (2.11). ■

3. Determining Equations

For those equations (1.1) which satisfy the conditions of the previous lemma, the component ξ of the operator (2.1) does not depend on u. Therefore it follows from (2.12) that $\partial_u^2 \eta = 0$. Consequently the equations $\partial_u \xi = 0$, $\partial_u^2 \eta = 0$ are included in the system of determining equations and, according to Section 7.14, the full Lie algebra of equation (1.1) has the structure $L^r \oplus L^\infty$, where the subalgebra L^r consists of the operators

$$\zeta \cdot \partial = \xi(x)\partial_x + \sigma(x)u\partial_u, \tag{3.1}$$

and L^∞ is generated by operators of the form $v(x)\partial_u$, where v is any solution of equation (1.1). Thus it is possible to limit consideration to the system *DE* of determining equations of the Lie algebra L^r. If it is assumed that $\lambda_0 = \sigma + \mu$, then finally the system *DE*, in which nontrivial contributions are made only by equations (2.8), (2.13), and (2.14), is

$$a^{ik}\partial_k \xi^j + a^{jk}\partial_k \xi^i - \xi^k \partial_k a^{ij} = \mu a^{ij} \qquad (i, j = 1, \ldots, n), \tag{3.2}$$

$$2a^{ik}\partial_k \sigma = a^{jk}\partial_j \partial_k \xi^i + b^k \partial_k \xi^i - \xi^k \partial_k b^i - \mu b^i \qquad (i = 1, \ldots, n), \tag{3.3}$$

$$a^{ij}\partial_i \partial_j \sigma + b^i \partial_i \sigma = -\xi^k \partial_k c - \mu c, \tag{3.4}$$

where the function σ is determined up to a constant term. This *DE* system is used to establish the following important fact.

Theorem. *Equation* (1.1) *admits a nonzero operator of the form* (3.1) *if and only if it is equivalent to an equation all of whose coefficients are independent of one of the coordinates.*

Proof. It is evident that if all the coefficients do not depend (for example) on x^1, then equation (1.1) admits a group G^1 of translations relative to the coordinate x^1; i.e., it admits transformations with the operator $\zeta \cdot \partial = \partial_1$. It is easy to establish that this operator satisfies the equations *DE*.

Conversely, because of the theorem in Section 3.8 it is always possible to establish, using transformations (α) in Section 27.1, that the coordinates of the component ξ of the operator (3.1) would have values $\xi^i = \delta_1^i$. In this case equations *DE* (3.2)–(3.4) assume the form

$$\partial_1 a^{ij} = -\mu a^{ij}, \qquad \partial_1 b^i = -\mu b^i - 2a^{ik}\partial_k \sigma, \qquad \partial_1 c = -\mu c - a^{ij}\partial_i \partial_j \sigma - b^i \partial_i \sigma. \tag{3.5}$$

Upon introducing the functions

$$\omega = \int \sigma\, dx^1, \qquad \nu = -\int \mu\, dx^1,$$

assuming

$$a^{ij} = e^{\nu}\tilde{a}^{ij}, \qquad b^i = e^{\nu}(\tilde{b}^i - 2\tilde{a}^{ij}\omega_j), \qquad c = e^{\nu}(\tilde{c} - \tilde{a}^{ij}\omega_{ij} - \tilde{b}^i\omega_i + \tilde{a}^{ij}\omega_i\omega_j), \tag{3.6}$$

where $\omega_i = \partial_i\omega$, $\omega_{ij} = \partial_i\partial_j\omega$, and substituting these expressions into equations (3.5) there results the following simplification:

$$\partial_1\tilde{a}^{ij} = 0, \qquad \partial_1\tilde{b}^i = 0, \qquad \partial_1\tilde{c} = 0 \qquad (i, j = 1, \ldots, n). \tag{3.7}$$

On the other hand if, one performs a transformation of the type (β) in Section 27.1, with $\theta = \omega$, followed by a transformation of the type (γ) in Section 27.1, with $2\theta = \nu$, and then uses (3.6), a simple calculation shows that (1.1) will transform into the equivalent equation

$$\tilde{a}^{ij}u_{ij} + \tilde{b}^i u_i + cu = 0,$$

whose coefficients, from (3.7), do not depend on x^1. ■

The independence of the coefficients of (1.1) from one of the coordinates allows the construction of a class of (invariant, or partially invariant) solutions for it by separation of variables (not complete), and also a reduction of the problem of finding the general solution to an equation of lower dimensionality (by the Fourier method). Consequently, the previous result establishes a connection between admitted operators and classical methods of constructing solutions.

4. Associated Riemannian Space

A further analysis of the determining equations (3.2)–(3.4) will be done for the case when *the matrix* (a^{ij}) *is not degenerate*, i.e., *when* $\det(a^{ij}) \neq 0$. This assumption essentially simplifies the investigation, which can be seen in the example of group classification of a linear equation discussed in Section 9 for the case $n = 2$. For degenerate equations, when $n > 2$, there does not appear to be complete results as yet.

Equations (1.3) show that the coefficients a^{ij} are components of a contravariant tensor. If it is not degenerate, i.e., $\det(a^{ij}) \neq 0$, then the covariant components a_{ij} of the same tensor are determined from the equations

$$a_{jk}a^{ij} = \delta^i_k \qquad (i, k = 1, \ldots, n). \tag{4.1}$$

The tensors a_{ij} and a^{ij} are symmetric and a fundamental metric quadratic form

$$ds^2 = a_{ij}\,dx^i\,dx^j \tag{4.2}$$

can be introduced into the considerations.

The Riemannian space V_n with the matric (4.2), for which (4.1) holds, will be called a *space associated with the differential equation* (1.1).

Equations (1.3) and (1.5) show that the transformations (α) and (β) in Section 27.1 of equation (1.1) do not change the space V_n. Consequently, the Riemannian space associated with equation (1.1) is an invariant object relative to these transformations. Application of a transformation of the form (γ) in Section 27.1 leads to an equation whose associated space is conformal with respect to the space V_n.

It should be noted that the component ξ of the operator (3.1) is a contravariant vector of the associated space V_n and that σ is a scalar. Below it will be shown that the determining equations (3.2)–(3.4) can be converted into a *covariant form* relative to this V_n, i.e., they can be written using values and operations defined by the space V_n.

One important result follows directly from the determining equations. First it is necessary to note the following. Let G^r be the group of transformations of the space $Z = X \times Y$ with corresponding Lie algebra L^r, admitted by equation (1.1). Obviously, the subspace $X \times \{0\} \subset Z$, which is given by the equation $u = 0$ and can be identified with the space X, is invariant relative to G^r. Thus the group $G^r | X$, induced by the group G^r on its invariant manifold X (see Section 18.3), is determined. The Lie algebra $L^r | X$ of the group $G^r | X$ is obtained if it is assumed that $u = 0$ in (3.1). It is easy to verify that the Lie algebras L^r and $L^r | X$ (which also means the groups G^r and $G^r | X$) are isomorphic. Indeed if $\xi = 0$, then (3.2) implies that $\mu = 0$ and then from (3.3) $\partial_i \sigma = 0$ $(i = 1, \ldots, n)$; i.e., $\sigma = 0$ as the function σ is determined up to a constant term.

Theorem. *If a nondegenerate equation* (1.1) *admits the group* G^r, *then the group* $G^r | X$ *is a subgroup of the conformal transformation group of the associated Riemannian space.*

Proof. Multiplication of equation (3.2) by $a_{is}a_{jt}$, summing with respect to i, j, and using (4.1), leads to the equations

$$a_{ik}\partial_l \xi^i + a_{il}\partial_k \xi^i + \xi^i \partial_i a_{kl} = \mu a_{kl} \qquad (k, l = 1, \ldots, n). \tag{4.3}$$

As is known from Riemannian geometry these equations are the determining equations for the operators $\zeta \cdot \partial = \xi^i \partial_i$ of the Lie algebra of the Riemannian space conformal transformations, with the metric (4.2). Thus any operator (3.1), being a solution of the system (3.2)–(3.4), induces some operator of the conformal transformation group of the space V_n. ■

5. Some Formulas of Riemannian Geometry

The formulas to be given of Riemannian geometry, where all indices assume the values $1, \ldots, n$, will be used to transform equations (3.2)–(3.4).

The Christoffel symbols of the second order are defined by

$$\Gamma^i_{kl} = \tfrac{1}{2}a^{ij}(\partial_l a_{kj} + \partial_k a_{lj} - \partial_j a_{kl}). \tag{5.1}$$

For any tensor $T^{i_1 \cdots i_r}_{k_1 \cdots k_s}$ its covariant derivatives relative to the fundamental tensor a_{ij} are denoted with a comma and are given by

$$T^{i_1 \cdots i_r}_{k_1 \cdots k_s, j} = \partial_j T^{i_1 \cdots i_r}_{k_1 \cdots k_s} + \sum_{\alpha=1}^{r} T^{i_1 \cdots i_{\alpha-1} l i_{\alpha+1} \cdots i_r}_{k_1 \cdots k_s} \cdot \Gamma^{i_\alpha}_{lj}$$

$$- \sum_{\beta=1}^{s} T^{i_1 \cdots i_r}_{k_1 \cdots k_{\beta-1} l k_{\beta+1} \cdots k_s} \cdot \Gamma^l_{k_\beta j}. \tag{5.2}$$

In particular, for a contravariant vector with components ξ^i

$$\xi^i_{,j} = \partial_j \xi^i + \xi^l \Gamma^i_{lj}, \tag{5.3}$$

and for its covariant components $\xi_k = a_{ik}\xi^i$

$$\xi_{k,j} = \partial_j \xi_k - \xi_l \Gamma^l_{kj}. \tag{5.4}$$

The components of the tensors a_{ik} and a^{ik} in covariant differentiation behave as constants, i.e.,

$$a_{ik,j} = 0, \qquad a^{ik}_{,j} = 0. \tag{5.5}$$

The tensor components for the Riemannian curvature of the space V_n are determined by the formulas

$$R^l_{ijk} = \partial_j \Gamma^l_{ik} - \partial_k \Gamma^l_{ij} + \Gamma^s_{ik}\Gamma^l_{sj} - \Gamma^s_{ij}\Gamma^l_{sk} \tag{5.6}$$

and satisfy the identities

$$R^l_{ijk} = -R^l_{ikj}, \qquad R^l_{ijk} + R^l_{jki} + R^l_{kij} = 0. \tag{5.7}$$

Ricci's tensor $R_{ij} = R^k_{ijk}$ is symmetric. The scalar

$$R = a^{ij}R_{ij} = a^{ij}R^k_{ijk} \tag{5.8}$$

is called the *scalar curvature* of the space V_n.

The tensor $R_{ijkl} = a_{is}R^s_{jkl}$ satisfies the identities

$$R_{ijkl} = R_{klij} = -R_{lkij}. \tag{5.9}$$

The tensor $R^i_k = a^{ij}R_{jk}$ and the scalar (5.8) are related by

$$\partial_k R = 2R^i_{k,i}. \tag{5.10}$$

For any vector a^i the Ricci identities

$$a^i_{,kl} - a^i_{,lk} = a^s R^i_{slk} \tag{5.11}$$

are satisfied.

If the determining equations (4.3), for the conformal transformations of the space V_n, are expressed in the covariant components $\xi_k = a_{ik}\xi^i$ of the vector ξ, then these equations assume the form

$$\xi_{i,j} + \xi_{j,i} = \mu a_{ij}. \tag{5.12}$$

The compatibility conditions of the last equations are

$$\xi_{i,jk} = \tfrac{1}{2}(a_{ij}\mu_k + a_{ik}\mu_j - a_{jk}\mu_i) + \xi_s R^s_{kji}, \tag{5.13}$$

where $\mu_k = \partial_k \mu$. Also the relation

$$a^{ij}\mu_{i,j} = \frac{2}{n-1}(R^i_{k,i}\xi^k + R^i_k \xi^k_{,i}) \tag{5.14}$$

holds, where $R^i_k = a^{ij}R_{jk}$. Relation (5.14) can be written in an equipollent form using (5.10) and the equality $R^i_k \xi^k_{,i} = \frac{1}{2}R\mu$ from (5.12):

$$a^{ij}\mu_{i,j} = \frac{1}{n-1}(\xi^k \partial_k R + R\mu). \tag{5.15}$$

The Christoffel symbols are not components of a tensor. Under transformation (α) in Section 27.1 they become

$$\Gamma^s_{ij} p^k_s = p^k_i + \Gamma'^k_{sl}\, p^s_i p^l_j. \tag{5.16}$$

6. Covariant Form of Determining Equations

Equation (3.2) is already written in the covariant form (5.12). To transform equations (3.3) and (3.4) it should be noted that the coefficients b^i of equation (1.1) are not components of a vector. But the quantities

$$a^i = b^i + a^{kl}\Gamma^i_{kl} \qquad (i = 1, \ldots, n) \tag{6.1}$$

form a contravariant vector. Indeed, under the transformation (α) in Section 27.1, using (5.16) and (1.3), the following relations hold:

$$\begin{aligned} a'^k &= b'^k + a'^{sl}\Gamma'^k_{sl} = a^{ij}p^k_{ij} + b^i p^k_i + a^{ij}p^s_i p^l_j \Gamma'^k_{sl} \\ &= a^{ij}p^k_{ij} + b^i p^k_i + a^{ij}\Gamma^s_{ij}p^k_s - a^{ij}p^k_{ij} = (b^i + a^{sj}\Gamma^i_{sj})p^k_i = a^i p^k_i. \end{aligned}$$

Transformation of equation (3.3). The derivative $\partial_j \partial_k \xi^i$ on the right-hand side is expressed in covariant derivatives using (5.2) and (5.3). The result is

$$a^{jk}\partial_j \partial_k \xi^i = a^{jk}\xi^i_{,jk} - a^{jk}\Gamma^i_{lj}\xi^l_{,k} - a^{jk}\Gamma^i_{lk}\xi^l_{,j} + a^{jk}\Gamma^l_{jk}\xi^i_{,l} - a^{jk}(\partial_k \Gamma^i_{lj} - \Gamma^s_{lk}\Gamma^i_{sj})\xi^l.$$

Further, because of (6.1), simple transformations generate the relations

$$b^k\partial_k \xi^i - \xi^k \partial_k b^i - \mu b^i = a^k \xi^i_{,k} - a^{jk}\Gamma^l_{jk}\xi^i_{,l} - a^i_{,k}\xi^k + a^{jk}(\partial_l \Gamma^i_{jk} + \Gamma^s_{jk}\Gamma^i_{ls} - \Gamma^s_{jl}\Gamma^i_{sk} - \Gamma^s_{kl}\Gamma^i_{js})\xi^l - \mu a^i + \mu a^{kl}\Gamma^i_{kl}.$$

Adding these expressions, using the definitions (5.6) and equation (5.12), reduces the right-hand side of equation (3.3) to the form

$$a^{jk}\xi^i_{,jk} + a^k\xi^i_{,k} - a^i_{,k}\xi^k - \mu a^i + a^{jk}R^i_{jlk}\xi^l.$$

But from (5.13)

$$a^{jk}\xi^i_{,jk} = \frac{2-n}{2}a^{ik}\mu_k + a^{jk}a^{is}R^l_{kjs}\xi_l \tag{6.2}$$

and then, according to (5.9),

$$R^i_{jlk}\xi^l = a^{is}R_{sjlk}\xi^l = -a^{is}R_{lkjs}\xi^l = -a^{is}R^l_{kjs}\xi_l.$$

Thus the right-hand side of (3.3) becomes

$$\frac{2-n}{2}a^{ik}\mu_k + a^k\xi^i_{,k} - a^i_{,k}\xi^k - \mu a^i.$$

Finally, if the covariant components $a_k = a_{ik}a^i$ are introduced and equation (5.12) is taken into consideration, then the previous expression becomes

$$a^{ik}\left(\frac{2-n}{2}\mu_k - a_j\xi^j_{,k} - a_{k,j}\xi^j\right).$$

Thus equation (3.3) has the covariant form

$$2\partial_k \sigma = \frac{2-n}{2}\mu_k - a_j\xi^j_{,k} - a_{k,j}\xi^j. \tag{6.3}$$

It is convenient to introduce the skew symmetric tensor

$$K_{ij} = a_{i,j} - a_{j,i}, \tag{6.4}$$

with the aid of which equation (3.3) takes its final form

$$2\partial_i \sigma = \frac{2-n}{2}\mu_i - (a_k\xi^k)_{,i} - K_{ij}\xi^j. \tag{6.5}$$

From (6.5) the function σ is uniquely determined by the given ξ and μ if the compatibility conditions

$$(K_{is}\xi^s)_{,j} = (K_{js}\xi^s)_{,i} \tag{6.6}$$

are satisfied. These are equations of the first order for the unknown vector ξ, additional to (5.12).

Transformation of equation (3.4). If the left-hand side of equation (3.4) is denoted by $\frac{1}{2}N$, then using (3.3) and the definition (6.1) of the vector a^i, the expression

$$N = 2a^{ij}(\partial_i\sigma)_{,j} + 2a^i\partial_i\sigma$$

is obtained. Substituting the expressions for $2\partial_i\sigma$ from (6.3) gives

$$\begin{aligned} N &= \frac{2-n}{2}a^{ij}\mu_{i,j} - \left(a_k a^{ij}\xi^k_{,ij} - \frac{2-n}{2}a^i\mu_i + a^i_{,ki}\xi^k\right) \\ &\quad - (a^{ij}a_{k,j}\xi^k_{,i} + a^j_{,k}\xi^k_{,j}) - (a^i a_k\xi^k_{,i} + a^i a_{i,k}\xi^k) \\ &= N_1 - N_2 - N_3 - N_4. \end{aligned}$$

Here N_1 is defined by (5.15). To transform N_2, the relations (6.2), (5.11), and (5.9) are used to find

$$N_2 = a^i_{,ik}\xi^k.$$

By (5.12), the term N_3 becomes

$$N_3 = a^i_{,i}\mu.$$

Finally, N_4 by (5.12) becomes the form

$$N_4 = \tfrac{1}{2}a^i a_i\mu + \tfrac{1}{2}(a^i a_i)_{,k}\xi^k.$$

Therefore, the expression for N is found to be

$$N = \left(\frac{2-n}{2n-2}R - a^i_{,i} - \frac{1}{2}a^i a_i\right)_{,k}\xi^k + \left(\frac{2-n}{2n-2}R - a^i_{,i} - \frac{1}{2}a^i a_i\right)\mu.$$

Consequently, equation (3.4), with the form $N + 2c_{,k}\xi^k + 2c\mu = 0$, is finally rewritten in the covariant form

$$\xi^k H_{,k} + \mu H = 0, \tag{6.7}$$

where the scalar H is defined by the formula

$$H = -2c + a^i_{,i} + \frac{1}{2}a^i a_i + \frac{n-2}{2(n-1)}R. \tag{6.8}$$

So *in covariant form the system of determining equations consists of the equations* (5.12), (6.5), (6.6) *and* (6.7). *These equations are invariant under any transformations of the kind* (α) *in Section* (27.1).

7. Equation Invariants

The tensor K_{ij} and scalar H, which are included in the system of determining equations, will be called the *equation* (1.1) *invariants*. This term is justified since the aforementioned terms are analogues of the Laplace invariants of a nondegenerate equation for the case $n = 2$ (see Section 9.2). That is, the following statement holds.

Theorem 1. *Equations having the same associated Riemannian spaces (the same coefficients a^{ij}) are equivalent relative to the transformation (β) in Section* 27.1 *if and only if they have the same tensors $K_{i,j}$ and scalars H.*

Proof. If (1.4) holds, then the relations (1.6), written in the covariant components a_i of the vector (6.1) for b^i, have the form

$$a_i' = a_i + 2\theta_i \qquad (i = 1, \ldots, n), \tag{7.1}$$

from which, because of definition (5.4), the equalities

$$K_{ij}' = K_{ij} \qquad (i, j = 1, \ldots, n) \tag{7.2}$$

follow. Also, by relations (1.6) and (7.1) the equality

$$-2c' + a'^{i}_{,i} + \tfrac{1}{2}a'^{i}a_i' = -2c + a^{i}_{,i} + \tfrac{1}{2}a^{i}a_i$$

is established by a simple calculation, from which the equality

$$H' = H \tag{7.3}$$

follows because of definition (6.8).

On the contrary, let equalities (7.2) and (7.3) be satisfied. From (7.2) it follows that $a_i' - a_i$ is the gradient of some function, for example, 2θ. Then equalities (7.1) hold, and so does the first series of equalities (1.6) which are equipollent to (7.1). Also, it is possible to invert the calculation, which leads to equality (7.3), and the expression (1.6) for c'. Thus, with the function θ chosen in this way, the equality

$$e^{-\theta}F[e^{\theta}u] = a^{ij}u_{ij} + b'^{i}u_i + c'u = F'[u]$$

is found, and this means that transformation (β) in Section 27.1 applies to equation (1.1). ■

It is also important to determine the behavior of the invariants under the transformations (γ) in Section 27.1, i.e., under multiplication of the left-hand side of equation (1.1) by an exponential. From that the associated Riemannian space V_n transforms into its conformal Riemannian space $\overline{V}_n$ with the fundamental form $\bar{a}_{ij}\,dx^i\,dx^j$.

Theorem 2. *Under the transformation of the form* (γ) *in Section* 27.1 *the invariant* K_{ij} *is unchanged and the invariant* H *obtains a multiplier, namely,*

$$\overline{H} = e^{-2\theta}H. \tag{7.4}$$

Proof. From Riemannian geometry the following relation between the Christoffel symbols of the spaces V_n and $\overline{V}_n$ is known:

$$\Gamma^k_{i,j} = \overline{\Gamma}^k_{i,j} - \delta^k_i\theta_j - \delta^k_j\theta_i + \bar{a}_{ij}\bar{a}^{kl}\theta_l. \tag{7.5}$$

From (1.8), (7.5), and the definition of the vector $\bar{a}^k$ for the transformed equation, the relation

$$a^k = e^{2\theta}[\bar{a}^k + (n-2)\bar{a}^{kl}\theta_l] \tag{7.6}$$

is found. For the change to covariant components both sides of equation (7.6) are multiplied by $a_{ik} = e^{-2\theta}\bar{a}_{ik}$ and summed with respect to k. This gives

$$a_i = \bar{a}_i + (n-2)\theta_i, \tag{7.7}$$

from which the equality $K_{ij} = \overline{K}_{ij}$ follows. Further, denoting covariant differentiation in V_n (as it used to be) by a comma in the subscript and covariant differentiation in $\overline{V}_n$ by a semicolon in the subscript, and using (7.5) and (7.6), easily gives the relation

$$a^i_{,i} = e^{2\theta}[\bar{a}^i_{;i} + (n-2)\bar{a}^{ij}\theta_{i;j} - (n-2)\bar{a}^i\theta_i - (n-2)^2\bar{a}^{ij}\theta_i\theta_j].$$

Also, by the multiplication of (7.6) and (7.7) the relation

$$a^i a_i = e^{2\theta}[\bar{a}^i\bar{a}_i + 2(n-2)\bar{a}^i\theta_i + (n-2)^2\bar{a}^{ij}\theta_i\theta_j]$$

is found. From this

$$a^i_{,i} + \tfrac{1}{2}a^i a_i = e^{2\theta}[\bar{a}^i_{;i} + \tfrac{1}{2}\bar{a}^i\bar{a}_i + (n-2)\bar{a}^{ij}\theta_{i;j} - \tfrac{1}{2}(n-2)^2\bar{a}^{ij}\theta_i\theta_j].$$

On the other hand, from Riemannian geometry the relation

$$R = e^{2\theta}[\overline{R} - 2(n-2)\bar{a}^{ij}\theta_{i;j} + (n-1)(n-2)\bar{a}^{ij}\theta_i\theta_j]$$

is known. Using these two equations, (7.4) is obtained from (6.8). ■

8. Group of Motions

It is known from Riemannian geometry that the space V_n admits a group of motions, with operators of the kind $\xi^i(x)\partial_i$, if and only if the Killing's equations

$$\xi_{i,j} + \xi_{j,i} = 0 \qquad (i, j = 1, \ldots, n) \tag{8.1}$$

are satisfied. It is known also that the dimensionality of a motions group is not more than the number $\frac{1}{2}n(n + 1)$. Using these facts further properties of the group admitted by equation (1.1) are determined.

Theorem. *If for equation* (1.1) *the invariant* $H \neq 0$, *then the group* $G^r|X$, *isomorphic to the group* G^r *admitted by this equation, is a subgroup of the* motions groups *of the Riemannian space* V_n *associated with* (1.1).

Proof. If the equations $F'[u] = 0$ and $F[u] = 0$ are related by the transformation (γ) in Section 27.1, then they admit the same group. But when $H \neq 0$ it is possible to choose the function θ such that $H = e^{2\theta}$. But then, because of (7.4), the equation $F'[u] = 0$ will have the invariant $\overline{H} = 1$. Consequently, the determining equation (6.7) will assume the form $\bar{\mu} = 0$ and equations (5.12) reduce to the Killing's equations (8.1). ■

Consequence. *The order of the group* G^r *admitted by equation* (1.1), *with the invariant* $H \neq 0$, *satisfies the inequality*

$$r \leq \tfrac{1}{2}n(n + 1). \tag{8.2}$$

9. The Group with Maximum Order

According to the theorem in Section 27.8, equation (1.1), for which the invariant $H \neq 0$, is equivalent to an equation in which $H = \text{const}$. Here it will be shown that if such an equation admits the group with maximal possible order (8.2), then it is equivalent to a *standard* equation.

This result is obtained using the known fact from Riemannian geometry that the Riemannian space V_n, with a motions group of maximal order (8.2), necessarily has a constant curvature K_0 connected with the scalar curvature by the relation

$$R = -n(n - 1)K_0. \tag{9.1}$$

Also, in a space V_n of constant curvature there is a coordinate system in which the fundamental metric form has the Riemannian form

$$ds^2 = P^{-1} \sum_{i=1}^{n} \varepsilon_i (dx^i)^2, \tag{9.2}$$

where

$$P = (1 + \tfrac{1}{4}K_0 \rho^2)^2, \qquad \rho^2 = \sum_{i=1}^{n} \varepsilon_i (x^i)^2,$$

and $\varepsilon_i = \pm 1$ $(i = 1, \ldots, n)$. If all the $\varepsilon_i = 1$, then ρ is distance from the point x to the coordinate origin. The general case (when not all $\varepsilon_i = 1$) is led to this one by formally considering X to be a complex space.

Theorem. *Equation* (1.1) *admitting the group* G^r *of order* $r = n(n+1)/2$, $n > 2$, *with the invariant* $H \neq 0$, *is equivalent to the equation*

$$(1 + \tfrac{1}{4}K_0\rho^2)^2\,\Delta u = \tfrac{1}{2}Hu, \tag{9.3}$$

where $H = \text{const}$ *and* Δ *is the generalized Laplace operator*

$$\Delta = \sum_{i=1}^{n} \varepsilon_i \partial_i^2, \tag{9.4}$$

and the ε_i *have the same values as in* (9.2).

Proof. It is supposed that equation (1.1) is already such that $H = \text{const}$. First, it is determined that for the considered equation the tensor $K_{ij} = 0$. Indeed, as was shown in the proof in Section 27.8, here $\mu = 0$. Then equations (5.12) reduce to equations (8.1), and besides them there must not be any equations relating the values of the components ξ_i with the derivatives $\xi_{i,j}$ (otherwise, the order r will be less than the number $n(n+1)/2$). This means that from equations (6.6), rewritten in the form

$$K_{is}\xi^s_{,j} - K_{js}\xi^s_{,i} + (K_{is,j} - K_{js,i})\xi^s = 0,$$

there must follow the relation $K_{is}\xi^s_{,j} - K_{js}\xi^s_{,i} = 0$, or in the covariant components of the vector ξ

$$K_{is}a^{sl}\xi_{l,j} - K_{js}a^{sl}\xi_{l,i} = 0.$$

From here, because of (8.1), the identities

$$K_{is}a^{sl}\delta^k_j - K_{js}a^{sl}\delta^k_i - K_{is}a^{sk}\delta^l_j + K_{js}a^{sk}\delta^l_i = 0$$

follow. If here it is assumed that $k = j$ and the sum is taken with respect to j, then the equality $(n-2)K_{is}a^{sl} = 0$ is found, from which it follows that $K_{ij} = 0$.

From this fact it follows that the vector a_i is the gradient of some function ψ, i.e., $a_i = \partial_i\psi$. Because of (9.2) the considered equation can be put in the form

$$P\,\Delta u + b^i u_i + cu = 0. \tag{9.5}$$

Now it is confirmed that this is equivalent to an equation in which all the $b^i = 0$. Indeed, a simple calculation of the Christoffel symbols leads to the formula

$$a^{kl}\Gamma^i_{kl} = \frac{n-2}{2}\,\delta^{is}\partial_s P.$$

Thus from definition (6.1) it follows that the b^i are components of the vector whose covariant components are

$$b_i = a_i - \frac{n-2}{2}\partial_i \ln P = \partial_i\left(\psi - \frac{n-2}{2}\ln P\right) \qquad (i = 1, \ldots, n).$$

Consequently, transformation (β) in Section 27.1 with the function

$$\theta = \frac{n-2}{4}\ln P - \tfrac{1}{2}\psi,$$

because of formula (1.6), leads to the equation in which the $b'^i = 0$. Therefore the invariant H does not change and, in particular, remains constant.

So it is possible to assume in equation (9.5) that the coefficients $b^i = 0$. In that case an elementary calculation of the invariant H, by formula (6.8), leads to the value $H = -2c$, because of which (9.5) coincides with (9.3). ■

Consequence. *Any equation* (1.1) *admitting the group G^r of maximal order, with the invariant $H \neq 0$, is equivalent to one of the following standard equations: If $K_0 = 0$*

$$\Delta u = \pm u, \tag{9.6}$$

and if $K_0 \neq 0$

$$(1 \pm \rho^2)^2\,\Delta u = \lambda u \qquad (\lambda = \text{const}). \tag{9.7}$$

Proof. When $K_0 = 0$, through a suitable extension of the vector x the equality $H = \pm 1$ can be achieved. When $K_0 \neq 0$, an analogous transformation gives $K_0 = \pm 4$ and so equation (9.7) with $\lambda = 2H/K_0$. ■

10. Groups Admitted by Standard Equations

For simplicity the Lie algebras L^r, admitted by standard equations in the elliptic case when all $\varepsilon_i = 1$, will be found here. The general case is obtained from this by the choice of $x'^i = \sqrt{-1}\,x^i$ for those values of i for which $\varepsilon_i = -1$, if $K_0 > 0$ (or for $\varepsilon_i = 1$, if $K_0 < 0$).

The group G^r, admitted by equation (9.6), is seen directly to be generated by n translations along the coordinate axes and by $n(n-1)/2$ rotations around these axes, and the quantity u remains invariant. The basis of the Lie algebra for this group consists of the translation operators

$$\zeta_i \cdot \partial = \partial_i \qquad (i = 1, \ldots, n) \tag{10.1}$$

and the rotation operators

$$\zeta_{ij} \cdot \partial = x^i\partial_j - x^j\partial_i \qquad (i, j = 1, \ldots, n;\ i < j). \tag{10.2}$$

Equation (9.7) also admits the operators (10.2). To find the remainder of the n operators it is convenient to rewrite the equation in the form

$$\Delta u - \lambda(1 + \rho^2)^{-2} u = 0. \tag{10.3}$$

Here $a^{ij} = \delta^{ij}$, $a_{ij} = \delta_{ij}$, and all $\Gamma^i_{kl} = 0$, because of which the covariant derivatives coincide with the usual ones ∂_i. Also, the curvature tensor $R^l_{kij} = 0$. An application of equations (5.13) shows that when $n > 2$ the function μ is linear and any other coordinate ξ^i is quadratic in the $\{x^i\}$. Considering this fact for equations (5.12) it is easy to find the following general solution (the prime at the summation sign means that the term with $s = i$ should be omitted):

$$\xi^i = A_i\left[(x^i)^2 - \sum_{s=1}^{n}{}'(x^s)^2\right] + 2\sum_{s=1}^{n}{}' A_s x^s x^i + Bx^i + \sum_{s=1}^{n}{}' C_{is} x^s + C_i,$$
$$\mu = 4A_k x^k + 2B, \tag{10.4}$$

where A_i, B, C_{ij}, and C_i are arbitrary constants and $C_{ij} = -C_{ji}$.

From (10.4) and the fact that $a_i = 0$ and $K_{ij} = 0$, equation (6.5) can be easily integrated, with the result

$$\sigma = -(n-2)A_k x^k, \tag{10.5}$$

and equation (6.6) is satisfied uniquely. It is left to satisfy the last of the determining equations (6.7), which here assumes the form

$$\xi^k \partial_k c + \mu c = 0,$$

where $c = -\lambda(1 + \rho^2)^{-2}$. Substitution of the general solution (10.4) into this equation shows that it is satisfied if and only if $B = 0$ and $C_k = A_k$ $(k = 1, \ldots, n)$.

Finally, the general solution of the determining equations of the Lie algebra L^r of operators (3.1), which are admitted by equation (10.3), is

$$\xi^i = A_i\left[(x^i)^2 - \sum_{s=1}^{n}{}'(x^s)^2 + 1\right] + 2\sum_{s=1}^{n}{}' A_s x^s x^i + \sum_{s=1}^{n} C_{is} x^s,$$
$$\sigma = -(n-2)A_k x^k. \tag{10.6}$$

Here the constants A_i are arbitrary and the C_{ij} satisfy the relations $C_{ij} = -C_{ji}$. The prime at the summation sign has the same meaning as in (10.4).

Evidently, the constants C_{ij} give the basis operators (10.2). The constants A_i correspond to new operators for equation (10.3), which replace the translation operators (10.1). These new operators are such that (as the associated space V_n with the equation (10.3) is Euclidean, $x^i = x_i$)

$$\zeta_i \cdot \partial = (2x_i z^k + (1 - \rho^2)\delta^k_i)\partial_k - (n-2)x_i u\partial_u \qquad (i = 1, \ldots, n). \tag{10.7}$$

The groups G^r, *admitted by equations* (9.6) *and* (10.3), *have the same dimensionality, but they are not isomorphic since the Lie algebra of operators with the basis* (10.1), (10.2) *contains a solvable* (*Abelian*) *ideal with the basis* (10.1), *and the Lie algebra* (10.2), (10.7) *is semisimple.* It is not difficult to verify the last fact using the criterion of Cartan in Section 14.6.

§28. TANGENTIAL TRANSFORMATION

1. Special Space Expansion

To formulate the definition and different properties of tangential transformations, the following abstract construction of the space expansion is introduced.

Let the Banach space $Z = X \times Y$ be given and let $Y_1 = \mathscr{L}(X; Y)$ be the first continuation of Y with respect to X (see Section 4.2). The continuation space $Z_1 = Z \times Y_1$ can be considered as the product of the spaces X and $Y^{(1)} = Y \times Y_1$. In this case it will also be denoted by $Z^{(1)}$. This representation $Z^{(1)} = X \times Y^{(1)}$ is the basis of a special expansion. For vectors of the spaces X and Y the usual notations are used, namely, $x \in X$, $y \in Y$, and a new notation $p \in Y_1$ is introduced; then a pair (y, p) is a general vector of the space $Y^{(1)}$.

As in Section 4.2, the continuations $Y_k^{(1)}$ of the space $Y^{(1)}$ with respect to X and the continuations $Z_k^{(1)}$ of the space $Z^{(1)}$ corresponding to them are considered. It is clear that every continuation $Y_k^{(1)}$ is a product of the corresponding continuations Y_k and $(Y_1)_k$, i.e., $Y_k^{(1)} = Y_k \times (Y_1)_k$, because of which a general vector of the space $Y_k^{(1)}$ is written in the form of the pair $(\underset{k}{y}, \underset{k}{p})$. Consequently, the vectors $\underset{k}{z}^{(1)} \in Z_k^{(1)}$ have the representation

$$\underset{k}{z}^{(1)} = (x, (y, p), (\underset{1}{y}, \underset{1}{p}), \ldots, (\underset{k}{y}, \underset{k}{p})). \tag{1.1}$$

In the continuation theory an important role is played by the total differentiation operator of the mappings $Z_k^{(1)} \to P$, which, in accordance with Section 4.3, has the form

$$D = \partial_x + \underset{1}{y} \cdot \partial_y + \underset{1}{p} \cdot \partial_p + \underset{2}{y} \cdot \partial_{\underset{1}{y}} + \underset{2}{p} \cdot \partial_{\underset{1}{p}} + \cdots. \tag{1.2}$$

2. Fundamental Definition

The space $Z^{(1)}(x, y, p)$ is that of transformations of class $C_\infty(Z^{(1)})$, with the components $f: Z^{(1)} \to X$, $g: Z^{(1)} \to Y$, and $h: Z^{(1)} \to Y_1$, defined by

$$x' = f(x, y, p), \qquad y' = g(x, y, p), \qquad p' = h(x, y, p). \tag{2.1}$$

Every such transformation, satisfying a condition analogous to (4.5.2), can be continued according to the rules given in Section 4.5, up to the transformation $Z_1^{(1)} \to Z_1^{(1)}$. From this the additional components of the continuation transformation are given by (see (4.5.4))

$$\underset{1}{y}' = (Dg) \circ (Df)^{-1}, \qquad \underset{1}{p}' = (Dh) \circ (Df)^{-1}. \tag{2.2}$$

Among all transformations (2.1) the tangential transformations are distinguished by the following definition.

The transformation (2.1) is called a *tangential transformation of the space* $Z^{(1)}$ if in $Z_1^{(1)}$ a manifold, given by the equation

$$\underset{1}{y} = p, \tag{2.3}$$

is invariant relative to the continuous transformation (2.1), (2.2). This means that the identity

$$\underset{1}{y}' = p' \tag{2.4}$$

holds on the manifold (2.3).

The geometric meaning of this definition is given by an analytic interpretation of the manifold (2.3) on which the operation of transformation (2.1) on the mapping $X \to Y^{(1)}$, of the form $y = u(x)$, $p = v(x)$, is considered. According to Section 4.4 the continuation of such a mapping is given by $\underset{1}{y} = \partial u(x)$, $\underset{1}{p} = \partial v(x)$, because of which $v(x) = \partial u(x)$ holds on the manifold (2.3). Therefore, the assignment of the manifold (2.3) is equipollent to considering a special class of mappings $X \to Y^{(1)}$, operating by the formulas

$$y = u(x), \qquad p = \partial u(x). \tag{2.5}$$

If the first of these equations is considered as an equation of the manifold $U \subset Z$, then the second can be considered as giving a tangential space to the manifold U in the sense that for any $a \in X$, the vector $p\langle a\rangle = \partial u(x)\langle a\rangle$ is tangent to U at the point x. Consequently, the given definition of a tangential transformation means that *for any manifold $U \subset Z$, with the equation $y = u(x)$, the following holds: If U transforms into U', then at the same time the tangential mapping to U transforms into the tangential mapping to U'.*

In particular, when the mappings f and g in (2.1) do not depend on p, the first equation of (2.2) gives the already known (from (4.5.4)) continuation of transformations of the space Z, and, because of (2.4), it coincides with the last equation of (2.1). Such a tangential transformation is called a *continuous dot transformation.*

In connection with the given definition the questions arise: Are there dot transformations which are not continuous, and are there other, possibly simpler, representations of any tangential transformation? In what follows these and other questions will be studied for *Lie groups of tangential transformations.* Owing to the construction introduced previously, which is connected with consideration of the invariant manifold (2.3), the answers to those questions can be obtained from an infinitesimal study (i.e. in terms of the Lie algebra of operators), discussed in the previous chapters.

3. Infinitesimal Operator Representation

The operator for the one-parameter group $G^1(\tau)$ of tangential transformations is given by the vector field $\tau\colon Z^{(1)} \to Z^{(1)}$ and is written in the form

$$\tau \cdot \partial = \xi \cdot \partial_x + \eta \cdot \partial_y + \zeta \cdot \partial_p, \tag{3.1}$$

for some mappings $\xi\colon Z^{(1)} \to X$, $\eta\colon Z^{(1)} \to Y$, and $\zeta\colon Z^{(1)} \to Y_1$ of the class $C_\infty(Z^{(1)})$. The first continuation of the operator (3.1) has the form

$$\underset{1}{\tau} \cdot \partial = \tau \cdot \partial + \alpha \cdot \partial_{\underset{1}{y}} + \beta \cdot \partial_{\underset{1}{p}}$$

and is obtained from (4.8.4) using the total differentiation operator (1.2), whence

$$\alpha = D\eta - \underset{1}{y} \cdot D\xi, \qquad \beta = D\zeta - p \cdot D\xi. \tag{3.2}$$

From the fundamental definition in Section 28.2 and the invariance criterion (3.12.1) of the manifold, it follows that (3.1) is the operator of the group $G'(\tau)$ of tangential transformations if and only if the identity $\alpha = \zeta$ holds on the manifold (2.3) and, because of (3.2), it has the expanded form

$$\partial_x \eta + p \cdot \partial_y \eta + p_1 \cdot \partial_p \eta - p \cdot (\partial_x \xi + p \cdot \partial_y \xi + p_1 \cdot \partial_p \xi) = \zeta.$$

As the mappings ξ, η, ζ do not depend upon p_1, this relation can be decomposed relative to p_1. The following determining equations of the tangential transformations result:

$$\partial_p \eta = p \cdot \partial_p \xi, \tag{3.3}$$

$$\zeta = \partial_x \eta + p \cdot \partial_y \eta - p \cdot (\partial_x \xi + p \cdot \partial_y \xi). \tag{3.4}$$

Lemma. *If* $\dim Y > 1$, *then necessarily* $\partial_p \xi = \partial_p \eta = 0$, *i.e., any tangential mapping is a continuous dot one.*

Proof. This fact follows from the integrability condition of equation (3.3) as an equation for the unknown η. Using an arbitrary constant (*tentatively*) vector $A \in Y_1$, equation (3.3) is rewritten in the form

$$\partial_p \eta \langle A \rangle = p \langle \partial_p \xi \langle A \rangle \rangle.$$

Upon repeated differentiation with respect to p, taking a derivative of the tentative vector $B \in Y_1$ and writing symmetry condition of the second derivatives $\partial_p^2 \eta\langle B, A\rangle$, the relation

$$A\langle \partial_p \xi\langle B\rangle\rangle = B\langle \partial_p \xi\langle A\rangle\rangle. \tag{3.5}$$

is obtained. In particular, if A and B take the forms $A = a \otimes c$ and $B = b \otimes d$, where $a, b \in Y$ and $c, d \in X^*$, then because of the formula $(a \otimes c)\langle e\rangle = a(c \cdot e)$, which holds for any vector $e \in X$, the previous relation assumes the form

$$a(\partial_p(c \cdot \xi)\langle B\rangle) = b(\partial_p(d \cdot \xi)\langle A\rangle),$$

where the multipliers $\partial_p(c \cdot \xi)\langle B\rangle$ are scalars. Therefore, if $\dim Y > 1$, then there exist linearly independent vectors $a, b \in Y$ so that this relation implies that $\partial_p(c \cdot \xi)\langle B\rangle = 0$ is equipollent to $\partial_p \xi = 0$. ■

Note. If $\dim Y = 1$, then vectors A and B of the space X^* can be taken in the relation (3.5). Then (3.5) can be rewritten as

$$A \cdot \partial_p \xi\langle B\rangle = B \cdot \partial_p \xi\langle A\rangle,$$

and this relation is equipollent to the complete integrability of the mapping $\xi\colon Y_1 \to X$, i.e., to existence of solutions for the equation

$$\partial_p W = \xi, \tag{3.6}$$

for an unknown mapping $W\colon Y_1 \to R$.

The general representation of operators of tangential transformations for the space $Z^{(1)}$ is given by the following theorem:

Theorem. *The group $G^1(\tau)$ of tangential transformations of the space $Z^{(1)}$ is not a group of continuous dot transformations of the space $Z = X \times Y$ only if* $\dim Y = 1$. *In the last case the group $G^1(\tau)$ is a group of tangential transformations if and only if there exists a function $W\colon Z^{(1)} \to R$ such that the representation*

$$\xi = \partial_p W, \qquad \eta = p \cdot \partial_p W - W, \qquad \zeta = -\partial_x W - p\partial_y W \tag{3.7}$$

holds for the components of the operator (3.1) *of this group.*

Proof. The first statement follows from the previous lemma. If $\dim Y = 1$, then for any function $W\colon Z^{(1)} \to R$ equations (3.7) give a solution of the determining equations (3.3), (3.4). Conversely, if such a solution is given, then the function W can be defined by the formula

$$W = p \cdot \xi - \eta, \tag{3.8}$$

for which it is not difficult to verify that equalities (3.7) are satisfied. ■

From this theorem it follows that the characteristic function W is in one-to-one correspondence with every tangential transformation operator, the components of which are given by equations (3.7). If there are two operators $\tau_1 \cdot \partial$ and $\tau_2 \cdot \partial$ of tangential transformations, then their commutator $[\tau_1, \tau_2] \cdot \partial$ is also an operator of tangential transformation. This follows from the lemma in Section 7.11, wherein the manifold which admits two operators also admits their commutator, and from the fact that tangential transformations are defined as transformations admitted by the manifold (2.3). Because of the previous theorem, the characteristic function, which must be defined by the characteristic functions W_1 and W_2 of the operators $\tau_1 \cdot \partial$ and $\tau_2 \cdot \partial$, corresponds to the operator $[\tau_1, \tau_2] \cdot \partial$. This function will be called the *commutator of the functions* W_1 *and* W_2 and will be denoted by the symbol $[W_1, W_2]$. A simple calculation gives an expression for this commutator

$$[W_1, W_2] = \partial_p W_1 \cdot \partial_x W_2 + (p \cdot \partial_p W_1 - W_1)\partial_y W_2 \\ - \partial_p W_2 \cdot \partial_x W_1 - (p \cdot \partial_p W_2 - W_2)\partial_y W_1. \tag{3.9}$$

The tangential transformation (2.1) is said to be *homogeneous* if $g(x, y, p) = y$ and the mappings f and h do not depend on y. The characteristic function W of the group $G^1(\tau)$ of homogeneous transformations does not depend on y and the operator of such a group has the form

$$\tau \cdot \partial = (\partial_p W) \cdot \partial_x - (\partial_x W) \cdot \partial_p. \tag{3.10}$$

With this it follows from (3.7) that $p \cdot \partial_p W = W$, i.e., because of Euler's theorem the characteristic function W is homogeneous of first order. Conversely, it is clear that the operator (3.10), with any homogeneous function W of first order, is an operator of the group $G^1(\tau)$ of homogeneous tangential transformations. The commutator (3.9) of such homogeneous characteristic functions reduces to the expression

$$[W_1, W_2] = \partial_p W_1 \cdot \partial_x W_2 - \partial_p W_2 \cdot \partial_x W_1, \tag{3.11}$$

which is usually called the Poisson bracket of the functions W_1 and W_2.

4. Applications

Many applications of tangential transformations in geometry and mechanics result from the following fact. On the space $Z^{(1)} = X \times Y \times Y_1$, where $\dim Y = 1$, let the first-order differential equation

$$W(x, y, \underset{1}{y}) = 0 \tag{4.1}$$

be given. Then *the group* $G^1(\tau)$ *of tangential transformations of the space* $Z^{(1)}$, *for which* W *is a characteristic function, possesses the property that the orbits found by its operation on the points of the space* $Z^{(1)}$ *coincide with the characteristics of equation* (4.1). Indeed, because of (3.7), the Lie equations (which are orbit differential equations) of the group $G^1(\tau)$ have the form

$$\partial_a x = \partial_p W, \qquad \partial_a y = p\partial_p W - W, \qquad \partial_a p = -\partial_x W - p\partial_y W, \quad (4.2)$$

and these differential equations are the characteristics of equation (4.1).

In the case of the group $G^1(\tau)$ of homogeneous tangential transformations, equations (4.2) assume the form of Hamilton's system

$$\partial_a x = \partial_p W, \qquad \partial_a p = -\partial_x W,$$

which provides the possibility of using group analysis in the mechanics of dynamic systems.† It is also interesting to use tangential mappings for the group analysis of differential mappings for the group analysis of differential equations. For example, the group classification of the second-order general (nonlinear) equation with two independent variables, with respect to tangential transformations, was given by S. Lie. The complete exposition of this result takes too much space. As an example, only the Monge–Ampère equation will be considered. It is already clear in this example that transition to tangential transformations significantly expands the admitted group.

Let $Z = X \times Y$, where $X = R^2(x^1, x^2)$ and $Y = R(y)$. The simplest Monge–Ampère equation is

$$y_{11}y_{22} - y_{12}^2 = 0, \quad (4.3)$$

where $y_{ij} = \partial_i\partial_j y$. A group of homogeneous tangential transformations, which is admitted by equation (4.3), is to be found. The admitted operator is written in the form

$$\underset{1}{\tau}\cdot\partial = W_p\partial_1 + W_q\partial_2 - W_1\partial_p - W_2\partial_q, \quad (4.4)$$

where $p = y_1$, $q = y_2$, and the derivatives $\partial_i = \partial_{x^i}$, ∂_p, and ∂_q are denoted by subscripts of W. The continuation of the operator (4.4) to the derivatives $r = y_{11}$, $s = y_{12}$, $t = y_{22}$ (Monge's notation) is

$$\underset{1}{\tau}\cdot\partial = \tau\cdot\partial - \pi\partial_r - \rho\partial_s - \sigma\partial_t,$$

where π, ρ, σ are found from (3.2) and have the expressions

$$\pi = W_{11} + 2rW_{1p} + 2sW_{1q} + r^2W_{pp} + 2rsW_{pq} + s^2W_{qq},$$
$$\rho = W_{12} + rW_{2p} + s(W_{1p} + W_{2q}) + tW_{1q} + rsW_{pp} + (rt + s^2)W_{pq} + stW_{qq},$$
$$\sigma = W_{22} + 2sW_{2p} + 2tW_{2q} + s^2W_{pp} + 2stW_{pq} + t^2W_{qq}.$$

† Further details about applications in geometry and mechanics may be found in Elsenhart's *Continuous Groups of Transformations*, Dover, 1961.

The result of operating by the operator $\underset{1}{\tau} \cdot \partial$ on the function $rt - s^2$, transition to the manifold $rt = s^2$, and decomposition relative to the variables r, t gives the determining equations

$$W_{1p} + W_{2q} = 0, \qquad W_{11} = W_{12} = W_{22} = 0.$$

Construction of the general solution for these equations is not difficult, with the result

$$W = x^1\phi_q - x^2\phi_p + \psi, \tag{4.5}$$

where the arbitrary functions $\phi, \psi: R^2(p, q) \to R$ must be homogeneous with ϕ of the second degree and ψ of the first degree.

The enlargement of the admitted group is evident here since the operator (4.4) will be an operator of the continuous dot transformations group, with the characteristic function (4.5), only if ϕ is a homogeneous polynomial of the second degree and ψ of the first degree.

Generally, it is useful to keep in mind that the groups of tangential transformations are Lie groups of transformations. Hence all the facts, conclusions, and algorithms considered in the previous chapters hold for them. In particular, using those results it is possible to organize the search for those classes of differential equations which have partial solutions with one unknown function, to realize the group classification and group splitting, and so on.

5. Continuation of Tangential Transformations

In connection with the fundamental definition of the tangential transformation given in Section 28.2 it is necessary to determine what role is played there by a "duplicate" of the continuation spaces Y_k. Is the second copy of this space "unnecessary"? It appears this is indeed so and that this second copy can be removed at the expense of a translation on an invariant manifold. To simplify the account here the corresponding results will be obtained for the case of Lie groups of tangential transformations. [It is sufficient to restrict attention to the groups $G^1(\tau)$.]

The continuation spaces $Z_k^{(1)}$ of the terms (1.1), already introduced in Section 28.1, and the manifold $T_k \subset Z_k^{(1)}$ given by the equations

$$\underset{1}{y} = p, \qquad \underset{2}{y} = \underset{1}{p}, \qquad \ldots, \qquad \underset{k}{y} = \underset{k-1}{p}, \tag{5.1}$$

are considered. Let $G^{(1)}(\tau)$ be the group of tangential transformations of the space $Z^{(1)}$ with the operator (3.1) and $\underset{k}{G^{(1)}}(\tau)$ be its continuation on the space $Z_k^{(1)}$.

Lemma. *For any $k \geq 1$, the manifold T_k is invariant with respect to the group $\underset{k}{G}^{(1)}(\tau)$. The operator induced by the continuation operator $\underset{k}{\tau} \cdot \partial$ on the manifold T_k coincides with the $(k+1)$th continuation of the operator induced on T_1 by the operator $\xi \cdot \partial_x + \eta \cdot \partial_y$.*

Proof. It is evident that the manifold T_k is a continuous manifold obtained by continuation of the manifold T_1 in the manner described in Section 19.5. It is easy to verify from (3.3) and (3.4) that the representation

$$(\underset{1}{\tau} \cdot \partial)(\underset{1}{y} - p) = F \cdot (\underset{1}{y} - p)$$

holds for some mapping $F: Z_1^{(1)} \to Y_1$. Thus, here the conditions of the lemma in Section 19.5 are satisfied and the first statement is a particular case of that lemma.

Further, the kth continuation of the operator (3.1) is

$$\begin{aligned} \underset{k}{\tau} \cdot \partial = \xi \cdot \partial_x + \eta \cdot \partial_y + (\zeta \cdot \partial_p + \underset{1}{\eta} \cdot \partial_{\underset{1}{y}}) + \cdots \\ + (\underset{k-1}{\zeta} \cdot \partial_{\underset{k-1}{p}} + \underset{k}{\eta} \cdot \partial_{\underset{k}{y}}) + \underset{k}{\zeta} \cdot \partial_{\underset{k}{p}}, \end{aligned} \tag{5.2}$$

and the condition of invariance of the manifold (5.1) consists in the requirement that

$$\zeta = \underset{1}{\eta}, \qquad \underset{1}{\zeta} = \underset{2}{\eta}, \qquad \ldots, \qquad \underset{k-1}{\zeta} = \underset{k}{\eta} \tag{5.3}$$

on T_k. As a result of the transformation $(\underset{s}{y}, \underset{s-1}{p}) \to (\underset{s}{y}, \underset{s-1}{\theta})$, where $\underset{s-1}{\theta} = \underset{s}{y} - \underset{s-1}{p}$ $(s = 1, \ldots, k)$, the operator, induced by the operator (5.2) on T_k because of (5.3) and the fact that $\underset{k}{\zeta} = \underset{k+1}{\eta}$, $\underset{k}{p} = \underset{k+1}{y}$, is

$$\underset{k+1}{\bar{\tau}} \cdot \partial = \xi \cdot \partial_x + \eta \cdot \partial_y + \underset{1}{\eta} \partial_{\underset{1}{y}} + \cdots + \underset{k}{\eta} \cdot \partial_{\underset{k}{y}} + \underset{k+1}{\eta} \cdot \partial_{\underset{k+1}{y}}. \tag{5.4}$$

In addition, by a similar approach on the manifold (5.1) the operator of total differentiation D becomes

$$\bar{D} = \partial_x + \underset{1}{y} \cdot \partial_y + \underset{2}{y} \cdot \partial_{\underset{1}{y}} + \cdots.$$

Thus operator (5.4) coincides with the $(k+1)$th continuation of the operator $\bar{\tau} \cdot \partial = \xi \cdot \partial_x + \eta \cdot \partial_y$, taken on the manifold $p = \underset{1}{y}$. ■

This lemma shows that in operations with tangential transformations the variable p actually appears to be "unnecessary" and does not need to be considered. This variable fulfills its function in writing the tangential transformation.

6. Higher Order Tangential Transformations

The previous work related tangential transformations and first-order tangential transformation groups. A natural question arises. Would the consideration of higher order tangential transformations lead to an expansion of a group class?

The concept of such transformations is formulated in a manner analogous to that one which is presented in Section 28.1, 28.2 and is its generalization. The continuous space Z_l is given in the product form $Z^{(l)} = X \times Y^{(l)}$, where $Y^{(l)} = Y \times Y_1 \times \cdots \times Y_l$ and the continuation space $Z_1^{(l)}$, obtained by continuation of $Y^{(l)}$ using X, is considered. Then $Z_1^{(l)} = X \times Y^{(l)} \times Y_1^{(l)}$, where $Y_1^{(l)} = \mathscr{L}(X;\ Y^{(l)})$. In accordance with the representation of $Y^{(l)}$ and $Y_1^{(l)}$ in the form of a product, vectors of these spaces are written as

$$(y, p^{(1)}, \ldots, p^{(l)}) \in Y^{(l)}, \qquad (\underset{1}{y}, \underset{1}{p}^{(1)}, \ldots, \underset{1}{p}^{(l)}) \in Y_1^{(l)}.$$

The space $Z^{(l)}$ of class C_∞ transformations and continuations on the space $Z_1^{(l)}$ are considered. The main definition here is

A transformation of the space $Z^{(l)}$ is said to be *an order l tangential transformation* if the manifold in $Z_1^{(l)}$

$$\underset{1}{y} = p^{(1)}, \qquad \underset{1}{p}^{(1)} = p^{(2)}, \qquad \ldots, \qquad \underset{1}{p}^{(l-1)} = p^{(l)} \tag{6.1}$$

is invariant with respect to the corresponding continuous transformation.

Evidently, the lth continuation of any first-order tangential transformation can be considered as an $(l + 1)$th-order tangential transformation. It appears that from this the introduced transformations are exhausted in the general case. That is, the following statement holds:

Any higher order tangential transformation is the continuation of a first-order tangential transformation.

This result is easily established for the Lie groups $G^1(\tau)$ of lth-order tangential transformations. In that case, when $\dim Y_1 > 1$ it is sufficient to repeat, with suitable changes, the reasoning used in the proof of the lemma in Section 18.5. The variable $\underset{1}{p}^{(l)}$, which appears in the components of the continuous operator $\underset{1}{\tau} \cdot \partial$, plays the role of a "limiter," when the operator $\tau \cdot \partial$ itself does not depend on it. This causes a decomposition of the invariance condition of the manifold (6.1) relative to this variable, which leads to the aforementioned result. The case of $\dim Y_1 = 1$, when all the spaces X, Y, and Y_k are one dimensional, is a little bit more complicated. Here it is essential to use the invariance of the whole manifold (6.1).

7. Lie–Bäcklund Groups

Removal of the limitations which impeded the expansion of classes of groups, by means of a transition to tangential transformations, is possible if the space $Z^{(l)}$ "does not terminate," i.e., by use of the infinite continuation $Z^{(\infty)}$. In this case $Z^{(\infty)} = X \times Y^{(\infty)}$ and $Z_1^{(\infty)} = X \times Y^{(\infty)} \times Y_1^{(\infty)}$, where $Y_1^{(\infty)}$ is obtained by continuation of $Y^{(\infty)}$ with respect to X. In the previous notation the vectors of these spaces are written as the infinite sets

$$(y, p^{(1)}, p^{(2)}, \ldots) \in Y^{(\infty)}, \quad (\underset{1}{y}, p_1^{(1)}, p_1^{(2)}, \ldots) \in Y_1^{(\infty)}, \quad (x, y, p^{(1)}, p^{(2)}, \ldots) \in Z^{(\infty)},$$

and the corresponding manifold $T_\infty \subset Z_1^{(\infty)}$ is given by an infinite sequence of equations

$$\underset{1}{y} = p^{(1)}, \qquad p_1^{(1)} = p^{(2)}, \qquad \underset{1}{p}^{(2)} = p^{(3)}, \qquad \ldots. \tag{7.1}$$

The transformations of the space $Z^{(\infty)}$

$$\omega' = F(\omega), \tag{7.2}$$

subject to the following definition, where $\omega = (x, y, p^{(1)}, p^{(2)}, \ldots)$, are considered.

The transformation (7.2) is said to be a *Lie–Bäcklund transformation* if the manifold T_∞, given by equation (7.1), is invariant with respect to the corresponding continuation space.

As in the case of tangential transformations, the duplicate of the spaces Y_k is eliminated by a translation of the invariant space T_∞. As a result transformations of the points

$$\underset{\infty}{z} = (x, y, \underset{1}{y}, \underset{2}{y}, \ldots) \tag{7.3}$$

of the infinite continuation space Z_∞, operating by

$$x' = f(\underset{\infty}{z}), \qquad y' = g(\underset{\infty}{z}), \qquad \underset{1}{y}' = \underset{1}{v}(\underset{\infty}{z}), \qquad \underset{2}{y}' = \underset{2}{v}(\underset{\infty}{z}), \qquad \ldots, \tag{7.4}$$

are obtained if the functions $\underset{k}{v}$ are expressed in terms of f, g by the continuation formulas (4.5.3), (4.5.4), which should be written with the "infinite" operator of total differentiation D.

Let the group $G^1(\tau)$ be a group of transformations (7.4). Then its infinitesimal operator is

$$\tau \cdot \partial = \xi \cdot \partial_x + \eta \cdot \partial_y + \underset{1}{\eta} \cdot \underset{1}{\partial} + \underset{2}{\eta} \cdot \underset{2}{\partial} + \cdots, \tag{7.5}$$

where the $\underset{k}{\eta}$ are defined by (4.8.5), namely,

$$\underset{k}{\eta} = D^k(\eta - \underset{1}{y} \cdot \xi) + \underset{k+1}{y} \cdot \xi \qquad (k = 1, 2, \ldots). \tag{7.6}$$

Based upon these formulas it is possible to give the following definition of a Lie–Bäcklund group:

The group $G^1(\tau)$ of transformations of the space Z_∞ with the operator (7.5) is called a *Lie–Bäcklund group* if this operator's components satisfy relations (7.6).

In connection with this formal definition are a number of problems about the topological properties of the considered transformations (thus far no topology was introduced in Z_∞) and about possible applications (even formal) of these groups to the solution of problems of differential equations by group analysis. At the present time these problems are being studied, and it is possible only to note that the available (sparse) results indicate some possibilities for the applications of Lie–Bäcklund groups.

§29. BOUNDARY VALUE PROBLEMS

1. General Considerations

In the general sense a boundary value problem for a given differential equation E (system of equations) is to find the solution U satisfying some additional condition Γ (system of conditions) which is necessary to determine one concrete solution. Usually, such a condition is given on the boundary of the domain of definition (boundary conditions). The question arises, Can the solution of a boundary value problem be assisted by means of group analysis, and in what way can such assistance be rendered?

Examples already available in this field indicate that the technique of group analysis turns out to be useful in the study of boundary value problems, in some respects. It finds applications in the transformation of a given problem (the equation E and boundary condition Γ) to a simpler and more convenient form, the determination and investigation of peculiarities of the solution, the approximate analysis of the problem by determining simplified equations and boundary conditions (simulating the boundary value problem), determining whether the unknown solution belongs to a simpler class of partial solutions, and so on. Substantial examples of these applications of group analysis are available. But because of a shortage of space only the problem of invariant (or partially invariant) solutions of some boundary value problems will be considered below.

If a boundary value problem has a unique solution, then the invariance of the solution U of the boundary value problem follows from the invariance of E and Γ with respect to some (the same) transformation group. This statement is obvious in those cases where E and Γ can be considered as manifolds in the same space (Γ is interpreted as a boundary manifold) and U is a solution of the boundary value problem that can be written by the inclusion statement

$$\Gamma \subset U_k \subset E \tag{1.1}$$

with a suitable continuation U_k of the manifold solution U.

To evaluate correctly the possibilities of applying the group analysis method to boundary value problems, it is necessary to take into account the fact that Lie transformation group theory considers local transformations, while a number of boundary value problems consider a "global" boundary condition (for example, elliptic equations). Thus the developed technique of group analysis can be applied, without additional limitations, to boundary value problems admitting local consideration, such as Cauchy's problem, Goursat's problem (in the case of hyperbolic equations), and some others.

2. First-Order Systems

The invariance problem will be considered for a specific (see Section 23.1) first-order system, given on the space $Z = X \times Y$, where $X = R(t) \times X'$, $X' = R^n(x)$, $Y = R^m(y)$. The representation $\underset{1}{y} = (y_t, p)$, the components of which belong to the spaces Y and $Y'_1 = \mathscr{L}(X'; Y)$ and are defined in terms of the mapping $u: X \to Y$ by $y_t = \partial_t u$, $p = \partial_x u$, holds for the vector $\underset{1}{y} \in Y_1$. Also let $Z'_1 = Z \times Y'_1$. In this notation the considered system is written as the equation

$$y_t = F(t, x, y, p), \tag{2.1}$$

with the given mapping $F: Z'_1 \to Y$ of the class $C_\infty(Z'_1)$.

A typical boundary value problem for equation (2.1) is Cauchy's problem, where the boundary (or initial) value manifold has the form

$$t = 0, \qquad y = \phi(x), \tag{2.2}$$

where $\phi: X' \to Y$ is a given mapping.

Equation (2.1) is called quasi-linear (see also Section 23.1) if there exist mappings $A: Z \to \mathscr{L}(Y'_1; Y)$ and $B: Z \to Y$ such that F is given by

$$F = A\langle p\rangle + B. \tag{2.3}$$

Let $u: X \to Y$ be a solution of equation (2.1) considered as the manifold $U \subset Z$ with the equation $y = u(t, x)$. The following construction leads to a new equation derived from (2.1) using U. A change of variable $y \to y'$ is made with $y = u + ay'$, with a real parameter a, a Taylor's expansion of the right part is taken with respect to the parameter a, and the terms of first order are compared. This leads to a linear equation for the unknown $y' \in Y$ (U_1 is the first continuation of the manifold U)

$$y'_t = (\partial_p F | U_1)\langle p' \rangle + (\partial_y F | U_1)\langle y' \rangle. \tag{2.4}$$

Equation (2.4) is said to be *the equation obtained by linearization of equation* (2.1) *with respect to its solution U* or for short, *the linearized* (*on U*) *equation of* (2.1).

In what follows it will be assumed that *Cauchy's problem* (2.2) *has a unique solution both for equation* (2.1) *and* (2.4).

The operator $\zeta \cdot \partial$ of some Lie group $G^1(\zeta)$ of transformations of the space Z is written in the form

$$\zeta \cdot \partial = \tau \partial_t + \xi \cdot \partial_x + \eta \cdot \partial_y, \tag{2.5}$$

with mappings $\tau: Z \to R$, $\xi: Z \to X'$, and $\eta: Z \to Y$.

Lemma. *If equation* (2.1) *admits the operator* (2.5), *then whatever the solution* $U\{y = u(t, x)\}$ *of equation* (2.1) *is, the mapping* $u': X \to Y$, *operating by*

$$u'(t, x) = (\zeta \cdot \partial)(y - u(t, x))|_U, \tag{2.6}$$

is a solution of the linear equation (2.4).

Proof. Let the generating mapping of the group $G^1(\zeta)$ be

$$\bar{t} = f(z, a), \qquad \bar{x} = g(z, a), \qquad \bar{y} = h(z, a). \tag{2.7}$$

Since the group $G^1(\zeta)$ is admitted by equation (2.1), under the transformations (2.7) the solution U transforms into the solution $\bar{U}$, expressed by the equation $\bar{y} = \bar{u}(\bar{t}, \bar{x}, a)$, where $\bar{u}$ is determined from

$$h(t, x, u(t, x), a) = \bar{u}(f(t, x, u(t, x), a), g(t, x, u(t, x), a), a). \tag{2.8}$$

It is clear that the mapping $\bar{u}$ can be given in the form

$$\bar{u}(t, x, a) = u(t, x) + au'(t, x) + a^2\psi(t, x, a),$$

with a smooth mapping $\psi: X \times R \to Y$. From the construction of the linearized equation it follows that u' must be a solution of equation (2.4). On the other hand, the equation

$$u'(t, x) = \partial_a \bar{u}(t, x, 0),$$

from which u' can be calculated by differentiation of the identity (2.8) with respect to the parameter a evaluated at $a = 0$, is also correct. The operator (2.5), using the generating mapping (2.7), becomes

$$\eta | U = u' + (\tau \partial_t u) | U + (\xi \cdot \partial_x u) | U,$$

from which equation (2.6) is obtained. ■

Note. The differential structure of equation (2.1) was never used in the proof. Therefore, *an analogous lemma holds for any differential equation E.*

3. Invariant Solutions of the Cauchy Problem

Using the previous lemma it is easy to develop an invariance criterion for the solution of the Cauchy problem. Let H be the Lie group admitted by equation (2.1) and let L be its Lie algebra of operators (2.5).

Theorem. *The solution of the Cauchy problem* (2.1), (2.2) *is an invariant H-solution if and only if the mapping ϕ satisfies the relation*

$$\eta(0, x, \phi) - \partial_x \phi\langle\xi(0, x, \phi)\rangle - \tau(0, x, \phi)F(0, x, \phi, \partial_x \phi) = 0 \qquad (3.1)$$

for any operator $\zeta \cdot \partial \in L$.

Proof. Application of the invariance criterion in Section 18.2 to the manifold U shows that U is invariant if and only if the mapping u' defined by equation (2.6) is zero. Because of the lemma in Section 29.2 the mapping u' is a solution of the linear homogeneous equation (2.4) and it is equal to zero when and only when $u'(0, x) = 0$. It is left to note that the calculation of $u'(0, x)$, using (2.1), (2.2), and (2.5), gives an expression for it which coincides with the left-hand side of equation (3.1). ■

It is interesting to note that *relation* (3.1) *does not imply invariance of the initial manifold* Γ *given by equations* (2.2). According to the criterion in Section 18.2, such an invariance is equipollent to satisfaction of the relations

$$\tau(0, x, \phi) = 0, \qquad \eta(0, x, \phi) - \partial_x \phi\langle\xi(0, x, \phi)\rangle = 0, \qquad (3.2)$$

which are sufficient for (3.1) to hold but, conversely, are not necessary. But the situation where $\tau(0, x, \phi) \neq 0$ (even for one operator from L) is special, in a way, as the following result shows.

Consequence. *If condition* (3.1) *is satisfied and $\tau(0, x, \phi) \neq 0$, then the solution U of the Cauchy problem* (2.1), (2.2) *coincides with the orbit* (Γ, H) *of the manifold* (2.2).

Proof. The point $z = (0, x, \phi(x))$ becomes $\bar{z} = (\bar{t}, \bar{x}, \bar{y})$ under the transformation (2.7) and, because of the previous lemma, is on U for any a. It remains to show that the points $(\bar{t}, \bar{x})$ cover some neighborhood of the point $(0, x_0)$ in the space X. For that it is sufficient to consider a subgroup $G^1(\zeta) \subset H$ such that $\tau(0, x_0, \phi(x_0)) \neq 0$ and to show that the mapping $(a, x) \to (\bar{t}, \bar{x})$, given by

$$\bar{t} = f(0, x, \phi(x), a), \qquad \bar{x} = g(0, x, \phi(x), a),$$

is invertible in a neighborhood of the point $(0, x_0)$. This follows from the inverse function theorem because the expression

$$\det(\partial_{(a,x)}(\bar{t}, \bar{x})(0, x_0)) = \tau(0, x_0, \phi(x_0))$$

is the Jacobian of the mapping $(a, x) \to (\bar{t}, \bar{x})$. ■

So if (3.1) holds and $\tau(0, x, \phi(x)) \neq 0$, then a solution of the Cauchy problem (2.1), (2.2) can be found by applying the transformation (2.7) to the initial manifold (2.2). One can also note that *in this case* Γ *is a partially invariant manifold of the group* H *with the defect* $\delta(\Gamma, H) = 1$.

4. Invariance of Characteristic

The quasi-linear system (2.3) is considered here. The hypersurface $V \subset X$, with the equation $v(t, x) = \text{const.}$, is a characteristic surface of the solution U (see Section 23.1) if

$$\det(v_t I_Y - q \cdot A) | U = 0, \tag{4.1}$$

where $v_t = \partial_t v$ and $q = \partial_x v$. It is well known that, under group transformations of the space Z, characteristics transform into characteristics. But then, generally speaking, equation (2.3), the solution U, and the surface V also change. If it is assumed that equation (2.3) admits some transformation, then the question arises, How does the manifold given by the equation (4.1) behave? To formulate this problem exactly, the space Y should be extended to the space $\bar{Y} = Y \times R(v)$ and the space of transformations (2.7) of $\bar{Z} = X \times Y$ added, with the equality $\bar{v} = v$, should be considered. Then one should distinguish between the invariance property of characteristic equations "on any solution," i.e., equations

$$v = \text{const.}, \qquad \det(v_t I_Y - q \cdot A) = 0 \tag{4.2}$$

with respect to the group $G^1(\zeta)$ of transformations of the space Z admitted by equation (2.3), and the invariance property of characteristic equations on a given invariant solution U, i.e., equation (4.1). The calculation of corresponding results is based on the determining equations for the operators (2.5) admitted by equation (2.3). To derive these equations the first continuation of the operator (2.5) on the space $\bar{Z}_1$ is

$$\underset{1}{\zeta} \cdot \partial = \zeta \cdot \partial + \sigma \cdot \partial_{y_t} + \theta \cdot \partial_p + \omega \cdot \partial_{v_t} + \lambda \cdot \partial_q,$$

where

$$\begin{aligned}
\sigma &= \eta_t + \eta_y\langle y_t\rangle - y_t\tau_t - y_t(y_t \cdot \tau_y) - p\langle \xi_t\rangle - p\langle \xi_y\langle y_t\rangle\rangle, \\
\theta &= \eta_x + \eta_y \circ p - y_t \otimes \tau_x - y_t \otimes (p \cdot \tau_y) - p \circ \xi_x - p \circ \xi_y \circ p, \\
\omega &= -v_t(\tau_t + y_t \cdot \tau_y) - q \cdot (\xi_t + \xi_y\langle y_t\rangle), \\
\lambda &= -v_t(\tau_x + p \cdot \tau_y) - q \cdot (\xi_x + \xi_y \circ p).
\end{aligned}$$

(The derivatives $\eta_x = \partial_x \eta$, $\xi_y = \partial_y \xi$, etc., are denoted by subscripts; the symbol $a \otimes q$ denotes a tensor multiplication of the vectors $a \in Y$ and $q \in X^*$.) In this notation the invariance condition for equation (2.3) becomes

$$\sigma = A\langle\theta\rangle + A'\langle p\rangle + B',$$

where, for short, it is assumed that $A' = (\zeta \cdot \partial)A$ and $B' = (\zeta \cdot \partial)B$. After the first transition to the manifold (2.3) this condition decomposes with respect to the vector $p \in Y_1'$ and leads to the equations (the vector p remains, here it plays the role of a "tentative" vector; for short, $Ap = A\langle p\rangle$ is written)

$$Ap((Ap)\cdot\tau_y) + p\langle\xi_y\langle Ap\rangle\rangle = A((Ap)\otimes(p\cdot\tau_y)) + A(p\circ\xi_y\circ p); \quad (4.3)$$

$$\begin{aligned}\eta_y\langle Ap\rangle - Ap\tau_t - Ap(B\cdot\tau_y) - B((Ap)\cdot\tau_y) - p\langle\xi_t\rangle - p\langle\xi_y\langle B\rangle\rangle \\ = A(\eta_y\circ p) - A((Ap)\otimes\tau_x) - A(B\otimes(p\cdot\tau_y)) - A(p\circ\xi_x) + A'p; \quad (4.4)\end{aligned}$$

$$\eta_t - \eta_y\langle B\rangle - B\tau_t - B(B\cdot\tau_y) = A\eta_x - A(B\otimes\tau_x) + B'. \quad (4.5)$$

The following simple fact from linear algebra is used in the proof of the following theorem. Let d be a (scalar) linear differential operator on some space T and let the mapping $M: T \to \mathscr{L}(Y)$ be given. Then if there exist mappings $P, Q \in \mathscr{L}(Y)$ such that

$$dM = P\circ M + M\circ Q, \quad (4.6)$$

then the relation

$$d|M| = \operatorname{tr}(P + Q)|M| \quad (4.7)$$

holds for the determinant $\det(M) = |M|$.

Theorem. *If equation* (2.3) *admits the operator* (2.5), *then this operator is also admitted by the system of equations* (4.2).

Proof. Evidently, it is sufficient to prove equality (4.6) for $M = v_t I_Y - q\cdot A$ and $d = \underset{1}{\zeta}\cdot\partial$, as then the result will follow from the invariance criterion in Section 18.2. From the previous continuation formulas the expression

$$\begin{aligned}dM = &-v_t(\tau_t + y_t\cdot\tau_y)I_Y - q\cdot(\xi_t + \xi_y\langle y_t\rangle)I_Y \\ &+ v_t(\tau_x + p\cdot\tau_y)\cdot A + (q\cdot(\xi_x + \xi_y\circ p))\cdot A - qA'.\end{aligned}$$

is obtained. After elminating v_t and y_t with $v_t I_Y = M + q\cdot A$ and $y_t = Ap + B$ this expression is reduced to the form

$$\begin{aligned}dM = &[-(\tau_t + B\cdot\tau_y + A\langle p, \tau_y\rangle)I_Y + (\tau_x + p\cdot\tau_y)\cdot A]\circ M \\ &+ q\cdot[-(\tau_t + B\cdot\tau_y)A - (\xi_t + \xi_y\langle B\rangle)I_Y + (\tau_x\cdot A)\circ A + \xi_x\cdot A - A'] \\ &+ q\cdot[((p\cdot\tau_y)\cdot A)\circ A - A\langle p, \tau_y\rangle A - \xi_y\langle Ap\rangle I_Y + (\xi_y\circ p)\cdot A].\end{aligned}$$

Here the last term $q \cdot [\ \ ]$ is equal to zero because of (4.3) and the second term $q \cdot [\ \ ]$ can be transformed using the determining equation (4.4). To do that it is necessary to assume $p = a \otimes q$ in (4.4), where $a \in Y$, $q \in X^*$, and to transform every term to the result of some linear mapping operating on the vector a. After "elimination" of the vector a, equation (4.4) becomes the following

$$\begin{aligned} q \cdot [-(\tau_t + B \cdot \tau_y)A - (\xi_t + \xi_y\langle B\rangle)I_Y + (\tau_x \cdot A) \circ A + \xi_x \cdot A - A'] \\ = -\eta_y \circ (q \cdot A) + (q \cdot A) \circ \eta_y + (B \otimes \tau_y) \circ (q \cdot A) - (q \cdot A) \circ (B \otimes \tau_y). \end{aligned}$$

Here the left-hand side coincides with the second term $q \cdot [\ \ ]$ in the expression for dM, and the right-hand side, because of the equality $q \cdot A = v_t I_Y - M$, is equal to

$$\begin{aligned} \eta_y \circ (M - v_t I_Y) + (v_t I_Y - M) \circ \eta_y + (B \otimes \tau_y) \circ (v_t I_Y - M) \\ + (M - v_t I_Y) \circ (B \otimes \tau_y) \\ = \eta_y \circ M - M \circ \eta_y - (B \otimes \tau_y) \circ M + M \circ (B \otimes \tau_y) \\ = (\eta_y - B \otimes \tau_y) \circ M - M \circ (\eta_y - B \otimes \tau_y). \end{aligned}$$

Finally, the required representation is obtained:

$$\begin{aligned} dM = [\eta_y - B \otimes \tau_y - (\tau_t + B \cdot \tau_y + A\langle p, \tau_y\rangle)I_Y \\ + (\tau_x + p \cdot \tau_y) \cdot A] \circ M + M \circ (B \otimes \tau_y - \eta_y). \quad \blacksquare \end{aligned}$$

The analogous result, relating to equation (4.1), is actually a consequence of the theorem. It is necessary to note that if U is an invariant $G^1(\zeta)$-solution, then the expression (2.6) is equal to zero and the operator, induced on U by the operator (2.5), has the form

$$\zeta' \cdot \partial = (\tau \,|\, U) \cdot \partial_t + (\xi \,|\, U) \cdot \partial_x. \tag{4.8}$$

Consequence. *If U is an invariant $G^1(\zeta)$-solution, then the characteristic equations* (4.1) *admit the operator* (4.8) *on this solution.*

Proof. The intersection between the invariant manifold (4.2) and the invariant manifold U_1 admits the induced group $G^1(\zeta) | U$, i.e., it is invariant with respect to the first continuation of the operator (4.8). ■

5. Invariance of a Strong Discontinuity

The operator of total differentiation D' on the space $X' \times Y$ is a vector of the space X^* (see Section 4.3) and thus for any mapping $\Phi: Z \to \mathscr{L}(X^*; Y)$ the expression $\Phi\langle D'\rangle = D' \cdot \Phi$ is defined. A special case of the quasi-linear system (2.3), which can be written in the form

$$y_t = D' \cdot \Phi + \Psi, \tag{5.1}$$

with the mappings $\Phi: Z \to \mathscr{L}(X^*; Y)$ and $\Psi: Z \to Y$, is called a *differential law of conservation*. As $D' = \partial_x + p \cdot \partial_y$ it follows that equation (5.1) is like (2.3) with

$$A = \partial_y \cdot \Phi, \qquad B = \partial_x \Phi + \Psi. \tag{5.2}$$

The concept of a *generalized solution*, a particular case of which is a solution with a strong discontinuity, is well defined for equation (5.1).

The hypersurface V with the equation $v(t, x) = 0$ is called a *surface of a strong discontinuity for the solution* $U\{y = u(t, x)\}$ if on every side of it there are continuous limiting values $u|V$ which have a discontinuity of the first class ("jump" $u_1 - u_2$) in transition from side "1" to side "2." Using that definition the equation of a strong discontinuity

$$(v_t y - q \cdot \Phi)_1 = (v_t y - q \cdot \Phi)_2 \tag{5.3}$$

is satisfied. Here v_t and q are the same as in Section 29.4, and only $y = u(x)$ has the jump discontinuity $y_1 - y_2 = u_1(t, x) - u_2(t, x)$. The problem of invariance for equation (5.3) becomes complicated by the fact that the initial equation is written in the form of the differential law of conservation (5.1), generally speaking *not uniquely*, because of which the same surface V may be a surface of a strong discontinuity in one form and not in another form. The exact result here can be found only for transformation groups of the space Z whose operators have the form

$$\zeta \cdot \partial = \tau \partial_t + \xi \cdot \partial_x + (\alpha\langle y\rangle + \beta) \cdot \partial_y, \tag{5.4}$$

where $\tau: R(t) \to R$, $\xi: X \to X$, $\alpha: X \to \mathscr{L}(Y)$, and $\beta: X \to Y$.

As in the case of characteristics one should distinguish the invariance properties of the strong discontinuity equation (5.3), considered as an equation of a manifold in some expanded space, or for a given solution U. Expansion of the space Z to $\tilde{Z}$ is carried out here by "doubling" the space Y (see Section 17.6) and adding the component $R(v)$, where $\bar{v} = v$, so that $\tilde{Z} = X \times R(v) \times Y \times Y$. Then the vectors $\tilde{z} \in \tilde{Z}$ have the representation $\tilde{z} = (t, x, v, y_1, y_2)$, in accordance with which the "doubled" and expanded operator, corresponding to the operator (5.4), is

$$\tilde{\zeta} \cdot \partial = \tau \partial_t + \xi \cdot \partial_x + (\alpha\langle y_1\rangle + \beta) \cdot \partial_{y_1} + (\alpha\langle y_2\rangle + \beta) \cdot \partial_{y_2}. \tag{5.5}$$

Let $\mathscr{D}$ be a manifold in $\tilde{Z}_1$, given by the equations $v = 0$ and (5.3).

Theorem. *If equation* (5.1) *admits the operator* (5.4) *then the manifold* $\mathscr{D}$ *admits the corresponding operator* (5.5).

Proof. Using the special form of the operator (5.4) considerably simplifies the continuation operator, and also the determining equations. In particular, equation (4.3) is identically satisfied and (4.4) assumes the form

$$\alpha\langle Ap\rangle - \tau_t Ap - p\langle \xi_t\rangle = A(\alpha \circ p) - A(p \circ \xi_x) + A'p. \tag{5.6}$$

Let $w = v_t y - q \cdot \Phi$ and $w' = (\underset{1}{\zeta} \cdot \partial)w$. The expression

$$w' = -(v_t \tau_t + q \cdot \xi_t)y + v_t(\alpha\langle y\rangle + \beta) + (q \cdot \xi_x) \cdot \Phi + q \cdot \Phi'$$

holds, where $\Phi' = (\zeta \cdot \partial)\Phi$. Differentiating with respect to y and introducing (5.2) gives

$$\partial_y w' = -(v_t \tau_t + q \cdot \xi_t)I_Y + v_t \alpha + (q \cdot \xi_x) \cdot A - q \cdot \partial_y \Phi'.$$

This expression can be very much simplified. First of all, since

$$\partial_y \Phi' = \partial_y(\zeta \cdot \partial)\Phi = (\zeta \cdot \partial)\partial_y \Phi + (\partial_y \Phi) \circ \alpha = A' + A \circ \alpha,$$

it follows that

$$\partial_y w' = (\alpha - \tau_t I_Y)v_t - (q \cdot \xi_t)I_Y + (q \cdot \xi_x) \cdot A - q \cdot A' - (q \cdot A) \circ \alpha.$$

On the other hand, if it is assumed that $p = a \otimes q$ in (5.6) and then a factor a is "canceled," the equation will be

$$\alpha \circ (q \cdot A) - \tau_t q \cdot A = (q \cdot \xi_t)I_Y + (q \cdot A) \circ \alpha - (q \cdot \xi_x)A + q \cdot A'.$$

Adding the two last equations leads to an expression which is easily transformed to

$$\partial_y w' = (\alpha - \tau_t I_Y) \circ (v_t I_Y - q \cdot A).$$

But $v_t I_Y - q \cdot A = \partial_y(v_t y - q \cdot \Phi) = \partial_y w$, so

$$\partial_y w' = (\alpha - \tau_t I_Y) \circ \partial_y w.$$

This relation shows that the vector $w' - (\alpha - \tau_t I_Y)\langle w\rangle$ does not depend on y, because of which its change in the transition $y_1 \to y_2$ (t, x are not changed) is equal to zero, i.e.,

$$w_1' - w_2' = (\alpha - \tau_t I_Y)\langle w_1 - w_2\rangle,$$

or, finally,

$$(\underset{1}{\tilde{\zeta}} \cdot \partial)(w_1 - w_2) = (\alpha - \tau_t I_Y)\langle w_1 - w_2\rangle. \tag{5.7}$$

From (5.7) it follows that the statement of theorem is correct because of the criterion in Section 18.2 for the invariance of a manifold. ■

These theorems about the invariance of characteristics or strong discontinuity equations find application in the problem about invariant continuation of invariant solutions through an invariant characteristic or a plane of a strong discontinuity.

6. The Navier–Stokes Equations

In the notation accepted in hydrodynamics, the Navier–Stokes equations are given on the space $Z = X \times Y$, where $X = R(t) \times R^3(x)$, $Y = R^3(u) \times R(p)$, and in reduced form they are

$$(\partial_t + u \cdot \partial_x)u + \partial_x p = \Delta u, \qquad \operatorname{div} u = 0, \tag{6.1}$$

where Δ is the Laplacian operator. Here u is the velocity vector and p is pressure (constant density ρ and constant viscosity ν have numerical values equal to unity in suitable measurement units).

The full group of system (6.1) appears to be infinite. Its Lie algebra can be presented in the form of the direct sum $L^\infty \oplus L^{11}$, where the infinite-dimensional ideal L^∞ is generated by the operators

$$\zeta_\phi \cdot \partial = \phi \cdot \partial_x + \phi' \cdot \partial_u - (\phi'' \cdot x)\partial_p, \qquad \zeta_\psi \cdot \partial = \psi \partial_p, \tag{6.2}$$

with the arbitrary functions $\phi\colon R(t) \to R^3$ and $\psi\colon R(t) \to R$ (here $\phi' = \partial_t \phi$, $\phi'' = \partial_t^2 \phi$). In the coordinates $x = (x^1, x^2, x^3)$, $u = (u^1, u^2, u^3)$ the subalgebra L^{11} has the following basis:

$$\begin{aligned} &\xi_0 \cdot \partial = \partial_t, \qquad \xi_i \cdot \partial = \partial_{x^i}, \qquad \eta_i \cdot \partial = t\partial_{x^i} + \partial_{u^i} \qquad (i = 1, 2, 3),\\ &\zeta_{ik} \cdot \partial = x^i \partial_{x^k} - x^k \partial_{x^i} + u^i \partial_{u^k} - u^k \partial_{u^i} \qquad (i < k),\\ &\tau \cdot \partial = 2t\partial_t + x^i \partial_{x^i} - u^i \partial_{u^i} - 2p\partial_p. \end{aligned} \tag{6.3}$$

If the dilatation operator $\tau \cdot \partial$ is neglected, then the remaining operators of (6.3) will generate the basis of the algebra L^{10}, which is the Lie algebra of the *Galilean group* G^{10}, whose transformations operate in seven-dimensional space $Z' = R(t) \times R^3(x) \times R^3(u)$. It is useful to note that the invariance property of the equations of mechanics, relative to the Galilean group, is indeed one of the main axioms of Newton's mechanics. Thus any mathematical models (differential equations) describing general laws of motion of physical bodies within Newtonian dynamics, must admit the Galilean group. This is confirmed for the Navier–Stokes equation. Another example is given by the equations of gas dynamics, considered in Section 11, for which the kernel of the full Lie algebra (11.3.4) is an extension of the Lie algebra of the Galilean group G^{10}, found by adding one dilatation operator to it.

7. Free Boundary Problems

Among boundary value problems, which are usually considered for equations (6.1), the problem of a free boundary value is discussed in the following statement. Two conditions, *kinematic* (V consists of the same liquid particles all the time),

$$\partial_t v + (u \cdot \partial_x)v = 0,$$

and *dynamic* (the component of the stress vector normal to V and operating on V is equal to zero),

$$(-pI_X + \partial_x u + (\partial_x u)^*)\langle \partial_x v \rangle = 0,$$

are given on the hypersurface $V \subset X$ with the equation $v(t, x) = 0$. It is convenient to rewrite these conditions, assuming $v_t = \partial_t v$, $q = \partial_x v$, $r = \partial_x u$, and introduce the notation $r\langle q \rangle = r \cdot q$, $r^*\langle q \rangle = q \cdot r$. Then they assume the form of the equations

$$v = 0, \qquad v_t + u \cdot q = 0, \qquad pq = r \cdot q + q \cdot r, \tag{7.1}$$

for the manifold Γ.

Theorem. *The manifold* Γ, *given by equations* (7.1), *admits the group* G^{11} *whose Lie algebra has the basis of operators* (6.3).

Proof. Any operator from L^{11} has the form

$$\zeta \cdot \partial = \alpha(t)\partial_t + \beta(t, x) \cdot \partial_x + \eta(u) \cdot \partial_u + \pi(p)\partial_p. \tag{7.2}$$

Continuation of such an operator on the space $\tilde{Z}_1 = (Z \times R(v))_1$, considering that the transformed $\bar{v} = v$, is

$$\bar{\zeta} \cdot \partial = \zeta \cdot \partial + (\eta_u \circ r - r \circ \beta_x) \cdot \partial_r - (v_t \alpha_t + q \cdot \beta_t)\partial_{v_t} - (q \cdot \beta_x) \cdot \partial_q + \cdots,$$

where the dots denote the coordinates which are not included in the result of operating with the operator (7.2) on equations (7.1). Application of this operator to equations (7.1) and transition to Γ give the following invariance conditions for the manifold Γ:

$$\begin{aligned} v_t \alpha_t + q \cdot \beta_t - \eta \cdot q - u \cdot (q \cdot \beta_x) &= 0, \\ \pi q = (\eta \circ r - r \circ \beta_x) \cdot q + q \cdot (\eta_u \circ r &- \beta_x \circ r). \end{aligned} \tag{7.3}$$

By direct substitution of any components of the operator (6.3) into equation (7.3) it is easy to verify that both equations become identities on Γ. ■

This theorem can be applied to determine invariant and partially invariant solutions of problems concerning viscous liquid flow with a free boundary value.

§30. CONSERVATION LAWS

1. Conservation Laws

The conservation laws which can be satisfied by solutions of the given differential equation E play an essential role in a number of problems of mathematical physics. They are used both for theoretical analysis of the properties of solutions of the equation E (a priori estimations, theorems of existence and uniqueness, and so on) and to obtain physical quantities which are invariant in time. Conservation laws can be presented in differential or integral form. In what follows, a definition, given in the notation introduced in Sections 4 and 5, is expounded.

Let the equation E be given on the space $Z = X \times Y$; the relation

$$D \cdot A = 0 \tag{1.1}$$

(for some mapping $A: Z_p \to X$), which is identically satisfied for any solution of equation E, is called a *conservation law*.

From the geometrical point of view the conservation law (1.1) can be considered as an equation of a manifold in the continuation space Z_{p+1}, embedded in a suitably continuous manifold E.

One of the general ways to construct conservation laws is to add the relation (1.1), with an undetermined vector A, to the equation E and then to investigate the problem of compatibility of the resulting overdefined system. But this approach leads to cumbersome calculations, which can be completed only in a few cases. Thus it can hardly be considered as a constructive method.

On the other hand, for those equations which arise during the solutions of variational problems the conservation laws can be found using the group admitted by them. In its general form this problem has been studied by E. Noether, who proceeded from the idea of invariance of a functional given in the form of a variational integral. In particular, this method, which has been developed recently and appears to be fruitful, will be discussed.

It is evident that for the given equation E the set of vectors A, describing the conservation law (1.1), generates a vector space (over the field R). Hence the concept of *linearly dependent or independent conservation laws* is meaningful.

2. Invariance of the Functional

Let the space $Z = X \times Y$, the function $\Phi: Z_k \to R$ of the class $C_\infty(Z_k)$, and some region $V \subset X$ be considered. On the class of mappings $u: X \to Y$, $u \in C_\infty(X)$, the functional

$$l(u, V) = \int_V (\Phi \circ \underset{k}{u})\, dx \tag{2.1}$$

is defined, where the kth continuation of the mapping $\underset{k}{u}$ is defined in Section 4.4.

The problem of determining extremal values of the functional (2.1), when the region V is fixed, in the subclass of mappings u which satisfy some boundary conditions on ∂V, is called a *variational problem*. The necessary condition for the extremum is reduced to a boundary value problem, with the boundary conditions on ∂V, for Euler's differential equation, which is equipollent to the first variation of this functional becoming zero. This equation has the form

$$E(\Phi) \equiv \sum_{l=0}^{k} (-1)^l D^l \cdot \partial_{\underset{l}{y}} \Phi = 0. \tag{2.2}$$

In this form the derivative $\partial_{\underset{l}{y}}\Phi$ is considered as an element of the space $\mathscr{L}_s(X^{*l}; Y^*)$ and its value after the l times repeated vector $D \in X^*$ is denoted as follows:

$$D^l \cdot \partial_{\underset{l}{y}} \Phi = \partial_{\underset{l}{y}} \Phi\langle \underbrace{D, \ldots, D}_{l} \rangle.$$

The case $k = 1$, in which Euler's equation (in component form) is

$$\partial_{y^s}\Phi - D_i(\partial_{y_i^s}\Phi) = 0 \qquad (s = 1, \ldots, m),$$

where the D_i are the components of the total differentiation operator (see Section 4.3). This form is met most often.

Let the transformation h of the space Z depending on the parameter a and operating by $z' = h(z, a)$, or in component form

$$x' = f(x, y, a), \qquad y' = g(x, y, a), \tag{2.3}$$

be given. It is assumed that this transformation satisfies condition (4.5.2). The transformed functional (2.1), obtained using the transformation h, is called $l(u', V')$, where $V' = (f \circ \underset{0}{u})(V)$ and the mapping u' is calculated by operations in Section 4.5 (see also Section 5.6). This quantity $l(u', V')$ is given by

$$l(u', V') = \int_{V'} (\Phi \circ \underset{k}{u'})\, dx'.$$

The functional (2.1) is said to be *invariant with respect to the transformation* (2.3) if for any mapping $u: X \to Y$ and for any region $V \in X$ the relation

$$l(u, V) = l(u', V') \tag{2.4}$$

holds.

The definition of functional invariance (2.4) can be presented in another form which is more convenient for group analysis. To do this it is necessary to change the variables of integration $x' \to x$ in the integral $l(u', V')$ by the formula $x' = (f \circ \underset{0}{u})(x)$. Then from the lemma in Section 4.5 and equation (4.5.3) there results $\underset{k}{u'}(x') = \underset{k}{h} \circ \underset{k}{u}(x)$ and the transformed functional takes the form

$$l(u', V') = \int_V (\Phi \circ \underset{k}{h} \circ \underset{k}{u})\,|\partial_x(f \circ \underset{0}{u})|\, dx,$$

where $|\partial_x(f \circ \underset{0}{u})|$ is the Jacobian of the mapping $x \to x'$. Thus the definition of functional invariance (2.4) is equipollent to

$$\Phi \circ \underset{k}{u} = (\Phi \circ \underset{k}{h} \circ \underset{k}{u})|\partial_x(f \circ \underset{0}{u})|$$

because of the arbitrariness of the region V. The latter equation, in its turn, is satisfied for any mapping u if and only if, for any $\underset{k}{z} \in Z_k$, the equation

$$\Phi = (\Phi \circ \underset{k}{h})|Df|, \tag{2.5}$$

obtained from the previous one by noting that $\partial_x(f \circ \underset{0}{u}) = (Df) \circ \underset{1}{u}$, is correct. Consequently, *identity* (2.5) *is equipollent to invariance of the functional* (2.1), *with respect to the mapping h, operating by* (2.3).

Identity (2.5) does not have the form of an equation of invariance for some function. This can be achieved by expansion of the space Z_k to the space $\tilde{Z}_k = Z_k \times R(v)$ by introducing an auxiliary variable v whose transformation is defined by

$$v' = |Df|v. \tag{2.6}$$

From the foregoing a conclusion follows from relations (2.5) and (2.6): *The invariance condition of the functional is equipollent to invariance of the function* Φv *with respect to the transformation* $\underset{k}{\tilde{h}}$ *of the space* $\tilde{Z}_k$.

The multiplication rule for Jacobians shows that, under the composition of transformations (2.3), the composition of the expanded transformations is equal to the expansion of the initial transformations. Thus the expansion of any Lie group $G(h)$ up to the Lie group $G(\tilde{h})$ is defined by (2.6).

Let a be a scalar parameter and $G^1(\zeta)$ be a one-parameter Lie group of transformations (2.3) of the space Z with the operator

$$\zeta \cdot \partial = \xi \cdot \partial_x + \eta \cdot \partial_y. \tag{2.7}$$

Differentiation of expression (2.6) with respect to a at the point $a = 0$ gives

$$\partial_a v'(0) = (D \cdot \xi)v.$$

Consequently, the dilatation operator of the continuation group $\underset{k}{G^1}(\zeta)$ is

$$\underset{k}{\tilde{\zeta}} \cdot \partial = \underset{k}{\zeta} \cdot \partial + (D \cdot \xi)v\partial_y. \tag{2.8}$$

Applying the invariance criterion in Section 2.4 to the function Φv gives the *infinitesimal condition of invariance of the functional* (2.1), *with respect to the group* $G^1(\zeta)$ *whose operator is* (2.7),

$$(\underset{k}{\zeta} \cdot \partial)\Phi + (D \cdot \xi)\Phi = 0, \tag{2.9}$$

which is equipollent to (2.4). In this form the invariance property of the functional is of particular use in the group analysis of conservation laws.

It is necessary to note that the *invariance of some functional* (2.1), *under the transformation* (2.3), *causes invariance of the corresponding Euler equation* (2.2). Indeed, if the functional $l(u, V)$ has an extremal value for some mapping u, then, because of equation (2.4), the transformed functional $l(u', V')$ has the same extremal value for the transformed mapping u'. But the converse statement is not correct, i.e., invariance of the functional does not follow, generally speaking, from invariance of Euler's equation, since the functional (2.1) is not uniquely associated with a given Euler's equation (2.2).

3. Main Lemma

The relation existing between the left-hand sides of equations (2.2) and (2.9) has considerable importance.

Lemma. *The identity*

$$(\underset{k}{\zeta} \cdot \partial)\Phi + (D \cdot \xi)\Phi = D \cdot A + (\eta - \underset{1}{y} \cdot \xi) \cdot E(\Phi) \tag{3.1}$$

is satisfied, where

$$A = \xi\Phi + \sum_{s=1}^{k} \sum_{i=1}^{s} (-1)^{i-1}(D^{s-i}(\eta - \underset{1}{y} \cdot \xi)) \cdot D^{i-1} \cdot \partial_{\underset{s}{y}} \Phi. \tag{3.2}$$

Proof. Let $w = \eta - \underset{1}{y} \cdot \xi$. From (4.8.5) the following relations hold:

$$\begin{aligned}(\underset{k}{\zeta} \cdot \partial)\Phi &= (\zeta \cdot \partial)\Phi + \sum_{s=1}^{k} \underset{s}{\eta} \cdot \partial_{\underset{s}{y}}\Phi \\ &= (\zeta \cdot \partial)\Phi + \sum_{s=1}^{k} (D^s w) \cdot \partial_{\underset{s}{y}}\Phi + \sum_{s=1}^{k} \underset{s+1}{y} \cdot \xi \cdot \partial_{\underset{s}{y}}\Phi \\ &= (\zeta \cdot \partial)\Phi + (\xi \cdot D)\Phi - \xi \cdot \partial_x\Phi - \xi \cdot \underset{1}{y} \cdot \partial_y\Phi + \sum_{s=1}^{k} (D^s w) \cdot \partial_{\underset{s}{y}}\Phi. \end{aligned} \tag{3.3}$$

The expression before the summation sign can be rewritten in the form

$$D \cdot (\xi\Phi) - (D \cdot \xi)\Phi + w \cdot \partial_y\Phi. \tag{3.4}$$

The term under the summation sign in (3.3) is transformed by the product differentiation rule, whose application gives the following sequence of identities:

$$\sum_{s=i}^{k}(D^{s-i+1}\omega)\cdot D^{i-1}\cdot\partial_{\underset{s}{y}}\Phi = D\cdot\sum_{s=i}^{k}(D^{s-i}w)\cdot D^{i-1}\cdot\partial_{\underset{s}{y}}\Phi$$

$$-\sum_{s=i+1}^{k}(D^{s-i}w)\cdot D^{i}\cdot\partial_{\underset{s}{y}}\Phi - w\cdot D^{i}\cdot\partial_{\underset{i}{y}}\Phi \qquad (i = 1,\ldots,k-1),$$

$$\sum_{s=k}^{k}(D^{s-k+1}w)\cdot D^{k-1}\cdot\partial_{\underset{s}{y}}\Phi = D\cdot\sum_{s=k}^{k}(D^{s-k}w)\cdot D^{k-1}\cdot\partial_{\underset{s}{y}}\Phi - w\cdot D^{k}\cdot\partial_{\underset{k}{y}}\Phi.$$

The linear combination of these identities, where the ith row is multiplied by $(-1)^{i-1}$, using the notation of (2.2) and (3.2), becomes the form

$$\sum_{s=1}^{k}(D^{s}w)\cdot\partial_{\underset{s}{y}}\Phi$$

$$= D\cdot\sum_{i=1}^{k}(-1)^{i-1}\sum_{s=i}^{k}(D^{s-i}w)\cdot D^{i-1}\cdot\partial_{\underset{s}{y}}\Phi + w\cdot\sum_{i=1}^{k}(-1)^{i}D^{i}\cdot\partial_{\underset{i}{y}}\Phi$$

$$= D\cdot A - D\cdot(\xi\Phi) + w\cdot E(\Phi) - w\cdot\partial_{y}\Phi. \tag{3.5}$$

Evidently, the substitution of (3.4) and (3.5) into (3.3) gives the identity (3.1). ■

A new form can be given to the invariance condition for the functional (2.9) because of this theorem.

Consequence. *A necessary and sufficient condition for invariance of the functional* (2.1), *with respect to the group* $G^1(\zeta)$ *whose operator is* (2.7), *is that*

$$D\cdot A + (\eta - \underset{1}{y}\cdot\xi)\cdot E(\Phi) = 0, \tag{3.6}$$

where the vector A *is defined by equation* (3.2) *and* $E(\Phi)$ *is defined in* (2.2).

4. The Noether Theorem

The identity (3.1) and the invariance criterion (2.9) for the functional (2.1) are the sources for a number of statements about conservation laws for Euler's differential equation (2.2). First of all they lead to the following classical result, where *independence of Euler's (scalar) equations is assumed* in the sense that their left-hand sides are linearly independent over the field of functions $Z_1 \to R$.

Theorem. *If the functional* (2.1) *is invariant with respect to the Lie group* $G^r(\zeta)$, *then for Euler's equations* (2.2) *there are r linearly independent conservation laws* (1.1) *with the vectors* A_α *defined by* (3.2) *using the basis operators* $\zeta_\alpha \cdot \partial$ ($\alpha = 1, \ldots, r$) *of the Lie algebra* $L^r(\zeta)$.

Proof. From (3.6) it follows that for any solution of the equation $E(\Phi) = 0$ the conservation laws $D \cdot A_\alpha = 0$ are satisfied. Their linear independence follows from the fact that if $A = 0$, then (3.6) assumes the form

$$(\eta - \underset{1}{y} \cdot \xi) \cdot E(\Phi) = 0.$$

From this, because of the independence of Euler's equations, the identity $\eta - \underset{1}{y} \cdot \xi = 0$ follows, and therefore, finally, $\xi = \eta = 0$. ■

It is clear that the Noether theorem gives only a sufficient condition for the existence of conservation laws for equation (2.2). It consists of the requirement of the absolute functional invariance of (2.1), as expressed by the identity (2.9). Examples show that this condition, in general, is not necessary: Conservation laws with vectors (3.2) can exist when the *Lagrangian* Φ does not satisfy the identity (2.9). But it is evident that because of (3.1), relation (2.9) must necessarily be satisfied for solutions of the equation $E(\Phi) = 0$ if there is a conservation law with the vector (3.2). Therefore, the following strengthening of the Noether theorem, noted by N. Ibragimov, is obtained.

A necessary and sufficient condition for the existence of a conservation law (1.1), *with the vector* (3.2), *for the Euler equation* (2.2) *admitting the group* $G^1(\zeta)$, *consists in the invariance of the functional* (2.1) *with respect to the group* $G^1(\zeta)$ *on the manifold given by equation* (2.2), *i.e., satisfaction of the relation*

$$((\underset{k}{\zeta} \cdot \partial)\Phi + (D \cdot \xi)\Phi)|_{E(\Phi)=0} = 0. \tag{4.1}$$

It should be noted that *the resulting conservation laws must not be linearly independent, in contrast to those which are given by the Noether theorem.* At any rate, since the group admitted by equation (2.2), without fail, contains the group admitted by the functional (2.1) (see the note at the end of Section 30.2), *all the conservation laws, obtained from the Noether theorem, are surely obtained using the stronger theorem.*

5. Other Conservation Laws

The main identity (3.1) shows that for Euler's equation (2.2) it is possible to find a conservation law even when the functional (2.1) is not invariant with respect to the group $G^1(\zeta)$, i.e., when the Langrangian Φ does not satisfy the identity (2.9). This will be so in case the more general identity

$$(\underset{k}{\zeta} \cdot \partial)\Phi + (D \cdot \xi)\Phi = D \cdot B \tag{5.1}$$

is satisfied, with some vector mapping $B: Z_p \to X$, instead of (2.9). From (3.1) it follows that the conservation law

$$D \cdot (A - B) = 0 \tag{5.2}$$

holds for equation (2.2), where the vector A is still given by equation (3.2).

To understand this situation correctly, it is necessary to note that an expression of "divergence type" $D \cdot \Psi$, with the mapping $\Psi: Z_p \to X$, can be added to the Lagrangian Φ without changing Euler's equation (2.2). This well known fact follows from the easily verified identity

$$E(\Phi + D \cdot \Psi) = E(\Phi).$$

Together with this, the invariance condition for the functional changes under the transition from the Lagrangian Φ to $\Phi_1 = \Phi + D \cdot \Psi$, since the identity

$$(\underset{\infty}{\zeta} \cdot \partial)\Phi_1 + (D \cdot \xi)\Phi_1 = (\underset{\infty}{\zeta} \cdot \partial)\Phi + (D \cdot \xi)\Phi + D \cdot ((\underset{\infty}{\zeta} \cdot \partial)\Psi) \tag{5.3}$$

holds.

So *if the functional* (2.1) *is not invariant with respect to the group* $G^1(\zeta)$, *but becomes invariant after the transition from* Φ *to* Φ_1, *then for Euler's equation there is a conservation law* (5.2), *with the vector* A *from* (3.2), *and with the vector* $B = -(\underset{\infty}{\zeta} \cdot \partial)\Psi$. But the equation

$$(\underset{\infty}{\zeta} \cdot \partial)\Psi = -B, \tag{5.4}$$

with a given B and unknown Ψ, surely has a solution, and thus equation (5.1), indeed, does not give an essential generalization of the previous result.

It is also useful to note the obvious possibility of finding new higher order conservation laws which are given by differentiation of the conservation law (2.1) by applying the operator D. Since the operators D_j are permutable,

$$D_j(D \cdot A) = D \cdot (D_j A) \qquad (j = 1, \ldots, n). \tag{5.5}$$

Consequently, one conservation law (2.1), which has some order p, gives n conservation laws of order $p + 1$.

6. Inversion of the Noether Theorem

That problem for which *all* the conservation laws for some Euler's equations (2.2) can be obtained by Noether's theorem, i.e., when the invariance of the functional (2.1) is necessary to give the conservation laws with vectors (3.2), is of interest. If this property holds, then it is said that the converse Noether theorem is true.

To answer this question it is necessary to restrict Euler's equation (2.2) by some condition which guarantees the independence of the invariance condition of the functional (2.9) from equation (2.2). More precisely, it is necessary that (2.9) should not be an algebraic consequence of equations (2.2), but be satisfied identically on the space Z_k. Apparently, the corresponding criterion has yet been found. So, here only some indication sufficient for this condition's satisfaction can be given. One of them is the following criterion, in which the facts that $k \geq 1$ and, in the general case, Euler's equation has order $2k$, are used.

(α) *None of the algebraic consequences of Euler's equation* (2.2) *has order less than or equal to k.*

Theorem. *If property* (α) *holds and Euler's equation* (2.2) *of the functional* (2.1) *admits the group* $G^1(\zeta)$, *then the invariance of the functional* (2.1) *with respect to the group* $G^1(\zeta)$ *follows from the existence of a conservation law, with the vector* (3.2), *for the equation.*

Proof. Because of the identity (3.1), equation (2.9) is satisfied for any solution of equation (2.2). Condition (α) means that the derivatives $\underset{l}{u}$ of order $l \leq k$ are not connected with any relation obtained from the solutions u of equation (2.2), i.e., they remain sufficiently arbitrary. Thus equation (2.9), having order k, must be satisfied identically. ■

Sometimes the functional (2.1) for which property (α) holds, is said to be *nondegenerate* (in the opposite case it is *degenerate*). The majority of the functionals appearing in mechanics and physics are nondegenerate. The *Dirac equations* are an example with a degenerate functional.

7. Example

The equation for the unsteady potential transonic flow of a gas (Lin–Reissner–Tsien equation) in two spatial dimensions is considered on the space $Z = R^3(t, x, y) \times R(\phi)$,

$$2\phi_{tx} + \phi_x \phi_{xx} - \phi_{yy} = 0, \tag{7.1}$$

where $\phi_t = \partial_t \phi$, $\phi_x = \partial_x \phi$, and so on. It is easy to verify that (7.1) is Euler's equation for the functional (2.1) with the Lagrangian

$$\Phi = -\phi_t \phi_x - \tfrac{1}{6}\phi_x^3 + \tfrac{1}{2}\phi_y^2. \tag{7.2}$$

The full group of equation (7.1) is known. It is an infinite Lie group, where the basis operators of its Lie algebra contain five arbitrary functions α, β, γ,

δ, ε: $R(t) \to R$ and have the form (the derivatives ∂_t are denoted by primes):

$$\begin{aligned} \zeta_\alpha \cdot \partial &= 3\alpha\partial_t + (\alpha' x + \alpha'' y^2)\partial_x \\ &\quad + 2\alpha' y \partial_y + (-\alpha'\phi + \alpha'' x^2 + 2\alpha''' x y^2 + \tfrac{1}{3}\alpha^{(4)} y^4)\partial_\phi; \\ \zeta_\beta \cdot \partial &= \beta' y \partial_x + \beta \partial_y + (2\beta'' x y + \tfrac{2}{3}\beta''' y^3)\partial_\phi; \\ \zeta_\gamma \cdot \partial &= \gamma \partial_x + (2\gamma' x + 2\gamma'' y^2)\partial_\phi; \\ \zeta_\delta \cdot \partial &= \delta y \partial_\phi; \\ \zeta_s \cdot \partial &= \varepsilon \partial_\phi; \\ \zeta_0 \cdot \partial &= 2t\partial_t + y\partial_y - 2\phi\partial_\phi. \end{aligned} \tag{7.3}$$

An application of the rules expounded above for constructing conservation laws shows that all these operators, with the exception of the dilatation operator $\zeta_0 \cdot \partial$, generate independent conservation laws. Consequently, the space of first-order conservation laws for equation (7.1) is infinite dimensional. Now the functional (2.1), with the Lagrangian (7.2), is not invariant and must be "enlarged" by adding the term $D \cdot \Psi$ described in Section 30.5, leading to the conservation law (5.2).

For example, if the operator $\zeta_\delta \cdot \partial$ is taken, then its first continuation is

$$\underset{1}{\zeta_\delta} \cdot \partial = \delta y \partial_\phi + \delta' y \partial_{\phi_t} + \delta \partial_{\phi_y},$$

where $D \cdot \xi_\delta = 0$. Thus here

$$(\underset{1}{\zeta_\delta} \cdot \partial)\Phi + (D \cdot \xi_\delta)\Phi = -\delta' y \phi_x + \delta \phi_y.$$

Obviously, this is a relation of the type (5.1) with the vector

$$B = (0, -\delta' y \phi, \delta \phi).$$

Here equation (3.2) can be rewritten in the form

$$A = \xi\Phi + (\eta - (\phi_t, \phi_x, \phi_y) \cdot \xi)(\partial_{\phi_t}\Phi, \partial_{\phi_x}\Phi, \partial_{\phi_y}\Phi),$$

and for the operator $\zeta_\delta \cdot \partial$ the vector

$$A = \delta y(-\phi_x, -\phi_t - \tfrac{1}{2}\phi_x^2, \phi_y)$$

is obtained. According to Section 30.5, the conservation law (5.2) with the vector $A - B$ holds for equation (7.1):

$$D_t(-\delta y \phi_x) + D_x(\delta' y \phi - \delta y \phi_t - \tfrac{1}{2}\delta y \phi_x^2) + D_y(-\delta\phi + \delta y \phi_y) = 0. \tag{7.4}$$

This conservation law, for every fixed function δ, can be obtained by the Noether theorem if $\Phi_1 = \Phi + D \cdot \Psi$ is taken instead of the Lagrangian (7.2), where the vector Ψ should be determined from equation (5.4). That calculation gives

$$\Psi = \left(0, -\frac{\delta'}{2\delta}\phi^2, -\frac{1}{2y}\phi^2\right).$$

So equation (7.1) is Euler's equation of the functional (2.1), invariant with respect to the operator $\zeta_\delta \cdot \partial$ with the Lagrangian

$$\Phi_1 = -\phi_t\phi_x - \frac{1}{6}\phi_x^3 + \frac{1}{2}\phi_y^2 - \frac{\delta'}{\delta}\phi\phi_x + \frac{1}{2y^2}\phi^2 - \frac{1}{4}\phi\phi_y,$$

and the conservation law (7.4) is obtained by the Noether theorem. It is clearly seen that for every concrete function $\delta: R(t) \to R$ there is a Lagrangian. Thus there is no unique functional for the entire infinite Lie group with the Lie algebra of operators $\zeta_\delta \cdot \partial$.

REFERENCES

Investigations of linear second order differential equations from the group point of view were started by S. Lie (see Ref. [3] in Chapter II) and his close associates [1]. The covariant form of determining equations was given in Ref. [2] for the first time. The group classification by means of a weak nonlinearity was given in Ref. [3]. The important physical idea of a wave equation in a Riemann space was introduced in Ref. [4]. Classification problems for linear second-order equations are closely connected with the classification problems of Riemann spaces by motion groups, to which the monograph [5] is devoted. Relatively recently there has been discovered a remarkable connection between the conformal invariance of a Riemann space and the Huygens principle for hyperbolic equations [6–8]. The work [9] is attached here. Recently there was begun a systematic investigation of standard equations admitting a maximally large group of motions [10]. The paper [11] is devoted to problems of separation of variables.

The idea of tangential mappings was treated in detail by S. Lie (see Ref. [1] in Chapter I). A formal consideration of infinitely continuous space transformations was initiated by Bäcklund [12]. This part of the theory was revived only recently in studies [13] and [14].

There are a few papers concerning the invariance problem of boundary value problems. Problems of characteristics and strong discontinuity invariance were studied in [15] and [16]. The group analysis of the Navier–Stokes equations was done in Refs. [17] and [18]. Invariance of the conditions on a free boundary was investigated in Ref. [19].

The main work by E. Noether [20] was essentially supplemented by the results obtained in Ref. [21], and also in Ref. [7] in Chapter I, where many examples are given. The statement of these results, given in Section 30, differs by the fact that here they are derived for any order functional. During application to the hydrodynamics equations new conservation laws were found in Ref. [22].

[1] Medolaghi P. Classificazione delle equationi alle derivate parziali del secondo ordine, che ammetono up gruppo infinito di transformationi puntuali. — Ann. mat. pura ed appl. 1 (1898), 229—263.
[2] Овсянников Л. В. Об отыскании группы линейного дифференциального уравнения второго порядка. — ДАН СССР **132**, № 1 (1960), 44—47.
[3] Ибрагимов Н. Х. К групповой классификации дифференциальных уравнений второго порядка. — ДАН СССР, **183**, № 2 (1968), 274—277.
[4] Ибрагимов Н. Х. Волновое уравнение в римановом пространстве. — В сб. «Динамика сплошной среды», вып. 1 (1969), Новосибирск, 36—47.
[5] Петров А. З. Новые методы в общей теории относительности. М., «Наука», 1966.
[6] Ибрагимов Н. Х. Конформная инвариантность и принцип Гюйгенса. — ДАН СССР, **194**, № 1 (1970), 24—27.
[7] Ибрагимов Н. Х. Принцип Гюйгенса. — В сб. «Некоторые проблемы математики и механики». Л., «Наука», 1970, 159—170.
[8] Ибрагимов Н. Х., Мамонтов Е. В. Sur le probléme de J. Hadamard relatif à la diffusion des ondes. — C. r. Acad. sci. (1970), 456—458.
[9] Чупахин А. П. Нелинейные конформно-инвариантные уравнения в пространстве V_n с нетривиальной конформной группой. — В сб. «Динамика сплошной среды», вып. 25 (1976), Новосибирск, 122—123.
[10] Чиркунов Ю. А. О линейных дифференциальных уравнениях второго порядка, допускающих группу максимальной размерности. — В сб. «Динамика сплошной среды», вып. 24 (1976), Новосибирск, 124—137.
[11] Шаповалов В. Н. К групповым свойствам линейных уравнений.— Известия вузов, Физика, № 6 (1968), 75—80.
[12] Bäcklund A. V. Ueber flächentransformationen. — Math. Annalen. B. 9 (1876), 297—320.
[13] Ибрагимов Н. Х., Андерсон Р. Д. Группы касательных преобразований Ли—Бэклунда. — ДАН СССР **227**, № 3 (1976), 539—542.
[14] Ибрагимов Н. Х. Группы Ли—Бэклунда и законы сохранения. — ДАН СССР, **270**, № 1 (1976).
[15] Меньщиков В. М. О продолжении инвариантных решений уравнений газовой динамики через ударную волну. — В сб. «Динамика сплошной среды», вып. 4 (1969), Новосибирск, 163—169.
[16] Меньщиков В. М. О непрерывном сопряжении инвариантных решений. — В сб. «Динамика сплошной среды», вып. 10 (1972), Новосибирск, 70—84.
[17] Пухначев В. В. Групповые свойства уравнений Навье—Стокса в плоском случае. — Ж. прикл. мех. и техн. физ. № 1 (1960), 83—90.
[18] Бытев В. О. Групповые свойства уравнений Навье—Стокса. — В сб. «Численные методы механики сплошной среды», 3, № 5 (1972), Новосибирск, ВЦ СО АН СССР, 13—17.
[19] Пухначев В. В. Инвариантные решения уравнений Навье—Стокса, описывающие движения со свободной границей. — ДАН СССР, **202**, № 2 (1972), 302—305.
[20] Нётер Э. Инвариантные вариационные задачи. — В сб. «Вариационные принципы механики». М., Физматгиз, 1959, 611—630.
[21] Ибрагимов Н. Х. Инвариантные вариационные задачи и законы сохранения. — Теор. и матем. физ. **1** № 3 (1969), 350—359.
[22] Ибрагимов Н. Х. Законы сохранения в гидродинамике. — ДАН СССР **210**, № 6 (1973), 1307—1309.

APPENDIX

Tables of Calculated Full Groups

In addition to examples given in the main text, the full groups or group classification results obtained by various authors for concrete systems are given here. There are many results. The list provided here has been selected to show a variety of possible applications. It should be emphasized that every such result is a theorem stating that the given system of differential equations admits the indicated Lie group and does not admit any other operators.

The usually accepted name of the system of equations and a reference to the bibliography in the corresponding chapter are given for every example. Then the spaces $X = R^n(t, x, \ldots)$ of independent variables and $Y = R^m$ $(u, p, \ldots)$ of unknown functions are indicated and the system of equations is given, where the derivatives are denoted by subscripts ($\partial_t u = u_t$, $\partial_{x^i} u = u_i$, and so on). An operator component is denoted by the same letter as the variable but with a star over it, so an operator has the form

$$\zeta \cdot \partial = \overset{*}{t}\partial_t + \overset{*}{x}\partial_x + \cdots + \overset{*}{u}\partial_u + \overset{*}{p}\partial_p + \cdots.$$

Components of the operator, linearly generating the admitted Lie algebra, are given in the tables. In the case of trivial equations, the trivial ideal (see Section 7.14) is not given. The arbitrary element is described separately and its specializations, leading to expansion of the admitted Lie algebra, are given in the first column.

§A1. EULER EQUATIONS OF AN IDEAL INCOMPRESSIBLE LIQUID†

$X = R^4(t, x^1, x^2, x^3)$, $Y = R^4(v^1, v^2, v^3, p)$. The system E,

$$v_i^i = 0, \qquad v_t^k + v^i v_i^k = -p_k \qquad (k = 1, 2, 3)$$

admits an infinite group. The full Lie algebra LE is given in terms of the arbitrary functions $\alpha\colon R(t) \to R$ and $\beta\colon R(t) \to R^3$.

In Table 1, A is an arbitrary antisymmetric linear mapping $R^3 \to R^3$ and $\beta' = \partial_t \beta$.

TABLE 1

	$\overset{*}{t}$	$\overset{*}{x}$	$\overset{*}{v}$	$\overset{*}{p}$
ζ_1	1	0	0	0
ζ_2	t	x	0	0
ζ_3	0	x	v	$2p$
ζ_A	0	$A\langle x \rangle$	$A\langle v \rangle$	0
ζ_α	0	0	0	α
ζ_β	0	β	β'	$-\beta'' \cdot x$

§A2. EINSTEIN EQUATIONS IN VACUUM‡

$X = R^n(x)$, $Y = \mathscr{L}_s(X; X^*) = R^{n(n+1)/2}(y)$, where $y = \{y_{ij}\}$ and $y_{ij} = y_{ji}$. The fundamental metric is given using the mapping $g\colon X \to Y$, operating by the formula $y = g(x)$ or in component form $y_{ij} = g_{ij}(x)$, as

$$ds^2 = g_{ij}(x)\, dx^i\, dx^j.$$

With the Ricci tensor R_{ik}, the system E is

$$R_{ik} = 0 \qquad (i, k = 1, \ldots, n).$$

The full Lie algebra LE is infinite dimensional and is given in Table 2 using the arbitrary vector field $\xi\colon X \to X$.

TABLE 2

	$\overset{*}{x}$	$\overset{*}{y}$
ζ_1	0	y
ζ_ξ	ξ	$-y \circ \partial\xi - \partial\xi \circ y$

† From Ref. [16] in Chapter III.
‡ Ref. [13] in Chapter III.

§A3. MAXWELL EQUATIONS IN VACUUM†

$X = R^4(x)$, $Y = \mathscr{L}_s(X; X^*)$. The unknown is the tensor of the electromagnetic field $u: X \to Y$, operating by the formula $y = u(x)$ or in the components $y_{kl} = u_{kl}(x)$. With the notation $u_{kl,m} = \partial_m u_{kl}$ the system E is (all the indices assume the values 1, 2, 3, 4)

$$u_{kl,m} + u_{lm,k} + u_{mk,l} = 0, \qquad \sum_{l=1}^{4} u_{kl,l} = 0.$$

It is convenient to present the full group by means of the group G^{15} of conformal transformations of the Minkowski's four-dimensional space with the metric $-(dx^1)^2 + (dx^2)^2 + (dx^3)^2 + (dx^4)^2$ and with the operators $\xi \cdot \partial = \overset{*}{x}{}^i \partial_i$. The Lie algebra, generated by the basis is given in Table 3.

TABLE 3

	$\overset{*}{x}{}^1$	$\overset{*}{x}{}^2$	$\overset{*}{x}{}^3$	$\overset{*}{x}{}^4$
ξ_0	x^1	x^2	x^3	x^4
ξ_1	1	0	0	0
ξ_2	0	1	0	0
ξ_3	0	0	1	0
ξ_4	0	0	0	1
ξ_5	x^2	x^1	0	0
ξ_6	x^3	0	x^1	0
ξ_7	x^4	0	0	x^1
ξ_8	0	x^3	$-x^2$	0
ξ_9	0	$-x^4$	0	x^2
ξ_{10}	0	0	x^4	$-x^3$
ξ_{11}	r_1^2	$2x^1x^2$	$2x^1x^3$	$2x^1x^4$
ξ_{12}	$2x^1x^2$	r_2^2	$2x^2x^3$	$2x^2x^4$
ξ_{13}	$2x^1x^3$	$2x^2x^3$	r_3^2	$2x^3x^4$
ξ_{14}	$2x^1x^4$	$2x^2x^4$	$2x^3x^4$	r_4^2

Here $r_i^2 = \sum_{k=1}^4 \varepsilon_{ik}(x^k)^2$, where ε_{ik} are the elements of the matrix

$$(\varepsilon_{ik}) = \begin{bmatrix} 1 & 1 & 1 & 1 \\ 1 & 1 & -1 & -1 \\ 1 & -1 & 1 & -1 \\ 1 & -1 & -1 & 1 \end{bmatrix}.$$

The Lie algebra L^{15} is the full Lie algebra of the wave equation in R^4. The full group of Maxwell's equations is G^{17} with the Lie algebra of operators

† From Ref. [13] in Chapter III.

$\zeta \cdot \partial = \xi \cdot \partial_x + \overset{*}{u} \cdot \partial_u$ as given in Table 4, where $\xi \cdot \partial_x$ is any operator from Table 3 with the components $\xi^r = \overset{*}{x}{}^r$, $\xi^r_i = \partial_i \xi^r$.

TABLE 4

	$\overset{*}{x}$	$\overset{*}{u}_{kl}$
ζ	ξ	$-u_{kr}\xi^r_l - u_{rl}\xi^r_k$
ζ_{16}	0	u_{kl}
ζ_{17}	0	$\sum_{r,s=1}^{4} e_{klrs} u_{rs}$

§A4. DIRAC EQUATION (WITH ZERO MASS)†

$X = R^4(x)$, $Y = R^4(\psi)$. The system E is written in the form

$$\gamma^k \partial_k \psi = 0,$$

where the constant matrices $\gamma^k \in \mathscr{L}(R^4)$ are expressed in terms of the elementary matrices e_{ik} (in which at some intersection of the ith row and kth column there is unity and the rest of the elements are zero) by the formulas

$$\gamma^1 = \sqrt{-1}(-e_{14} - e_{23} + e_{32} + e_{41}), \qquad \gamma^2 = -e_{14} + e_{23} + e_{32} - e_{41},$$
$$\gamma^3 = \sqrt{-1}(-e_{13} + e_{24} + e_{31} - e_{42}), \qquad \gamma^4 = e_{11} + e_{22} - e_{33} - e_{44}.$$

The full Lie algebra of operators $\zeta \cdot \partial = \overset{*}{x} \cdot \partial_x + \overset{*}{\psi} \cdot \partial_\psi$ is given in Table 5.

TABLE 5

	$\overset{*}{x}$	$\overset{*}{\psi}$
ζ	ξ	$S\langle\psi\rangle$
ζ_{16}	0	$(e_{13} + e_{31} + e_{24} + e_{42})\langle\psi\rangle$
ζ_{17}	0	ψ

Here $\xi \cdot \partial_x$ is any operator from L^{15} of Example A3 and the matrix S is

$$S = \frac{1}{8} \sum_{k,l=1}^{4} \partial_l \xi^k (\gamma^k \gamma^l - \gamma^l \gamma^k - 3\delta^{kl}).$$

† From Ref. [15] in Chapter III.

§A5. STATIONARY HEAT CONVECTION EQUATIONS†

$X = R^2(x, y)$, $Y = R^4(u, v, p, \theta)$. The system E is

$$u_x + v_y = 0$$

$$uu_x + vu_y = -p_x + u_{xx} + u_{yy},$$

$$uv_x + vv_y = -p_y + v_{xx} + v_{yy} + \theta,$$

$$u\theta_x + v\theta_y = \sigma(\theta_{xx} + \theta_{yy}) \qquad (\sigma = \text{const}).$$

The full Lie algebra is given in Table 6.

TABLE 6

	$\overset{*}{x}$	$\overset{*}{y}$	$\overset{*}{u}$	$\overset{*}{v}$	$\overset{*}{p}$	$\overset{*}{\theta}$
ζ_1	1	0	0	0	0	0
ζ_2	0	1	0	0	0	0
ζ_3	0	0	0	0	1	0
ζ_4	0	0	0	0	y	1
ζ_5	x	y	$-u$	$-v$	$-2p$	-3θ

§A6. HOPF EQUATION‡

$X = R^2(t, x)$, $Y = R(u)$. The equation E is

$$u_t + uu_x = (\phi(u)u_x)_x.$$

The arbitrary element is $\phi: R(u) \to R$, $\phi \neq 0$. The result of the group classification is given in Table 7.

TABLE 7

ϕ		$\overset{*}{t}$	$\overset{*}{x}$	$\overset{*}{u}$
$\sim$	ζ_1	1	0	0
	ζ_2	0	1	0
$\exp u$	ζ_3	t	$x + t$	1
u^{2m}	ζ_3	$2(m-1)t$	$(2m-1)x$	u
1	ζ_4	0	t	1
	ζ_5	t^2	tx	$x - tu$

† From Ref. [9] in Chapter V.
‡ From Ref. [8] in Chapter V.

TABLE 8

A		$\overset{*}{x}$	$\overset{*}{y}$	$\overset{*}{z}$	$\overset{*}{u}$	$\overset{*}{v}$	$\overset{*}{w}$	$\overset{*}{p}$
	ζ_1	0	y	0	$-2u$	$-v$	$-2w$	$-4p$
—	ζ_2	0	0	0	0	0	0	1
	ζ_3	0	f	0	0	$A^{-1}(f_x u + f_z w)$	0	0
$g(z)e^x$	ζ_4	1	0	0	u	0	w	$2p$
$g(z)$	ζ_4	1	0	0	0	0	0	0
z^α	ζ_5	x	0	z	$(\alpha + 1)u$	0	$(\alpha + 1)w$	$2(\alpha + 1)p$
βz^{-1}	ζ_6	$\frac{1}{2}(z^2 - x^2)$	0	$-xz$	zw	0	$-zu$	0
$\beta(\cosh z)^{-1}$	ζ_5	$-\cos x \cdot \sinh z$	0	$\sin x \cdot \cosh z$	$-(\cos x \cdot \cosh z)w$	0	$(\cos x \cdot \cosh z)u$	0
	ζ_6	$\sin x \cdot \sinh z$	0	$\cos x \cdot \cosh z$	$(\sin x \cdot \cosh z)w$	0	$-(\sin x \cdot \cosh z)u$	0
1	ζ_6	0	0	1	0	0	0	0
	ζ_7	z	0	$-x$	w	0	$-u$	0

§A7. EQUATIONS OF STATIONARY THREE-DIMENSIONAL BOUNDARY LAYER ON AN ARBITRARY PLANE†

$X = R^3(x, y, z)$, $Y = R^4(u, v, w, p)$. The system $E(A)$ is

$$u_x + w_z + Av_y + A^{-1}A_x u - A^{-1}A_z w = 0,$$

$$uu_x + wu_z + Avu_y + A^{-1}A_z uw - A^{-1}A_x w^2 + p_x = Au_{yy},$$

$$uw_x + ww_z + Avw_y + A^{-1}A_x uw - A^{-1}A_z u^2 + p_z = Aw_{yy},$$

$$p_y = 0.$$

The arbitrary element $A: R^2(x, z) \to R$ gives the form of a steamline surface through its fundamental metric $ds^2 = A(dx^2 + dy^2)$. The results of the group classification are given in Table 8. Here $f: R^2(x, z) \to R$ and $g: R(z) \to R$ are arbitrary functions and α, β are arbitrary constants.

§A8. EQUATION OF MOTION OF VISCOPLASTIC MEDIUM‡

$X = R^2(t, x)$, $Y = R(u)$. The equation $E(f)$ is

$$u_t = f(u_x)u_{xx}.$$

The arbitrary element $f: R(q) \to R$. The result of the group classification is given in Table 9.

TABLE 9

$\overset{*}{f}$		$\overset{*}{t}$	$\overset{*}{x}$	$\overset{*}{u}$
	ζ_1	1	0	0
~	ζ_2	0	1	0
	ζ_3	0	0	1
	ζ_4	$2t$	x	u
g^k	ζ_5	kt	0	$-u$
e^q	ζ_5	t	0	$-x$

† From Ref. [8] in Chapter III.
‡ From Ref. [17] in Chapter III.

§A9. EQUATIONS OF STATIONARY BOUNDARY LAYER WITH MOMENT STRESSES†

$X = R^2(x, y)$, $Y = R^4(u, v, \omega, p)$. The system E is

$$u_x + v_y = 0, \qquad p_y = 0$$

$$uu_x + vu_y + p_x = u_{yy} + \omega_y, \qquad u\omega_x + v\omega_y = \omega_{yy}.$$

The admitted Lie algebra is $L^8[\phi\colon R(x) \to R$ is an arbitrary function] and is given in Table 10. A number of variants of the system discussed here are also considered in the referenced work.

TABLE 10

	$\overset{*}{x}$	$\overset{*}{y}$	$\overset{*}{u}$	$\overset{*}{v}$	$\overset{*}{\omega}$	$\overset{*}{p}$
ζ_1	1	0	0	0	0	0
ζ_2	0	0	0	0	0	1
ζ_3	0	0	0	0	1	0
ζ_4	x	0	u	0	ω	$2p$
ζ_5	0	$-y$	$2u$	v	3ω	$4p$
ζ_ϕ	0	ϕ	0	$\phi' u$	0	0

§A10. EQUATIONS OF LIQUID MOTION IN LONG PIPES‡

$X = R^2(t, x)$, $Y = R^2(u, v)$. The system $E(n, m)$ is $u_t + v_x = 0$, $u_x + v^n u^{-m} = 0$. The arbitrary element is a pair of numbers (n, m) and $n \neq 0$. The result of the group classification is given in Table 11. Here $A, B\colon R^2(t, u) \to R$ are arbitrary functions, satisfying the equation $\partial_t F - \partial_u(u^{-m}\partial_u F) = 0$.

TABLE 11

(n, m)		$\overset{*}{t}$	$\overset{*}{x}$	$\overset{*}{u}$	$\overset{*}{v}$
	ζ_1	1	0	0	0
$\sim$	ζ_2	0	1	0	0
	ζ_3	$(n + 1)t$	nx	0	$-v$
	ζ_4	0	$(m - n + 1)x$	$(n + 1)u$	$(m + 2)v$
$(1, -\frac{3}{4})$	ζ_5	0	$-x^2$	$3xu$	$xv - 3u^{-1/3}$
$(-1, \sim)$	ζ_4	$(m + 2)t$	0	u	$-(m + 1)v$
	ζ_B	0	B	0	$u^{-m}\partial_u B$
$(-1, -2)$	ζ_A	0	xA	0	$vA + xu^2\partial_u A$
	ζ_6	$2t$	0	$(t + \ln u)u$	$(t - 1 + \ln u)v$

† From Ref. [20] in Chapter III.

‡ From Ref. [11] in Chapter V.

TABLE 12

	$\overset{*}{t}$	$\overset{*}{x}{}^1$	$\overset{*}{x}{}^2$	$\overset{*}{x}{}^3$	$\overset{*}{v}{}^1$	$\overset{*}{v}{}^2$	$\overset{*}{v}{}^3$	$\overset{*}{E}_1$	$\overset{*}{E}_2$	$\overset{*}{E}_3$	$\overset{*}{H}_1$	$\overset{*}{H}_2$	$\overset{*}{H}_3$	ρ
ζ_1	1	0	0	0	0	0	0	0	0	0	0	0	0	0
ζ_2	0	1	0	0	0	0	0	0	0	0	0	0	0	0
ζ_3	0	0	1	0	0	0	0	0	0	0	0	0	0	0
ζ_4	0	0	0	1	0	0	0	0	0	0	0	0	0	0
ζ_5	t	x^1	x^2	x^3	0	0	0	$-E_1$	$-E_2$	$-E_3$	$-H_1$	$-H_2$	$-H_3$	-2ρ
ζ_6	0	$-x^2$	x^1	0	$-v^2$	v^1	0	$-E_2$	E_1	0	$-H_2$	H_1	0	0
ζ_7	0	$-x^3$	0	x^1	$-v^3$	0	v^1	$-E_3$	0	E_1	$-H_3$	0	H_1	0
ζ_8	0	0	$-x^3$	x^2	0	$-v^3$	v^2	0	$-E_3$	E_2	0	$-H_3$	H_2	0
ζ_9	x^1	t	0	0	$1-(v^1)^2$	$-v^1v^2$	$-v^1v^3$	0	H_3	$-H_2$	0	$-E_3$	E_2	ρv^1
ζ_{10}	x^2	0	t	0	$-v^1v^2$	$1-(v^2)^2$	$-v^2v^3$	$-H_3$	0	H_1	E_3	0	$-E_1$	ρv^2
ζ_{11}	x^3	0	0	t	$-v^1v^3$	$-v^2v^3$	$1-(v^3)^2$	H_2	$-H_1$	0	$-E_2$	E_1	0	ρv^2

§A11. EQUATIONS OF A RELATIVISTIC PENCIL OF CHARGED PARTICLES†

$X = R^4(t, x)$, where $x \in V_3$ with the metric $ds^2 = g_{ij}\, dx^i\, dx^j$, $Y = R^{10}(v, E, H, \rho)$, where $v, E, H \in R^3$, $\rho \in R$. The system E is written in covariant form with respect to V_3 (all the indices assume the values 1, 2, 3):

$$v_t^i + v^k v^i{}_{,k} = g^{il}\sqrt{1 - g_{jq} v^j v^q}(E_l + \sqrt{g}\, e_{mnl} v^m H^n - g_{rs} v^r E^s v^l)$$

$$\frac{1}{\sqrt{g}} e^{ikl} \partial_i E_k + H_t^l = 0, \qquad \frac{1}{\sqrt{g}} e^{ikl} \partial_i H_k - E_t^l = \rho v^l,$$

$$\partial_i(\sqrt{g}\, g^{ik} E_k) = \sqrt{g}\, \rho, \qquad \partial_i(\sqrt{g}\, g^{il} H_k) = 0.$$

This system admits the system G^{11}. The basis for L^{11} is given in Table 12. Also, the invariants of the system E obtained for simplifications are considered in the quoted works.

§A12. EQUATIONS OF A ONE-DIMENSIONAL TWO-COMPONENT HYDRODYNAMIC MODEL OF A PLASMA‡

$X = R^2(t, x)$, $Y = R^5(\eta_e, \eta_i, v_e, v_i, E)$. In the accepted notation the system, which contains the four parameters e, m, M, Z and the two arbitrary functions $f, g\colon R(s) \to R$ characterizing the pressure dependence on concentration, is

$$\partial_t n_e + \partial_x(n_e v_e) = 0, \qquad \partial_x E = 4\pi e(Zn_i - n_e), \qquad \partial_t E = 4\pi e(n_e v_e - Zn_i v_i),$$

$$\partial_t v_e + v_e \partial_x v_e + \frac{e}{m} E + f(n_e)\partial_x \eta_e = 0,$$

$$\partial_t v_i + v_i \partial_x v_i - \frac{Ze}{m} E + g(n_i)\partial_x \eta_i = 0.$$

The arbitrary element is a pair of functions (f, g). The result of group classification (a, b, γ are arbitrary constants) is given in Table 13.

TABLE 13

(f, g)		$\overset{*}{t}$	$\overset{*}{x}$	$\overset{*}{n}_e$	$\overset{*}{n}_i$	$\overset{*}{v}_e$	$\overset{*}{v}_i$	$\overset{*}{E}$
	ζ_1	1	0	0	0	0	0	0
~	ζ_2	0	1	0	0	0	0	0
	ζ_3	0	t	0	0	1	1	0
$(as\gamma^{-2}, bs\gamma^{-2})$	ζ_4	t	$(2-\gamma)x$	$-2n_e$	$-2n_i$	$(1-\gamma)v_e$	$(1-\gamma)v_i$	$-\gamma E$
(0, 0)	ζ_5	0	x	0	0	v_e	v_i	E

† From Refs. [9, 10] in Chapter V.
‡ From Ref. [18] in Chapter V.

§A13. EQUATIONS OF LONGITUDINAL ONE-DIMENSIONAL MOTIONS OF A COLLISIONLESS PLASMA†

$X = R^3(t, x, v)$, $Y = R^2(f, E)$. The integrodifferential system

$$f_t + vf_x - Ef_v = 0$$

$$E_t = \int vf\,dv, \qquad E_x = 1 - \int f\,dv, \qquad E_v = 0$$

is considered, where the integration is carried out over all $R(v)$. The admitted group is calculated through a formal approach to the infinite system of differential equations for the moments

$$M_k = \int v^k f\,dv \qquad (k = 0, 1, 2, \ldots), \qquad \partial_t M_k + \partial_x M_{k+1} + kEM_{k-1} = 0,$$

$$\partial_t E = M_1, \qquad \partial_x E = 1 - M_0.$$

For that system the group G^5 is obtained as given in Table 14.

TABLE 14

	$\overset{*}{t}$	$\overset{*}{x}$	$\overset{*}{v}$	$\overset{*}{E}$	$\overset{*}{f}$
ζ_1	1	0	0	0	0
ζ_2	0	1	0	0	0
ζ_3	0	x	v	E	$-f$
ζ_4	0	$\cos t$	$-\sin t$	$\cos t$	0
ζ_5	0	$\sin t$	$\cos t$	$\sin t$	0

§A14. EQUATIONS OF FLOW IN BAROCLINIC LAYER OF THE OCEAN‡

$X = R^3(x, y, z)$, $Y = R^5(u, v, w, p, \rho)$. The system E is

$$u_x + v_y + w_z = 0, \qquad u\rho_x + v\rho_y + w\rho_z = \rho_{xx} + \rho_{yy} + \rho_{zz},$$

$$p_x = (1 - x)v, \qquad p_y = -(1 - x)v, \qquad p_z = \rho.$$

The full group G^5 is given in Table 15.

† From Ref. [19] in Chapter V.
‡ From Ref. [13] in Chapter V.

TABLE 15

	$\overset{*}{x}$	$\overset{*}{y}$	$\overset{*}{z}$	$\overset{*}{u}$	$\overset{*}{v}$	$\overset{*}{w}$	$\overset{*}{p}$	$\overset{*}{\rho}$
ζ_1	0	1	0	0	0	0	0	0
ζ_2	0	0	1	0	0	0	0	0
ζ_3	0	0	0	0	0	0	1	0
ζ_4	0	0	0	0	0	0	z	1
ζ_5	$x - 1$	y	z	$-u$	$-v$	$-w$	p	0

§A15. LAMÉ EQUATIONS†

$X = R^4(t, x)$, $Y = R^3(w)$. The system E in the components

$$x = (x^1, x^2, x^3),\ w = (w^1, w^2, w^3) \text{ is}$$

$$\partial_t^2 w^k = \sum_{i=1}^{3} (\partial_i \partial_k w^i + \alpha \partial_i^2 w^k) \qquad (k = 1, 2, 3;\ \alpha = \text{const}).$$

The full (nontrivial) group is G^8 [here I_X is an identity transformation and A is an arbitrary antisymmetric mapping of the space $R^3(x)$] as given in Table 16.

TABLE 16

	$\overset{*}{t}$	$\overset{*}{x}$	$\overset{*}{w}$
ζ_1	1	0	0
ζ_I	0	I_X	0
ζ_A	0	$A\langle x \rangle$	$A\langle w \rangle$
ζ_8	t	x	0

† From Ref. [22] in Chapter III.

Supplementary References

1. S. Lie, *Theorie der Transformationgruppen*, 2nd ed. Chelsea, New York, 1970.
2. S. Lie, Über die Integration durch bestimmte Integrale von einer Klasse linearer partieller Differentialgleichungen, *Arch. Math. Naturvidensk.* **6,** 328–368 (1881).
3. E. Vessiot, Sur l'intégration des systémes différentiels qui admettent des groupes continus de transformations, *Acta Math.* **28,** 307–349 (1904).
4. L. E. Dickson, Differential equations from the group standpoint, *Ann. of Math.* **25,** 287–378 (1924).
5. A. Cohen, *An Introduction to the Lie Theory of One Parameter Groups*, Heath, Boston, Massachusetts, 1911.
6. L. P. Eisenhart, *Continuous Groups of Transformations*, Princeton Univ. Press, Princeton, New Jersey, 1933.
7. G. Birkhoff, *Hydrodynamics*, 2nd ed. Princeton Univ. Press, Princeton, New Jersey, 1960.
8. A. J. A. Morgan, The reduction by one of the number of independent variables in some systems of partial differential equations, *Quart. J. Math.* **3,** 250–259 (1952).
9. L. V. Ovsiannikov, *Group Properties of Differential Equations* (in Russian), Novosibirsk, 1962; English translation by G. W. Bluman, 1967.
10. R. E. Boisvert, *Group Analysis of the Navier-Stokes Equations*, Ph.D. Dissertation, Georgia Institute of Technology, Atlanta, Georgia, 1982.
11. G. W. Bluman and J. D. Cole, *Similarity Methods for Differential Equations*, Appl. Math. Sci., Vol. 13. Springer-Verlag, New York, 1974.
12. H. S. Woodard, *Similarity Solutions for Partial Differential Equations Generated by Finite and Infinitesimal Groups*, Ph.D. Dissertation, Univ. of Iowa, Iowa City, 1971.
13. W. F. Ames, *Nonlinear Partial Differential Equations in Engineering*, Vol. II. Academic Press, New York, 1972.
14. W. F. Ames, R. J. Lohner, and E. Adams, Group properties of $u_{tt} = [f(u)u_x]_x$, *Internat. J. Non-Linear Mech.* **16,** 439–447 (1981).

15. N. S. Zakharov and V. P. Korobeinikov, Group analysis of the generalized Korteweg–DeVries–Burgers equations, *J. Appl. Math. Mech.* **44,** 668–671 (1981).
16. M. J. Moran and R. A. Gaggioli, A new systematic formalism for similarity analysis, *J. Engrg. Math.* **3,** 151–162 (1969).
17. M. S. Klamkin, On the transformation of a class of boundary value problems into initial value problems for ordinary differential equations, *SIAM Rev.* **4,** 43–47 (1962).
18. T. Y. Na, Transforming boundary conditions to initial conditions for ordinary differential equations, *SIAM Rev.* **9,** 204–210 (1967).
19. T. Y. Na, Further extension on transforming from boundary value to initial value problems, *SIAM Rev.* **10,** 85–87 (1968).
20. M. S. Klamkin, Transformation of boundary value problems into initial value problems, *J. Math. Anal. Appl.* **32,** 308–330 (1970).
21. W. F. Ames and E. Adams, Exact shooting and eigenparameter problems, *Nonlinear Anal. Theory Meth. Appl.* **1,** 75–82 (1976).
22. W. F. Ames, *Numerical Methods for Partial Differential Equations*, 2nd ed. Academic Press, New York, 1977.
23. N. N. Janenko and Ju. I. Sokin, Group classification of difference schemes for a system of one-dimensional equations of gas dynamics, *Amer. Math. Soc. Transl.* **104,** 259–265 (1976).
24. W. F. Ames and N. H. Ibragimov, *Utilization of Group Properties in Computation*, Paper presented at the International Joint IUTAM/IMU Symposium, Novosibirsk, 1978.
25. V. L. Katov and N. I. Kostukova, in Dynamics of Fluids (in Russian), Vol. 1, pp. 48–60. Novosibirsk, 1969.
26. P. Rosenau and J. L. Schwarzmeier, *Similarity Solutions of Systems of Partial Differential Equations Using MACSYMA*. Courant Inst. of Math. Sci. Report No. COO-3077-160/MF-94, 1979.
27. G. W. Bluman and J. D. Cole, The general similarity solution of the heat equation, *J. Math. Mech.* **18,** 1025–1042 (1969).
28. C. Chevalley, *Theory of Lie Groups*, Princeton Univ. Press, Princeton, New Jersey, 1946.
29. N. Jacobson, *Lie Algebras*. Wiley (Interscience), New York, 1962.
30. G. A. Nariboli, Self-similar solutions of some nonlinear equations, *Appl. Sci. Res.* **22,** 449–461 (1970).
31. V. V. Puhnachev, Group properties of the equations of Navier–Stokes in the plane (in Russian), *J. Appl. Mech. Tech. Phys.* **1,** 83–90 (1960).
32. V. O. Bitev, Group properties of the equations of Navier–Stokes, in *Numerical Methods of the Mechanics of Fluids* (in Russian). Novosibirsk, 1972.
33. S. P. Lloyd, The infinitesimal group of the Navier–Stokes equations, *Acta Mech.* **38,** 85–98 (1981).
34. C. L. M. H. Navier, "Mémoire sur les lois du mouvement des fluides," *Mem. Acad. Sci. Inst. Fr.* **6,** 389–440 (1823).
35. G. G. Stokes, On the theories of the internal friction of fluids in motion, and of the equilibrium and motion of elastic solids, *Camb. Phil. Soc. Trans.* **8,** 287–319 (1849).
36. W. F. Ames, Ad hoc techniques for nonlinear partial differential equations, in *Nonlinear Partial Differential Equations* (W. F. Ames, ed.), pp. 55–72. Academic Press, New York, 1967.
37. G. Hamel, Spiralförmige Bewegungen zäher Flüssigkeiten, *Jber. Deutsch. Math.-Verein.* **25,** 34–60 (1916).
38. H. Görtler and K. Wieghardt, Über eine gewisse Klasse von Strömungenzäher Flüssigkeiten und eine Kennzeichnung der Poiseuille-Strömung, *Math. Z.* **48,** 247–250 (1942).
39. A. Strumpf, *On a Class of Transformations Leading to "Similar" Solutions of the Steady,*

Two-Dimensional, Navier–Stokes Equations, Ph.D. Dissertation, Stevens Inst. of Tech., Hoboken, New Jersey, 1964.

40. P. Moon and D. E. Spencer, *Field Theory Handbook*, Springer-Verlag, Berlin and New York, 1961.
41. G. B. Jeffrey, The two-dimensional steady motion of a viscous fluid, *Phil. Mag.* **29,** 455–465 (1915).
42. M. D. Riabouchinski, Quelques considérations sur les mouvements plans rotationnels d'un liquide, *C. R. Acad. Sci.* **179,** 1133–1136 (1924).
43. R. Berker, *Sur quelques cas d'intégration des équations du mouvement d'un fluide visqueux incompressible*, Paris-Lille, 1936 (see Ref. 47).
44. G. I. Taylor, On the decay of vortices in a viscous fluid, *Phil. Mag.* **46,** 671–674 (1923).
45. H. L. Dryden, F. D. Murnaghan, and H. Bateman, Report of the Committee on Hydrodynamics, *Bull. Nat. Res. Council* No. 84, Washington, 1932.
46. R. Berker, An exact solution of the Navier–Stokes equations, *Internat. J. Engng. Sci.* **20,** 217–230 (1982).
47. R. Berker, Intégration des équations du mouvement d'un fluide visqueux incompressible, in *Handbuch der Physik*, Bd. VIII/2, pp. 1–384. 1963.

Index

D

H

I